331.708

24/2

£22.50

AF869605

The Use of Integral Transforms

The Use of Integral Transforms

Ian N. Sneddon

Simson Professor of Mathematics,
University of Glasgow

McGraw-Hill Book Company

New York · St Louis · San Francisco · Düsseldorf
Johannesburg · Kuala Lumpur · London · Mexico
Montreal · New Delhi · Panama · Rio de Janeiro
Singapore · Sydney · Toronto

1198711

BIRMINGHAM UNIVERSITY LIBRARY

The Use of Integral Transforms

Copyright © 1972 by McGraw-Hill, Inc. All rights reserved. Printed in the United States of America. No part of this publication may be reproduced, stored in a retrieval system, or transmitted, in any form or by any means, electronic, mechanical, photocopying, recording, or otherwise, without the prior written permission of the publisher.

Library of Congress Catalog Card Number 79-155885

ISBN 07-059436-8

67890 MAMM 79876543

This book was set in Times Roman, and was printed and bound by The Maple Press Company. The designer was Alan Chapman. The editors were Lee W. Peterson and Alan Chapman. John A. Sabella supervised production.

Contents

CHAPTER 3

CHAPTER 9

Preface

The aim of this book is to provide an introduction to the use of integral transforms for students of applied mathematics, physics and engineering. It is based upon courses of lectures given at various times in the University of Glasgow, North Carolina State University and the State University of New York at Stony Brook to postgraduate students of these subjects. Some sections have also been used in lectures to undergraduates reading for "honours" degrees in mathematics or engineering in the University of Glasgow.

Since this book has been written for students whose primary interest is in the applications of the theory rather than in the theory itself, it does not attempt to present the basic theorems relating to integral transforms in their most general forms. The treatment is restricted to classes of functions which are sufficiently wide to embrace the functions which normally occur in physical applications. It is hoped that most of the book is accessible to a student well grounded in the methods of advanced calculus and of the solution of elementary differential equations.

An attempt has been made to cover the commonly used properties of all those integral transforms which are currently in use in the solution of boundary value problems. There are chapters on the Fourier transforms and those of Lapace, Mellin, Hankel, Mehler and Fock, Kontorovich and Lebedev. Closely related topics such as finite transforms, dual integral equations, the Wiener-Hopf procedure and the elementary parts of the theory of generalized functions are also considered. In an effort to make the book more useful as a textbook for students of applied mathematics each chapter contains an account of applications of the theory of integral transforms to the solution of problems arising in mathematical physics. In this way is illustrated the use of transform techniques in finding the solution of boundary and initial value problems for ordinary and partial differential equations, of integral equations and difference equations. Problems are included at the end of each chapter so that the reader may further develop his skill in the application of integral transforms.

The critical comments of colleagues and of students who attended the lectures on which this book is based were of great value and are gratefully acknowledged. In particular I wish to thank M. Lowengrub, R. P. Srivastav, M. P. Stallybrass, J. Tweed and M. S. Alala for their comments on the final manuscript and for their help in reading the proofs.

My thanks are due also to Mrs. Dorothy Long for her assistance in the preparation of the author and subject indices.

Ian N. Sneddon

1

Introduction

Since, in this book, we are concerned primarily with the *use* of integral transforms, we begin by outlining briefly the circumstances in which integral transforms arise in the analysis of some boundary value and initial value problems of classical physics. This provides some motivation for the mathematical problems with which the remainder of the book is concerned.

In sec. 1-1 we discuss in a general kind of way how the use of the principle of linear superposition in constructing integral representations of solutions of linear partial differential equations can lead in a natural way to a consideration of the properties of integral transforms. In secs. 1-3 through 1-5 this method is used to introduce the Fourier transform, the Fourier sine and cosine transforms, and the Hankel transform. An alternative approach is developed in sec. 1-2 and in secs. 1-6, 1-7 this is used to introduce the Laplace transform and the Mellin transform.

We conclude this introductory chapter by saying something about the kinds of functions we shall be discussing in the subsequent chapters.

1-1 INTEGRAL TRANSFORMS

Many of the phenomena of classical physics may be described by partial differential equations of a particularly simple type. For example, problems in potential theory (which embraces gravitational theory, electrostatics, magnetostatics, and the irrotational flow of a perfect fluid) may be discussed with reference either to *Laplace's equation*

$$\frac{\partial^2 u}{\partial x^2} + \frac{\partial^2 u}{\partial y^2} + \frac{\partial^2 u}{\partial z^2} = 0$$

or to *Poisson's equation*

$$\frac{\partial^2 u}{\partial x^2} + \frac{\partial^2 u}{\partial y^2} + \frac{\partial^2 u}{\partial z^2} = f(x,y,z).$$

In these equations u is the potential function whose value at an arbitrary point (x,y,z) is to be found; in Poisson's equation the function $f(x,y,z)$ is prescribed. In terms of the three-dimensional Laplace operator Δ_3 defined by the equation

$$\Delta_3 = \frac{\partial^2}{\partial x^2} + \frac{\partial^2}{\partial y^2} + \frac{\partial^2}{\partial z^2}$$

and the position vector $\mathbf{r} = (x,y,z)$ of a typical point in E_3 we may write Laplace's equation in the form

$$\Delta_3 u(\mathbf{r}) = 0,$$

and Poisson's equation in the form

$$\Delta_3 u(\mathbf{r}) = f(\mathbf{r}).$$

In many problems of potential theory, the potential function u does not depend on one of the cartesian coordinates, z say, in which case Laplace's equation and Poisson's equation reduce to

$$\frac{\partial^2 u}{\partial x^2} + \frac{\partial^2 u}{\partial y^2} = 0$$

and

$$\frac{\partial^2 u}{\partial x^2} + \frac{\partial^2 u}{\partial y^2} = f(x,y)$$

respectively, or in terms of the two-dimensional Laplace operator, Δ_2, defined by

$$\Delta_2 = \frac{\partial^2}{\partial x^2} + \frac{\partial^2}{\partial y^2}$$

to

$$\Delta_2 u(\mathbf{r}) = 0$$

and

$$\Delta_2 u(\mathbf{r}) = f(\mathbf{r}),$$

where, of course, $\mathbf{r}$ is now the position vector (x,y) of a typical point in E_2.

Similarly the propagation of waves in E_3 is described by the *wave equation*

$$\Delta_3 u = \frac{1}{c^2}\frac{\partial^2 u}{\partial t^2},$$

where t denotes the time and c the velocity of propagation of the waves, and diffusion phenomena, such as the conduction of heat, by the *diffusion equation*

$$\Delta_3 u = \frac{1}{\kappa}\frac{\partial u}{\partial t},$$

where κ denotes the thermal diffusivity of the medium.

If we take $\mathbf{r} = (x,y,z,t)$ to be the position vector of a point of space-time, we can write these equations in the forms

$$\left(c^2\Delta_3 - \frac{\partial^2}{\partial t^2}\right) u(\mathbf{r}) = 0$$

and

$$\left(\kappa\Delta_3 - \frac{\partial}{\partial t}\right) u(\mathbf{r}) = 0$$

respectively.

It will be observed that these equations are linear, of second order, and that in cartesian coordinates they have constant coefficients, but in spite of their somewhat special character, these equations are of fundamental importance in many branches of physics and engineering and since the eighteenth century have played an important role in the development of pure mathematics as well as of that of mathematical physics.

Perhaps the oldest systematic technique for the solution of the partial differential equations of mathematical physics is the method of *separation of variables*. Introduced by d'Alembert, Daniel Bernoulli, and Euler in the middle of the eighteenth century it remains a method of great value today and lies at the heart of the use of integral transforms in the solution of problems in applied mathematics.

To illustrate the method we shall consider its use in the solution of Laplace's two-dimensional equation in the half-plane $y \geqslant 0$ when it is assumed that the potential function $u(x,y)$ tends to zero as $\rho \to \infty$, where $\rho = \sqrt{(x^2 + y^2)}$ denotes distance from the origin.

The method of separation of variables consists of trying to find solutions of $\Delta_2 u(x,y) = 0$ which are of the form

$$u(x,y) = X(x)Y(y),$$

where X is a function of x alone and Y is a function of y alone. Since for this form of $u(x,y)$

$$\frac{\partial^2 u}{\partial x^2} = \frac{d^2 X}{dx^2} Y$$

and

$$\frac{\partial^2 u}{\partial y^2} = X \frac{d^2 Y}{dy^2},$$

it follows that $\Delta_2 u = 0$ will have solutions of this form if we can find functions X and Y such that

$$\frac{d^2 X}{dx^2} Y + X \frac{d^2 Y}{dy^2} = 0$$

for all (x,y) in the half-plane $y \geqslant 0$. Rewriting this equation in the form

$$\frac{1}{X}\frac{d^2 X}{dx^2} = -\frac{1}{Y}\frac{d^2 Y}{dy^2}$$

and noting that whereas the left-hand side is a function of x alone, the right-hand side is a function of y alone, we see that both sides must be equal to a constant, $-\xi^2$ say. Hence X and Y are determined by the equations

$$\frac{d^2 X}{dx^2} = -\xi^2 X, \quad \frac{d^2 Y}{dy^2} = \xi^2 Y$$

So far we have only said that ξ is a constant. If we wish to obtain solutions u tending to zero as $\rho \to \infty$ we must choose functions Y which tend to zero as $y \to \infty$, so *we must choose ξ to be real* and take

$$Y(y) = e^{-|\xi|y}.$$

(We take $e^{-|\xi|y}$ since if we took $e^{-\xi y}$, Y would not tend to zero as $y \to \infty$ for negative values of ξ.) We also take

$$X(x) = e^{\pm i\xi x}$$

This in turn shows us that

$$u(x,y) = e^{i\xi x - |\xi|y}, \quad \xi \in \mathbf{R}$$

is *a* solution of $\Delta_2 u = 0$.

In a physical problem the field variable is determined not only by the partial differential equation but by the boundary values or the initial values assumed by the function. The function derived above certainly satisfies Laplace's equation but it is the solution only of the boundary value problem

$$\begin{aligned} \Delta_2 u(x,y) &= 0, \qquad -\infty < x < \infty, \qquad y \geqslant 0 \\ u(x,y) &= e^{i\xi x}, \qquad -\infty < x < \infty \\ u(x,y) &\to 0 \qquad \text{as} \qquad \sqrt{(x^2 + y^2)} \to \infty. \end{aligned}$$

For other forms of $u(x,0)$ we have to construct other solutions. We do this by building up complicated solutions from the simple one we have just found. Since

$$\Delta_2 e^{i\xi x - |\xi| y} = 0$$

for *all* real ξ and Δ_2 is a linear operator, we see that for certain forms of the function $F(\xi)$ the function

$$u(x,y) = \int_{-\infty}^{\infty} F(\xi) e^{i\xi x - |\xi| y} \, d\xi$$

would also satisfy Laplace's equation in the half-plane $y \geqslant 0$ and hence give us a new solution.[1] For example, if we take $F(\xi) = \frac{1}{2} e^{-|\xi| a}$, we obtain the solution

$$u(x,y) = \frac{a + y}{x^2 + (a + y)^2},$$

which satisfies the boundary condition $u(x,0) = a(a^2 + x^2)^{-1}$.

The problem which naturally arises is: can we find the solution corresponding to the boundary condition $u(x,0) = f(x)$, $-\infty < x < \infty$? In other words: can we find a function $F(\xi)$ such that

$$f(x) = \int_{-\infty}^{\infty} F(\xi) e^{i\xi x} \, d\xi$$

for all real values of x?

This problem can be generalized in an obvious way.

Suppose that we wish to find the solution in a domain D of a homogeneous partial differential operator of the form

$$\mathbf{L} u(\mathbf{r}) = 0, \tag{1-1-1}$$

where $\mathbf{r}$ is the position vector, with components $(x_1, x_2, \ldots, x_n)$, of a field point in E_n, the Euclidean space of n dimensions and $\mathbf{L}$ is a linear differential operator in the variables $x_1, x_2, \ldots, x_n$. For instance $\mathbf{L}$ could be Δ_3 or Δ_2 or $\kappa\Delta_3 - \partial/\partial t$.

[1] This is known as the *superposition principle*.

We can often find a simple solution $\eta(\mathbf{r};\xi)$ of the equation (1-1-1) involving an arbitrary constant ξ whose values, which may be real or complex, are taken from a set Ω. If

$$\mathbf{L}\eta(\mathbf{r};\xi) = 0 \qquad \mathbf{r} \in D, \qquad \xi \in \Omega$$

then, since the operator $\mathbf{L}$ is linear, it is probable that the function

$$u(\mathbf{r}) = \int_\Omega F(\xi)\eta(\mathbf{r};\xi)\, d\xi, \qquad \mathbf{r} \in D$$

will also be a solution of (1-1-1) for any arbitrary function F of ξ—at least if it is such that the integral on the right side of this last equation converges uniformly for all $\mathbf{r} \in D$. It may indeed turn out that the boundary conditions imposed in the problem are sufficient to enable us to determine the form of the function $F(\xi)$.

To illustrate the kind of situation which may arise we consider the case in which $u(x,y)$ is a function of two independent variables (x,y). We then have a solution of the form

$$u(x,y) = \int_\Omega F(\xi)\eta(x,y;\xi)\, d\xi, \qquad (x,y) \in D.$$

If, in addition, it is postulated that $u(x,0) = f(x)$, where the function f is prescribed, we have the relation

$$f(x) = \int_\Omega F(\xi)K(x;\xi)\, d\xi, \qquad x \in D_1, \tag{1-1-2}$$

where D_1 is the domain of x and the function $K(x;\xi)$ is defined by the equation

$$K(x;\xi) = \eta(x,0;\xi). \tag{1-1-3}$$

When a function f is defined in terms of a function F by means of an integral relation of the type (1-1-2), we say that $f(x)$ is the *integral transform* of the function $F(\xi)$ by the *kernel* $K(x,\xi)$, and we employ some such notation as

$$f(x) = \mathscr{T}[F(\xi);x]. \tag{1-1-4}$$

For a particular choice of kernel we use not a general symbol $\mathscr{T}$ for the operator of the transform but some specific symbol which reflects the nature of the kernel. For instance, for kernels traditionally associated with the name of Fourier we use $\mathscr{F}$ instead of $\mathscr{T}$, and for those associated with the names of Laplace, Mellin, and Hankel we use the symbols $\mathscr{L}$, $\mathscr{M}$, and $\mathscr{H}$, respectively.

We note that the operator $\mathscr{T}$ defined by equations (1-1-2) and (1-1-4) is a linear operator, i.e., that if

$$\mathscr{T}[F_1(\xi);x] = f_1(x), \qquad \mathscr{T}[F_2(\xi);x] = f_2(x)$$

and c_1 and c_2 are arbitrary constants, then

$$\mathscr{T}[c_1 F_1(\xi) + c_2 F_2(\xi); x] = c_1 f_1(x) + c_2 f_2(x). \tag{1-1-5}$$

For example, if Ω is the entire real line $\mathbf{R}$ and $K(x;\xi)$ is $(2\pi)^{-\frac{1}{2}} e^{i\xi x}$, we say that $f(x)$ is the *exponential Fourier transform* of $F(\xi)$ or, simply, the *Fourier transform* of $F(\xi)$. We therefore have as the definition of the Fourier transform

$$f(x) = \frac{1}{\sqrt{(2\pi)}} \int_{-\infty}^{\infty} F(\xi) e^{i\xi x}\, d\xi. \tag{1-1-6}$$

It is sometimes convenient to write the integral on the right as

$$f(x) = \frac{1}{\sqrt{(2\pi)}} \int_{\mathbf{R}} F(\xi) e^{i\xi x}\, d\xi.$$

In sec. 1-3 we shall see how this transform arises in a natural way in the analysis of a simple boundary value problem in potential theory.

Similarly, if Ω is the positive real line $\mathbf{R}^+$ and $K(x;\xi)$ is $(2/\pi)^{\frac{1}{2}} \cos(\xi x)$, we say that the function $f(x)$ defined by the equation

$$f(x) = \sqrt{\frac{2}{\pi}} \int_0^{\infty} F(\xi) \cos(\xi x)\, d\xi \equiv \sqrt{\frac{2}{\pi}} \int_{\mathbf{R}^+} F(\xi) \cos(\xi x)\, d\xi \tag{1-1-7}$$

is the *Fourier cosine transform* of $F(\xi)$, and if $K(\xi;x)$ is $(2/\pi)^{\frac{1}{2}} \sin(\xi x)$, we say that

$$f(x) = \sqrt{\frac{2}{\pi}} \int_0^{\infty} F(\xi) \sin(\xi x)\, d\xi \equiv \sqrt{\frac{2}{\pi}} \int_{\mathbf{R}^+} F(\xi) \sin(\xi x)\, d\xi \tag{1-1-8}$$

is the *Fourier sine transform* of $F(\xi)$. We shall see in sec. 1-4 how such transforms arise naturally in the derivation of certain solutions of the one-dimensional wave equation.

Another integral transform which is of frequent use in mathematical physics is obtained by taking Ω to be the positive real line $\mathbf{R}^+$ and $K(x;\xi)$ to be defined by the equation

$$K(x;\xi) = \xi J_\nu(\xi x),$$

where $J_\nu(z)$ denotes the Bessel function of the first kind of order ν and argument z. We then say that $f(x)$, defined by the equation

$$f(x) = \int_0^{\infty} \xi F(\xi) J_\nu(\xi x)\, d\xi \tag{1-1-9}$$

is the Hankel transform[1] of order ν of the function $F(\xi)$. In sec. 1-5 we shall

[1] In some books $K(x;\xi)$ is chosen to be $(x\xi)^{\frac{1}{2}} J_\nu(x\xi)$ to give a kernel which is a function of the single variable $x\xi$. The choice made here is more appropriate in the analysis of problems in applied mathematics.

see how such a transform arises in a simple problem.

In the discussion of boundary value problems for partial differential equations, the basic problem often turns out to be the determination of the function $F(\xi)$ when the function $f(x)$ is prescribed, i.e., it reduces to the solution of the integral equation (1-1-2). In many cases it turns out that we can find a solution of an integral equation of this type in the form

$$F(\xi) = \int_{D_1} f(x)H(x;\xi)\,dx, \qquad \xi \in \Omega \tag{1-1-10}$$

where D_1 is the domain of x, i.e., in the problem considered $x \in D_1$. Such a result is called an *inversion formula* for the transform (1-1-2) and equations (1-1-2) and (1-1-10) taken together constitute what is called an *inversion theorem* for the transform (1-1-2).

When it happens, as it frequently does, that the sets D_1 and Ω are identical and $K(x;\xi) = H(\xi;x)$, so that tne relation between f and F is symmetrical, we say that $K(x;\xi)$ is a *Fourier kernel.*[1]

If $K(x;\xi) = K(\xi;x)$, we say—as in the theory of integral equations—that the kernel K is a *symmetrical kernel.* The simplest type of symmetric kernel is one of the form $K(\xi x)$, which is a function of the single variable ξx. In problem 1-1 a simple necessary and sufficient condition for a kernel to be a Fourier kernel is stated; later—in sec. 4-5—this is used as the basis for a useful criterion by means of which we can determine whether or not a function $K(t)$ defines a Fourier kernel.

1-2 AN ALTERNATIVE APPROACH

The concept of an integral transform arises naturally in another way which we introduce through a simple example.

Suppose we wish to solve the two-dimensional Laplace equation $\Delta_2 u = 0$ in the positive quadrant $x \geqslant 0$, $y \geqslant 0$ subject to the boundary conditions

$$u(x,0) = f(x), \qquad \frac{\partial u(0,y)}{\partial x} = g(y)$$

and the condition that as $\sqrt{(x^2 + y^2)} \to \infty$, $u(x,y)$ and its first order derivatives tend to zero. Applying the rule for integrating by parts twice we find that

$$\int_0^\infty \frac{\partial^2 u}{\partial x^2}\cos(x\xi)\,dx$$
$$= \left[\frac{\partial u}{\partial x}\cos(x\xi) + \xi u \sin(x\xi)\right]_0^\infty - \xi^2 \int_0^\infty u(x,y)\cos(x\xi)\,dx$$

[1] It turns out that the kernel of the Fourier transform $\mathscr{F}$ is **not** a Fourier kernel! This is an unfortunate consequence of using what is now the standard term "Fourier" kernel for a kernel with the defined property; it would be more appropriate to call it an "idempotent" kernel.

so that for the solution of our problem

$$\int_0^\infty \frac{\partial^2 u}{\partial x^2} \cos(x\xi)\, dx = -\xi^2 \bar{u}(\xi,y) - g(y)$$

where the function $\bar{u}(\xi,y)$ is defined by the equation

$$\bar{u}(\xi,y) = \int_0^\infty u(x,y) \cos(x\xi)\, dx$$

Since, in general

$$\frac{\partial^2 \bar{u}}{\partial y^2} = \int_0^\infty \frac{\partial^2 u(x,y)}{\partial y^2} \cos(x\xi)\, dx$$

we see that

$$\int_0^\infty \Delta_2 u(x,y) \cos(x\xi)\, dx = \frac{\partial^2 \bar{u}}{\partial y^2} - \xi^2 \bar{u} - g(y)$$

and hence that $\Delta_2 u = 0$ is equivalent to the equation

$$\frac{\partial^2 \bar{u}}{\partial y^2} - \xi^2 \bar{u} = g(y).$$

Also the conditions $u(x,0) = f(x)$ and $u(x,y) \to 0$ as $y \to \infty$, $x > 0$ give us the boundary conditions $\bar{u}(\xi,0) = \bar{f}(\xi)$, $\bar{u}(\xi,y) \to 0$ as $y \to \infty$, where

$$\bar{f}(\xi) = \int_0^\infty f(x) \cos(x\xi)\, dx.$$

The differential equation governing $\bar{u}(\xi,y)$ is in effect an ordinary differential (since derivatives of $\bar{u}$ with respect to ξ nowhere occur in the problem) so that the boundary value problem for determining $\bar{u}(\xi,y)$ is much simpler than that for finding $u(x,y)$. $\bar{u}(\xi,y)$ is an integral transform of $u(x,y)$.

To show the general situation, suppose that we wish to solve the equation (1-1-1) for $\mathbf{r} \in D$ and that, when $\mathbf{r} \in D$, $x_1 \in D_1$. Then for a wide class of functions $K(x_1,\xi)$ the partial differential equation (1-1-1) is equivalent to the relation

$$\int_{D_1} K(x_1,\xi) \mathbf{L} u(r)\, dx_1 = 0. \tag{1-2-1}$$

Now it may happen that, with a suitable choice of $K(x_1,\xi)$, the integral on the left side of this equation can be written in the form

$$\int_{D_1} K(x_1,\xi) \mathbf{M} u(r)\, dx_1 - g_1(\xi, x_2, \ldots, x_n)$$

where $\mathbf{M}$ is a linear differential operator in the variables $x_2, \ldots, x_n$ whose coefficients may be functions of ξ as well as of $x_2, \ldots, x_n$. (For example, in the problem considered above $\mathbf{L} = \Delta_2$, $\mathbf{M} = \partial^2/\partial y^2 - \xi^2$, $K(x,\xi) = \cos(x\xi)$ and $g_1(\xi,y) = g(y)$.) Since the operator $\mathbf{M}$ does not depend on x_1 we can, if the integral is uniformly convergent, take it outside the integral sign in which case the equation (1-2-1)—and hence the equation (1-1-1)—is equivalent to the partial differential equation

$$\mathbf{M}\bar{u}(\xi,x_2, \ldots ,x_n) = g_1(\xi,x_2, \ldots ,x_n) \tag{1-2-2}$$

where the function $\bar{u}(\xi,x_2, \ldots ,x_n)$ is defined by the equation

$$\bar{u}(\xi,x_2, \ldots ,x_n) = \int_{D_1} u(x_1,x_2, \ldots ,x_n)K(x_1,\xi)\, dx_1 \tag{1-2-3}$$

As before, we say that $\bar{u}(\xi,x_2, \ldots ,x_n)$ is the integral transform of $u(\mathbf{r})$ by the kernel $K(x_1,\xi)$ and we use a notation of the form

$$\bar{u}(\xi,x_2, \ldots ,x_n) = \mathscr{T}[u(\mathbf{r});x_1 \to \xi],$$

where in a special case the symbol $\mathscr{T}$ would give an indication of the nature of the kernel K, and the "arrow" notation is used to indicate unambiguously which of the n variables $x_1,x_2, \ldots ,x_n$ is "integrated out." Thus, if the domain of x_2 were the same as that of x_1, we should denote the transform

$$\int_{D_1} u(x_1,x_2, \ldots ,x_n)K(x_2,\xi)\, dx_2$$

by the symbol

$$\mathscr{T}[u(\mathbf{r});x_2 \to \xi].$$

It will be observed that (1-2-2) is a partial differential equation in $n - 1$ independent variables whereas equation (1-1-1) is in n independent variables. The effect of the transform is to reduce the number of independent variables by 1. In particular if $n = 2$, a partial differential equation in two independent variables is reduced to what is effectively an ordinary differential equation. This procedure is illustrated in secs. 1-6 and 1-7 below.

If we take D_1 to be the positive real line and $K(x_1,\xi)$ to be $e^{-x_1\xi}$, we get the *Laplace transform*; i.e., the function $\bar{u}(p,y)$, defined by the equation

$$\bar{u}(p,y) = \int_0^\infty u(x,y)e^{-px}\, dx, \tag{1-2-4}$$

is called the Laplace transform of the function $u(x,y)$ with respect to the variable x and

$$\bar{u}(x,p) = \int_0^\infty u(x,y)e^{-py}\, dy$$

its Laplace transform with respect to the variable y.

Similarly, if we take D_1 to be the positive real line and $K(x_1,\xi)$ to be

x_1^{s-1}, we get the *Mellin transform*; i.e.,

$$u^*(s,y) = \int_0^\infty u(x,y)x^{s-1}\,dx \tag{1-2-5}$$

is the Mellin transform of the function $u(x,y)$ with respect to the variable x.

We shall show in secs. 1-6 and 1-7 below how such transforms arise naturally in the solution of simple problems in mathematical physics. It may appear rather mysterious how we know what function $K(x_1,\xi)$ will transform the original differential equation (1-2-1) to give the required simplification. Some indication of how we know what kernels we should choose is given in Problems 1-2 and 1-3 at the end of this chapter.

1-3 SOLUTION OF LAPLACE'S EQUATION IN A HALF-PLANE

We shall now illustrate, by means of a simple example, how the Fourier transform defined in equation (1-1-6) arises in practical applications.

We consider the problem of determining the function $u(x,y)$ satisfying Laplace's equation

$$\Delta_2 u(x,y) = 0 \tag{1-3-1}$$

in the half-plane $y \geqslant 0$ subject to the boundary conditions

$$u(x,0) = f(x), \qquad -\infty < x < \infty \tag{1-3-2}$$

$$u(x,y) \to 0, \qquad \text{as} \qquad \rho \to \infty \tag{1-3-3}$$

where $\rho = (x^2 + y^2)^{\frac{1}{2}}$.

We saw above (sec. 1-1) that the method of separation of variables yields a simple solution of equation (1-3-1) of the form

$$\eta(x,y;\xi) = (2\pi)^{-\frac{1}{2}} e^{i\xi x - |\xi| y}$$

where ξ is a real separation constant. At this stage it may seem strange to introduce a factor $(2\pi)^{-\frac{1}{2}}$; it is done to define a transform whose inversion theorem (it will emerge later) has a more symmetrical form. This solution has the property that $\eta(x,y;\xi) \to 0$ as $y \to \infty$. If $F(\xi)$ is an arbitrary function of the real variable ξ, we should therefore expect from a simple use of the superposition principle that the function

$$u(x,y) = (2\pi)^{-\frac{1}{2}} \int_{-\infty}^{\infty} F(\xi) e^{i\xi x - |\xi| y}\,d\xi \tag{1-3-4}$$

would satisfy Laplace's equation if $F(\xi)$ were such that the integral on the right were uniformly convergent for all (x,y) in the upper half-plane $y > 0$. We should also expect it to satisfy the condition $u(x,y) \to 0$ as $y \to \infty$.

The remaining conditions will be satisfied if we can find a function $F(\xi)$ such that

$$f(x) = (2\pi)^{-\frac{1}{2}} \int_{-\infty}^{\infty} F(\xi)e^{i\xi x}\, d\xi \tag{1-3-5}$$

and such that

$$\int_{-\infty}^{\infty} F(\xi)e^{i\xi x - |\xi|y}\, d\xi \to 0, \qquad \text{as} \qquad |x| \to \infty. \tag{1-3-6}$$

From (1-3-5) we see that the function $f(x)$ is the Fourier transform of the function $F(\xi)$, so that the problem of finding $F(\xi)$ is equivalent to that of finding a function whose Fourier transform is prescribed, i.e., of finding the inversion theorem for the Fourier transform. This latter problem is discussed in sec. 2-3. All that then remains is to verify that the function $F(\xi)$ obtained as a result of this inversion, satisfies the condition (1-3-6).

To illustrate the second method of approach outlined in sec. 1-2, we multiply both sides of (1-3-1) by $\exp(i\xi x)$ and integrate with respect to x over the whole real line to obtain the equivalent relation

$$\int_{-\infty}^{\infty} \Delta_2 u(x,y)e^{i\xi x}\, dx = 0.$$

Now, if both u and $\partial u/\partial x$ tend to zero as $|x| \to \infty$ it follows as a result of two integrations by parts that

$$\int_{-\infty}^{\infty} \frac{\partial^2 u}{\partial x^2} e^{i\xi x}\, dx = -\xi^2 \int_{-\infty}^{\infty} u(x,y)e^{i\xi x}\, dx.$$

Hence if we introduce the Fourier transform

$$U(\xi,y) = (2\pi)^{-\frac{1}{2}} \int_{-\infty}^{\infty} u(x,y)e^{i\xi x}\, dx \tag{1-3-7}$$

of the function $u(x,y)$ with respect to the variable x, we find that it is a solution of the differential equation

$$\left(\frac{\partial^2}{\partial y^2} - \xi^2\right) U(\xi,y) = 0, \tag{1-3-8}$$

which is, in effect, an ordinary differential equation. From (1-3-2) we see that

$$U(\xi,0) = (2\pi)^{-\frac{1}{2}} \int_{-\infty}^{\infty} f(x)e^{i\xi x}\, dx \tag{1-3-9}$$

i.e., that

$$U(\xi,0) = F(\xi), \tag{1-3-10}$$

where $F(\xi)$ is the Fourier transform of the prescribed function $f(x)$. Similarly the limiting condition (1-3-3) is equivalent to the condition

$$U(\xi,y) \to 0, \qquad y \to \infty. \tag{1-3-11}$$

Solving the equation (1-3-8) subject to the conditions (1-3-9) and (1-3-11) we find that the transform $U(\xi,y)$ is given by the simple formula

$$U(\xi,y) = F(\xi)e^{-|\xi|y}. \tag{1-3-12}$$

From the equation (1-3-7) we see that the problem of determining the function $u(x,y)$ is equivalent to that of finding the function whose Fourier transform with respect to x is the function on the right side of equation (1-3-12). As before, therefore, we once again discover that we need to have an inversion formula for Fourier transforms; as pointed out above, this is discussed in sec. 2-3.

1-4 SOLUTION OF THE WAVE EQUATION FOR A SEMI-INFINITE LINE

As an example of how the Fourier cosine and the Fourier sine transform can be introduced to solve problems in mathematical physics, we consider the solution of the one-dimensional wave equation

$$\frac{\partial^2 u}{\partial x^2} = \frac{1}{c^2}\frac{\partial^2 u}{\partial t^2} \tag{1-4-1}$$

for $x > 0$, $t > 0$.

In the first instance we shall consider the initial conditions

$$u(x,0) = f(x), \qquad u_t(x,0) = 0 \tag{1-4-2}$$

and the boundary conditions

$$u(0,t) = 0, \tag{1-4-3}$$

$$u(x,t) \to 0 \quad \text{as} \quad x \to \infty. \tag{1-4-4}$$

By the method of separation of variables we see that for any positive real separation constant ξ,

$$\eta(x,t) = (2/\pi)^{\frac{1}{2}} \sin(\xi x)\cos(\xi ct)$$

is a solution of the wave equation (1-4-1) satisfying the conditions

$$\eta_t(x,0) = 0, \qquad \eta(0,t) = 0.$$

Using the superposition principle we see that, for any arbitrary function $F_s(\xi)$ which leads to the integral being uniformly convergent, the function

$$u(x,t) = \left(\frac{2}{\pi}\right)^{\frac{1}{2}} \int_0^\infty F_s(\xi)\sin(\xi x)\cos(\xi ct)\,d\xi \tag{1-4-5}$$

is a solution of the wave equation satisfying the second of the initial conditions (1-4-2) and the boundary condition (1-4-3). It will satisfy the remaining initial condition of the pair (1-4-2) if we can find a function $F_s(\xi)$ such that

$$f(x) = \left(\frac{2}{\pi}\right)^{\frac{1}{2}} \int_0^\infty F_s(\xi) \sin(\xi x)\, d\xi. \tag{1-4-6}$$

From equation (1-8) we see that the right side of equation (1-4-6) is the Fourier sine transform of the function $F_s(\xi)$. Our problem is therefore shown to be equivalent to that of finding a function $F_s(\xi)$ whose Fourier sine transform is $f(x)$. This problem is discussed in sec. 2-5 below.

Having determined $F_s(\xi)$ we still have to show that the function defined by equation (1-4-5) satisfies the condition (1-4-4). Since we can write equation (1-4-5) in the form

$$u(x,t) = (2\pi)^{-\frac{1}{2}} \int_0^\infty F_s(\xi)\{\sin \xi(x + ct) - \sin \xi(x - ct)\}\, d\xi,$$

we see that this condition will be satisfied if we can show that

$$\int_0^\infty F_s(\xi) \sin(\lambda\xi)\, d\xi \to 0$$

as $\lambda \to \infty$. This problem is considered in sec. 2-2.

If, instead of the condition (1-4-3), we have the condition

$$u_x(0,t) = 0, \tag{1-4-8}$$

the other conditions remaining unaltered, we can easily show by a similar method that the corresponding solution can be written in the form

$$u(x,t) = \left(\frac{2}{\pi}\right)^{\frac{1}{2}} \int_0^\infty F_c(\xi) \cos(\xi x) \cos(\xi ct)\, d\xi \tag{1-4-9}$$

provided that we can find a function $F_c(\xi)$ such that the integral on the right side of this equation is uniformly convergent and that the conditions

$$f(x) = \left(\frac{2}{\pi}\right)^{\frac{1}{2}} \int_0^\infty F_c(\xi) \cos(\xi x)\, d\xi, \tag{1-4-10}$$

$$\int_0^\infty F_c(\xi) \cos(\xi\lambda)\, d\xi \to 0, \quad \text{as} \quad \lambda \to \infty \tag{1-4-11}$$

are satisfied.

We therefore need to look at the problem of determining a function when its Fourier cosine transform is prescribed; this is considered in sec. 2-4 below. The second problem, expressed by the relation (1-4-11) is considered in sec. 2-2.

1-5 AN AXISYMMETRIC PROBLEM IN POTENTIAL THEORY

To show the manner in which Hankel transforms may enter into the analysis of boundary value problems we consider the axisymmetric solution $u(\rho,z)$ of Laplace's equation

$$\frac{\partial^2 u}{\partial \rho^2} + \frac{1}{\rho}\frac{\partial u}{\partial \rho} + \frac{\partial^2 u}{\partial z^2} = 0 \tag{1-5-1}$$

in the half-space $\rho > 0$, $z > 0$, which satisfies the boundary conditions

$$u(\rho,0) = f(\rho), \qquad \rho > 0, \tag{1-5-2}$$

$$u(\rho,z) \to 0, \qquad \text{as} \qquad r \to \infty, \tag{1-5-3}$$

where $f(\rho)$ is a prescribed function of ρ, and $r = (\rho^2 + z^2)^{\frac{1}{2}}$.

It is easily shown (Cf. Appendix, sec. A.1) that a simple solution of the equation (1-5-1) is

$$\eta(\rho,z;\xi) = J_0(\xi\rho)e^{-\xi z},$$

where J_0 denotes the zero-order Bessel function of the first kind. This solution has the property that, if $\xi > 0$, $\eta(\rho,z;\xi) \to 0$ as $r \to \infty$. By simple superposition we can therefore construct solutions of the form

$$u(\rho,z) = \int_0^\infty \xi\psi(\xi)J_0(\xi\rho)e^{-\xi z}\, d\xi, \tag{1-5-4}$$

where $\psi(\xi)$ is any arbitrary function of ξ such that the integral on the right is uniformly convergent. The condition (1-5-2) will be satisfied if we can find a function $\psi(\xi)$ such that

$$f(\rho) = \int_0^\infty \xi\psi(\xi)J_0(\xi\rho)\, d\xi, \tag{1-5-5}$$

i.e., if we can find a function $\psi(\xi)$ whose Hankel transform of order zero is $f(\rho)$, and which is such that the integral on the right of (1-5-4) tends to zero as $r \to \infty$.

The properties of Hankel transforms are discussed in Chapter 5.

1-6 DIFFUSION ALONG A SEMI-INFINITE LINE

The occurrence of Laplace transforms can be demonstrated by considering the simple problem of determining the solution $u(x,t)$ of the one-dimensional diffusion equation

$$\frac{\partial^2 u}{\partial x^2} = \frac{1}{\kappa}\frac{\partial u}{\partial t} \tag{1-6-1}$$

in the region $x > 0$, $t > 0$ subject to the initial condition

$$u(x,0) = 0 \tag{1-6-2}$$

and the boundary conditions

$$u(0,t) = f(t), \tag{1-6-3}$$

$$u(x,t) \to 0, \qquad \text{as} \qquad x \to \infty. \tag{1-6-4}$$

We shall also assume that $u(x,t)$ remains finite as $t \to \infty$.

We use the method of sec. 1-2.

If we multiply both sides of (1-6-1) by e^{-pt} and integrate with respect to t from 0 to ∞, we find that

$$\int_0^\infty \frac{\partial^2 u}{\partial x^2} e^{-pt}\,dt = \frac{1}{\kappa}\int_0^\infty \frac{\partial u}{\partial t} e^{-pt}\,dt.$$

Integrating by parts we see that the integral on the right side of this equation can be written as

$$\frac{1}{\kappa}[u(x,t)e^{-pt}]_0^\infty + \frac{p}{\kappa}\bar{u}(x,p),$$

where

$$\bar{u}(x,p) = \int_0^\infty u(x,t)e^{-pt}\,dt \tag{1-6-5}$$

is the Laplace transform of $u(x,t)$ with respect to the variable t. Because of the initial condition (1-6-2) and the fact that u remains finite as $t \to \infty$, we see that

$$\frac{1}{\kappa}\int_0^\infty \frac{\partial u}{\partial t} e^{-pt}\,dt = \frac{p}{\kappa}\bar{u}(x,p),$$

so that

$$\frac{\partial^2 \bar{u}}{\partial x^2} = \frac{p}{\kappa}\bar{u}(x,p). \tag{1-6-6}$$

Also, from equation (1-6-3), we have the condition

$$\bar{u}(0,p) = \int_0^\infty u(0,t)e^{-pt}\,dt = \int_0^\infty f(t)e^{-pt}\,dt,$$

showing that

$$\bar{u}(0,p) = \bar{f}(p), \tag{1-6-7}$$

where, comparing with equation (1-2-4), we see that $\bar{f}(p)$ is the Laplace transform of the prescribed function $f(t)$. Similarly we see that the condition (1-6-4) is equivalent to the relation

$$\bar{u}(x,p) \to 0 \qquad \text{as} \qquad x \to \infty. \tag{1-6-8}$$

We recall that $(p/\kappa)^{\frac{1}{2}}$ is a complex number. If we take the branch of $(p/\kappa)^{\frac{1}{2}}$ with the positive real part we see that the solution of the differential equation (1-6-6) satisfying the boundary conditions (1-6-7) and (1-6-8) is

$$\bar{u}(x,p) = \bar{f}(p)e^{-x\sqrt{(p/k)}}.$$

This function can be found once we know the inversion formula for the Laplace transform; this is discussed in sec. 3-10. Other properties of the Laplace transform are discussed in Chapter 3.

1-7 A PROBLEM IN POTENTIAL THEORY CONCERNING AN INFINITE WEDGE

Mellin transforms arise in a natural fashion in the discussion of solutions $u(\rho,\phi)$ of the two-dimensional Laplace equation appropriate to the infinite wedge $\rho > 0$, $|\phi| < \alpha$, where ρ and ϕ are plane polar coordinates. In terms of these coordinates, Laplace's equation can be written

$$\frac{\partial^2 u}{\partial \phi^2} + \frac{1}{\rho}\frac{\partial u}{\partial \rho} + \frac{1}{\rho^2}\frac{\partial^2 u}{\partial \phi^2} = 0. \tag{1-7-1}$$

If we use the formula for integrating by parts we see that

$$\begin{aligned}\int_0^\infty \rho^{s+1}\left(\frac{\partial^2 u}{\partial \rho^2} + \frac{1}{\rho}\frac{\partial u}{\partial \rho}\right) d\rho &= \int_0^\infty \rho^s \frac{\partial}{\partial \rho}\left(\rho \frac{\partial u}{\partial \rho}\right) d\rho \\ &= \left[\rho^{s+1}\frac{\partial u}{\partial \rho} - s\rho^s u(\rho,\phi)\right]_0^\infty + s^2 \int_0^\infty u(\rho,\phi)\rho^{s-1}\, d\rho\end{aligned}$$

so that if the functions

$$\rho^{s+1}\frac{\partial u}{\partial \rho}, \qquad \rho^s u(\rho,\phi)$$

tend to zero as $\rho \to 0$ and as $\rho \to \infty$ we find that

$$\int_0^\infty \rho^{s+1}\left(\frac{\partial^2 u}{\partial \rho^2} + \frac{1}{\rho}\frac{\partial u}{\partial \rho}\right) d\rho = s^2 u^*(s,\phi) \tag{1-7-2}$$

where

$$u^*(s,\phi) = \int_0^\infty \rho^{s-1} u(\rho,\phi)\, d\rho. \tag{1-7-3}$$

Comparing equation (1-7-3) with equation (1-2-5) we see that $u^*(s,\phi)$ is the Mellin transform of the function $u(\rho,\phi)$ with respect to the variable ρ.

Equation (1-7-1) is therefore equivalent to the ordinary differential equation

$$\frac{\partial^2 u^*}{\partial \phi^2} + s^2 u^*(s,\phi) = 0. \tag{1-7-4}$$

If the conditions on the faces of the wedge are specified by the equations

$$u(\rho, \pm\alpha) = f(\rho),$$

we see that

$$u^*(s, \pm\alpha) = f^*(s), \tag{1-7-5}$$

where $f^*(s)$ is the Mellin transform of the function $f(\rho)$. The solution of equation (1-7-4) subject to the boundary conditions (1-7-5) is

$$u^*(s,\phi) = f^*(s)\,\frac{\cos(s\phi)}{\cos(s\alpha)}.$$

Hence we see that the required solution is the function $u(\rho,\phi)$, whose Mellin transform with respect to ρ is

$$f^*(s)\,\frac{\cos(s\phi)}{\cos(s\alpha)},$$

where $f^*(s)$ is the Mellin transform of the prescribed function $f(\rho)$. The properties of Mellin transforms, and, in particular, the inversion theorem are considered in Chapter 4.

1-8 CLASSES OF FUNCTIONS

The inversion theorems for integral transforms of functions of a very general type are lengthy and somewhat intricate. Since we are interested primarily in applications we shall consider only functions of the type which normally arise in the analysis of physical problems.

If a function $f(x)$, of a single independent variable x, is continuous on an open interval (a,b), we say that $f(x)$ belongs to the class of functions $\mathscr{C}(a,b)$, and we write $f \in \mathscr{C}(a,b)$; if the interval on which the function is continuous is closed we write $f \in \mathscr{C}[a,b]$. If the function f is continuous over the whole real line we write $f \in \mathscr{C}(\mathbf{R})$ or $f \in \mathscr{C}(-\infty,\infty)$.

A function $f(x)$ is said to be *piecewise continuous* in an interval (a,b) if the interval can be partitioned into a *finite* number of nonintersecting intervals

$$(a,a_1), (a_1, a_2), \ldots, (a_{n-1},b),$$

in each of which the function is continuous and has finite limits as x approaches the endpoints of each of the subintervals. A function of this kind is shown in Fig. 1-1. When a function is piecewise continuous in (a,b) we write $f \in \mathscr{P}(a,b)$.

If a_r is an endpoint of one of the subintervals into which (a,b) is partitioned, then by $f(a_r-)$ we mean

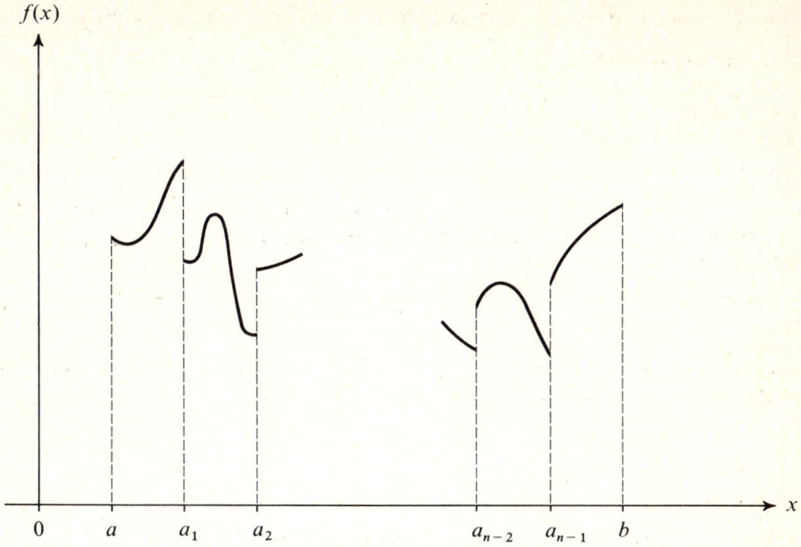

Fig. 1-1

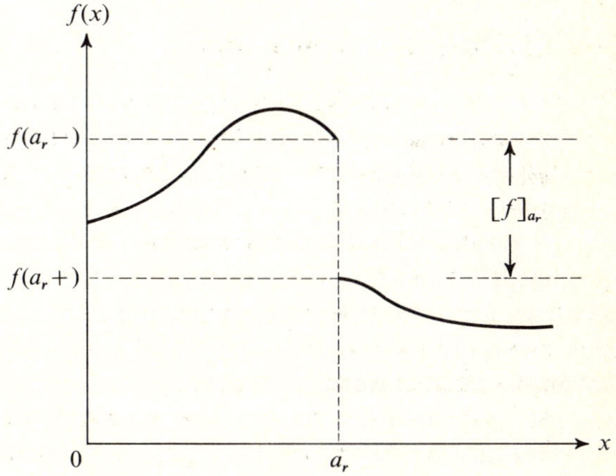

Fig. 1-2

$$\lim_{\varepsilon \to 0} f(a_r - \varepsilon)$$

where ε tends to zero through *positive* values, and by $f(a_r+)$ we mean

$$\lim_{\varepsilon \to 0} f(a_r + \varepsilon),$$

where, again, ε tends to zero through *positive* values. These values in a particular case are shown graphically in Fig. 1-2. The magnitude of the finite

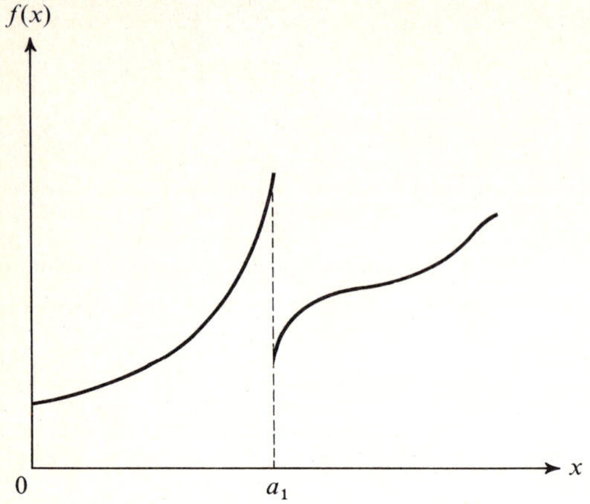

Fig. 1-3

discontinuity in the function at the point a_r is written $[f]_{a_r}$ so that

$$[f]_{a_r} = f(a_r+) - f(a_r-). \tag{1-8-1}$$

When a function $f(x)$ and each of its first m derivatives $f'(x), \ldots, f^{(m)}(x)$ is continuous on the interval (a,b) we say that f belongs to the class $\mathscr{C}^m(a,b)$ and we write $f \in \mathscr{C}^m(a,b)$. If *all* the derivatives of f are continuous on (a,b), i.e., if $f \in \mathscr{C}^k(a,b)$ for all positive integral values of k, we say that $f \in \mathscr{C}^\infty(a,b)$, or that f is *infinitely smooth* or, simply, *smooth*.

When a function and each of its first m derivatives are piecewise continuous on the interval (a,b) we say that $f \in \mathscr{P}^m(a,b)$.

It should be noted that f can belong to $\mathscr{P}(a,b)$ but not to $\mathscr{P}^1(a,b)$. For instance, the function $f(x)$ shown in Fig. 1-3 is piecewise continuous but not piecewise-continuously differentiable since $f'(x)$ is infinite at the point $x = a_1$.

It should be noted also that even if $f \in \mathscr{P}^1(a,b)$, the points of finite discontinuity of $f(x)$ need not coincide with those of its derivative $f'(x)$. For example, the function whose graph is shown in Fig. 1-4 has a point of finite discontinuity at $x = a_1$ but its derivative is continuous there (in the sense that the function possesses both a left-derivative and a right-derivative and their values are equal) while its derivative has a point of discontinuity at $x = b_1$, and $a_1 \neq b_1$.

If $f(x,y)$ is a function of two independent variables x and y which is continuous for all values (x,y) in the set $\{(x,y): x \in K_1,\ y \in K_2\}$, we write $f \in \mathscr{C}(K_1 \times K_2)$.

We have a similar notation for functions of several variables. If $x = (x_1, x_2, \ldots, x_n)$ is a point of the n-dimensional real Euclidean space $\mathbf{R}^n$, and if the function $f(x)$ is continuous at every point x of a domain D of $\mathbf{R}^n$,

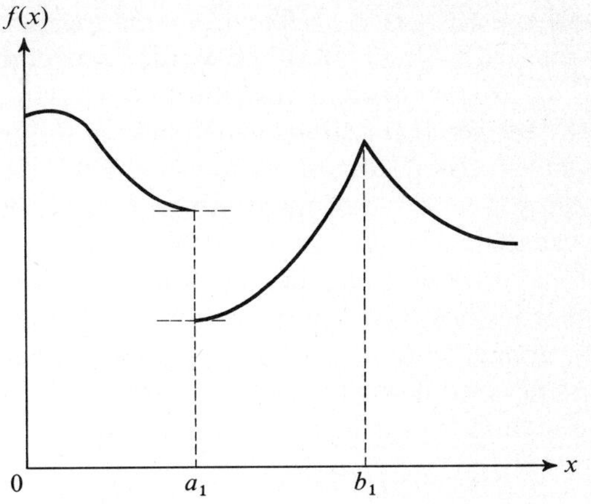

Fig. 1-4

we write $f \in \mathscr{C}(D)$. If, in addition, each of the partial derivatives of f up to and including all those of the m-th order are continuous on D we write $f \in \mathscr{C}^m(D)$. If f has continuous derivatives of all orders at all points of the domain D, we say that it is *infinitely smooth*, or simply that it is *smooth*, on D; we write $f \in \mathscr{C}^\infty(D)$.

Similarly, if D can be partitioned into the union of a finite number N of nonintersecting open sets $D_1, D_2, \ldots, D_N$ in each of which the function f and each of its partial derivatives up to and including all those of order m are continuous, we say that f is piecewise continuous on D and we write $f \in \mathscr{P}^m(D)$.

We now introduce the idea of the support of a function of a single real variable. If a function $f(x)$ is zero for values of x lying outside a set of points K on the real line, we say that K is the *support of f*; in other words the support of f is the closure of the set of points x at which $f(x) \neq 0$. We write

$$K = \operatorname{supp} f$$

so that

$$\operatorname{supp} f = \overline{\{x \in \mathbf{R} : f(x) \neq 0\}}.$$

If supp f consists of a closed bounded set on the real line we say that the function f has *compact support*.[1]

For example, if the function $g(x)$ is continuous on any interval which includes the open interval $(-a,a)$, the function

$$f(x) = g(x)H(a - |x|),$$

[1] This is because in **R**, a compact set is simply a bounded closed set. It will be recalled that in more general types of topological vector spaces a closed bounded set is not necessarily compact.

where $H(x)$ is the *Heaviside unit function*,[2] has as its support the closed interval $[-a,a]$. In other words f has compact support.

In the theory of distributions (Cf. Chapter 9) an important role is played by the set of functions which belong to the class $\mathscr{C}^\infty(\Omega)$, $\Omega \subset \mathbf{R}$, and which have a compact support K contained in Ω. A function $\phi \in \mathscr{C}^\infty(\Omega)$ such that $\operatorname{supp} \phi = K$ is a compact subset of Ω is said to belong to the class $\mathscr{D}(\Omega)$. The elements of $\mathscr{D}(\Omega)$ are also called *test functions* on Ω.

For example, the function e^{-x^2} belongs to $\mathscr{C}^\infty(\mathbf{R})$ but not to $\mathscr{D}(\mathbf{R})$ since it does not have compact support. On the other hand $e^{-x^2}H(1-|x|)$ has compact support $[-1,1]$ but does not belong to $\mathscr{C}^\infty(\mathbf{R})$ since its derivative is discontinuous at the points $x = \pm 1$. The prototype test function is the function $\zeta(x)$ defined by the equation

$$\zeta(x) = H(1 - |x|)\exp\{(x^2 - 1)^{-1}\}, \tag{1-8-2}$$

where H again denotes the Heaviside unit function. This function obviously has compact support $[-1,1]$ and it is easy to show that it is infinitely differentiable. By means of this function it is a simple matter to construct a test function whose support lies within the closed interval $[-a,a]$; $\zeta(x/b)$ would be such a function for any positive b satisfying the inequality $b < a$.

It should be noticed that the definition of $\mathscr{D}$ does not demand that all the elements of $\mathscr{D}$ should have the same support; for example, $\zeta(x)$ and $\zeta(x-3)$ are both elements of $\mathscr{D}(\mathbf{R})$ although their supports are respectively $[-1,1]$ and $[2,4]$ respectively.

These ideas are easily generalized to functions of several real variables. If Ω is a subset of $\mathbf{R}^n$, we say that ϕ is a test function and write $\phi \in \mathscr{D}(\Omega)$ if $\phi \in \mathscr{C}^\infty(\Omega)$ and if $\operatorname{supp} \phi$ is a compact subset of Ω. As an example of a test function in R^n, we have

$$\phi(x) = \zeta(|x|), \tag{1-8-3}$$

where $\zeta(x)$ is defined by equation (1-8-2) and $|x|$ by the equation

$$|x| = \sqrt{(x_1^2 + x_2^2 + \cdots + x_n^2)}. \tag{1-8-4}$$

Another space of test functions which enters into the theory of distributions is the space $\mathscr{S}$ of *test functions of rapid descent*. $\mathscr{S}$ is also called the *Schwartz space*. A function ϕ is said to be a test function of rapid descent if $\phi \in \mathscr{C}^\infty(R^n)$ and if ϕ and all its partial derivatives decrease to zero faster than every power of $|x|^{-1}$. In the case of $n = 1$ this latter condition can be written in the form

$$|x^m\phi^{(k)}(x)| \leqslant C_{mk}, \tag{1-8-5}$$

for all non-negative integers m and k and for all values of x.

[2]The function $H(x)$ takes the value 1 when $x > 0$, and the value 0 when $x < 0$.

For the corresponding expression in $\mathbf{R}^n$ it is convenient to denote the vector $(k_1,k_2, \ldots ,k_n)$, whose components are non-negative integers k_i, by k and the partial derivative

$$\frac{\partial^{k_1+k_2+\cdots+k_n}}{\partial x_1^{k_1}\,\partial x_2^{k_2}\cdots\partial x_n^{k_n}}\,\phi(x_1,x_2, \ldots ,x_n)$$

by $D^k\phi(x)$. The counterpart of (1-8-5) is then

$$|x|^m\left|D^k\phi(x)\right| \leqslant C_{m,k} \tag{1-8-6}$$

For some purposes it is also convenient to introduce the space whose elements are called *functions of slow growth*. We say that $f \in \mathscr{N}(\mathbf{R})$ if $f \in \mathscr{C}^\infty(\mathbf{R})$ and if there exists a positive integer N with the property that

$$|f^{(r)}(x)| = 0(|x|^N), \qquad |x| \to \infty \tag{1-8-7}$$

for all non-negative integers r. This can be written in the alternative form: every $f \in \mathscr{N}$ satisfies the set of inequalities

$$|x^{-N}f^{(r)}(x)| \leqslant c_r, \tag{1-8-8}$$

for some fixed integer N and for all non-negative integers r. The set $\mathscr{N}(\mathbf{R}^n)$ can be defined in a similar way with (1-8-8) replaced by

$$|x|^{-N}|D^k f(x)| \leqslant c_k, \tag{1-8-9}$$

with the notation of (1-8-6).

The function e^{-x^2} is obviously an element of $\mathscr{S}$ while any polynomial is an element of $\mathscr{N}$.

It should also be observed that Lighthill (1958) calls the elements of $\mathscr{S}$ *good functions* and those of $\mathscr{N}$ *fairly good functions*.

The concept of convergence in these spaces is important from the point of view of the theory of distributions (Cf. sec. 9-3).

A sequence of test functions $\{\phi_r(x)\}_{r=1}^\infty$, $x \in \mathbf{R}$, is said to *converge in* $\mathscr{D}$ if

(a) $\phi_r(x) \in \mathscr{D}, \qquad r = 1,2, \ldots ,$
(b) each of the functions $\phi_r(x)$ vanishes outside some fixed interval I of the real line;
(c) for each fixed positive integer k, the sequence $\{\phi_r^{(k)}(x)\}_{r=1}^\infty$ converges uniformly for all real values of x.

If $\phi(x)$ is the limit function of the sequence $\{\phi_r(x)\}_{r=1}^\infty$, it is easily shown that $\phi(x) \in \mathscr{D}$; we describe this property by saying that *the space* $\mathscr{D}$ *is closed under convergence*.

The definition of convergence in $\mathscr{D}(\mathbf{R}^n)$ can be extended in an obvious way.

Similarly, a sequence of functions $\{\phi_r(x)\}_{r=1}^\infty$ is said to *converge in* $\mathscr{S}$

if

(a) $\phi_r(x) \in \mathscr{S}, \qquad r = 1,2, \ldots;$

(b) for each set of non-negative integers $k = (k_1, k_2, \ldots, k_n)$ and for each positive integer m, the sequence $|x|^m\{D^k\phi_r(x)\}_{r=1}^{\infty}$ converges uniformly over $\mathbf{R}^n$.

In condition (b) the convergence must be uniform for any given set k and each non-negative integer m, but it is not required to be uniform over all such sets.

If the limit function of the sequence $\{\phi_r(x)\}_{r=1}^{\infty}$ of functions of rapid descent is $\phi(x)$, then again it is obvious that $\phi(x) \in \mathscr{S}$. Again we express this property in the words: *the space $\mathscr{S}$ is closed under convergence.*

Finally we introduce the set of functions $\mathscr{A}_1(\Omega)$. We say that $f \in \mathscr{A}_1(\Omega)$ if f is *absolutely integrable* over Ω, i.e., if

$$\int_\Omega |f(x)|\, dx$$

is finite. Similarly the statement $f \in \mathscr{A}_p(\Omega)$ implies that

$$\int_\Omega |f(x)|^p\, dx$$

is finite.

PROBLEMS

1-1 Show that a necessary and sufficient condition for $K(\xi x)$ to be a Fourier kernel is that

$$\int_0^\infty u^{-1}K(xu)K_1(tu)\, du = H(x) - H(x - t)$$

where H denotes the Heaviside unit function and

$$K_1(t) = \int_0^t K(s)\, ds.$$

1-2 If $u(x,y)$ satisfies the partial differential equation

$$a(x)\frac{\partial^2 u}{\partial x^2} + b(x)\frac{\partial u}{\partial x} + c(x)u + \frac{\partial^2 u}{\partial y^2} = 0$$

in the quadrant $x \geqslant 0$, $y \geqslant 0$, and the boundary conditions

(*a*) $u(0,y) = 0,$

(*b*) $u(x,0) = f(x),$

(*c*) $u(x,y) \to 0 \qquad \text{as} \qquad (x^2 + y^2)^{\frac{1}{2}} \to \infty,$

show that the transform $U(\xi,y)$ defined by the equation

$$U(\xi,y) = \int_0^\infty u(x,y)K(x,\xi)\, dx$$

whose kernel $K(x,\xi)$ satisfies the equation

$$\frac{\partial^2}{\partial x^2}\{a(x)K(x,\xi)\} - \frac{\partial}{\partial x}\{b(x)K(x,\xi)\} + \{c(x) + \xi^2\}K(x,\xi) = 0$$

and the condition $a(0)K(0,\xi) = 0$, is given by

$$U(\xi,y) = F(\xi)e^{-|\xi|y}$$

where

$$F(\xi) = \int_0^\infty f(x)K(x,\xi)\,dx.$$

1-3 If, in the above problem, the condition (a) is replaced by the condition $u_x(0,y) = 0$, show that the above result still holds except that now $K(x,\xi)$ satisfies the initial condition

$$\left[b(x)K(x,\xi) - \frac{\partial}{\partial x}\{a(x)K(x,\xi)\}\right]_{x=0} = 0.$$

1-4 The function $\delta_a(x)$, $x \in \mathbf{R}$, is defined by the equation

$$\delta_a(x) = \frac{\zeta(x/a)}{a\int_{-\infty}^{\infty}\zeta(t)\,dt}, \qquad (a > 0)$$

where the function ζ is defined by equation (1-8-2). Show that

$$\int_{-\infty}^{\infty}\delta_a(x)\,dx = 1, \qquad \operatorname{supp}\delta_a = [-a,a]$$

and sketch the graphs of the functions δ_a with $a = 0.25, 0.5, 1.0$.

1-5 Suppose that $f(x)$ is continuous for all x and that the function $\phi_a(x)$ is defined by the equation

$$\phi_a(x) = \int_{-\infty}^{\infty}\delta_a(x-\xi)f(\xi)\,d\xi, \qquad (a > 0)$$

where δ_a is the function defined in Problem 1-4.

Show that if $\operatorname{supp} f = [\alpha,\beta]$, then $\operatorname{supp}\phi_a = [\alpha - a, \beta + a]$, and that $\phi_a \in \mathscr{D}$.

Show also that, if we are given a prescribed positive number ε, we can find a positive number A such that, for all x,

$$|f(x) - \phi_a(x)| < \varepsilon,$$

whenever $a < A$.

1-6 If the functions ϕ_r and ψ_r are defined in terms of the function ζ— cf. equation (1-8-2)— through the equations

$$\phi_r(x) = r^{-1}\zeta(x), \qquad \psi_r(x) = r^{-1}\zeta(x/r),$$

show that the sequence $\{\phi_r(x)\}$ converges in $\mathscr{D}$ to zero but that the sequence $\{\psi_r(x)\}$ does not.

1-7 If the function $\xi(x)$ is defined by the equation

$$\xi(x) = H(1 - |x|) \exp \frac{x^2}{x^2 - 1}, \qquad x \in \mathbf{R}$$

show that the sequence $\{\xi(x/r)\}$ converges to 1, and that $\xi(x/r)$ is uniformly bounded with respect to r over the whole of $\mathbf{R}$.

Show also that if $\phi(x) \in \mathscr{S}$, and $\phi_r(x)$ is defined by the equation

$$\phi_r(x) = \phi(x)\xi(x/r), \qquad (r = 1,2, \ldots),$$

then $\phi_r \in \mathscr{D}$ and the sequence $\{\phi_r(x)\}$ converges in $\mathscr{S}$ to $\phi(x)$.†

†This result is usually expressed by saying that *the space $\mathscr{D}$ is dense in the space $\mathscr{S}$* (Cf. Zemanian 1965, Theorem 4.2-1, p. 101.)

2

Fourier Transforms

2-1 SOLUTION OF LAPLACE'S EQUATION FOR A LONG STRIP

We shall begin by deriving a *formal* solution of the problem stated in sec. 1-3 by finding the limiting form (as $a \to \infty$) of the solution of the corresponding problem in the semi-infinite strip $|x| < a, y > 0$. (Cf. Fig. 2-1.)

Our problem is first to determine a function $u(x,y)$ satisfying Laplace's equation

$$\frac{\partial^2 u}{\partial x^2} + \frac{\partial^2 u}{\partial y^2} = 0 \tag{2-1-1}$$

in the strip $|x| < a$, $y > 0$, and satisfying the boundary conditions

$$u(x,0) = f(x), \qquad -a \leqslant x \leqslant a \tag{2-1-2}$$

$$u(x,y) \to 0, \qquad \text{as} \qquad y \to \infty. \tag{2-1-3}$$

By a simple procedure based on the method of separation of variables we have a solution of the equation (2-1-1) of the form

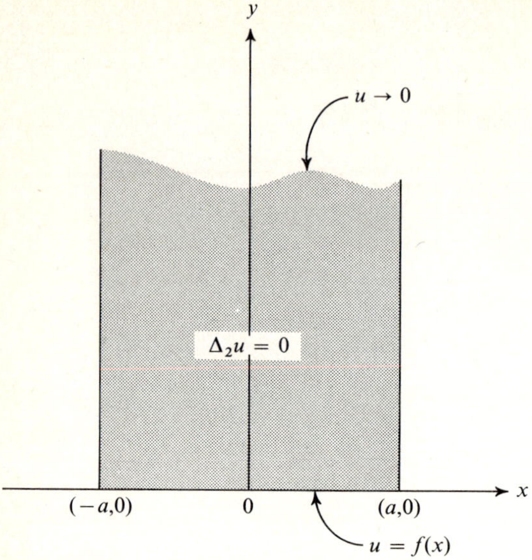

Fig. 2-1

$$u(x,y) = \sum_{n=-\infty}^{\infty} a_n \exp\left(-in\pi x/a - |n|\pi y/a\right) \tag{2-1-4}$$

where, to satisfy the boundary condition (2-1-2) we must choose the constants a_n to be such that

$$f(x) = \sum_{n=-\infty}^{\infty} a_n e^{-in\pi x/a}, \qquad -a < x < a. \tag{2-1-5}$$

From the elementary theory of Fourier series we know that if $f(x) \in \mathscr{P}^1(-a,a)$, the constants a_n are calculated from the formula

$$a_n = \frac{1}{2a}\int_{-a}^{a} f(u)e^{in\pi u/a}\,du \tag{2-1-6}$$

so that the required potential function is given by the equation

$$u(x,y) = \frac{1}{2a}\sum_{n=-\infty}^{\infty}\int_{-a}^{a} f(u)e^{in\pi(u-x)/a-|n|\pi y/a}\,du. \tag{2-1-7}$$

We now consider the form taken by these expressions as we let $a \to \infty$.

If we substitute from (2-1-6) into (2-1-5) we obtain the identity

$$f(x) = \frac{1}{2a}\sum_{n=-\infty}^{\infty} e^{-in\pi x/a}\int_{-a}^{a} f(u)e^{in\pi u/a}\,du. \tag{2-1-8}$$

Proceeding purely formally, we find on writing $\delta\xi = \pi/a$, that

$$f(x) = \frac{1}{2\pi} \sum_{n=-\infty}^{\infty} \delta\xi \int_{-\pi/\delta\xi}^{\pi/\delta\xi} f(u) e^{i(u-x)n\delta\xi}\, du$$

and we note that $\delta\xi \to 0$ as $a \to \infty$. By the definition of the Riemann integral as the limit of a sum we see that as $\delta\xi \to 0$, the series

$$\sum_{n=-\infty}^{\infty} \delta\xi \int_{-\pi/\delta\xi}^{\pi/\delta\xi} f(u) e^{i(u-x)n\delta\xi}\, du$$

tends to the integral

$$\int_{-\infty}^{\infty} d\xi \int_{-\infty}^{\infty} f(u) e^{i(u-x)\xi}\, du,$$

so that, in the limit as $a \to \infty$ the identity (2-1-8) would appear to take the form

$$f(x) = \frac{1}{2\pi} \int_{-\infty}^{\infty} e^{-ix\xi}\, d\xi \int_{-\infty}^{\infty} f(u) e^{iu\xi}\, du. \tag{2-1-9}$$

This can be written in another way. If we put

$$F(\xi) = (2\pi)^{-\frac{1}{2}} \int_{-\infty}^{\infty} f(u) e^{iu\xi}\, du, \tag{2-1-10a}$$

then equation (2-1-9) states that

$$f(x) = (2\pi)^{-\frac{1}{2}} \int_{-\infty}^{\infty} F(\xi) e^{-ix\xi}\, d\xi. \tag{2-1-10b}$$

In a similar way if we let $a \to \infty$ in (2-1-7) we find that the formal solution of this potential problem is

$$u(x,y) = \frac{1}{2\pi} \int_{-\infty}^{\infty} e^{-ix\xi - y|\xi|}\, d\xi \int_{-\infty}^{\infty} f(u) e^{iu\xi}\, du,$$

which can be written in the form

$$u(x,y) = (2\pi)^{-\frac{1}{2}} \int_{-\infty}^{\infty} F(\xi) e^{-ix\xi - y|\xi|}\, d\xi, \tag{2-1-11}$$

where $F(\xi)$ is defined by equation (2-1-10*a*).

2-2 FOURIER'S INTEGRAL THEOREM

We saw in the last section by a purely formal argument that in some sense the identity

$$f(x) = \frac{1}{2\pi} \int_{-\infty}^{\infty} e^{-ix\xi}\, d\xi \int_{-\infty}^{\infty} f(u) e^{iu\xi}\, du$$

might be valid. This identity is known as *Fourier's integral theorem*. In this section we shall show that Fourier's integral theorem holds for functions $f(x)$ which are piecewise-continuously differentiable and absolutely integrable on the whole real line.

Fourier's integral has been proved under much more general conditions on the behavior of the function $f(x)$. We shall look at some of these results in sec. 2-11, but we shall here confine our attention to establishing the theorem for the class of functions stated since that class is sufficiently wide to embrace most of the functions which arise in problems in mathematical physics.

The central result is:

The Riemann–Lebesgue lemma *If* $f(t) \in \mathscr{C}[a,b]$, *where* $0 < a < b < \infty$,

$$\int_a^b f(t) \sin(\lambda t)\, dt \to 0, \qquad \int_a^b f(t) \cos(\lambda t)\, dt \to 0$$

as $\lambda \to \infty$.

We shall begin by proving the first result.

By a suitable change of variable we find that

$$\int_a^b f(t) \sin(\lambda t)\, dt = -\int_{a-\pi/\lambda}^{b-\pi/\lambda} f(\tau + \pi/\lambda) \sin(\lambda\tau)\, d\tau$$

and hence that

$$2\int_a^b f(t) \sin(\lambda t)\, dt = \int_a^b f(t) \sin(\lambda t)\, dt - \int_{a-\pi/\lambda}^{b-\pi/\lambda} f(t + \pi/\lambda) \sin(\lambda t)\, dt.$$

We may now write this result in the form

$$\int_a^b f(t) \sin(\lambda t)\, dt = \tfrac{1}{2}(I_1 + I_2 - I_3)$$

where

$$I_1 = \int_{b-\pi/\lambda}^{b} f(t) \sin(\lambda t)\, dt, \quad I_2 = \int_a^{b-\pi/\lambda} \{f(t) - f(t + \pi/\lambda)\} \sin(\lambda t)\, \mathrm{dt}$$

$$I_3 = \int_{a-\pi/\lambda}^{a} f(t + \pi/\lambda) \sin(\lambda t)\, dt.$$

Since $f(t)$ is continuous in $[a,b]$ it is bounded in that interval; suppose that $|f(t)| < M$ in $[a,b]$. Then given any arbitrary positive constant ε, we can choose a positive constant η_1 such that

$$\frac{\pi}{\lambda} < \frac{\varepsilon}{2M}, \qquad (\lambda > \eta_1)$$

so that, if $\lambda > \eta_1$,

$$|I_1| < M\pi/\lambda < \tfrac{1}{2}\varepsilon, \qquad |I_3| < M\pi/\lambda < \tfrac{1}{2}\varepsilon.$$

If $f(t)$ is continuous in a finite closed interval $[a,b]$ then it is uniformly continuous in $[a,b]$ so that, given ε we can find a positive number η_2 such that

$$|f(t + \pi/\lambda) - f(t)| < \varepsilon/(b - a), \qquad (\lambda > \eta_2).$$

Hence if $\lambda > \eta_2$,

$$|I_2| < \int_a^{b-\pi/\lambda} |f(t + \pi/\lambda) - f(t)|\, dt < \frac{b - a - \pi/\lambda}{b - a}\varepsilon$$

showing that $|I_2| < \varepsilon$.

Collecting these results together we see that if we write $\eta = \max(\eta_1,\eta_2)$, then

$$\left|\int_a^b f(t)\sin(\lambda t)\, dt\right| < 2\varepsilon, \qquad \lambda > \eta,$$

and the first result follows.

In proving this result the only properties of $\sin(\lambda t)$ which we have used are that it is continuous and that for real values of t, $\sin(\lambda t + \pi) = -\sin(\lambda t)$ and $|\sin(\lambda t)| \leqslant 1$. The function $\cos(\lambda t)$ has the same properties so that the second result can be established in exactly the same way.

Corollary 1 *If $f(t) \in \mathscr{P}[a,b]$, where $0 < a < b < \infty$,*

$$\int_a^b f(t) \begin{matrix}\sin\\ \cos\end{matrix} (\lambda t)\, dt \to 0,$$

as $\lambda \to \infty$.

We apply the Riemann–Lebesgue lemma to each of the integrals over the intervals $[a,a_1),(a_1,a_2), \ldots ,(a_{n-1},a_n),(a_n,b]$ in which $f(t)$ is continuous, and the result follows immediately.

Corollary 2 *If $f(t) \in \mathscr{P}[a,\infty)$, $a > 0$, and if $f(t) \in \mathscr{A}_1[a,\infty)$, then*

$$\int_a^\infty f(t) \begin{matrix}\sin\\ \cos\end{matrix} (\lambda t)\, dt \to 0$$

as $\lambda \to \infty$.

To prove this we may assume, without loss of generality, that $f(t)$ is continuous in $[a,\infty)$. Since $f(t)$ is absolutely integrable in the same interval, we can find a positive number N such that

$$\int_N^\infty |f(t)|\,dt < \tfrac{1}{2}\varepsilon$$

for any prescribed positive number ε.

Since

$$\left|\int_a^\infty f(t)\frac{\sin}{\cos}(\lambda t)\,dt\right| = \left|\int_a^N f(t)\frac{\sin}{\cos}(\lambda t)\,dt + \int_N^\infty f(t)\frac{\sin}{\cos}(\lambda t)\,dt\right|$$
$$\leqslant \left|\int_a^N f(t)\frac{\sin}{\cos}(\lambda t)\,dt\right| + \int_N^\infty |f(t)|dt$$

and given ε and N we can, by the Riemann–Lebesgue lemma, find a positive number η such that

$$\left|\int_a^N f(t)\frac{\sin}{\cos}(\lambda t)\,dt\right| < \tfrac{1}{2}\varepsilon, \qquad \lambda > \eta$$

so that the result follows immediately.

We next prove:

The localization lemma *If $f(t) \in \mathscr{P}^1[0,a]$, where a is finite,*

$$\int_0^a f(t)\frac{\sin(\lambda t)}{t}\,dt \to \tfrac{1}{2}\pi f(0+)$$

as $\lambda \to \infty$.

It is sufficient to prove the result for $f(t) \in \mathscr{C}^1[0,a]$.

We may write

$$\int_0^a f(t)\frac{\sin(\lambda t)}{t}\,dt = I_1 + I_2$$

where

$$I_1 = f(0+)\int_0^a \frac{\sin(\lambda t)}{t}\,dt$$

$$I_2 = \int_0^a \{f(t) - f(0+)\}\frac{\sin(\lambda t)}{t}\,dt.$$

Since $f'(t)$ is continuous in $[0,a]$ it follows that $t^{-1}\{f(t) - f(0+)\}$ is continuous in $[0,a]$ and hence, by the Riemann–Lebesgue lemma, we see that given any prescribed positive number ε we can find a positive number η_1 such that

$$|I_2| < \tfrac{1}{2}\varepsilon, \qquad \text{if} \qquad \lambda > \eta_1.$$

If $f(0+) = 0$, the result is now proved. If $f(0+) \neq 0$ we need a further result. By a trivial change of the variable of integration we can write

$$\int_0^a \frac{\sin(\lambda t)}{t}\,dt = \int_0^{\lambda a} \frac{\sin u}{u}\,du.$$

Now the integral

$$\int_0^\infty \frac{\sin u}{u}\,du$$

is convergent and has the value $\frac{1}{2}\pi$. Hence for the given value of ε and for $f(0+) \neq 0$, we can always find a positive number η_2 such that

$$\left|\int_0^{\lambda a} \frac{\sin \xi}{\xi}\,d\xi - \tfrac{1}{2}\pi\right| < \frac{\varepsilon}{2|f(0+)|}, \qquad \lambda > \eta_2.$$

It follows that if we take η to be $\max(\eta_1, \eta_2)$, then for $\lambda > \eta$,

$$\left|\int_0^a f(t)\frac{\sin(\lambda t)}{t}\,dt - \tfrac{1}{2}\pi f(0+)\right| \leqslant |f(0+)|\left|\int_0^{\lambda a} \frac{\sin \xi}{\xi}\,d\xi - \tfrac{1}{2}\pi\right| + |I_2|$$

and the lemma is established.

Corollary 1 *If $f(t) \in \mathscr{P}^1[0,\infty)$, and if $t^{-1}f(t) \in \mathscr{A}_1[a,\infty)$, where a is an arbitrary positive number,*

$$\int_0^\infty f(t)\frac{\sin(\lambda t)}{t}\,dt \to \tfrac{1}{2}\pi f(0+)$$

as $\lambda \to \infty$.

Since $t^{-1}f(t)$ is absolutely integrable in $[a,\infty)$ we can find a positive number N such that for any prescribed positive number ε,

$$\int_N^\infty |t^{-1}f(t)|\,dt < \tfrac{1}{2}\varepsilon$$

and

$$\left|\int_0^\infty f(t)\frac{\sin(\lambda t)}{t}\,dt - \tfrac{1}{2}\pi f(0+)\right| \leqslant \left|\int_0^N f(t)\frac{\sin(\lambda t)}{t}\,dt - \tfrac{1}{2}\pi f(0+)\right| + \int_N^\infty |t^{-1}f(t)|\,dt.$$

From the localization lemma we know that corresponding to the given ε, N there exists a positive number η such that

$$\left| \int_0^N f(t) \frac{\sin(\lambda t)}{t} dt - \tfrac{1}{2}\pi f(0+) \right| < \tfrac{1}{2}\varepsilon, \quad (\lambda > \eta),$$

and the result follows.

Corollary 2 *If $f(t) \in \mathscr{P}^1(\mathbf{R})$, $f(t) \in \mathscr{A}_1(\mathbf{R})$, where $\mathbf{R}$ is the whole real line,*

$$\int_0^\infty f(x + t) \frac{\sin(\lambda t)}{t} dt \to \tfrac{1}{2}\pi f(x+), \qquad (x > 0),$$

as $\lambda \to \infty$.

This is a trivial extension of Corollary 1, obtained by a simple change of the variable of integration and the fact that if $f(t) \in \mathscr{A}_1(0,\infty)$, then $t^{-1}f(t) \in \mathscr{A}_1[a,\infty)$ for any $a > 0$.

Corollary 3 *If $f(t) \in \mathscr{P}^1(\mathbf{R})$, $f(t) \in \mathscr{A}_1(\mathbf{R})$, where $\mathbf{R}$ is the whole real line,*

$$\int_{-\infty}^\infty f(x + t) \frac{\sin(\lambda t)}{t} dt \to \tfrac{1}{2}\pi[f(x+) + f(x-)], \qquad (x \in \mathbf{R}),$$

as $\lambda \to \infty$.

This follows immediately since

$$\int_{-\infty}^0 f(x + t) \frac{\sin(\lambda t)}{t} dt = \int_0^\infty f(x - t) \frac{\sin(\lambda t)}{t} dt$$

We are now in a position to prove the key result:

Fourier's integral theorem *If $f(t) \in \mathscr{P}^1(\mathbf{R})$, $f(t) \in \mathscr{A}_1(\mathbf{R})$, then for all $x \in \mathbf{R}$,*

$$\frac{1}{\pi} \int_0^\infty d\xi \int_{-\infty}^\infty f(t) \cos\{\xi(x - t)\}\, dt = \tfrac{1}{2}[f(x+) + f(x-)].$$

To prove this result we begin by noting that if $k > 0$,

$$\begin{aligned}
\int_0^\infty f(t)\, dt \int_0^\lambda \cos\{\xi(t - x)\}\, d\xi - \int_0^\lambda d\xi \int_0^\infty f(t) \cos\{\xi(t - x)\}\, dt \\
= \int_0^k f(t)\, dt \int_0^\lambda \cos\{\xi(t - x)\}\, d\xi \\
+ \int_k^\infty f(t)\, dt \int_0^\lambda \cos\{\xi(t - x)\}\, d\xi \\
- \int_0^\lambda d\xi \int_0^k f(t) \cos\{\xi(t - x)\}\, dt \\
- \int_0^\lambda d\xi \int_k^\infty f(t) \cos\{\xi(t - x)\}\, dt
\end{aligned}$$

$$= \int_k^\infty f(t)\,dt \int_0^\lambda \cos\{\xi(t-x)\}\,d\xi$$
$$- \int_0^\lambda d\xi \int_k^\infty f(t)\cos\{\xi(t-x)\}\,dt.$$

Since $f(t)$ is absolutely convergent, we know that if we are given an arbitrary positive number ε, we can find a number K_1 such that

$$\int_k^\infty |f(t)|\,dt < \frac{\varepsilon}{2\lambda}, \qquad k > K_1.$$

From this it follows that if $k > K_1$,

$$\left|\int_0^\lambda d\xi \int_k^\infty f(t)\cos\{\xi(t-x)\}\,dt\right| < \int_0^\lambda d\xi \int_k^\infty |f(t)|\,dt < \tfrac{1}{2}\varepsilon.$$

Also, since

$$\int_0^\lambda \cos\{\xi(t-x)\}\,d\xi = \frac{\sin\{\lambda(t-x)\}}{t-x}$$

it follows that

$$\left|\int_k^\infty f(t)\,dt \int_0^\infty \cos\{\xi(t-x)\}\,d\xi\right| = \left|\int_{k-x}^\infty f(t+x)\frac{\sin(\lambda t)}{t}\,dt\right|.$$

By Corollary 1 to the localization lemma, we can find a number K_2 such that if $k > K_2 + x$,

$$\left|\int_{k-x}^\infty f(t+x)\frac{\sin(\lambda t)}{t}\,dt\right| < \tfrac{1}{2}\varepsilon.$$

Thus no matter how large λ is, $K = \max(K_1,K_2)$ can be chosen so large that

$$\left|\int_k^\infty f(t)\,dt \int_0^\lambda \cos\{\xi(t-x)\}\,d\xi - \int_0^\lambda d\xi \int_k^\infty f(t)\cos\{\xi(t-x)\}\,dt\right| < \varepsilon,$$
$$(k > K).$$

In other words

$$\lim_{\lambda\to\infty} \int_0^\infty f(t)\,dt \int_0^\lambda \cos\{\xi(t-x)\}\,d\xi$$
$$= \lim_{\lambda\to\infty} \int_0^\lambda d\xi \int_0^\infty f(t)\cos\{\xi(t-x)\}\,dt.$$

In an exactly similar way we can show that

$$\lim_{\lambda\to\infty} \int_{-\infty}^0 f(t)\,dt \int_0^\lambda \cos\{\xi(t-x)\}\,d\xi$$
$$= \lim_{\lambda\to\infty} \int_0^\lambda d\xi \int_{-\infty}^0 f(t)\cos\{\xi(t-x)\}\,dt$$

and, by addition, that

$$\lim_{\lambda\to\infty}\int_{-\infty}^{\infty} f(t)\,dt\int_0^{\lambda}\cos\{\xi(t-x)\}\,d\xi$$
$$=\lim_{\lambda\to\infty}\int_0^{\lambda}d\xi\int_{-\infty}^{\infty} f(t)\cos\{\xi(t-x)\}\,dt.$$

Now, by Corollary 3 of the localization lemma, we have that

$$\begin{aligned}\tfrac{1}{2}\{f(x+)+f(x-)\} &= \lim_{\lambda\to\infty}\frac{1}{\pi}\int_{-\infty}^{\infty} f(x+u)\frac{\sin(\lambda u)}{u}\,du\\ &= \lim_{\lambda\to\infty}\frac{1}{\pi}\int_{-\infty}^{\infty} f(t)\frac{\sin\{\lambda(t-x)\}}{t-x}\,dt\\ &= \lim_{\lambda\to\infty}\frac{1}{\pi}\int_{-\infty}^{\infty} f(t)\,dt\int_0^{\lambda}\cos\{\xi(t-x)\}\,d\xi\\ &= \lim_{\lambda\to\infty}\frac{1}{\pi}\int_0^{\lambda}d\xi\int_{-\infty}^{\infty} f(t)\cos\{\xi(t-x)\}\,dt.\end{aligned}$$

Hence we obtain the result

$$\tfrac{1}{2}[f(x+)+f(x-)] = \frac{1}{\pi}\int_0^{\infty}d\xi\int_{-\infty}^{\infty} f(t)\cos\{\xi(t-x)\}\,dt.$$

If we replace the cosine function by its exponential form we see that we may rewrite this equation in the form

$$\tfrac{1}{2}[f(x+)+f(x-)] = \frac{1}{2\pi}\int_{-\infty}^{\infty} e^{-i\xi x}\,d\xi\int_{-\infty}^{\infty} f(t)e^{i\xi t}\,dt.$$

Corollary *If, in addition, $f(t)$ is continuous at the point $t = x$,*

$$\frac{1}{2\pi}\int_{-\infty}^{\infty} e^{-i\xi x}\,d\xi\int_{-\infty}^{\infty} f(t)e^{i\xi t}\,dt = f(x).$$

2-3 FOURIER TRANSFORMS

If we write

$$F(\xi) = (2\pi)^{-\frac{1}{2}}\int_{-\infty}^{\infty} f(t)e^{i\xi t}\,dt, \tag{2-3-1}$$

then by the corollary to Fourier's integral theorem we see that if $f(t)$ is piecewise-continuously differentiable and absolutely integrable on **R**, then at a point at which it is continuous

$$f(t) = (2\pi)^{-\frac{1}{2}} \int_{-\infty}^{\infty} F(\xi)e^{-i\xi t}\, d\xi. \tag{2-3-2}$$

We say that the function $F(\xi)$ is the *Fourier transform* of $f(t)$. The pair of formulae (2-3-1) and (2-3-2) embody what is called the *Fourier inversion theorem*.

We also write $\mathscr{F}[f(t);\xi]$ or $\mathscr{F}[f]$ for the Fourier transform of $f(t)$. For example,

$$\mathscr{F}[H(a - |t|);\xi] = (2\pi)^{-\frac{1}{2}} \int_{-a}^{a} e^{i\xi t}\, dt = \left(\frac{2}{\pi}\right)^{\frac{1}{2}} \frac{\sin(\xi a)}{\xi}.$$

When the function f depends on more than one independent variable, we make use of a less ambiguous notation. For instance if f depends on both t and x, both of them ranging over all real values, we write $\mathscr{F}[f(t,x);t \to \xi]$ to denote the Fourier transform

$$(2\pi)^{-\frac{1}{2}} \int_{-\infty}^{\infty} f(t,x)e^{i\xi t}\, dt,$$

and $\mathscr{F}[f(t,x);x \to \xi]$ to denote the Fourier transform

$$(2\pi)^{-\frac{1}{2}} \int_{-\infty}^{\infty} f(t,x)e^{i\xi x}\, dx.$$

From (2-3-1) we see that

$$|F(\xi)| \leqslant (2\pi)^{-\frac{1}{2}} \int_{-\infty}^{\infty} |f(t)|\, dt$$

so that if $f(t) \in \mathscr{A}_1(\mathbf{R})$, there is a finite positive number J such that

$$\int_{-\infty}^{\infty} |f(t)|\, dt = J$$

and hence such that

$$|F(\xi)| \leqslant (2\pi)^{-\frac{1}{2}} J$$

showing that $F(\xi)$ *is bounded*. We note also that in terms of this notation, the Riemann–Lebesgue lemma can be written

$$\lim_{|\xi| \to \infty} F(\xi) = 0.$$

Also from the definition (2-3-1), we find that

$$F(\xi + h) - F(\xi) = I(\xi,h)$$

where

$$I(\xi,h) = (2\pi)^{-\frac{1}{2}} \int_{-\infty}^{\infty} f(t)e^{i\xi t}(e^{i\xi h} - 1)\, dt.$$

Obviously,

$$|I(\xi,h)| < (2/\pi)^{\frac{1}{2}} \int_{-\infty}^{\infty} |f(t)| \, dt$$

so that $I(\xi,h)$ is uniformly convergent and we can write

$$\lim_{h\to 0} I(\xi,h) = (2\pi)^{-\frac{1}{2}} \int_{-\infty}^{\infty} f(t)e^{i\xi t} \left\{ \lim_{h\to 0} (e^{i\xi h} - 1) \right\} dt = 0$$

showing that, *if f is piecewise-continuously differentiable and absolutely integrable on the whole real line, its Fourier transform F is continuous.*

If $f(t)$ is a *real* function of t, the complex conjugate of its Fourier transform is

$$\mathscr{F}^*[f] = (2\pi)^{-\frac{1}{2}} \int_{-\infty}^{\infty} f(t)e^{-i\xi t} \, dt$$

so that

$$\mathscr{F}^*[f(t);\xi] = \mathscr{F}[f(t);-\xi]. \tag{2-3-3}$$

When $f(t)$ is no longer real we use this equation to *define* the operator $\mathscr{F}^*$.

With this notation we can rewrite (2-3-2) as

$$f = \mathscr{F}^*F$$

and since $F = \mathscr{F}f$, this is equivalent to the operator equation

$$\mathscr{F}^*\mathscr{F} = I \tag{2-3-4}$$

in which I denotes the identity operator. This can be written in the alternative form

$$\mathscr{F}^{-1} = \mathscr{F}^* \tag{2-3-5}$$

$\mathscr{F}^{-1}$ denoting the operator which is the inverse of the Fourier operator $\mathscr{F}$.

By the argument which established equation (1-1-5) we see that $\mathscr{F}$ is a linear operator, i.e., that

$$\mathscr{F}(\lambda f + \mu g) = \lambda\mathscr{F}(f) + \mu\mathscr{F}(g) \tag{2-3-6}$$

where λ and μ are constants.

If $a > 0$, then, by the definition of a Fourier transform

$$\mathscr{F}[f(at);\xi] = (2\pi)^{-\frac{1}{2}} \int_{-\infty}^{\infty} f(at)e^{i\xi t} \, dt$$

$$= a^{-1}(2\pi)^{-\frac{1}{2}} \int_{-\infty}^{\infty} f(y)e^{i(\xi/a)y} \, dy$$

showing that

$$\mathscr{F}[f(at);\xi] = a^{-1}\mathscr{F}[f(t);\xi/a], \qquad (a > 0).$$

On the other hand, if $a < 0$,

$$\mathscr{F}[f(at);\xi] = a^{-1}(2\pi)^{-\frac{1}{2}} \int_{+\infty}^{-\infty} f(y)e^{i(\xi/a)y}\, dy$$

showing that

$$\mathscr{F}[f(at);\xi] = -a^{-1}\mathscr{F}[f(t);\xi/a], \qquad (a < 0).$$

In general, therefore, we find that if a is real,

$$\mathscr{F}[f(at);\xi] = \frac{1}{|a|}\mathscr{F}[f(t);\xi/a]. \tag{2-3-7}$$

Similarly we find, directly from the definition, that

$$\mathscr{F}[f(t-a);\xi] = (2\pi)^{-\frac{1}{2}} \int_{-\infty}^{\infty} f(t-a)e^{i\xi t}\, dt$$

$$= (2\pi)^{-\frac{1}{2}} \int_{-\infty}^{\infty} f(u)e^{i\xi(u+a)}\, du$$

from which it follows that

$$\mathscr{F}[f(t-a);\xi] = e^{i\xi a}\mathscr{F}[f(t);\xi]. \tag{2-3-8}$$

Also, if λ is a constant,

$$\mathscr{F}[e^{i\lambda t}f(t);\xi] = (2\pi)^{-\frac{1}{2}} \int_{-\infty}^{\infty} f(t)e^{i\xi t}e^{i\lambda t}\, dt$$

$$= (2\pi)^{-\frac{1}{2}} \int_{-\infty}^{\infty} f(t)e^{i(\xi+\lambda)t}\, dt$$

so that

$$\mathscr{F}[e^{i\lambda t}f(t);\xi] = \mathscr{F}[f(t);\xi+\lambda]. \tag{2-3-9}$$

If $f(t)$ is a complex-valued function, we see that

$$\mathscr{F}[f^*(-t);\xi] = (2\pi)^{-\frac{1}{2}} \int_{-\infty}^{\infty} f^*(-t)e^{i\xi t}\, dt$$

$$= (2\pi)^{-\frac{1}{2}} \int_{-\infty}^{\infty} f^*(u)e^{-iu\xi}\, du$$

and hence that

$$\mathscr{F}[f^*(-t);\xi] = F^*(\xi), \tag{2-3-10}$$

where $F^*(\xi)$ denotes the complex conjugate of $F(\xi)$, the Fourier transform of $f(t)$.

In the application of the Fourier transform to the solution of partial differential equations, we need to know the Fourier transform of the derivative of a function. If we proceed formally, then, if $f(t)$ is continuous

$$\mathscr{F}[f'(t);\xi] = (2\pi)^{-\frac{1}{2}} \int_{-\infty}^{\infty} f'(t)e^{i\xi t}\,dt$$

$$= (2\pi)^{-\frac{1}{2}}[f(t)e^{i\xi t}]_{-\infty}^{\infty} - i\xi(2\pi)^{-\frac{1}{2}} \int_{-\infty}^{\infty} f(t)e^{i\xi t}\,dt$$

Now for the Fourier transform on the right to exist we must have $f(t) \in \mathscr{A}_1(\mathbf{R})$ and this implies that

$$\lim_{|t|\to\infty} f(t) = 0,$$

so that the term in the square bracket vanishes. Hence if f belongs to both $\mathscr{C}(\mathbf{R})$ and $\mathscr{A}_1(\mathbf{R})$ and f' to both $\mathscr{P}^1(\mathbf{R})$ and $\mathscr{A}_1(\mathbf{R})$, we obtain the relation

$$\mathscr{F}[f'(t);\xi] = -i\xi\mathscr{F}[f(t);\xi]. \tag{2-3-11}$$

Continuing this process we see that $f, f', \ldots, f^{(n-1)}$ are continuous on $\mathbf{R}$, and $f^{(n)}$ is piecewise-continuously differentiable, and each of the derivatives $f^{(r)}$, $(r = 0,1,2, \ldots, n)$ is absolutely integrable on $\mathbf{R}$, then

$$\mathscr{F}[f^{(n)}(t);\xi] = (-i\xi)^n\mathscr{F}[f(t);\xi]. \tag{2-3-12}$$

In particular we note that (2-3-12) is valid if f belongs to either of the classes $\mathscr{D}$, $\mathscr{S}$ defined in sec. 1-8.

In a certain sense, then, we can say that the operator d/dt is mapped by the Fourier transform into the complex number $(-i\xi)$.

If $f(t)$ has finite discontinuities the result (2-3-11) has to be modified. For simplicity we shall suppose initially that $f(t)$ has a finite discontinuity at the single point $t = a$. Then we write

$$\mathscr{F}[f'(t);\xi] = (2\pi)^{-\frac{1}{2}} \int_{-\infty}^{a-} f'(t)e^{i\xi t}\,dt + (2\pi)^{-\frac{1}{2}} \int_{a+}^{\infty} f'(t)e^{i\xi t}\,dt.$$

Integrating by parts we therefore obtain the equation

$$\mathscr{F}[f'(t);\xi] = (2\pi)^{-\frac{1}{2}}[f(t)e^{i\xi t}]_{-\infty}^{a-} + (2\pi)^{-\frac{1}{2}}[f(t)e^{i\xi t}]_{a+}^{\infty}$$

$$- i\xi(2\pi)^{-\frac{1}{2}} \int_{-\infty}^{a-} f(t)e^{i\xi t}\,dt - i\xi(2\pi)^{-\frac{1}{2}} \int_{a+}^{\infty} f(t)e^{i\xi t}\,dt$$

which can be written in the form

$$\mathscr{F}[f'(t);\xi] = -i\xi\mathscr{F}[f(t);\xi] - (2\pi)^{-\frac{1}{2}}e^{i\xi a}[f]_a \tag{2-3-13}$$

where

$$[f]_a = f(a+) - f(a-). \tag{2-3-14}$$

If we now suppose that $f(t)$ has n points of finite discontinuity, denoted by $a_1, a_2, \ldots, a_n$, we see that (2-3-13) is modified to the form

$$\mathscr{F}[f'(t);\xi] = -i\xi\mathscr{F}[f(t);\xi] - (2\pi)^{-\frac{1}{2}} \sum_{r=1}^{n} [f]_{a_r} e^{i\xi a_r}. \tag{2-3-15}$$

We know that if $f(t) \in \mathscr{C}^1(\mathbf{R})$, then regarded as a function of t and ξ $f(t)e^{i\xi t} \in \mathscr{C}^1(\mathbf{R} \times \mathbf{R})$, so that if the integral

$$F(\xi) = (2\pi)^{-\frac{1}{2}} \int_{-\infty}^{\infty} f(t)e^{i\xi t}\,dt,$$

is convergent and the integral

$$i(2\pi)^{-\frac{1}{2}} \int_{-\infty}^{\infty} tf(t)e^{i\xi t}\,dt = (2\pi)^{-\frac{1}{2}} \int_{-\infty}^{\infty} \frac{\partial}{\partial \xi} f(t)e^{i\xi t}\,dt$$

is uniformly convergent, then

$$F'(\xi) = i\mathscr{F}[tf(t);\xi]. \tag{2-3-16}$$

We can, of course, continue this process: If $f \in \mathscr{C}^1(\mathbf{R})$ and $t^r f(t) \in \mathscr{A}_1(\mathbf{R})$ for $r = 0,1,2, \ldots, n$, then

$$F^{(r)}(\xi) = i^r \mathscr{F}[t^r f(t);\xi], \qquad (r = 0,1, \ldots, n). \tag{2-3-17}$$

In particular, if $f \in \mathscr{D}$ then (2-3-17) holds for all positive integral values of r. If we combine (2-3-12) and (2-3-17) we find that

$$\begin{aligned}\mathscr{F}[t^k f^{(m)}(t);\xi] &= (-i)^k \frac{d^k}{d\xi^k} \mathscr{F}[f^{(m\cdot}(t);\xi] \\ &= (-i)^{k+m} \frac{d^k}{d\xi^k} \xi^m F(\xi)\end{aligned} \tag{2-3-18}$$

where $F(\xi) = \mathscr{F}[f(t);\xi]$.

This result can be written in a slightly different way. Since

$$F^{(m)}(\xi) = i^m \mathscr{F}[t^m f(t);\xi]$$

it follows that

$$\xi^n F^{(m)}(\xi) = i^{m+n} \mathscr{F}\left[\frac{d^n}{dt^n} t^m f(t);\xi\right].$$

Using Leibnitz's theorem for the n-th derivative of a product, we see that

$$\frac{d^n}{dt^n}[t^m f(t)] = \sum_{r=0}^{p} \frac{n!m!}{r!(n-r)!(m-r)!} t^{m-r} f^{(n-r)}(t)$$

where $p = \min(m,n)$, from which we deduce that

$$\xi^n F^{(m)}(\xi) = i^{m+n} \sum_{r=0}^{p} \frac{n!m!}{r!(n-r)!(m-r)!} \mathscr{F}[t^{m-r} f^{(n-r)}(t); \xi] \quad (2\text{-}3\text{-}19)$$

From this result it follows that if $f \in \mathscr{S}$, then $F \in \mathscr{S}$, a result which we may write symbolically as

$$\mathscr{F}[\mathscr{S}] = \mathscr{S}. \quad (2\text{-}3\text{-}20).$$

The Fourier transform $\mathscr{F}$ therefore maps the class of functions $\mathscr{S}$ into itself.

It should be noticed that the Fourier transform of a function of compact support is *not* itself a function of compact support. For example, the function $f(t) = H(a - |t|)$, $(a > 0)$, has the compact support $[-a,a]$ but its Fourier transform $F(\xi) = (2/\pi)^{\frac{1}{2}}\xi^{-1}\sin(\xi a)$ does not have compact support.

2-4 FOURIER COSINE TRANSFORMS

When the function $f(t)$ is defined only for *positive* values of the real variable t, we may consider a function $f_+(t)$ defined over the whole real line by the equations

$$f_+(t) = \begin{cases} f(t), & t \geqslant 0, \\ f(-t), & t < 0, \end{cases}$$

(cf. Fig. 2-2) to be the "extension" of $f(t)$. Then, since

$$\int_{-\infty}^{\infty} f_+(t)e^{i\xi t}\,dt = \int_0^{\infty} f(t)e^{i\xi t}\,dt + \int_{-\infty}^{0} f(-t)e^{i\xi t}\,dt$$

$$= \int_0^{\infty} f(t)\{e^{i\xi t} + e^{-i\xi t}\}\,dt$$

$$= 2\int_0^{\infty} f(t)\cos(\xi t)\,dt$$

we find that

$$\mathscr{F}[f_+(t);\xi] = \left(\frac{2}{\pi}\right)^{\frac{1}{2}} \int_0^{\infty} f(t)\cos(\xi t)\,dt.$$

If we write

$$F_c(\xi) \equiv \mathscr{F}_c[f(t);\xi] = \left(\frac{2}{\pi}\right)^{\frac{1}{2}} \int_0^{\infty} f(t)\cos(\xi t)\,dt \quad (2\text{-}4\text{-}1)$$

we see that $F_c(-\xi) = F_c(\xi)$ so that $F_c(\xi)$ is an *even* function of ξ, and equation (2-3-2) implies that

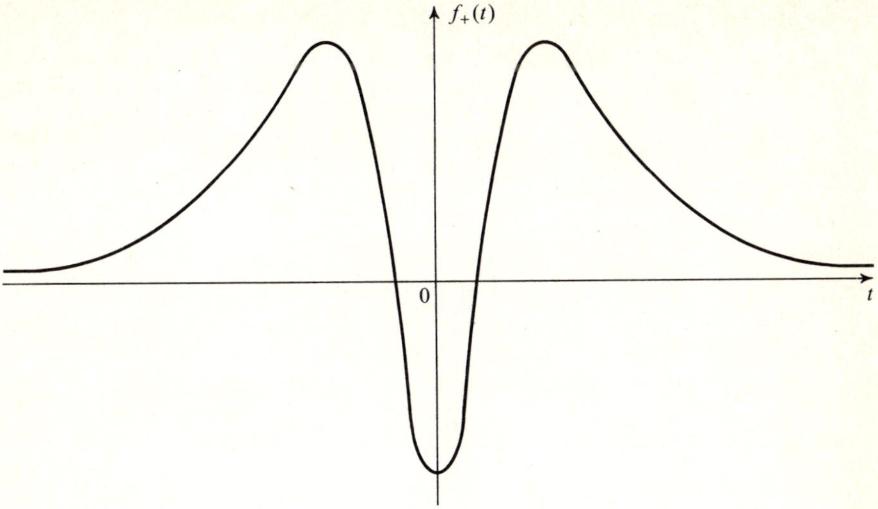

Fig. 2-2

$$f(t) = \left(\frac{2}{\pi}\right)^{\frac{1}{2}} \int_0^\infty F_c(\xi) \cos(\xi t)\, d\xi, \qquad (t \geqslant 0). \tag{2-4-2}$$

We say that $F_c(\xi)$ is the *Fourier cosine transform* of $f(t)$ and the result contained in the equations (2-4-1) and (2-4-2) is called the *Fourier cosine inversion theorem*. It should be noted that these formulae are symmetrical and also that

$$f(t) = \mathscr{F}_c[F_c(\xi); t] = \mathscr{F}_c[\mathscr{F}_c\{f(t); \xi\}; t],$$

so that if I denotes the identity operator

$$\mathscr{F}_c \mathscr{F}_c = I$$

that is,

$$\mathscr{F}_c^{-1} = \mathscr{F}_c. \tag{2-4-3}$$

These results may be stated simply by saying that $(2/\pi)^{\frac{1}{2}} \cos(\xi t)$ is a Fourier kernel in the sense defined at the end of sec. 1-1.

Obviously, $\mathscr{F}_c$ is a linear operator.

Fourier cosine transforms have many properties analogous to those of exponential Fourier transforms.

For example, since, if $a > 0$,

$$\mathscr{F}_c[f(at); \xi] = \left(\frac{2}{\pi}\right)^{\frac{1}{2}} \int_0^\infty f(at) \cos(\xi t)\, dt$$

$$= a^{-1} \left(\frac{2}{\pi}\right)^{\frac{1}{2}} \int_0^\infty f(x) \cos(\xi x/a)\, dx$$

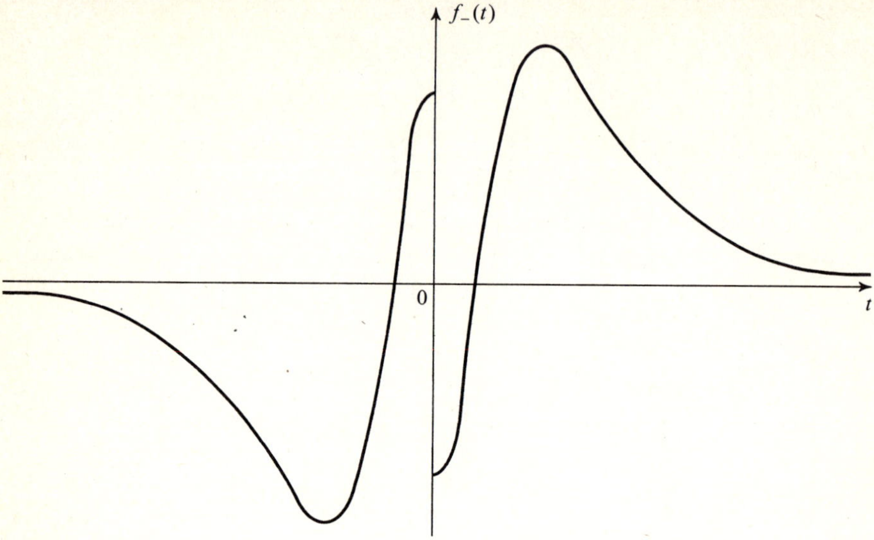

Fig. 2-3

we see that

$$\mathscr{F}_c[f(at);\xi] = a^{-1}\mathscr{F}_c[f(t);\xi/a], \qquad (a > 0). \tag{2-4-4}$$

It is also easily established that

$$\mathscr{F}_c[\cos(\omega t);\xi] = \tfrac{1}{2}\{F_c(\xi + \omega) + F_c(\xi - \omega)\}. \tag{2-4-5}$$

2-5 FOURIER SINE TRANSFORMS

If, instead of the function $f_+(t)$, we take as the "extension" of $f(t)$, defined on $(0,\infty)$, the *odd* function

$$f_-(t) = \begin{cases} f(t), & t > 0, \\ -f(-t), & t < 0, \end{cases}$$

(cf. Fig. 2-3), then, since

$$\begin{aligned}\int_{-\infty}^{\infty} f_-(t)e^{i\xi t}\,dt &= \int_0^{\infty} f(t)e^{i\xi t}\,dt - \int_{-\infty}^{0} f(-t)e^{i\xi t}\,dt \\ &= \int_0^{\infty} f(t)\{e^{i\xi t} - e^{-i\xi t}\}\,dt \\ &= 2i\int_0^{\infty} f(t)\sin(\xi t)\,dt\end{aligned}$$

we see that

$$\mathscr{F}[f_{-}(t);\xi] = i\left(\frac{2}{\pi}\right)^{\frac{1}{2}} \int_0^\infty f(t)\sin(\xi t)\,dt.$$

If we write

$$F_s(\xi) \equiv \mathscr{F}_s[f(t);\xi] = \left(\frac{2}{\pi}\right)^{\frac{1}{2}} \int_0^\infty f(t)\sin(\xi t)\,dt, \tag{2-5-1}$$

we see that $F_s(\xi)$ is an odd function of ξ and equation (2-3-2) implies that

$$f_{-}(t) = -i\left(\frac{2}{\pi}\right)^{\frac{1}{2}} \int_0^\infty \{iF_s(\xi)\}\sin(\xi t)\,dt$$

i.e., that

$$f(t) = \left(\frac{2}{\pi}\right)^{\frac{1}{2}} \int_0^\infty F_s(\xi)\sin(\xi t)\,d\xi, \qquad (t > 0). \tag{2-5-2}$$

We say that $F_s(\xi)$ is the *Fourier sine transform* of the function $f(t)$. The result expressed by (2-5-1) and (2-5-2) is called the *Fourier sine inversion theorem*. These formulae are symmetrical so that

$$\mathscr{F}_s^{-1} = \mathscr{F}_s; \tag{2-5-3}$$

also $\mathscr{F}_s$ is a linear operator.

It is very easy to show that

$$\mathscr{F}_s[\cos(\omega t)f(t);\xi] = \tfrac{1}{2}\{F_s(\xi+\omega) + F_s(\xi-\omega)\} \tag{2-5-4}$$

$$\mathscr{F}_c[\sin(\omega t)f(t);\xi] = \tfrac{1}{2}\{F_s(\omega+\xi) + F_s(\omega-\xi)\} \tag{2-5-5}$$

$$\mathscr{F}_s[\sin(\omega t)f(t);\xi] = \tfrac{1}{2}\{F_c(\xi-\omega) - F_c(\xi+\omega)\}. \tag{2-5-6}$$

Also

$$\mathscr{F}_s[f(at);\xi] = a^{-1}\mathscr{F}_s[f(t);\xi/a], \qquad (a > 0). \tag{2-5-7}$$

Another result is most useful in the actual calculation of Fourier sine transforms. If we can show that

$$\mathscr{F}_s[f(x);\xi] = \phi(\xi)$$

for $\xi > 0$, then for $\xi < 0$ we have

$$\mathscr{F}_s[f(x);\xi] = \left(\frac{2}{\pi}\right)^{\frac{1}{2}} \int_0^\infty f(x)\sin(-\eta x)\,dx$$

where we have written $\xi = -\eta$ with $\eta > 0$. Hence when $\xi < 0$,

$$\mathscr{F}_s[f(x);\xi] = -\phi(\eta) \equiv -\phi(-\xi)$$

and we have, in general,

$$\mathscr{F}_s[f(x);\xi] = \phi(|\xi|)\operatorname{sgn}\xi \tag{2-5-8}$$

where we have used the symbol

$$\operatorname{sgn} \xi = \begin{cases} +1, & \xi > 0, \\ -1, & \xi < 0. \end{cases} \tag{2-5-9}$$

2-6 FOURIER TRANSFORMS OF DERIVATIVES

When we make use of the theory of Fourier sine and cosine transforms to solve boundary value problems involving differential equations we need to know how to express the transforms of the derivatives of an unknown function in terms of the transform of the function itself.

Integrating by parts on the right-hand side of the equation

$$\mathscr{F}_c[f'(t);\xi] = \left(\frac{2}{\pi}\right)^{\frac{1}{2}} \int_0^\infty f'(t) \cos(\xi t)\, dt$$

and, assuming that $f(t) \in \mathscr{C}^2[0,\infty)$ and that $f(t)$, $f'(t)$ are both absolutely integrable, we find that

$$\mathscr{F}_c[f'(t);\xi] = -\left(\frac{2}{\pi}\right)^{\frac{1}{2}} f(0) + \xi \left(\frac{2}{\pi}\right)^{\frac{1}{2}} \int_0^\infty f(t) \sin(\xi t)\, dt$$

so that

$$\mathscr{F}_c[f'(t);\xi] = -(2/\pi)^{\frac{1}{2}} f(0) + \xi \mathscr{F}_s[f(t);\xi]. \tag{2-6-1}$$

Similarly, we can show that, for functions satisfying these same conditions,

$$\mathscr{F}_s[f'(t);\xi] = -\xi \mathscr{F}_c[f(t);\xi]. \tag{2-6-2}$$

Hence, from (2-6-1), we may deduce that if, in addition, $f''(t)$ belongs to $\mathscr{C}^1[0,\infty)$ and $\mathscr{A}_1[0,\infty)$,

$$\mathscr{F}_c[f''(t);\xi] = -(2/\pi)^{\frac{1}{2}} f'(0) + \xi \mathscr{F}_s[f'(t);\xi],$$

and hence from (2-6-2) that

$$\mathscr{F}_c[f''(t);\xi] = -(2/\pi)^{\frac{1}{2}} f'(0) - \xi^2 \mathscr{F}_c[f(t);\xi]. \tag{2-6-3}$$

Similarly from the same two equations applied in the reverse order we have for functions of the same type,

$$\mathscr{F}_s[f''(t);\xi] = (2/\pi)^{\frac{1}{2}} \xi f(0) - \xi^2 \mathscr{F}_s[f(t);\xi]. \tag{2-6-4}$$

In particular, if $f'(0) = 0$, we have the relation

$$\mathscr{F}_c[f''(t);\xi] = -\xi^2 F_c(\xi) \tag{2-6-5}$$

and if $f(0) = 0$, that we have the relation

$$\mathscr{F}_s[f''(t);\xi] = -\xi^2 F_s(\xi). \tag{2-6-6}$$

It should be noticed that these formulae give us some indication of which transform—sine or cosine—to employ in the solution of any problem. In any problem in which $f(0)$ is known but $f'(0)$ is not known, we should take the Fourier sine transform of $f''(t)$ and make use of equation (2-6-4). On the other hand, in any problem in which $f'(0)$ is known but $f(0)$ is not, we should take the Fourier cosine transform of $f''(t)$ and make use of equation (2-6-3).

These formulae have been derived on the assumption that $f \in \mathscr{C}^3[0,\infty)$; if f and f' have finite discontinuities, the formulae have to be modified in the same way as (2-3-15).

Similar results hold for the Fourier transforms of functions of several variables. For simplicity we shall state them for a function $f(x,y)$ of two independent variables x and y; the extension to a higher number of variables is obvious. If we write

$$F_c(\xi,y) = \mathscr{F}_c[f(x,y);x \to \xi]$$
$$F_s(\xi,y) = \mathscr{F}_s[f(x,y);x \to \xi]$$
$$f_r(y) = \sqrt{\frac{2}{\pi}}\left[\frac{\partial^r f}{\partial x^r}\right]_{x=0}$$

then the analogues of (2-6-1) and (2-6-2) for continuous functions are

$$\mathscr{F}_c\left[\frac{\partial f}{\partial x};x \to \xi\right] = -f_0(y) + \xi F_s(\xi,y) \tag{2-6-7}$$

$$\mathscr{F}_s\left[\frac{\partial f}{\partial x};x \to \xi\right] = -\xi F_c(\xi,y). \tag{2-6-8}$$

Applying these in succession we find that

$$\mathscr{F}_c\left[\frac{\partial^2 f}{\partial x^2};x \to \xi\right] = -\xi^2 F_c(\xi,y) - f_1(y) \tag{2-6-9}$$

$$\mathscr{F}_s\left[\frac{\partial^2 f}{\partial x^2};x \to \xi\right] = -\xi^2 F_s(\xi,y) + \xi f_0(y). \tag{2-6-10}$$

Now, by the definition of a cosine transform

$$\mathscr{F}_c\left[\frac{\partial f}{\partial y};x \to \xi\right] = \left(\frac{2}{\pi}\right)^{\frac{1}{2}}\int_0^\infty \frac{\partial f}{\partial y}\cos(\xi x)\,dx$$
$$= \frac{\partial}{\partial y}\left(\frac{2}{\pi}\right)^{\frac{1}{2}}\int_0^\infty f(x,y)\cos(\xi x)\,dx$$

so that

$$\mathscr{F}_c\left[\frac{\partial f}{\partial y};x \to \xi\right] = \mathrm{D}F_c(\xi,y)$$

where D denotes the differential operator $\partial/\partial y$. In general,

$$\mathscr{F}_c\left[\frac{\partial^r f}{\partial y^r};x \to \xi\right] = \mathrm{D}^r F_c(\xi,y) \tag{2-6-11}$$

$$\mathscr{F}_s\left[\frac{\partial^r f}{\partial y^r};x \to \xi\right] = \mathrm{D}^r F_s(\xi,y). \tag{2-6-12}$$

Hence for the transforms of

$$\Delta_2 f(x,y) = \frac{\partial^2 f}{\partial x^2} + \frac{\partial^2 f}{\partial y^2} \tag{2-6-13}$$

we have the relations

$$\mathscr{F}_c[\Delta_2 f(x,y);x \to \xi] = (\mathrm{D}^2 - \xi^2)F_c(\xi,y) - f_1(y), \tag{2-6-14}$$

$$\mathscr{F}_s[\Delta_2 f(x,y);x \to \xi] = (\mathrm{D}^2 - \xi^2)F_s(\xi,y) + \xi f_0(y). \tag{2-6-15}$$

In problems involving $\Delta_2 f$ in a region in which $x \in (0,\infty)$ and $f(0,y)$ is known but $f_x(0,y)$ is not known, we should take the Fourier sine transform of $f(x,y)$ with respect to x and make use of the formula (2-6-15). On the other hand, if $f_x(0,y)$ is known but $f(0,y)$ is not, we should take the Fourier cosine transform of $f(x,y)$ with respect to x and make use of the formula (2-6-14). This is illustrated in sec. 2-16(*c*) below.

We get similar results for higher derivatives:

$$\mathscr{F}_c\left[\frac{\partial^3 f}{\partial x^3};x \to \xi\right] = -\xi^3 F_s(\xi,y) + \xi^2 f_0(y) - f_2(y) \tag{2-6-16}$$

$$\mathscr{F}_s\left[\frac{\partial^3 f}{\partial x^3};x \to \xi\right] = \xi^3 F_c(\xi,y) + \xi f_1(y) \tag{2-6-17}$$

$$\mathscr{F}_c\left[\frac{\partial^4 f}{\partial x^4};x \to \xi\right] = \xi^4 F_c(\xi,y) + \xi^2 f_1(y) - f_3(y) \tag{2-6-18}$$

$$\mathscr{F}_s\left[\frac{\partial^4 f}{\partial x^4};x \to \xi\right] = \xi^4 F_s(\xi,y) - \xi^3 f_0(y) + \xi f_2(y) \tag{2-6-19}$$

and for mixed derivatives:

$$\mathscr{F}_c\left[\frac{\partial^4 f}{\partial x^2 \partial y^2};x \to \xi\right] = -\xi^2 \mathrm{D}^2 F_c(\xi,y) - f_1''(y) \tag{2-6-20}$$

$$\mathscr{F}_s\left[\frac{\partial^4 f}{\partial x^2 \partial y^2};x \to \xi\right] = -\xi^2 \mathrm{D}^2 F_s(\xi,y) + \xi f_0''(y). \tag{2-6-21}$$

Hence for the plane biharmonic operator $\Delta_2{}^2$ defined by the relation

$$\Delta_2{}^2 = \Delta_2\Delta_2 = \frac{\partial^4}{\partial x^4} + 2\frac{\partial^4}{\partial x^2 \partial y^2} + \frac{\partial^4}{\partial y^4}$$

we have

$$\mathscr{F}_c[\Delta_2{}^2 f(x,y); x \to \xi]$$
$$= (D^2 - \xi^2)^2 F_c(\xi,y) + (\xi^2 - 2D^2) f_1(y) - f_3(y) \qquad (2\text{-}6\text{-}22)$$

$$\mathscr{F}_s[\Delta_2{}^2 f(x,y); x \to \xi]$$
$$= (D^2 - \xi^2)^2 F_s(\xi,y) - \xi(\xi^2 + 2D^2) f_0(y) + \xi f_2(y). \qquad (2\text{-}6\text{-}23)$$

2-7 THE CALCULATION OF THE FOURIER TRANSFORMS OF SOME SIMPLE FUNCTIONS

In this section we shall consider the calculation of the Fourier transforms of some simple functions which occur frequently in applications.

We begin by considering the integrals

$$I = \int_0^\infty e^{-at} \cos(\xi t)\, dt, \qquad J = \int_0^\infty e^{-at} \sin(\xi t)\, dt$$

in which the constant a is taken to be real and positive. If we integrate by parts we find that

$$I = [-a^{-1} e^{-at} \cos(\xi t)]_0^\infty - \frac{\xi}{a} \int_0^\infty e^{-at} \sin(\xi t)\, dt$$

a relation which we may write in the form

$$I = \frac{1}{a} - \frac{\xi}{a} J.$$

On the other hand, if we integrate J by parts, we find that

$$J = [-a^{-1} e^{-at} \sin(\xi t)]_0^\infty + \frac{\xi}{a} \int_0^\infty e^{-at} \cos(\xi t)\, dt$$

which is equivalent to the relation

$$J = \frac{\xi}{a} I.$$

Solving these two equations for I and J we find that

$$I = \frac{a}{a^2 + \xi^2}, \qquad J = \frac{\xi}{a^2 + \xi^2}$$

from which we deduce immediately that

$$\mathscr{F}_c[e^{-at}; \xi] = \left(\frac{2}{\pi}\right)^{\frac{1}{2}} \frac{a}{a^2 + \xi^2}, \qquad (a > 0) \qquad (2\text{-}7\text{-}1)$$

$$\mathscr{F}_s[e^{-at}; \xi] = \left(\frac{2}{\pi}\right)^{\frac{1}{2}} \frac{\xi}{a^2 + \xi^2}, \qquad (a > 0). \qquad (2\text{-}7\text{-}2)$$

It will be observed, too, that (2-7-1) is equivalent to the formula

$$\mathscr{F}[e^{-a|t|};\xi] = \left(\frac{2}{\pi}\right)^{\frac{1}{2}} \frac{a}{a^2 + \xi^2}, \qquad (a > 0). \tag{2-7-3}$$

If we integrate both sides of (2-7-2) with respect to a from a to ∞ and make use of the result

$$\xi \int_a^\infty \frac{da}{a^2 + \xi^2} = \cot^{-1}\left(\frac{a}{\xi}\right)$$

we find that

$$\mathscr{F}_s[t^{-1}e^{-at};\xi] = \left(\frac{2}{\pi}\right)^{\frac{1}{2}} \cot^{-1}\left(\frac{a}{\xi}\right), \qquad (a > 0). \tag{2-7-3}$$

The limiting case of this formula as $a \to 0$ is of interest. From the definition of the Fourier sine transform we find that if $\xi > 0$

$$\mathscr{F}_s[t^{-1};\xi] = \left(\frac{2}{\pi}\right)^{\frac{1}{2}} \int_0^\infty \frac{\sin(\xi t)}{t}\,dt = (\tfrac{1}{2}\pi)^{\frac{1}{2}}.$$

Using (2-5-8) we see that, in general,

$$\mathscr{F}_s[t^{-1};\xi] = (\tfrac{1}{2}\pi)^{\frac{1}{2}} \operatorname{sgn} \xi. \tag{2-7-4}$$

This relation is, in turn, equivalent to the formula

$$\mathscr{F}[t^{-1};\xi] = (\tfrac{1}{2}\pi)^{\frac{1}{2}}\, i \operatorname{sgn} \xi. \tag{2-7-5}$$

If we differentiate both sides of (2-7-1) with respect to a we find that

$$\mathscr{F}_c[te^{-at};\xi] = \left(\frac{2}{\pi}\right)^{\frac{1}{2}} \frac{a^2 - \xi^2}{(a^2 + \xi^2)^2}, \qquad (a > 0) \tag{2-7-6}$$

and if we differentiate both sides of (2-7-2) with respect to a we obtain the formula

$$\mathscr{F}_s[te^{-at};\xi] = \left(\frac{2}{\pi}\right)^{\frac{1}{2}} \frac{2a\xi}{(a^2 + \xi^2)^2}, \qquad (a > 0). \tag{2-7-7}$$

These results can be written in the alternative forms

$$\mathscr{F}[te^{-a|t|};\xi] = \left(\frac{2}{\pi}\right)^{\frac{1}{2}} \frac{2ai\xi}{(a^2 + \xi^2)^2}, \qquad (a > 0) \tag{2-7-8}$$

$$\mathscr{F}[|t|e^{-a|t|};\xi] = \left(\frac{2}{\pi}\right)^{\frac{1}{2}} \frac{a^2 - \xi^2}{(a^2 + \xi^2)^2}, \qquad (a > 0). \tag{2-7-9}$$

Using the identity

$$t^2 - i\xi t = (t - \tfrac{1}{2}i\xi)^2 + \tfrac{1}{4}\xi^2,$$

we see that

$$\mathscr{F}[e^{-t^2};\xi] = (2\pi)^{-\frac{1}{2}} \int_{-\infty}^{\infty} e^{-(t-\frac{1}{2}i\xi)^2-\frac{1}{4}\xi^2}\,dt,$$

Since

$$\int_{-\infty}^{\infty} e^{-(t-\frac{1}{2}i\xi)^2}\,dt = \int_{-\infty}^{\infty} e^{-u^2}\,du = \pi^{\frac{1}{2}}$$

we see that

$$\mathscr{F}[e^{-t^2};\xi] = 2^{-\frac{1}{2}}e^{-\frac{1}{4}\xi^2}.$$

Making use of equation (2-3-7) we find that

$$\mathscr{F}[e^{-a^2t^2};\xi] = 2^{-\frac{1}{2}}a^{-1}e^{-\frac{1}{4}\xi^2/a^2}, \qquad (a > 0). \tag{2-7-10}$$

In particular, if we take $a = 2^{-\frac{1}{2}}$, we find that

$$\mathscr{F}[e^{-\frac{1}{2}t^2};\xi] = e^{-\frac{1}{2}\xi^2}. \tag{2-7-11}$$

A function $f(t)$ with the property that

$$\mathscr{F}[f(t);\xi] = f(\xi) \tag{2-7-12}$$

is said to be *self-reciprocal* under the Fourier transform. Equation (2-7-11) thus states that $e^{-\frac{1}{2}t^2}$ is a self-reciprocal function under the Fourier transform.

It also follows from (2-7-10) that

$$\mathscr{F}_c[e^{-a^2t^2};\xi] = 2^{-\frac{1}{2}}|a|^{-1}e^{-\frac{1}{4}\xi^2/a^2}. \tag{2-7-13}$$

If we differentiate both sides of (2-7-11) with respect to ξ we find that

$$\mathscr{F}_s[te^{-\frac{1}{2}t^2};\xi] = \xi e^{-\frac{1}{2}\xi^2}. \tag{2-7-14}$$

Hence $e^{-\frac{1}{2}t^2}$ is a self-reciprocal function under the Fourier cosine transform and $te^{-\frac{1}{2}t^2}$ is a self-reciprocal function under the Fourier sine transform.

An interesting property of Fourier transforms can be demonstrated from (2-7-10). It is easily shown that the function

$$\delta_n(x) = n\pi^{-\frac{1}{2}}e^{-n^2x^2}$$

has the property

$$\int_{-\infty}^{\infty} \delta_n(x)\,dx = 1,$$

and, from (2-7-10), we see that its Fourier transform is given by the expression

$$\Delta_n(\xi) = \mathscr{F}[\delta_n(x);\xi] = (2\pi)^{-\frac{1}{2}}e^{-\frac{1}{4}\xi^2/n^2}.$$

UNIVERSITY LIBRARY BIRMINGHAM

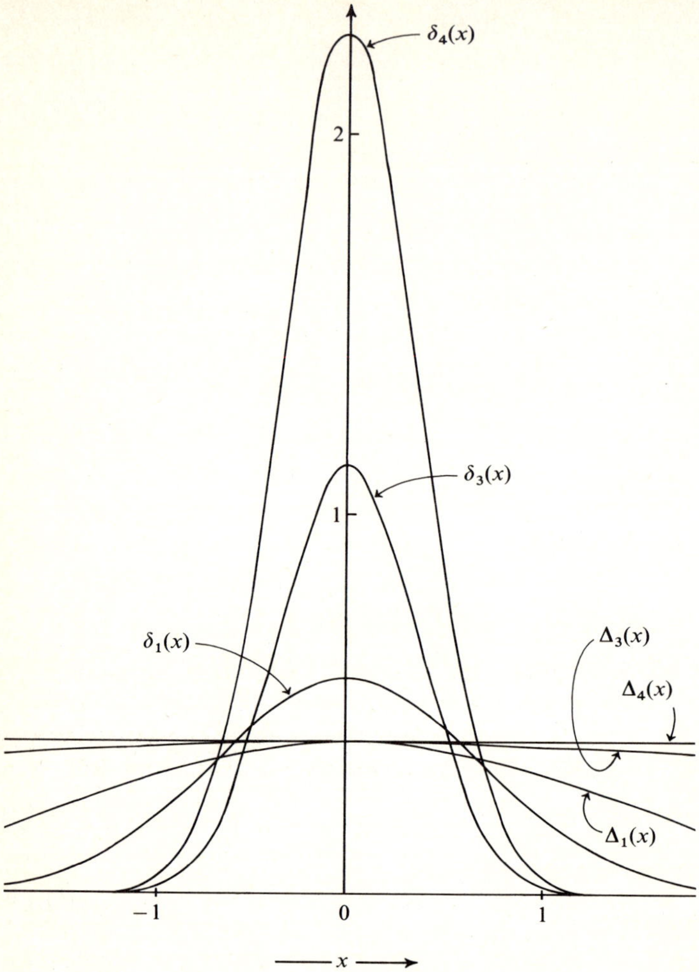

Fig. 2-4

The graphs of $\delta_n(x)$ and $\Delta_n(x)$ for various values of n are shown in Fig. 2-4. From these graphs we see that the more "concentrated" a function at the origin, the more "diffuse" its Fourier transform is.

Additional results of interest can be obtained by making the substitution

$$a = \frac{1}{\sqrt{2}} e^{-\frac{1}{4}\pi i} = \tfrac{1}{2}(1 - i)$$

in equation (2-7-10) and using the results

$$\frac{1}{a\sqrt{2}} = \frac{1 + i}{\sqrt{2}}, \qquad a^2 = -\tfrac{1}{2}i.$$

We find that

$$\mathscr{F}_c[e^{+\frac{1}{2}it^2};\xi] = \frac{1}{\sqrt{2}}(1+i)e^{-\frac{1}{2}i\xi^2}$$

from which, on equating real and imaginary parts, we deduce that

$$\mathscr{F}_c[\cos(\tfrac{1}{2}t^2);\xi] = \frac{1}{\sqrt{2}}\{\cos(\tfrac{1}{2}\xi^2) + \sin(\tfrac{1}{2}\xi^2)\}, \tag{2-7-15}$$

$$\mathscr{F}_c[\sin(\tfrac{1}{2}t^2);\xi] = \frac{1}{\sqrt{2}}\{\cos(\tfrac{1}{2}\xi^2) - \sin(\tfrac{1}{2}\xi^2)\}. \tag{2-7-16}$$

An alternative method of deriving these results is suggested in Problem 2-28.

We now evaluate some transforms by making use of the expansions

$$\cos z = \sum_{r=0}^{\infty}\frac{(-z^2)^r}{(2r)!}, \qquad \sin z = \sum_{r=0}^{\infty}\frac{(-1)^r z^{2r+1}}{(2r+1)!}.$$

From the duplication formula $\Gamma(\frac{1}{2})\Gamma(2z) = 2^{2z-1}\Gamma(z)\Gamma(z+\frac{1}{2})$ for the gamma function $\Gamma(z)$ (Cf. Widder, 1961, pp. 367–378) we find that

$$\frac{1}{(2r)!} = \frac{2^{-2r}\pi^{\frac{1}{2}}}{r!\Gamma(r+\frac{1}{2})}, \qquad \frac{1}{(2r+1)!} = \frac{2^{-2r-1}\pi^{\frac{1}{2}}}{r!\Gamma(r+\frac{3}{2})}$$

so that these expansions become

$$\cos z = \pi^{\frac{1}{2}}\sum_{r=0}^{\infty}\frac{2^{-2r}(-z^2)^r}{r!\Gamma(r+\frac{1}{2})}, \qquad \sin z = \pi^{\frac{1}{2}}\sum_{r=0}^{\infty}\frac{(-1)^r(\frac{1}{2}z)^{2r+1}}{r!\Gamma(r+\frac{3}{2})}.$$

From the first of these two expansions we see that

$$\mathscr{F}_c[(a^2-t^2)^{\nu-\frac{1}{2}}H(a-t);\xi] = \left(\frac{2}{\pi}\right)^{\frac{1}{2}}\int_0^a (a^2-t^2)^{\nu-\frac{1}{2}}\cos(\xi t)\,dt$$

$$= 2^{\frac{1}{2}}\sum_{r=0}^{\infty}\frac{(-\frac{1}{4}\xi^2)^r}{r!\Gamma(r+\frac{1}{2})}\int_0^a (a^2-t^2)^{\nu-\frac{1}{2}}t^{2r}\,dt.$$

Now, if $\nu > -\frac{1}{2}$, $r \geqslant 0$,

$$2\int_0^a (a^2-t^2)^{\nu-\frac{1}{2}}t^{2r}\,dt = a^{2\nu+2r}\frac{\Gamma(\nu+\frac{1}{2})\Gamma(r+\frac{1}{2})}{\Gamma(r+\nu+1)},$$

so that, if $\nu > -\frac{1}{2}$

$$\mathscr{F}_c[(a^2-t^2)^{\nu-\frac{1}{2}}H(a-t);\xi] = 2^{-\frac{1}{2}}\Gamma(\nu+\tfrac{1}{2})\sum_{r=0}^{\infty}\frac{(-1)^r(\frac{1}{2}\xi)^{2r}a^{2\nu+2r}}{r!\Gamma(\nu+r+1)}.$$

From the definition

$$J_\nu(z) = \sum_{r=0}^{\infty} \frac{(-1)^r(\frac{1}{2}z)^{\nu+2r}}{r!\Gamma(\nu + r + 1)} \tag{2-7-18}$$

of the Bessel function of the first kind of order ν we see that this result can be written in the form

$$\mathscr{F}_c[(a^2 - t^2)^{\nu-\frac{1}{2}}H(a - t);\xi] = 2^{\nu-\frac{1}{2}}\Gamma(\nu + \tfrac{1}{2})a^\nu\xi^{-\nu}J_\nu(a\xi). \tag{2-7-19}$$

Similarly, using the second of the two expansions we find that

$$\mathscr{F}_s[t(a^2 - t^2)^{\nu-\frac{3}{2}}H(a - t);\xi] = \left(\frac{2}{\pi}\right)^{\frac{1}{2}} \int_0^a t(a^2 - t^2)^{\nu-\frac{3}{2}} \sin(\xi t)\, dt$$

$$= 2^{\frac{1}{2}} \sum_{r=0}^{\infty} \frac{(-1)^r(\frac{1}{2}\xi)^{2r+1}}{r!\Gamma(r + \frac{3}{2})} \times$$

$$\times \int_0^a t^{2r+2}(a^2 - t^2)^{\nu-\frac{3}{2}}\, dt.$$

Now, if $\nu > \frac{1}{2}$,

$$2\int_0^a t^{2r+2}(a^2 - t^2)^{\nu-\frac{3}{2}}\, dt = \frac{\Gamma(r + \frac{3}{2})\Gamma(\nu - \frac{1}{2})}{\Gamma(r + \nu + 1)}\, a^{2r+2\nu}$$

so that

$$\mathscr{F}_s[t(a^2 - t^2)^{\nu-\frac{3}{2}}H(a - t);\xi] = 2^{-\frac{1}{2}} \sum_{r=0}^{\infty} \frac{(-1)^r(\frac{1}{2}\xi)^{2r+1}a^{2r+2}\Gamma(\nu - \frac{1}{2})}{r!\Gamma(r + \nu + 1)}$$

which is equivalent to the relation

$$\mathscr{F}_s[t(a^2 - t^2)^{\nu-\frac{3}{2}}H(a - t);\xi] = 2^{\nu-\frac{3}{2}}a^\nu\xi^{1-\nu}\Gamma(\nu - \tfrac{1}{2})J_\nu(\xi a),$$
$$(a > 0, \nu > \tfrac{1}{2}). \tag{2-7-20}$$

To illustrate the use of the calculus of residues in the calculation of Fourier transforms we shall derive the transform of t^{p-1}, where $0 < p < 1$.

If we write

$$f(z) = z^{p-1}e^{-\xi z}, \qquad \xi > 0, 0 < p < 1$$

then, by Cauchy's theorem,

$$\int_C f(z)\, dz = 0$$

where C is the contour shown in Fig. 2-5. Hence

$$\int_{C_1} f(z)\, dz + \int_\varepsilon^R x^{p-1}e^{-\xi x}\, dx + \int_{C_2} f(z)\, dz = e^{\frac{1}{2}\pi p i} \int_\varepsilon^R y^{p-1}e^{-i\xi y}\, dy.$$

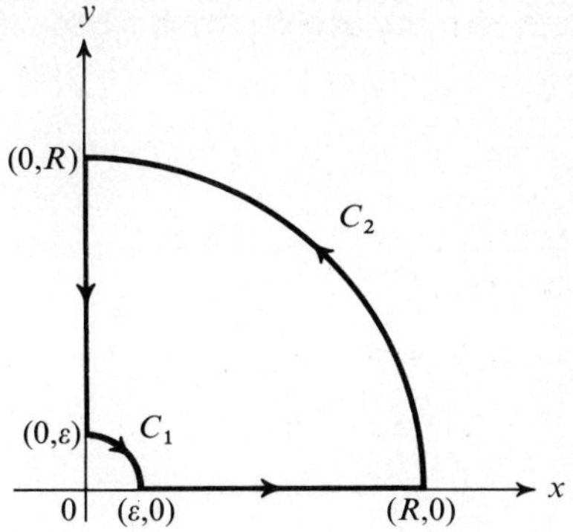

Fig. 2-5

To deal with the integral along the path C_1 we make use of:

Lemma 1 *If AB is the arc* $\theta_1 \leqslant \arg z \leqslant \theta_2$ *of the circle* $|z - a| = \varepsilon$ *and if* $(z - a)f(z) \to k$ *uniformly as* $z \to a$, *then*

$$\lim_{\varepsilon \to 0} \int_{AB} f(z)\, dz = ik(\theta_2 - \theta_1).$$

In particular if $(z - a)f(z) \to 0$ *as* $z \to a$,

$$\lim_{\varepsilon \to 0} \int_{AB} f(z)\, dz = 0.$$

Since, in our case,

$$\lim_{z \to 0} zf(z) = 0$$

if $p > 0$, it follows that as $\varepsilon \to 0$

$$\int_{C_1} f(z)\, dz \to 0.$$

To deal with the integral along the path C_2 we make use of:

Lemma 2 *If AB is the arc* $\theta_1 \leqslant \arg z \leqslant \theta_2$ *of the circle* $|z| = R$, *and if* $zf(z) \to K$, *uniformly, as* $R \to \infty$, *then*

$$\lim_{R \to \infty} \int_{AB} f(z)\, dz = iK(\theta_2 - \theta_1).$$

In particular if $f(z) \sim o(z^{-1})$ *as* $z \to \infty$,†

$$\lim_{R \to \infty} \int_{AB} f(z)\, dz = 0.$$

†It will be recalled that $f(z) \sim o(z^{-1})$ as $z \to \infty$, means simply that $\lim_{z \to \infty} zf(z) = 0$.

In the present case we see that if $\xi > 0$,

$$\lim_{z\to\infty} zf(z) = 0$$

so that, by Lemma 2, the integral

$$\int_{C_2} f(z)\,dz$$

tends to zero as $R \to \infty$.

The integral

$$\int_0^\infty y^{p-1}e^{-i\xi y}\,dy$$

is convergent if $0 < p < 1$, so that if we let $R \to \infty$, $\varepsilon \to 0$, we obtain the relation

$$\int_0^\infty y^{p-1}e^{-i\xi y}\,dy = e^{-\frac{1}{2}\pi pi}\int_0^\infty x^{p-1}e^{-\xi x}\,dx = e^{-\frac{1}{2}\pi pi}\xi^{-p}\Gamma(p).$$

Equating real and imaginary parts of both sides of this last equation we obtain the formulae

$$\mathscr{F}_c[x^{p-1};\xi] = \left(\frac{2}{\pi}\right)^{\frac{1}{2}}\xi^{-p}\Gamma(p)\cos\left(\tfrac{1}{2}\pi p\right), \qquad 0 < p < 1 \tag{2-7-21}$$

$$\mathscr{F}_s[x^{p-1};\xi] = \left(\frac{2}{\pi}\right)^{\frac{1}{2}}\xi^{-p}\Gamma(p)\sin\left(\tfrac{1}{2}\pi p\right), \qquad 0 < p < 1. \tag{2-7-22}$$

In particular by taking $p = \frac{1}{2}$ we obtain the special cases

$$\mathscr{F}_c[x^{-\frac{1}{2}};\xi] = \xi^{-\frac{1}{2}} \tag{2-7-23}$$

$$\mathscr{F}_s[x^{-\frac{1}{2}};\xi] = \xi^{-\frac{1}{2}} \tag{2-7-24}$$

showing that $x^{-\frac{1}{2}}$ is self-reciprocal under both the Fourier sine transform and the Fourier cosine transform.

2-8 THE FOURIER TRANSFORMS OF RATIONAL FUNCTIONS

The calculation of the Fourier transforms of rational functions can be effected easily by means of the calculus of residues. We shall suppose that the function $f(z)$, regarded as a function of the complex variable z, has the following properties:

(*a*) $f(z)$ has a finite number of singularities $a_1, a_2, \ldots, a_n$ in the upper half-plane;

(*b*) $f(z)$ is analytic at all points of the real axis except the points $b_1, b_2, \ldots, b_m$ which are simple poles;[1]

(*c*) $zf(z)e^{i\xi z} \to 0$ as $z \to \infty$, $\operatorname{Im} z > 0$.

[1] The extension to the case of multiple poles is trivial; we assume the poles to be simple merely for the sake of definiteness.

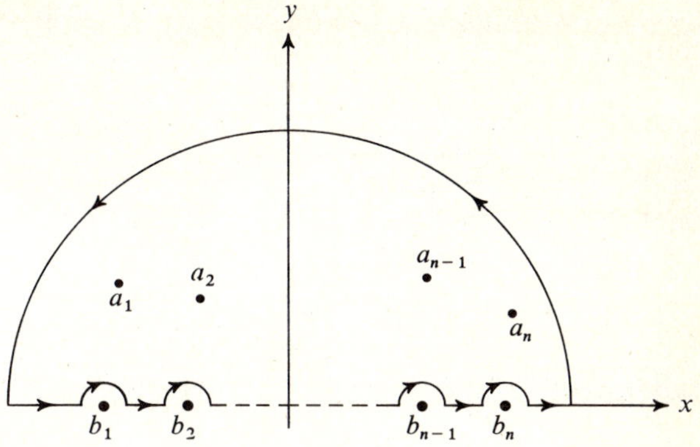

Fig. 2-6

If we integrate the function

$$e^{i\xi z}f(z), \qquad (\xi > 0),$$

round the contour shown in Fig. 2-6, and make use of Lemmas 1 and 2 of sec. 2-7, we find that

$$\frac{1}{2\pi i}\int_{-\infty}^{\infty} f(x)e^{i\xi x}\,dx = \tfrac{1}{2}\sum_{s=1}^{m} \operatorname{res}\{f(b_s)\}\,.\,e^{i\xi b_s} + \sum_{s=1}^{n} \operatorname{res}\{f(a_s)\}e^{i\xi a_s},$$

where we have used the notation $\operatorname{res}\{f(b_s)\}$ to denote the residue of the function $f(z)$ at the point $z = b_s$. This yields the simple formula

$$\mathscr{F}[f(x);\xi] = (2\pi)^{\frac{1}{2}}i\left\{\sum_{s=1}^{n} \operatorname{res}\{f(a_s)\}e^{i\xi a_s} + \tfrac{1}{2}\sum_{s=1}^{m} \operatorname{res}\{f(b_s)\}e^{i\xi b_s}\right\}, \qquad (\xi > 0). \qquad (2\text{-}8\text{-}1)$$

The simplest example is provided by taking

$$f(z) = z^{-1}.$$

This function has a simple pole at $z = 0$ and no other singularities and $\operatorname{res} f(0) = 1$, so that if $\xi > 0$ we have

$$\mathscr{F}[x^{-1};\xi] = i(\tfrac{1}{2}\pi)^{\frac{1}{2}}. \qquad (2\text{-}8\text{-}2)$$

Since x^{-1} is an odd function we see that this equation is equivalent to the formula

$$\mathscr{F}_s[x^{-1};\xi] = (\tfrac{1}{2}\pi)^{\frac{1}{2}}, \qquad (\xi > 0)$$

and we deduce immediately from (2-5-8) that, in general,

$$\mathscr{F}_s[x^{-1};\xi] = (\tfrac{1}{2}\pi)^{\frac{1}{2}} \operatorname{sgn} \xi. \qquad (2\text{-}8\text{-}3)$$

A more sophisticated example is given by

$$f(z) = \frac{1}{z(z^2 + a^2)}$$

with a real, which has a simple pole at $z = 0$ and at $z = ia$ with residues a^{-2} and $-\frac{1}{2}a^{-2}$ respectively. Hence if $\xi > 0$, we have

$$\mathscr{F}\left[\frac{1}{x(x^2 + a^2)}; \xi\right] = (\tfrac{1}{2}\pi)^{\frac{1}{2}} a^{-2} i(1 - e^{-\xi|a|})$$

from which it follows that

$$\mathscr{F}_s\left[\frac{1}{x(x^2 + a^2)}; \xi\right] = (\tfrac{1}{2}\pi)^{\frac{1}{2}} a^{-2}(1 - e^{-|\xi a|}) \operatorname{sgn} \xi. \tag{2-8-4}$$

2-9 THE CONVOLUTION INTEGRAL

The *convolution* of two integrable functions f and g defined by the equation

$$f \circ g = (2\pi)^{-\frac{1}{2}} \int_{-\infty}^{\infty} f(t - u)g(u)\, du \tag{2-9-1}$$

possesses many of the formal properties of an ordinary product. For example, if λ is a constant, it is obvious that

$$f \circ (\lambda g) = (\lambda f) \circ g = \lambda(f \circ g) \tag{2-9-2}$$

and that

$$f \circ (g + h) = f \circ g + f \circ h \tag{2-9-3}$$

We note that if both f and g belong to $\mathscr{C}^1(\mathbf{R})$ and $\mathscr{A}_1(\mathbf{R})$, then so does their convolution $h(t) = f \circ g$, for

$$\begin{aligned}(2\pi)^{\frac{1}{2}} \int_{-\infty}^{\infty} |h(t)|\, dt &= \int_{-\infty}^{\infty} dt \left| \int_{-\infty}^{\infty} f(u)g(t - u)\, du \right| \\ &< \int_{-\infty}^{\infty} dt \int_{-\infty}^{\infty} |f(u)g(t - u)|\, du \\ &= \int_{-\infty}^{\infty} |f(u)|\, du \int_{-\infty}^{\infty} |g(t - u)|\, dt\end{aligned}$$

so that

$$\int_{-\infty}^{\infty} |h(t)|\, dt < (2\pi)^{-\frac{1}{2}} \int_{-\infty}^{\infty} |f(u)|\, du \int_{-\infty}^{\infty} |g(v)|\, dv$$

and the result follows from the fact that both f and g belong to the class $\mathscr{A}_1(\mathbf{R})$.

By definition

$$g \circ f = (2\pi)^{-\frac{1}{2}} \int_{-\infty}^{\infty} g(t - v) f(v)\, dv.$$

Changing the variable of integration from v to u where $v = t - u$ we see that

$$g \circ f = (2\pi)^{-\frac{1}{2}} \int_{-\infty}^{\infty} g(u) f(t - u)\, du$$

so that we have the commutative law

$$f \circ g = g \circ f. \tag{2-9-4}$$

Similarly

$$f \circ (g \circ h) = (2\pi)^{-\frac{1}{2}} \int_{-\infty}^{\infty} f(t - u)\, du \cdot (2\pi)^{-\frac{1}{2}} \int_{-\infty}^{\infty} g(u - v) h(v)\, dv$$

$$= (2\pi)^{-1} \int_{-\infty}^{\infty} \int_{-\infty}^{\infty} f(t - u) g(u - v) h(v)\, dudv$$

and

$$(f \circ g) \circ h = (2\pi)^{-\frac{1}{2}} \int_{-\infty}^{\infty} dv \left\{ (2\pi)^{-\frac{1}{2}} \int_{-\infty}^{\infty} f(t - v - w) g(w)\, dw \right\} h(v).$$

Changing the variables of integration from v and w to v and u, where $w = u - v$, we find that

$$(f \circ g) \circ h = (2\pi)^{-1} \int_{-\infty}^{\infty} \int_{-\infty}^{\infty} f(t - u) g(u - v) h(v)\, dudv$$

from which it follows that

$$f \circ (g \circ h) = (f \circ g) \circ h \tag{2-9-5}$$

so that the associative law holds.

The important result we need is that concerning the Fourier transform of a convolution. If f and g both belong to the classes $\mathscr{C}^1(\mathbf{R})$ and $\mathscr{A}_1(\mathbf{R})$ then since (as was shown above) $f \circ g$ also belongs to $\mathscr{C}^1(\mathbf{R})$ and $\mathscr{A}_1(\mathbf{R})$, the Fourier transform of $f \circ g$ also exists. From the definitions of the Fourier transform and of the convolution of two functions we find that

$$\mathscr{F}[f \circ g; \xi] = (2\pi)^{-\frac{1}{2}} \int_{-\infty}^{\infty} e^{i\xi t}\, dt\, (2\pi)^{-\frac{1}{2}} \int_{-\infty}^{\infty} f(t - u) g(u)\, du$$

$$= (2\pi)^{-1} \int_{-\infty}^{\infty} \int_{-\infty}^{\infty} f(t - u) g(u) e^{i\xi t}\, dtdu.$$

Changing the variables of integration in the double integral from t and u to v and w, where

$$v = t - u, \qquad w = u,$$

we see that the right side of this last equation becomes

$$(2\pi)^{-1}\int_{-\infty}^{\infty}\int_{-\infty}^{\infty} f(v)g(w)e^{i\xi(v+w)}\,dvdw$$

which can be written as the product of two single integrals

$$(2\pi)^{-\frac{1}{2}}\int_{-\infty}^{\infty} f(v)e^{i\xi v}\,dv\,.\,(2\pi)^{-\frac{1}{2}}\int_{-\infty}^{\infty} g(w)e^{i\xi w}\,dw.$$

We immediately deduce that

$$\mathscr{F}[f\circ g;\xi] = F(\xi)G(\xi), \tag{2-9-6}$$

where $F(\xi)$ and $G(\xi)$ denote the Fourier transforms of f and g respectively.

Applying the operator $\mathscr{F}^{-1}$ to both sides of (2-9-6) and making use of (2-3-5) we obtain the result

$$\int_{-\infty}^{\infty} F(\xi)G(\xi)e^{-i\xi t}\,d\xi = \int_{-\infty}^{\infty} f(u)g(t-u)\,du. \tag{2-9-7}$$

The result (2-9-6) is known as the *convolution theorem* for Fourier transforms although this term is sometimes applied to the equivalent form (2-9-7). Putting $t = 0$ in equation (2-9-7) we obtain the relation

$$\int_{-\infty}^{\infty} F(\xi)G(\xi)\,d\xi = \int_{-\infty}^{\infty} f(u)g(-u)\,du \tag{2-9-8}$$

Taking the special case $g(u) = f^*(-u)$ and making use of equation (2-3-10) we obtain the relation

$$\int_{-\infty}^{\infty} F(\xi)F^*(\xi)\,d\xi = \int_{-\infty}^{\infty} f(u)f^*(u)\,du.$$

In other words, if $F(\xi)$ is the Fourier transform of $f(t)$

$$\int_{-\infty}^{\infty} |F(\xi)|^2\,d\xi = \int_{-\infty}^{\infty} |f(t)|^2\,dt \tag{2-9-9}$$

This relation, which is known as *Parseval's relation* for Fourier transforms, may be written in the alternative form

$$\|F\| = \|f\|. \tag{2-9-10}$$

2-10 PARSEVAL'S THEOREM FOR COSINE AND SINE TRANSFORMS

The convolution theorem for cosine and sine transforms is not so simple as (2-9-6) but the formulae corresponding to (2-9-8) and (2-9-9) are just as simple.

Inserting the definition of $G_c(\xi)$, we see that

$$\int_0^\infty F_c(\xi)G_c(\xi)\cos(\xi t)\,d\xi = \left(\frac{2}{\pi}\right)^{\frac{1}{2}}\int_0^\infty F_c(\xi)\cos(\xi t)\,d\xi\int_0^\infty g(u)\cos(\xi u)\,du.$$

Interchanging the order of the integrations we see that the right-hand side of this equation is equal to

$$\left(\frac{2}{\pi}\right)^{\frac{1}{2}}\int_0^\infty g(u)\,du\int_0^\infty F_c(\xi)\cos(\xi t)\cos(\xi u)\,d\xi = \left(\frac{2}{\pi}\right)^{\frac{1}{2}}\int_0^\infty g(u)\,du\int_0^\infty \tfrac{1}{2}[\cos\{\xi(t+u)\}+\cos(\xi|t-u|)]F_c(\xi)\,d\xi$$

showing that

$$\int_0^\infty F_c(\xi)G_c(\xi)\cos(\xi t)\,d\xi = \tfrac{1}{2}\int_0^\infty g(u)[f(t+u)+f(|t-u|)]\,du. \tag{2-10-1}$$

Putting $t = 0$ we obtain the equation

$$\int_0^\infty F_c(\xi)G_c(\xi)\,d\xi = \int_0^\infty f(u)g(u)\,du \tag{2-10-2}$$

which in the case in which g is identical with f^* gives the Parseval relation

$$\int_0^\infty |F_c(\xi)|^2\,d\xi = \int_0^\infty |f(u)|^2\,du \tag{2-10-3}$$

which can be stated in terms of the norm in the form

$$\|F_c\| = \|f\|.$$

We have similar results for the Fourier sine transform.

Replacing the Fourier sine transform G_s by its defining integral, we have the identity

$$\int_0^\infty F_s(\xi)G_s(\xi)\cos(\xi t)\,d\xi = \left(\frac{2}{\pi}\right)^{\frac{1}{2}}\int_0^\infty F_s(\xi)\cos(\xi t)\,d\xi\int_0^\infty g(u)\sin(\xi u)\,du.$$

Interchanging the order of the integrations we have the equation

$$\left(\frac{2}{\pi}\right)^{\frac{1}{2}} \int_0^\infty g(u)\,du \int_0^\infty F_s(\xi)\cos(\xi t)\sin(\xi u)\,d\xi$$

$$= \tfrac{1}{2}\int_0^\infty g(u)\,du \left(\frac{2}{\pi}\right)^{\frac{1}{2}} \int_0^\infty F_s(\xi)[\sin\{\xi(t+u)\} + \sin\{\xi(u-t)\}]\,d\xi$$

from which we deduce immediately that

$$\int_0^\infty F_s(\xi)G_s(\xi)\cos(\xi t)\,d\xi = \tfrac{1}{2}\int_0^\infty g(u)[f(u+t) + f(u-t)]\,du. \tag{2-10-4}$$

Giving t the particular value 0 we obtain the relation

$$\int_0^\infty F_s(\xi)G_s(\xi)\,d\xi = \int_0^\infty f(u)g(u)\,du \tag{2-10-5}$$

and if we take g to be identical with f^*, we get the Parseval relation

$$\int_0^\infty |F_s(\xi)|^2\,d\xi = \int_0^\infty |f(u)|^2\,du \tag{2-10-6}$$

that is,

$$\|F_s\| = \|f\|.$$

As a simple example of the use of these formulae we take

$$f(x) = e^{-ax},\ (a > 0), \qquad F_c(\xi) = (2/\pi)^{\frac{1}{2}} a(a^2+\xi^2)^{-1},$$
$$F_s(\xi) = (2/\pi)^{\frac{1}{2}}\xi(a^2+\xi^2)^{-1}$$

$$g(x) = e^{-bx},\ (b > 0), \qquad G_c(\xi) = (2/\pi)^{\frac{1}{2}} b(b^2+\xi^2)^{-1},$$
$$G_s(\xi) = (2/\pi)^{\frac{1}{2}}\xi(b^2+\xi^2)^{-1}.$$

Since

$$\int_0^\infty f(u)g(u)\,du = \frac{1}{a+b}$$

it follows from (2-10-2) that

$$\int_0^\infty \frac{d\xi}{(a^2+\xi^2)(b^2+\xi^2)} = \frac{\pi}{2ab(a+b)}, \qquad (a > 0, b > 0),$$

and from (2-10-4) that

$$\int_0^\infty \frac{\xi^2 d\xi}{(a^2+\xi^2)(b^2+\xi^2)} = \frac{\pi}{2(a+b)}.$$

These results can, of course, be easily obtained by other methods; they are derived here by means of the Parseval relations merely to illustrate the procedure.

We shall now use the method to derive a much more complicated result which we shall require later.

If we take

$$f(x) = x^{-s},\ (0 < s < 1), \qquad g(x) = (1 - x^2)^{\nu - \frac{1}{2}} H(1 - x),\ (\nu > -\tfrac{1}{2})$$

then by equations (2-7-21) and (2-7-19) we have

$$F_c(\xi) = \left(\frac{2}{\pi}\right)^{\frac{1}{2}} \xi^{s-1} \sin(\tfrac{1}{2}s\pi)\Gamma(1 - s), \quad G_c(\xi) = 2^{\nu - \frac{1}{2}}\Gamma(\nu + \tfrac{1}{2})\xi^{-\nu} J_\nu(\xi),$$

so that by equation (2-10-2)

$$2^\nu \pi^{-\frac{1}{2}} \sin(\tfrac{1}{2}s\pi)\Gamma(1 - s)\Gamma(\nu + \tfrac{1}{2}) \int_0^\infty \xi^{s-\nu-1} J_\nu(\xi)\, d\xi = \int_0^1 x^{-s}(1 - x^2)^{\nu - \frac{1}{2}}\, dx.$$

The simple change of variable $y = x^2$ shows that the integral on the right side of this equation has the value

$$\tfrac{1}{2} \int_0^1 y^{-\frac{1}{2}s - \frac{1}{2}}(1 - y)^{\nu - \frac{1}{2}}\, dy = \frac{\Gamma(\nu + \frac{1}{2})\Gamma(\frac{1}{2} - \frac{1}{2}s)}{2\Gamma(\nu - \frac{1}{2}s + 1)}.$$

Hence, since

$$\operatorname{cosec}(\tfrac{1}{2}s\pi) = \pi^{-1}\Gamma(\tfrac{1}{2}s)\Gamma(1 - \tfrac{1}{2}s),$$
$$\Gamma(\tfrac{1}{2})\Gamma(1 - s) = 2^{-s}\Gamma(\tfrac{1}{2} - \tfrac{1}{2}s)\Gamma(1 - \tfrac{1}{2}s)$$

we have the relation

$$\int_0^\infty \xi^{s-\nu-1} J_\nu(\xi)\, d\xi = \frac{2^{s-\nu-1}\Gamma(\frac{1}{2}s)}{\Gamma(\nu - \frac{1}{2}s + 1)}, \qquad (0 < s < 1, \nu > -\tfrac{1}{2}). \tag{2-10-7}$$

Another pair of formulae of which we shall make use is

$$\int_0^\infty F_s(\xi)G_c(\xi) \sin(\xi t)\, d\xi = \tfrac{1}{2} \int_0^\infty f(u)[g(|u - t|) - g(u + t)]\, du, \tag{2-10-8}$$

$$= \tfrac{1}{2} \int_0^\infty g(u)[f(t + u) - f(u - t)]\, du, \tag{2-10-9}$$

which can be established in a way similar to that used in the derivation of the relation (2-10-4).

2-11 OTHER FORMS OF THE FOURIER INVERSION THEOREM

We have proved (in sec. 2-2) Fourier's integral theorem for a very restricted class of functions, and on the basis of the Riemann integral (which is the one about which we learn in courses in calculus). Fourier integral theorems of the same type with very general conditions on $f(x)$ have been established. These are based on the concept of the integral introduced by Lebesgue.[1] At the end of the last century it became clear that the Riemann integral was not an instrument of sufficient power to deal with problems arising in mathematical analysis, and would have to be replaced by some other, more general, type of integral. There were attempts at such a generalization by Jordan, Borel, and W. H. Young as well as by Lebesgue, but it was Lebesgue's construction which proved to be the most successful.

It will be recalled that the fundamental idea of Riemann integration is that the integral of a function f over a finite interval $[a,b]$ is the limit as $n \to \infty$ of the sum

$$\sum_{i=1}^{n} f(x_i)m(E_i)$$

where $[a,b]$ is partitioned into n disjoint intervals $E_1, \ldots, E_n$ which cover the interval, $x_i \in E_i$ and $m(E_i)$ denotes the length of E_i.

General theories of integration are based on the concept of the *measure* of a set E, denoted by $m(E)$, which is a generalization of the idea of the length of an interval and which possesses obvious properties similar to those of length:

(*a*) $m(E) \geqslant 0$;
(*b*) $E_1 \subset E_2 \Rightarrow m(E_1) \leqslant m(E_2)$;
(*c*) $m(E_1 \cup E_2) = m(E_1) + m(E_2) - m(E_1 \cap E_2)$.

The *Lebesgue measure* of a point set is constructed as follows: Suppose that E is a set of points contained in the finite interval (a,b) of length $l = b - a$, and that we denote by E' the complement of E in (a,b), i.e., E' consists of all those points of (a,b) which do not belong to E. We further suppose that the set E is covered by a set of disjoint intervals

[1]The object of this section is to give only a very general idea of the nature of the Lebesgue integral—enough to enable an applied mathematician to read the statements of theorems in texts on the theory of Fourier transforms. A reader who does not wish to encounter considerations of this slight degree of subtlety, but wishes to have some idea of the nature of the theorems could interpret the statement "$f \in L^p(a,b)$" as meaning "$f(x)$ is such that the integral

$$\int_a^b |f(x)|^p \, dx$$

is finite," and "E is a set of measure zero" as meaning "E is an enumerable set of points." These are gross over-simplifications but they do give some idea of the points involved.

$\Lambda_1, \Lambda_2, \ldots$ of lengths $l_1, l_2, \ldots$ respectively so that

$$E \subset \bigcup_i \Lambda_i$$

and

$$\sum_i l_i \leqslant l.$$

The *exterior measure* of a set E, denoted by $m_e(E)$, is defined to be the smallest such sum $\sum_i l_i$.

In an exactly similar way we can cover E' by a collection of intervals $\Lambda_1', \Lambda_2', \ldots$ with lengths $l_1', l_2', \ldots$. The *interior measure* of the set E, denoted by $m_i(E)$, is then defined by the equation

$$m_i(E) = l - m_e(E'),$$

where $m_e(E')$ is the exterior measure of the complement E'. It can be shown that, in general,

$$0 \leqslant m_i(E) \leqslant m_e(E).$$

If it turns out that $m_e(E) = m_i(E)$, we say that the set E is *measurable* and we call the common value of $m_e(E)$ and $m_i(E)$ the *measure* $m(E)$ of E.

A special role is played in the theory by sets of measure zero. Among these is any enumerable set of points $\{x_n\}(n = 1,2, \ldots)$ contained within an interval of finite length. If two functions f and g are equal except on a set of measure zero, we say that they are *equal almost everywhere* and write $f = g$ (a.e.) or $f = g$ (p.p.).

The function $f(x)$ is said to be *measurable* on a set E if the sets E and $\{x : x \in E, f(x) > c\}$ are measurable for all finite values of c.

If the function $f(x)$ is a bounded measurable function defined on a measurable set E contained in a closed interval $[a,b]$, we choose a closed interval $[A,B]$ such that

$$A \leqslant f(x) \leqslant B, \qquad a \leqslant x \leqslant b,$$

and divide it into n parts by the points $y_0 = A, y_1, y_2, \ldots, y_n = B$ ordered in such a way that $y_0 < y_1 < \ldots y_{n-1} < y_n$. The sets $E_1, E_2, \ldots, E_n$ are now defined by the relation $E_i = \{x : x \in E, y_{i-1} \leqslant f(x) \leqslant y_i\}$, and the sums

$$s_n = \sum_{i=1}^n y_{i-1} m(E_i), \qquad S_n = \sum_{i=1}^n y_i m(E_i)$$

constructed. If, as $n \to \infty$ in such a way that each $y_i - y_{i-1} \to 0$, the sums s_n and S_n tend to a common limit s, the function f is said to be *integrable* (or *summable*) and the number s is called the *Lebesgue integral* of the function f over the set E.

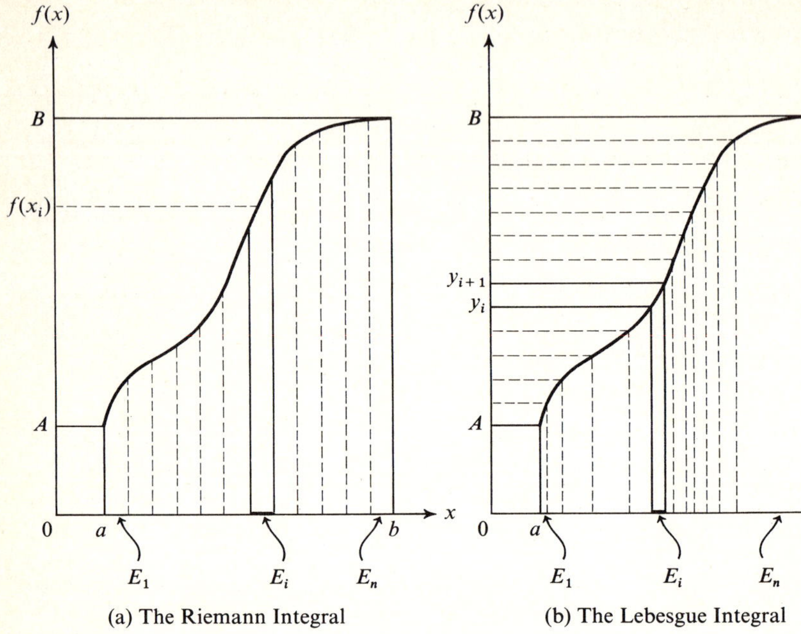

Fig. 2-7

The definition of the Lebesgue integral may be extended in an obvious way to cases in which the function f is unbounded and the set E is the whole real line.

The difference between Riemann integration and Lebesgue integration is shown graphically in Fig. 2-7. Lebesgue has described the difference in the following picturesque way: Suppose that we wish to calculate the value of a pile of coins of different face-value. The Riemann procedure consists of taking coins from the pile in succession in the order in which they lie and adding their values. The Lebesgue procedure consists of first of all sorting out the coins according to their values, each value then being multiplied by the relevant number of coins.

As a concrete example, we recall that the Dirichlet function

$$f(x) = \begin{cases} 1, & \text{if } x \text{ is rational,} \\ 0, & \text{if } x \text{ is irrational,} \end{cases}$$

does *not* possess a Riemann integral over $[0,1]$. On the other hand, it does have a Lebesgue integral since the set of all rational numbers in $[0,1]$ (being enumerable) has measure zero. Hence, the Lebesgue integral of the Dirichlet function over $[0,1]$ exists and has the value zero.

It can be shown that if a function $f(x)$ is integrable in the Riemann sense on the closed interval $[a,b]$, then it is integrable also in the Lebesgue sense and that the values of the two integrals coincide, so that any statement made about Riemann integrals is automatically true for Lebesgue integrals. It should be noticed, however, that the Lebesgue integral is a genuine generalization in the sense that there are functions which are Lebesgue-integrable on a closed interval but which are not integrable in the sense of Riemann.

We denote by $L^p(a,b)$ the class of functions for which $|f(x)|^p$ is integrable over (a,b). If the interval of integration is the whole real line we usually denote the class simply by L^p, i.e., we say that $f \in L^p$ if

$$\int_{-\infty}^{\infty} |f(x)|^p \, dx$$

exists. In dealing with a sequence of functions $\{f_n(x)\}$ and a given function $f(x)$, we are often interested not so much in the behavior of the difference $f(x) - f_n(x)$ as $n \to \infty$ as we are in the mean value of the difference. If $f_n(x) \in L^p(a,b)$, for all positive integers n, $f(x) \in L^p(a,b)$, and

$$\int_a^b |f_n(x) - f(x)|^p \, dx \to 0,$$

as $n \to \infty$ we say that $f_n(x)$ *converges in mean to* $f(x)$ *with index* p. The analogue of the "general principle of convergence" in Lebesgue theory is the

Riesz–Fischer theorem *If as m and n tend independently to infinity*

$$\int_a^b |f_n(x) - f_m(x)|^p \, dx \to 0,$$

then there exists a function $f \in L^p(a,b)$ *to which* $f_n(x)$ *converges in mean with index* p.

A theory of Fourier transforms which is completely symmetrical was first given by Plancherel (see Chapter III of Titchmarsh, 1948). It depends on mean convergence.[1] The key result is:

Plancherel's theorem *Suppose that* $f \in L^2$ *and that*

$$F(\xi,a) = (2\pi)^{-\frac{1}{2}} \int_{-a}^{a} f(x) e^{i\xi x} \, dx.$$

Then as $a \to \infty$, $F(\xi,a)$ *converges in mean over* $(-\infty,\infty)$ *to a function* $F(\xi) \in L^2$. *Reciprocally,*

[1] Instead of saying that a sequence converges in mean with index 2 we say merely that it "converges in mean."

$$f(x,a) = (2\pi)^{-\frac{1}{2}} \int_{-a}^{a} F(\xi) e^{-i\xi x}\, d\xi$$

converges in mean to $f(x)$.

Further, $f(x)$ *and* $F(\xi)$ *are related through the formulae*

$$F(\xi) = \frac{1}{\sqrt{(2\pi)}} \frac{d}{d\xi} \int_{-\infty}^{\infty} f(x) \frac{e^{i\xi x} - 1}{ix}\, dx,$$

$$f(x) = \frac{1}{\sqrt{(2\pi)}} \frac{d}{dx} \int_{-\infty}^{\infty} F(\xi) \frac{e^{-i\xi x} - 1}{-i\xi}\, d\xi.$$

Plancherel's theorem can be extended from functions of class L^2 to those of a general class L^p, $(p \geqslant 1)$, (see Chapter IV of Titchmarsh, 1948). It turns out that the Fourier transform maps the space L^p into the space L^q where

$$\frac{1}{p} + \frac{1}{q} = 1.$$

A typical theorem in this theory is:

Titchmarsh's theorem *Suppose that* $f \in L^p$, $(1 \leqslant p \leqslant 2)$, *and that*

$$F(\xi,a) = (2\pi)^{-\frac{1}{2}} \int_{-a}^{a} f(x) e^{i\xi x}\, dx.$$

Then as $a \to \infty$, $F(\xi,a)$ *converges in mean with index* q *to a function* $F(\xi)$, *where*

$$\int_{-\infty}^{\infty} |F(\xi)|^q\, d\xi \leqslant (2\pi)^{1-\frac{1}{2}q} \left\{ \int_{-\infty}^{\infty} |f(x)|^p\, dx \right\}^{1/(p-1)}.$$

Further, $f(x)$ *and* $F(\xi)$ *are related through the formulae*

$$F(\xi) = \frac{1}{\sqrt{(2\pi)}} \frac{d}{d\xi} \int_{-\infty}^{\infty} f(x) \frac{e^{i\xi x} - 1}{ix}\, dx,$$

$$f(x) = \frac{1}{\sqrt{(2\pi)}} \frac{d}{dx} \int_{-\infty}^{\infty} F(\xi) \frac{e^{i\xi x} - 1}{-i\xi}\, d\xi.$$

The convolution theorem reflects the same kind of reciprocity. If $f \in L^p$, $g \in L^{p'}$, with

$$\frac{1}{p} + \frac{1}{p'} > 1,$$

then $f \circ g \subset L^P$, where

$$P = \frac{pp'}{p + p' - pp'}$$

and $F(\xi)G(\xi) \in L^Q$, where

$$\frac{1}{P} + \frac{1}{Q} = 1$$

The existence of the integral defining the Fourier transform $F(\xi)$ imposes a certain restriction on the behavior of the function $f(x)$ at infinity. If we write $\zeta = \xi + i\eta$ with ξ and η real, then it is conceivable that in circumstances in which $F(\xi)$ does not exist for real values of ξ, the function

$$F_+(\zeta) = (2\pi)^{-\frac{1}{2}} \int_0^\infty f(x)e^{i\zeta x}\,dx$$

may exist for sufficiently large positive values of η and the function

$$F_-(\zeta) = (2\pi)^{-\frac{1}{2}} \int_{-\infty}^0 f(x)e^{i\zeta x}\,dx$$

may exist for sufficiently large negative values of η.

We now suppose $f(x)$ to be a complex-valued function of a real variable x; in addition, we shall say that such a function is continuous, integrable, of bounded variation,[1] etc., if its real and imaginary parts separately possess these properties.

If there exists a constant c such that $e^{-c|x|}|f(x)| \in L(-\infty,\infty)$, then it is readily shown that the function $F_+(\zeta)$ exists for $\eta \geqslant c$, and the function $F_-(\zeta)$ for $\eta \leqslant -c$. We also have:

Complex form of Fourier's theorem *If the function $f(t)$ is of bounded variation in a neighborhood of the point $t = x$, then*

$$\lim_{\lambda\to\infty} \left[(2\pi)^{-\frac{1}{2}} \int_{ia-\lambda}^{ia+\lambda} F_+(\zeta)e^{-ix\zeta}\,d\zeta + (2\pi)^{-\frac{1}{2}} \int_{ib-\lambda}^{ib+\lambda} F_-(\zeta)e^{-ix\zeta}\,d\zeta \right] = \tfrac{1}{2}[f(x+) + f(x-)],$$

where $a \geqslant c$, $b \leqslant -c$.

(For a proof of this theorem see p. 43 of Titchmarsh, 1948.)

In applications of this theorem[2] we often make use of the following theorem, which is proved in Vol. I (p. 463) of Morse and Feshbach (1953):

Suppose that the function $F(\zeta)$ is analytic in the strip $a_1 \leqslant \eta \leqslant a_2$ and that the function $G(\zeta)$ is analytic in the strip $b_1 \leqslant \eta \leqslant b_2 < a_1$ and that both $F(\zeta)$ and $G(\zeta)$ belong to the class $L(-\infty,\infty)$. If both $F(\zeta)$ and $G(\zeta)$ tend to zero as $|\xi| \to \infty$ in the region in which they are analytic, and

[1] We say that a function $f(x)$ is of *bounded variation* in an interval if, in that interval, it can be expressed in the form $u(x) - v(x)$ where each of the functions $u(x)$, $v(x)$ is nondecreasing and bounded.

[2] See, e.g. sec. 2-15 below.

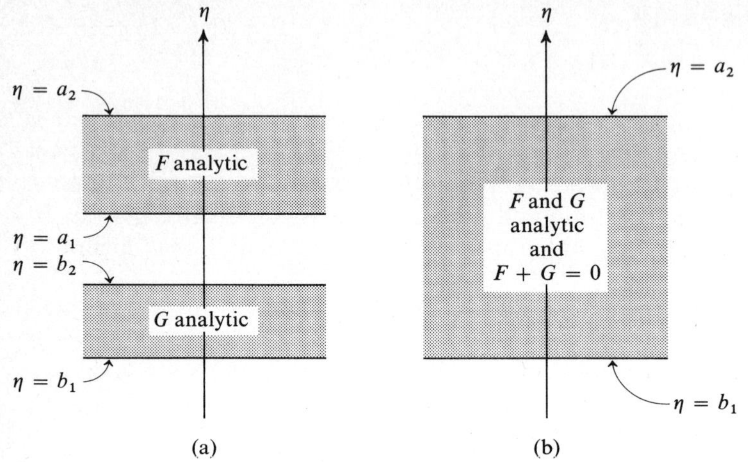

Fig. 2-8

$$\int_{ia-\infty}^{ia+\infty} F(\zeta)e^{-i\zeta x}\,d\zeta + \int_{ib-\infty}^{ib+\infty} G(\zeta)e^{-i\zeta x}\,d\zeta = 0,$$

$$(a_1 < a < a_2,\ b_1 < b < b_2)$$

then

(*a*) *$F(\zeta)$ and $G(\zeta)$ are both analytic in the strip $b_1 \leqslant \eta \leqslant a_2$;*
(*b*) *$F(\zeta) + G(\zeta) = 0$ in that strip.*

The hypotheses of this theorem are illustrated graphically in the diagram (*a*) of Fig. 2-8 and the conclusions are shown in (*b*).

2-12 A TRANSFORM WITH A MORE GENERAL TRIGONOMETRICAL KERNEL

We saw in sec. 2-6 that in any problem for a region $x \geqslant 0$, $(y,z) \in \Omega$, in which we require to calculate a Fourier transform of $\partial^2 f/\partial x^2$ when we know that

$$f(0,y,z) = \phi_1(y,z), \tag{2-12-1}$$

we may use the formula

$$\mathscr{F}_s\left[\frac{\partial^2 f}{\partial x^2}; x \to \xi\right] = (2/\pi)^{\frac{1}{2}}\phi_1(y,z) - \xi^2 F_s(\xi,y,z)$$

where $F_s(\xi,y,z) = \mathscr{F}_s[f(x,y,z); x \to \xi]$. (Cf. equation (2-6-4).)

On the other hand, if we are given that

$$f_x(0,y,z) = \phi_2(y,z) \tag{2-12-2}$$

it is appropriate to take a Fourier cosine transform for then we can use equation (2-6-3) to deduce that

$$\mathscr{F}_c\left[\frac{\partial^2 f}{\partial x^2}; x \to \xi\right] = -(2/\pi)^{\frac{1}{2}}\phi_2(y,z) - \xi^2 F_c(\xi,y,z),$$

where $F_c(\xi,y,z) = \mathscr{F}_c[f(x,y,z); x \to \xi]$.

It is natural now to seek a transform which can be used in the analysis of problems in which, instead of conditions either of the type (2-12-1) or (2-12-2) we have a condition of the type

$$f_x(0,y,z) - hf(0,y,z) = \psi(y,z), \tag{2-12-3}$$

where h is a constant and the function $\psi(y,z)$ is prescribed for all $(y,z) \in \Omega$. In the theory of the conduction of heat a condition of the type (2-12-3) is called a *radiation condition*. For that reason we shall denote the operator of the relevant transform by $\mathscr{R}$. A transform of this type was first considered by Churchill. (See sec. 71 of Churchill, 1944.)

Suppose that we consider a general transform defined by the equation

$$\tilde{f}(\xi,y,z) = \mathscr{T}[f(x,y,z); x \to \xi] = (2/\pi)^{\frac{1}{2}} \int_0^\infty f(x,y,z)K(x,\xi)\, dx.$$

Using the formula for integration by parts we find that, if $f \in \mathscr{C}^2(0,\infty)$ and f, f', f'' are all absolutely integrable over the positive real line,

$$\mathscr{T}\left[\frac{\partial^2 f}{\partial x^2}; x \to \xi\right] = (2/\pi)^{\frac{1}{2}}\phi(\xi,y,z) + (2/\pi)^{\frac{1}{2}} \int_0^\infty f(x,y,z)K_{xx}(x,\xi)\, dx$$

where the function $\phi(\xi,y,z)$ is defined by the equation

$$\phi(\xi,y,z) = f(0,y,z)K_x(0,\xi) - f_x(0,y,z)K(0,\xi). \tag{2-12-4}$$

Rearranging the terms on the right-hand side of this last equation we find that

$$\phi(\xi,y,z) = f(0,y,z)\{K_x(0,\xi) - hK(0,\xi)\} - K(0,\xi)\psi(y,z),$$

where the function $\psi(y,z)$ is the prescribed function occurring in the boundary condition (2-12-3). From this we deduce that if the kernel $K(x,\xi)$ satisfies the conditions

$$\frac{\partial^2 K}{\partial x^2} + \xi^2 K = 0, \qquad x > 0, \tag{2-12-5}$$

$$K_x(0,\xi) - hK(0,\xi) = 0, \tag{2-12-6}$$

then we have the relation

$$\mathscr{T}\left[\frac{\partial^2 f}{\partial x^2}; x \to \xi\right] = -\xi^2\tilde{f}(\xi,y,z) - (2/\pi)^{\frac{1}{2}}K(0,\xi)\psi(y,z), \tag{2-12-7}$$

for the transform of the second partial derivative of f with respect to x.

The solution of the equations (2-12-5) and (2-12-6) may be taken in the form

$$K(x,\xi) = \xi\cos(\xi x) + h\sin(\xi x)$$

so that the transform $\mathscr{R}$ defined by the relation

$$\begin{aligned}\tilde{f}(\xi,y,z) &= \mathscr{R}[f(x,y,z); x \to \xi] \\ &= (2/\pi)^{\frac{1}{2}} \int_0^\infty f(x,y,z)\{\xi\cos(\xi x) + h\sin(\xi x)\}\, dx \qquad (2\text{-}12\text{-}8)\end{aligned}$$

has the property that

$$\mathscr{R}\left[\frac{\partial^2 f}{\partial x^2}; x \to \xi\right] = -\xi^2\tilde{f}(\xi,y,z) - (2/\pi)^{\frac{1}{2}}\xi\psi(y,z) \qquad (2\text{-}12\text{-}9)$$

where $\psi(y,z)$ is the prescribed function occurring in the relation (2-12-3).

We can easily establish an inversion theorem for the $\mathscr{R}$-transform. If

$$\tilde{f}(\xi) = \mathscr{R}[f(x);\xi]$$

then it is obvious that

$$\tilde{f}(\xi) = \xi F_c(\xi) + hF_s(\xi) \qquad (2\text{-}12\text{-}10)$$

where $F_c(\xi)$ and $F_s(\xi)$ are respectively the Fourier cosine and the Fourier sine transform of the function $f(x)$. We note that a necessary condition for these transforms to exist is that

$$\lim_{x\to\infty} f(x) = 0. \qquad (2\text{-}12\text{-}11)$$

From equation (2-6-8) we see that we can write equation (2-12-10) in the alternative form

$$\tilde{f}(\xi) = -\mathscr{F}_s[f'(x) - hf(x);\xi]$$

and making use of the Fourier sine inversion theorem we find that

$$f'(x) - hf(x) = -\mathscr{F}_s[\tilde{f}(\xi);x]. \qquad (2\text{-}12\text{-}12)$$

Now the differential equation

$$y'(x) - hy(x) = \sin(\xi x) \qquad (2\text{-}12\text{-}13)$$

has the general solution

$$y(x) = ce^{hx} - \frac{\xi\cos(\xi x) + h\sin(\xi x)}{\xi^2 + h^2},$$

where c denotes an arbitrary constant. Hence if $h > 0$ and $f(x)$ has the property (2-12-11), it follows that we must take $c = 0$, i.e., that the equation (2-12-13) has solution

$$y(x) = -\frac{K(x,\xi)}{\xi^2 + h^2}$$

a result which can be written in the form

$$\left(\frac{d}{dx} - h\right)^{-1} \sin(\xi x) = -\frac{K(x,\xi)}{\xi^2 + h^2}.$$

Rewriting (2-12-12) in the form

$$f(x) = -(2/\pi)^{\frac{1}{2}} \int_0^\infty \tilde{f}(\xi)\left(\frac{d}{dx} - h\right)^{-1} \sin(\xi x)\, d\xi,$$

we see that

$$f(x) = \mathscr{R}^{-1}[\tilde{f}(\xi);x] = (2/\pi)^{\frac{1}{2}} \int_0^\infty \frac{K(x,\xi)}{\xi^2 + h^2}\tilde{f}(\xi)\, d\xi. \tag{2-12-14}$$

We can calculate $\mathscr{R}$-transforms easily from tables of Fourier sine and cosine transforms. For instance, since

$$\mathscr{F}_c[e^{-ax};\xi] = (2/\pi)^{\frac{1}{2}} a(a^2 + \xi^2)^{-1},$$

$$\mathscr{F}_s[e^{-ax}:\xi] = (2/\pi)^{\frac{1}{2}} \xi(a^2 + \xi^2)^{-1}$$

($a > 0$), it follows immediately from the definition that

$$\mathscr{R}[e^{-ax};\xi] = (2/\pi)^{\frac{1}{2}} \xi(a + h)(a^2 + \xi^2)^{-1}.$$

On the other hand, in working out inverse transforms it is sometimes more convenient to use the differential equation (2-12-12) than the inversion formula (2-12-14). To illustrate the procedure we shall calculate

$$\mathscr{R}^{-1}[\xi e^{-\xi^2 t};x].$$

If we write

$$\tilde{w}(\xi,t) = \xi e^{-\xi^2 t}, \qquad w(x,t) = \mathscr{R}^{-1}[\tilde{w}(\xi,t);\xi \to x], \tag{2-12-15}$$

then, by equation (2-12-12)

$$\left(\frac{\partial}{\partial x} - h\right) w(x,t) = -\mathscr{F}_s[\tilde{w}(\xi,t);\xi \to x].$$

Using the relation (2-5-4) we see that this is equivalent to

$$\left(\frac{\partial}{\partial x} - h\right) w(x,t) = -\frac{1}{2t}\mathscr{F}_s[\xi e^{-\frac{1}{2}\xi^2};\xi \to x(2t)^{-\frac{1}{2}}]$$

and using equation (2-7-14) to evaluate the expression on the right-hand side of this equation, we obtain the differential equation

$$\left(\frac{\partial}{\partial x} - h\right) w(x,t) = -x(2t)^{-\frac{3}{2}} e^{-x^2/4t}.$$

The solution of this equation is readily seen to be

$$w(x,t) = (2t)^{-\frac{3}{2}} e^{hx} \int_x^\infty u e^{-u^2/4t - hu} \, du.$$

Using the rule for integrating by parts on the integral on the right-hand side of this equation we obtain the solution in the form

$$w(x,t) = (2t)^{-\frac{1}{2}} e^{-x^2/4t} - h(2t)^{-\frac{1}{2}} e^{hx} \int_x^\infty e^{-hu - u^2/4t} \, du,$$

which is easily seen to be equivalent to the solution

$$w(x,t) = (2t)^{-\frac{1}{2}} e^{-x^2/4t} - (\tfrac{1}{2}\pi)^{\frac{1}{2}} h e^{hx + h^2 t} \operatorname{erfc}(\tfrac{1}{2} x t^{-\frac{1}{2}} + h t^{\frac{1}{2}}). \tag{2-12-16}$$

We shall now derive a result which is useful in deriving the form of the convolution integral which is most appropriate to a transform of this type. From the identity

$$f(x) = -\frac{d}{dx} \int_x^\infty f(v) \, dv$$

and equation (2-6-2) we obtain the relation

$$\mathscr{F}_s[f(x);\xi] = \xi \mathscr{F}_c\left[\int_x^\infty f(u) \, du;\, \xi\right].$$

From this we deduce that

$$\xi \mathscr{F}_c[f(x);\xi] + h\mathscr{F}_s[f(x);\xi] = \xi \mathscr{F}_c\left[f(x) + h \int_x^\infty f(v) \, dv; x \to \xi\right].$$

If we write

$$\phi(x) = f(x) + h \int_x^\infty f(v) \, dv$$

we see that this last equation can be put into the form

$$\xi^{-1} \tilde{f}(\xi) = \Phi_c(\xi) \tag{2-12-17}$$

where $\Phi_c(\xi) = \mathscr{F}_c[\phi(x);\xi]$.

To find the form of convolution appropriate to the $\mathscr{R}$-transform we write

$$f \circ g = \mathscr{R}^{-1}[\xi^{-1} \tilde{f}(\xi) \tilde{g}(\xi); x].$$

Then, by equation (2-12-12) we obtain

$$\left(\frac{\partial}{\partial x} - h\right) f \circ g = -\mathscr{F}_s[\Phi_c(\xi) \tilde{g}(\xi); x]$$

and, by the definition of $\tilde{g}(\xi)$, we can rewrite the right-hand side of this equation as

$$-\mathscr{F}_s[\xi\Phi_c(\xi)G_c(\xi);x] - h\mathscr{F}_s[\Phi_c(\xi)G_s(\xi);x].$$

Now by equation (2-10-8)

$$\mathscr{F}_s[\Phi_c(\xi)G_s(\xi);x] = (2\pi)^{-\frac{1}{2}}\int_0^\infty g(u)\{\phi(|x-u|) - \phi(x+u)\}\,du$$

and, by equations (2-6-2) and (2-10-1),

$$\begin{aligned}-\mathscr{F}_s[\xi\Phi_c(\xi)G_c(\xi);x]& \\ &= \frac{\partial}{\partial x}\mathscr{F}_c[\Phi_c(\xi)G_c(\xi);x] \\ &= \frac{\partial}{\partial x}(2\pi)^{-\frac{1}{2}}\int_0^\infty g(u)\{\phi(|x-u|) + \phi(x+u)\}\,du.\end{aligned}$$

We therefore find that

$$f\circ g = (2\pi)^{-\frac{1}{2}}\int_0^\infty g(u)\chi(x,u)\,du$$

where

$$\chi(x,u) = \phi(|x-u|) + \phi(x+u) + 2h\left(\frac{\partial}{\partial x} - h\right)^{-1}\phi(x+u).$$

Now it is easily shown that

$$\left(\frac{\partial}{\partial x} - h\right)^{-1}\phi(x+u) = -\int_{x+u}^\infty f(v)\,dv$$

so that

$$\chi(x,u) = f(|x-u|) + f(x+u) - h\int_{x+u}^\infty f(v)\,dv + h\int_{|x-u|}^\infty f(v)\,dv$$

and hence that, if the convolution $f\circ g$ of two functions f and g is defined by the equation

$$f\circ g = (2\pi)^{-\frac{1}{2}}\int_0^\infty g(u)\left\{f(|x-u|) + f(x+u) + h\int_{|x-u|}^{x+u} f(v)\,dv\right\}du, \tag{2-12-18}$$

then

$$\mathscr{R}^{-1}[\xi^{-1}\tilde{f}(\xi)\tilde{g}(\xi);x] = f\circ g. \tag{2-12-19}$$

From the symmetry of this result we deduce that for the convolution defined by equation (2-12-18),

$$f\circ g = g\circ f.$$

2-13 MULTIPLE FOURIER TRANSFORMS

If we have a function of n independent variables $x_1, x_2, \ldots, x_n$, each of which ranges over the whole real axis, we may take the Fourier transform of the function with respect to each of the variables in turn, i.e., we may take

$$F_{(1)}(\xi_1,x_2,\ldots,x_n) = \mathscr{F}_{(1)}[f(x_1,x_2,\ldots,x_n);x_1 \to \xi_1]$$

$$= \mathscr{F}[f(x_1,x_2,\ldots,x_n);x_1 \to \xi_1]$$

and we can define the double transform

$$F_{(2)}(\xi_1,\xi_2,\ldots,x_n) = \mathscr{F}_{(2)}[f(x_1,x_2,\ldots,x_n);x_1 \to \xi_1,x_2 \to \xi_2]$$

by the relation

$$F_{(2)}(\xi_1,\xi_2,\ldots,x_n) = \mathscr{F}[F_{(1)}(\xi_1,x_2,\ldots,x_n);x_2 \to \xi_2] \qquad (2\text{-}13\text{-}1)$$

and so on.

$$F_{(n)}(\xi_1,\xi_2,\ldots,\xi_{n-1},\xi_n)$$
$$= \mathscr{F}_{(n)}[f(x_1,x_2,\ldots,x_n);x_1 \to \xi_1,x_2 \to \xi_2,\ldots,x_n \to \xi_n] \qquad (2\text{-}13\text{-}2)$$

can be defined by the relation

$$F_{(n)}(\xi_1,\xi_2,\ldots,\xi_{n-1},\xi_n)$$
$$= \mathscr{F}[F_{(n-1)}(\xi_1,\xi_2,\ldots,\xi_{n-1},x_n);x_n \to \xi_n]. \qquad (2\text{-}13\text{-}3)$$

If we denote the n-vectors $(x_1,x_2,\ldots,x_n)$ and $(\xi_1,\xi_2,\ldots,\xi_n)$ by $\mathbf{x}$ and $\boldsymbol{\xi}$ respectively, and the inner product

$$x_1\xi_1 + x_2\xi_2 + \cdots + x_n\xi_n$$

by $(\mathbf{x}\cdot\boldsymbol{\xi})$ then the n-dimensional Fourier transform

$$F_{(n)}(\boldsymbol{\xi}) = \mathscr{F}_{(n)}[f(\mathbf{x});\mathbf{x} \to \boldsymbol{\xi}] \qquad (2\text{-}13\text{-}4)$$

of the function $f(\mathbf{x})$ is defined by the equation

$$F_{(n)}(\boldsymbol{\xi}) = (2\pi)^{-\frac{1}{2}n} \int_{E_n} f(\mathbf{x})e^{i(\boldsymbol{\xi}\cdot\mathbf{x})}\,d\mathbf{x} \qquad (2\text{-}13\text{-}5)$$

where E_n denotes the n-dimensional Euclidean space.

If we invert equation (2-13-3) by the Fourier inversion theorem we find that

$$F_{(n-1)}(\xi_1,\xi_2,\ldots,\xi_{n-1},x_n)$$
$$= \mathscr{F}^{-1}[F_{(n)}(\xi_1,\xi_2,\ldots,\xi_{n-1},\xi_n);\xi_n \to x_n]$$
$$= \mathscr{F}[F_{(n)}(\xi_1,\xi_2,\ldots,\xi_{n-1},\xi_n);\xi_n \to -x_n]$$

and so on until we reach

$$f(x_1,x_2,\ldots,x_n) = \mathscr{F}^{-1}[F_{(1)}(\xi_1,x_2,\ldots,x_n);\xi_1 \to x_1]$$

$$= \mathscr{F}[F_{(1)}(\xi_1,x_2,\ldots,x_n);\xi_1 \to -x_1]$$

$$= \mathscr{F}[\mathscr{F}\{F_{(2)}(\xi_1,\xi_2,\ldots,x_n);\xi_2 \to -x_2\};\ \xi_1 \to -x_1].$$

If we write out all the steps, we find that n applications of Fourier's integral theorem yields the inversion formula

$$f(x) = (2\pi)^{-\frac{1}{2}n} \int_{E_n} F_{(n)}(\boldsymbol{\xi})e^{-i(\boldsymbol{\xi}\cdot\mathbf{x})}\, d\boldsymbol{\xi}. \tag{2-13-6}$$

To show what is happening we shall write out the details of the three-dimensional case. Here

$$F_{(3)}(\boldsymbol{\xi}) = (2\pi)^{-\frac{3}{2}} \int_{-\infty}^{\infty}\int_{-\infty}^{\infty}\int_{-\infty}^{\infty} f(x_1,x_2,x_3)\exp\{i(\xi_1x_1 + \xi_2x_2 + \xi_3x_3)\}\, dx_1dx_2dx_3$$

which may be written in the form

$$F_{(3)}(\xi_1,\xi_2,\xi_3) = \mathscr{F}\left[\frac{1}{2\pi}\int_{-\infty}^{\infty}\int_{-\infty}^{\infty} f(x_1,x_2,x_3)\exp\{i(\xi_2x_2 + \xi_3x_3)\}\, dx_2dx_3;x_1 \to \xi_1\right]$$

This may be inverted to give

$$\frac{1}{2\pi}\int_{-\infty}^{\infty}\int_{-\infty}^{\infty} f(x_1,x_2,x_3)\exp\{i(\xi_2x_2 + \xi_3x_3)\}\, dx_2dx_3 = (2\pi)^{-\frac{1}{2}}\int_{-\infty}^{\infty} F_{(3)}(\xi_1,\xi_2,\xi_3)e^{-i\xi_1x_1}\, d\xi_1.$$

Writing this in turn as

$$\mathscr{F}\left[(2\pi)^{-\frac{1}{2}}\int_{-\infty}^{\infty} f(x_1,x_2,x_3)e^{i\xi_3x_3}\, dx_3;x_2 \to \xi_2\right] = (2\pi)^{-\frac{1}{2}}\int_{-\infty}^{\infty} F_{(3)}(\xi_1,\xi_2,\xi_3)e^{-i\xi_1x_1}\, d\xi_1$$

and using the Fourier inversion theorem we obtain the result

$$\mathscr{F}[f(x_1,x_2,x_3);x_3 \to \xi_3] = \frac{1}{2\pi}\int_{-\infty}^{\infty}\int_{-\infty}^{\infty} F_{(3)}(\xi_1,\xi_2,\xi_3)e^{-i(\xi_1x_1+\xi_2x_2)}\, d\xi_1d\xi_2$$

which in turn can be inverted to give

$$f(x_1,x_2,x_3) = (2\pi)^{-\frac{3}{2}} \int_{-\infty}^{\infty}\int_{-\infty}^{\infty}\int_{-\infty}^{\infty} F_{(3)}(\xi_1,\xi_2,\xi_3) \times$$

$$\times \exp\{-i(\xi_1 x_1 + \xi_2 x_2 + \xi_3 x_3)\}\, d\xi_1 d\xi_2 d\xi_3$$

which is the three-dimensional form of (2-13-6).

The n-dimensional analogues of the results of sec. 2-3 are easily derived. For example, the analogue of (2-3-8) is

$$\mathscr{F}_{(n)}[f(\mathbf{x} - \mathbf{a});\mathbf{x} \to \boldsymbol{\xi}] = e^{i(\boldsymbol{\xi}\cdot\mathbf{a})}\mathscr{F}_{(n)}[f(\mathbf{x});\mathbf{x} \to \boldsymbol{\xi}] \tag{2-13-7}$$

where **a** denotes a constant vector, while that of equation (2-3-11) is

$$\mathscr{F}_{(n)}\left[\frac{\partial f}{\partial x_s};\mathbf{x} \to \boldsymbol{\xi}\right] = -i\xi_s\mathscr{F}_{(n)}[f(\mathbf{x});\mathbf{x} \to \boldsymbol{\xi}] \tag{2-13-8}$$

which is valid if $f(x) \in \mathscr{C}^1(\mathbf{R}^n)$ and both f and $\partial f/\partial x_s$ are absolutely integrable in $\mathbf{R}^n$.

If $f(\mathbf{x})$ has finite discontinuities this last result has to be modified in the same way as in the one-dimensional case.

Derivatives of higher order can be dealt with in an obvious way; for example

$$\mathscr{F}_{(n)}\left[\frac{\partial^2 f}{\partial x_r \partial x_s};\mathbf{x} \to \boldsymbol{\xi}\right] = -\xi_r\xi_s\mathscr{F}_{(n)}[f(\mathbf{x});\mathbf{x} \to \boldsymbol{\xi}] \tag{2-13-9}$$

and so on. From this we deduce that, if L denotes the differential operator defined by the equation

$$\mathrm{L}f = \sum_{r,s=1}^{n} a_{rs}\frac{\partial^2 f}{\partial x_r \partial x_s} + \sum_{r=1}^{n} b_r\frac{\partial f}{\partial x_r} + cf \tag{2-13-10}$$

where the a_{rs}, b_r, and c are constants, then

$$\mathscr{F}_{(n)}[\mathrm{L}f;\mathbf{x} \to \boldsymbol{\xi}] = -P(\boldsymbol{\xi})\mathscr{F}_{(n)}[f(\mathbf{x});\mathbf{x} \to \boldsymbol{\xi}] \tag{2-13-11}$$

in which the quadratic form $P(\boldsymbol{\xi})$ is defined by the equation

$$P(\boldsymbol{\xi}) = \sum_{r,s=1}^{n} a_{rs}\xi_r\xi_s + i\sum_{r=1}^{n} b_r\xi_r - c. \tag{2-13-12}$$

In particular, if Δ_n denotes the n-dimensional Laplacian operator

$$\Delta_n = \frac{\partial^2}{\partial {x_1}^2} + \frac{\partial^2}{\partial {x_2}^2} + \cdots + \frac{\partial^2}{\partial {x_n}^2} \tag{2-13-13}$$

and $f \in \mathscr{C}^2(\mathbf{R}^n)$, then

$$\mathscr{F}_{(n)}[\Delta_n f(\mathbf{x});\mathbf{x} \to \boldsymbol{\xi}] = -\boldsymbol{\xi}^2 F_{(n)}(\boldsymbol{\xi}) \tag{2-13-14}$$

where

$$F_{(n)}(\boldsymbol{\xi}) = \mathscr{F}_{(n)}[f(\mathbf{x});\mathbf{x} \to \boldsymbol{\xi}]. \tag{2-13-15}$$

Similarly for the polyharmonic operator Δ_n^r defined by the recurrence relation

$$\Delta_n^r = \Delta_n \Delta_n^{r-1}$$

and the equation (2-13-13), we find that for functions $f(\mathbf{x}) \in \mathscr{C}^{2r}(\mathbf{R}^n)$,

$$\mathscr{F}_{(n)}[\Delta_n^r f(\mathbf{x});\mathbf{x} \to \boldsymbol{\xi}] = (-1)^r \xi^{2r} F_{(n)}(\boldsymbol{\xi}), \tag{2-13-16}$$

where, again, $F_{(n)}(\boldsymbol{\xi})$ is defined by equation (2-13-15).

The convolution of two functions $f(\mathbf{x})$, $g(\mathbf{x})$ of the vector $\mathbf{x}$ is defined by the equation

$$f \circ g(\mathbf{x}) = (2\pi)^{-\frac{1}{2}n} \int_{E_n} f(\mathbf{x} - \mathbf{u}) g(\mathbf{u})\, d\mathbf{u}. \tag{2-13-17}$$

Proceeding exactly as in the one-dimensional case (Cf. sec. 2-9) we find that

$$\begin{aligned}
\mathscr{F}_{(n)}&[f \circ g(\mathbf{x});\mathbf{x} \to \boldsymbol{\xi}] \\
&= (2\pi)^{-n} \int_{E_n} e^{i(\mathbf{x}\cdot\boldsymbol{\xi})}\, d\mathbf{x} \int_{E_n} f(\mathbf{x} - \mathbf{u}) g(\mathbf{u})\, d\mathbf{u} \\
&= (2\pi)^{-n} \int_{E_n} g(\mathbf{u})\, d\mathbf{u} \int_{E_n} f(\mathbf{x} - \mathbf{u}) e^{i(\mathbf{x}\cdot\boldsymbol{\xi})}\, d\mathbf{x} \\
&= (2\pi)^{-n} \int_{E_n} g(\mathbf{u}) e^{i(\mathbf{u}\cdot\boldsymbol{\xi})}\, d\mathbf{u} \int_{E_n} f(\mathbf{y}) e^{i(\mathbf{y}\cdot\boldsymbol{\xi})}\, d\mathbf{y} \\
&= F_{(n)}(\boldsymbol{\xi}) G_{(n)}(\boldsymbol{\xi})
\end{aligned}$$

where $F_{(n)}(\boldsymbol{\xi})$ and $G_{(n)}(\boldsymbol{\xi})$ are the Fourier transforms of $f(\mathbf{x})$ and $g(\mathbf{x})$ respectively. Hence we have the *convolution theorem*

$$\mathscr{F}_{(n)}[f \circ g(\mathbf{x});\mathbf{x} \to \boldsymbol{\xi}] = F_{(n)}(\boldsymbol{\xi}) G_{(n)}(\boldsymbol{\xi}). \tag{2-13-18}$$

This is most often used in its inverse form

$$\mathscr{F}_{(n)}^{-1}[F_{(n)}(\boldsymbol{\xi}) G_{(n)}(\boldsymbol{\xi});\boldsymbol{\xi} \to \mathbf{x}] = (2\pi)^{-\frac{1}{2}n} \int_{E_n} f(\mathbf{x} - \mathbf{u}) g(\mathbf{u})\, d\mathbf{u}. \tag{2-13-19}$$

2-14 FOURIER TRANSFORMS OF RADIALLY SYMMETRIC FUNCTIONS

In applications we often have to find the Fourier transform $F_{(n)}(\boldsymbol{\xi})$ of a function $f(r)$ which is a function of

$$r = |\mathbf{x}| = (x_1^2 + x_2^2 + \cdots + x_n^2)^{\frac{1}{2}} \tag{2-14-1}$$

only. We shall now show that in such a case $F_{(n)}(\boldsymbol{\xi})$ can be expressed simply in terms of a Hankel transform of the function $f(r)$. We begin with the two-dimensional case.

If we introduce plane polar coordinates (r,θ) in the x_1x_2-plane and (ρ,ϕ) in the $\xi_1\xi_2$-plane, we may write

$$x_1 = r\cos\theta,\ x_2 = r\sin\theta,\ \xi_1 = \rho\cos\phi,\ \xi_2 = \rho\sin\phi \tag{2-14-2}$$

from which it follows that

$$(\mathbf{x}\cdot\boldsymbol{\xi}) = r\rho\cos(\theta - \phi),$$

and hence that

$$\mathscr{F}_{(2)}[f(r); x_1 \to \xi_1, x_2 \to \xi_2] = \frac{1}{2\pi}\int_0^\infty r\,dr \int_0^{2\pi} f(r)e^{ir\rho\cos(\theta-\phi)}\,d\theta.$$

Now it is an elementary result from the theory of Bessel functions that

$$\frac{1}{2\pi}\int_0^{2\pi} e^{ir\rho\cos(\theta-\phi)}\,d\theta = \mathrm{J}_0(r\rho),$$

so that

$$\mathscr{F}_{(2)}[f(r); x_1 \to \xi_1, x_2 \to \xi_2] = \int_0^\infty rf(r)\mathrm{J}_0(r\rho)\,dr. \tag{2-14-3}$$

Comparing the integral on the right-hand side of this equation with that on the right-hand side of equation (1-1-9) we see that it is the Hankel transform of order zero of the function f. If we use the notation

$$\mathscr{H}_\nu[f(x);\xi]$$

to denote the Hankel transform of order ν of f

$$\int_0^\infty xf(x)\mathrm{J}_\nu(\xi x)\,dx$$

we see that equation (2-14-3) can be written in the alternative form

$$\mathscr{F}_{(2)}[f(r); x_1 \to \xi_1, x_2 \to \xi_2] = \mathscr{H}_0[f(r);\rho] \tag{2-14-4}$$

with $r = (x_1{}^2 + x_2{}^2)^{\frac{1}{2}}$, $\rho = (\xi_1{}^2 + \xi_2{}^2)^{\frac{1}{2}}$.

For example, if $k > 0$,

$$\mathscr{F}_{(2)}[r^{-1}e^{-kr}; x_1 \to \xi_1, x_2 \to \xi_2] = \mathscr{H}_0[r^{-1}e^{-kr};\rho].$$

Now the value of the integral

$$\int_0^\infty e^{-kr}\mathrm{J}_0(\rho r)\,dr$$

is known* to be $(k^2 + \rho^2)^{-\frac{1}{2}}$ and therefore we have the result

$$\mathscr{F}_{(2)}[r^{-1}e^{-kr}; x_1 \to \xi_1, x_2 \to \xi_2] = (k^2 + \rho^2)^{-\frac{1}{2}}, \quad (k > 0). \tag{2-14-5}$$

*This is a well-known result in the elementary theory of Bessel functions (Cf. sec. 3-7 below or sec. A-6).

The result (2-14-4) is readily generalized to the n-dimensional case. By the definition of the n-dimensional Fourier transform we have the equation

$$\mathscr{F}_{(n)}[f(r); x_1 \to \xi_1, \ldots, x_n \to \xi_n] = (2\pi)^{-\frac{1}{2}n} \int_{E_n} f(r) e^{i(\boldsymbol{\xi}\cdot\mathbf{x})}\, d\mathbf{x}. \tag{2-14-6}$$

We now make the transformations

$$\xi_i = \rho\alpha_i, \ (i = 1,2, \ldots, n), \qquad y_1 = \sum_{i=1}^{n} \alpha_i x_i$$

$$y_j = \sum_{i=1}^{n} \alpha_{ij} x_i, \ (j = 2, \ldots, n),$$

the coefficients α_{ij} being chosen in such a way as to make the transformation orthogonal. In this case we find that

$$r = (y_1^2 + y_2^2 + \cdots + y_n^2)^{\frac{1}{2}}$$

and that

$$d\mathbf{x} = d\mathbf{y}, \qquad (\boldsymbol{\xi}\cdot\mathbf{x}) = \sum_{i=1}^{n} \xi_i x_i = \rho \sum_{i=1}^{n} \alpha_i x_i = \rho y_1.$$

If we now put

$$\lambda^2 = y_2^2 + y_3^2 + \cdots + y_n^2,$$

we find that equation (2-14-6) reduces to

$$F_{(n)}(\boldsymbol{\xi}) = (2\pi)^{-\frac{1}{2}n} \int_{-\infty}^{\infty} dy_1 \int_0^{\infty} \Omega(\lambda) f\{(\lambda^2 + y_1^2)^{\frac{1}{2}}\} e^{i\rho y_1}\, d\lambda, \tag{2-14-7}$$

where we have written $F_{(n)}(\boldsymbol{\xi}) = \mathscr{F}_{(n)}[f(r); x_1 \to \xi_1, \ldots, x_n \to \xi_n]$, and $\Omega(\lambda)$ is defined to be such that, for any arbitrary function $g(\lambda)$,

$$\int_{E_{n-1}} g\{(y_2^2 + y_3^2 + \cdots + y_n^2)^{\frac{1}{2}}\}\, dy_2 dy_3 \ldots dy_n = \int_0^{\infty} \Omega(\lambda) g(\lambda)\, d\lambda. \tag{2-14-8}$$

To determine the form of the function $\Omega(\lambda)$ we note that, since $\Omega(\lambda)\, d\lambda$ is a volume element in E_{n-1}, it is of the form

$$\Omega(\lambda) = \omega_n \lambda^{n-2} \tag{2-14-9}$$

where ω_n is a function of n but not of λ. To find the value of ω_n, we consider the special case in which

$$g(\lambda) = e^{-\lambda^2}.$$

In this case

$$\int_{E_{n-1}} g\{(y_2{}^2 + y_3{}^2 + \cdots + y_n{}^2)^{\frac{1}{2}}\}\, dy_2 dy_3 \ldots dy_n$$

$$= \left\{\int_{-\infty}^{\infty} e^{-u^2}\, du\right\}^{n-1} = \pi^{\frac{1}{2}n-\frac{1}{2}}$$

and

$$\int_0^\infty \Omega(\lambda) g(\lambda)\, d\lambda = \omega_n \int_0^\infty \lambda^{n-2} e^{-\lambda^2}\, d\lambda = \tfrac{1}{2}\omega_n \Gamma(\tfrac{1}{2}n - \tfrac{1}{2})$$

so that we deduce the result

$$\omega_n = \frac{2\pi^{\frac{1}{2}n-\frac{1}{2}}}{\Gamma(\frac{1}{2}n - \frac{1}{2})}, \tag{2-14-10}$$

from equation (2-14-8).

Substituting from equations (2-14-9) and (2-14-10) into equation (2-14-7) we obtain the relation

$$F_{(n)}(\boldsymbol{\xi}) = \frac{1}{2^{\frac{1}{2}n-1}\pi^{\frac{1}{2}}\Gamma(\frac{1}{2}n - \frac{1}{2})} \int_0^\infty r^{n-1} f(r)\, dr \int_0^\pi \sin^{n-2}\theta\, e^{i\rho r\cos\theta}\, d\theta.$$

Making use of the well-known result[1]

$$\int_0^\pi \sin^{n-2}\theta\, e^{i\rho r\cos\theta}\, d\theta = \frac{2^{\frac{1}{2}n-1}\pi^{\frac{1}{2}}\Gamma(\frac{1}{2}n - \frac{1}{2})}{(\rho r)^{\frac{1}{2}n-1}} J_{\frac{1}{2}n-1}(\rho r), \tag{2-14-11}$$

we see that $F_{(n)}(\boldsymbol{\xi})$ is a function only of the variable ρ. If we write

$$F(\rho) = \mathscr{F}_{(n)}[f(r); x_1 \to \xi_1, \ldots, x_n \to \xi_n], \tag{2-14-12}$$

we find that

$$\rho^{\frac{1}{2}n-1} F(\rho) = \int_0^\infty r^{\frac{1}{2}n} f(r) J_{\frac{1}{2}n-1}(\rho r)\, dr. \tag{2-14-13}$$

Putting

$$\nu = \tfrac{1}{2}n - 1, \tag{2-14-14}$$

we see that this result may be written in the symmetrical form

$$\rho^\nu F(\rho) = \mathscr{H}_\nu[r^\nu f(r); \rho]. \tag{2-14-15}$$

The Fourier transform of a radially symmetrical function in E_n can therefore be reduced to a Hankel transform over the positive real line.

The case in which $n = 3$ is of special interest. Here $\nu = \frac{1}{2}$ and if we recall that

$$(\rho r)^{\frac{1}{2}} J_{\frac{1}{2}}(\rho r) = (2/\pi)^{\frac{1}{2}} \sin(\rho r)$$

[1] See, for example, p. 113 of Sneddon (1961).

we find that equation (2-14-15) reduces to the simple form

$$\mathscr{F}_{(3)}[f(r);\mathbf{x} \to \boldsymbol{\xi}] = \rho^{-1}\mathscr{F}_s[rf(r);\rho]. \tag{2-14-16}$$

In particular, if $k > 0$, we have the special results

$$\mathscr{F}_{(3)}[r^{-1}e^{-kr};\mathbf{x} \to \boldsymbol{\xi}] = \left(\frac{2}{\pi}\right)^{\frac{1}{2}} \frac{1}{\rho^2 + k^2}, \tag{2-14-17}$$

$$\mathscr{F}_{(3)}[e^{-kr};\mathbf{x} \to \boldsymbol{\xi}] = \left(\frac{2}{\pi}\right)^{\frac{1}{2}} \frac{2k}{(\rho^2 + k^2)^2}. \tag{2-14-18}$$

2-15 THE SOLUTION OF INTEGRAL EQUATIONS OF CONVOLUTION TYPE

In this section we shall make use of the theory of Fourier transforms to derive the solutions of certain simple types of integral equation.

The simplest integral equation to tackle by means of Fourier transform technique is the integral equation of convolution type

$$(2\pi)^{-\frac{1}{2}} \int_{-\infty}^{\infty} f(t)k(x - t)\,dt = g(x), \qquad -\infty < x < \infty, \tag{2-15-1}$$

in which the functions $g(x)$ and $k(x)$ are prescribed for all real values of x and are assumed to possess Fourier transforms $G(\xi)$ and $K(\xi)$ respectively. Taking the Fourier transform of each side of (2-15-1) and making use of the convolution we obtain *formally* the relation

$$F(\xi)K(\xi) = G(\xi), \tag{2-15-2}$$

connecting $F(\xi)$, the Fourier transform of the unknown function $f(x)$, with those of $g(x)$ and $k(x)$. We can write this equation in the form

$$F(\xi) = G(\xi)L(\xi)$$

where $L(\xi) = 1/K(\xi)$. If it did happen that

$$l(x) = \mathscr{F}^{-1}[L(\xi);x]$$

exists, then by applying the convolution theorem we should obtain the solution of (2-15-1) in the form

$$f(x) = \frac{1}{\sqrt{(2\pi)}} \int_{-\infty}^{\infty} g(t)l(x - t)\,dt.$$

The difficulty is that in problems of physical interest the function $l(x)$ usually does *not* exist. If, however, it so happens that for some positive integer n, the inverse transform

$$m(x) = \mathscr{F}^{-1}\left[\frac{1}{(-i\xi)^n K(\xi)};x\right] \tag{2-15-3}$$

exists, then writing (2-15-2) in the form

$$F(\xi) = (-i\xi)^n G(\xi) \cdot M(\xi),$$

we see that it yields the solution

$$f(x) = (2\pi)^{-\frac{1}{2}} \int_{-\infty}^{\infty} m(x - t) g^{(n)}(t)\, dt. \tag{2-15-4}$$

As an example, we consider the integral equation

$$(2\pi)^{-\frac{1}{2}} \int_{-\infty}^{\infty} |x - t|^{-\frac{1}{2}} f(t)\, dt = g(x).$$

This is of the type (2-15-1) with $k(x) = |x|^{-\frac{1}{2}}$ and hence, by equation (2-7-23), with $K(\xi) = |\xi|^{-\frac{1}{2}}$. Since

$$\mathscr{F}^{-1}\left[\frac{1}{(-i\xi)K(\xi)}; x\right] = \mathscr{F}^{-1}[i \operatorname{sgn} \xi \cdot |\xi|^{-\frac{1}{2}}; x] = |x|^{-\frac{1}{2}} \operatorname{sgn} x$$

the solution is given by (2-15-4) with $n = 1, m(x) = |x|^{-\frac{1}{2}} \operatorname{sgn} x$. The solution is therefore given by the equation

$$f(x) = \frac{1}{\sqrt{(2\pi)}} \int_{-\infty}^{x} \frac{g'(t)\, dt}{\sqrt{(x - t)}} - \frac{1}{\sqrt{(2\pi)}} \int_{x}^{\infty} \frac{g'(t)\, dt}{\sqrt{(t - x)}}.$$

In cases in which the inverse transform (2-15-3) does *not* exist it is sometimes still possible to derive a solution of equation (2-15-1) provided that the function $g(x)$ has derivatives of *all* orders. The method used is basically due to Eddington (1913) although interest in it has recently been revived by Kurth (1965). Eddington's method consists essentially in assuming a solution of the form

$$f(x) = \sum_{n=1}^{\infty} c_n g^{(n)}(x) \tag{2-15-5}$$

where the constants $c_n (n = 0,1,2, \ldots)$ are to be determined. For a solution of the type (2-15-5), we find that

$$F(\xi) = \sum_{n=0}^{\infty} c_n (-i\xi)^n G(\xi)$$

provided that the Fourier transforms of all the derivatives exist. If we substitute this form into equation (2-15-2) we find that

$$[K(\xi)]^{-1} = \sum_{n=0}^{\infty} c_n (-i\xi)^n$$

so that if $[K(\xi)]^{-1}$ can be expanded as a Maclaurin series in the neighborhood of the origin then $(-i)^n c_n$ is the coefficient of ξ^n in that expansion.

If there is an Eddington solution (2-15-5), then its coefficients $c_0, c_1, c_2, \ldots$ are uniquely determined by the function $k(x)$, but there *are* kernels $k(x)$ which exclude, for any infinitely differentiable function $g(x)$, all possibility that an Eddington solution exists.

As an illustration of a case in which there is an Eddington solution we consider the integral equation

$$\frac{1}{\sigma\sqrt{(2\pi)}}\int_{-\infty}^{\infty} f(t)e^{-\frac{1}{2}(x-t)^2/\sigma^2}\,dt = g(x)$$

where $g(x)$ has derivatives of all orders. In this equation

$$k(x) = \sigma^{-1}e^{-\frac{1}{2}x^2/\sigma^2}$$

so that

$$K(\xi) = e^{-\frac{1}{2}\xi^2\sigma^2}.$$

If, therefore, we assume a solution of the form (2-15-5), we obtain the identity

$$\sum_{n=0}^{\infty} c_n(-i\xi)^n = e^{\frac{1}{2}\xi^2\sigma^2}$$

by means of which we may determine the coefficients c_n. We deduce immediately that

$$c_{2m} = (-1)_m \frac{(\frac{1}{2}\sigma^{2m})}{m!}, \qquad c_{2m-1} = 0$$

and hence that the solution is given by

$$f(x) = \sum_{m=0}^{\infty} (-1)_m \frac{(\frac{1}{2}\sigma^{2m})}{m!} g^{(2m)}(x).$$

We have already noticed that it is possible that an Eddington solution does *not* exist; to illustrate this situation we show that the integral equation

$$\frac{\lambda}{\pi}\int_{-\infty}^{\infty} \frac{f(t)\,dt}{\lambda^2 + (x - t - \mu)^2} = g(x)$$

in which λ and μ are constants ($\lambda > 0$) does not possess an Eddington solution.

For this equation

$$k(x) = \left(\frac{2}{\pi}\right)^{\frac{1}{2}} \frac{\lambda}{\lambda^2 + (x - \mu)^2}, \qquad (\lambda > 0)$$

so that

$$K(\xi) = e^{i\mu\xi - \lambda|\xi|}.$$

Since this function cannot be expanded in a power series in any neighborhood of the origin, no Eddington solution exists. (It will be observed that this equation cannot be solved by the first method either, since the inverse transform (2-15-3) does not exist for any positive integer n.)

Kurth has shown that for an Eddington solution to exist it is necessary that both $k(x)$ and $g(x)$ possess derivatives of all orders which are absolutely integrable over the whole real line, and that $g(x)$ has moments[1] γ_0, γ_1, γ_2, . . . , of all orders such that the series

$$\sum_{n=0}^{\infty} \gamma_n \frac{(i\xi)^n}{n!}$$

converges for some (none-zero) value of ξ. Kurth also provides a set of sufficient conditions; he has shown that if

(*a*) $[K(\xi)]^{-1}$ has a Maclaurin expansion $\sum_{n=0}^{\infty} c_n(-i\xi)^n$, which converges for all values of ξ;

(*b*) $g(x)$ has derivatives of all orders such that for some suitable positive constant α and an integrable function $h(x)$,

$$|g^{(n)}(x)| \leqslant \alpha^n h(x), \qquad n = 0,1,2, \ldots;$$

then the series

$$\sum_{n=0}^{\infty} c_n g^{(n)}(x)$$

converges to a function which is integrable over the x-axis and is a solution of the integral equation (2-15-1).

The integral equation

$$(2\pi)^{-\frac{1}{2}} \int_{-\infty}^{\infty} f(t)k(x+t)\,dt = g(x), \qquad -\infty < x < \infty \tag{2-15-6}$$

can be treated in the same way by first of all observing that it can be written in the alternative form

$$(2\pi)^{-\frac{1}{2}} \int_{-\infty}^{\infty} f(-y)k(x-y)\,dy = g(x)$$

which is equivalent to the relation

$$F(-\xi)K(\xi) = G(\xi).$$

If we write

$$f(x) = \sum_{n=0}^{\infty} c_n \frac{d^n}{dx^n} g(-x), \tag{2-15-7}$$

[1]The moments are defined by

$$\gamma_r = \int_{-\infty}^{\infty} x^r g(x)\,dx.$$

then

$$F(\xi) = \sum_{n=0}^{\infty} c_n(-i\xi)^n G(-\xi)$$

and the coefficients $c_n(n = 0,1,2, \ldots)$ in the Eddington solution (2-15-7)—if it exists at all!—are given by the fact that c_n is the coefficient of $(i\xi)^n$ in the Maclaurin expansion of $[K(\xi)]^{-1}$.

We shall now consider the solution of the Fredholm integral equation

$$f(x) = g(x) + \lambda \int_{-\infty}^{\infty} k(x - t)f(t)\, dt \tag{2-15-8}$$

in which the functions $g(x)$ and $k(x)$ are assumed to be known, λ is a parameter and $f(x)$ is the unknown function which we wish to determine. The equation corresponding to the case in which $g(x) \equiv 0$:

$$f(x) = \lambda \int_{-\infty}^{\infty} k(x - t)f(t)\, dt \tag{2-15-9}$$

is said to be *homogeneous*.

We recall (Cf. Riesz and Nagy, 1955, Chapter IV), that if there exists a non-trivial solution $f_1(t)$ of the homogeneous equation (2-15-9) corresponding to the value λ_1 of λ, we say that λ_1 is an *eigenvalue* of the equation and f_1 an *eigenfunction* of that equation. The set of eigenvalues is known as the *spectrum* of the equation; if an equation has a discrete set of eigenvalues we say that it has a *line spectrum* and otherwise, a *band spectrum*. We begin our discussion by showing that a simple equation of the type (2-15-9) may have a band spectrum.

From equations (2-3-8) and (2-7-1) we have the relation

$$\mathscr{F}[e^{-|x-t|}; t \to \xi] = e^{i\xi x}\mathscr{F}[e^{-|t|}; \xi] = (2/\pi)^{\frac{1}{2}}(1 + \xi^2)^{-1}e^{i\xi x}.$$

Rewriting this in the form

$$e^{i\xi x} = \tfrac{1}{2}(1 + \xi^2) \int_{-\infty}^{\infty} e^{-|x-t| + i\xi t}\, dt$$

we see that the equation

$$f(x) = \int_{-\infty}^{\infty} e^{-|x-t|}f(t)\, dt \tag{2-15-10}$$

has an eigenfunction $e^{i\xi x}$ corresponding to the eigenvalue $\frac{1}{2}(1 + \xi^2)$ for all real values of ξ. In other words, the equation (2-15-10) has the band spectrum $\lambda > \frac{1}{2}$.

We return now to the discussion of the solution of equation (2-15-8). From the complex form of Fourier's theorem (see sec. 2-11), we see that we may represent the prescribed function $g(x)$ by a formula of the type

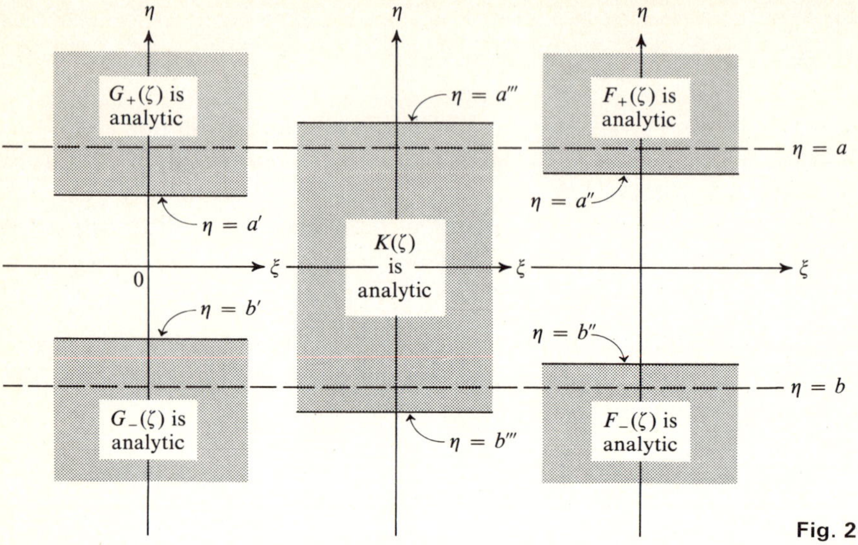

Fig. 2-9

$$g(x) = (2\pi)^{-\frac{1}{2}}\left\{\int_{ia'-\infty}^{ia'+\infty} G_+(\zeta)e^{-i\zeta x}\,d\zeta + \int_{ib'-\infty}^{ib'+\infty} G_-(\zeta)e^{-i\zeta x}\,d\zeta\right\} \tag{2-15-11}$$

where the function $G_+(\zeta)$ is analytic throughout the strip $\eta > a'$, and the function $G_-(\zeta)$ is analytic throughout the strip $\eta > b'$. For the method of the Fourier transform to be applicable, we must be able to represent the kernel $k(x)$ in a similar way, such that its Fourier transform $K(\zeta)$ is analytic throughout a strip $b''' < \eta < a'''$ which overlaps the region in which the functions $G_+(\zeta)$ and $G_-(\zeta)$ are analytic. In a similar way we have to assume that the unknown function $f(x)$ can be represented by a formula of the type (2-15-11) in which the function $F_+(\zeta)$ is analytic in a region $\eta > a''$, and $F_-(\zeta)$ is analytic in a region $\eta > b''$. For the method of complex Fourier transforms to be applicable we must have (Cf. Fig. 2-9)

$$a''' > \max(a', a''), \qquad b''' < \min(b', b'').$$

With these assumptions we may write equation (2-15-8) in the form

$$\begin{aligned}\int_{ia-\infty}^{ia+\infty} F_+(\zeta)e^{-i\zeta x}\,d\zeta &+ \int_{ib-\infty}^{ib+\infty} F_-(\zeta)e^{-i\zeta x}\,d\zeta \\ &= \int_{ia-\infty}^{ia+\infty} G_+(\zeta)e^{-i\zeta x}\,d\zeta + \int_{ib-\infty}^{ib+\infty} G_-(\zeta)e^{-i\zeta x}\,d\zeta \\ &\quad + \lambda(2\pi)^{\frac{1}{2}}\int_{ia-\infty}^{ia+\infty} F_+(\zeta)K(\zeta)e^{-i\zeta x}\,d\zeta \\ &\quad + \lambda(2\pi)^{\frac{1}{2}}\int_{ib-\infty}^{ib+\infty} F_-(\zeta)K(\zeta)e^{-i\zeta x}\,d\zeta\end{aligned}$$

where $a''' > a > \max(a', a'')$, and $b''' < b < \min(b', b'')$.

If we apply the theorem stated at the end of sec. 2-11, we see that the functions

$$\{1 - \lambda(2\pi)^{\frac{1}{2}}K(\zeta)\}F_+(\zeta) - G_+(\zeta),$$
$$\{1 - \lambda(2\pi)^{\frac{1}{2}}K(\zeta)\}F_-(\zeta) - G_-(\zeta)$$

are both analytic in the strip $a < \eta < b$. Furthermore, throughout that strip of the ζ-plane their sum is zero. We may therefore write

$$\{1 - \lambda(2\pi)^{\frac{1}{2}}K(\zeta)\}F_+(\zeta) - G_+(\zeta) = H(\zeta),$$
$$\{1 - \lambda(2\pi)^{\frac{1}{2}}K(\zeta)\}F_-(\zeta) - G_-(\zeta) = -H(\zeta),$$

where the function $H(\zeta)$ is analytic in the strip $a < \eta < b$.

Solving these equations, we find that

$$F_+(\zeta) = \frac{G_+(\zeta) + H(\zeta)}{1 - \lambda(2\pi)^{\frac{1}{2}}K(\zeta)}$$
$$F_-(\zeta) = \frac{G_-(\zeta) - H(\zeta)}{1 - \lambda(2\pi)^{\frac{1}{2}}K(\zeta)}.$$

Inserting these expressions into equation (2-11-1) we find that the solution of the integral equation (2-15-8) can be written in the form

$$\begin{aligned} f(x) &= (2\pi)^{-\frac{1}{2}} \int_{ia-\infty}^{ia+\infty} \frac{G_+(\zeta)e^{-i\zeta x}}{1 - \lambda(2\pi)^{\frac{1}{2}}K(\zeta)}\,d\zeta \\ &+ (2\pi)^{-\frac{1}{2}} \int_{ib-\infty}^{ib+\infty} \frac{G_-(\zeta)e^{-i\zeta x}}{1 - \lambda(2\pi)^{\frac{1}{2}}K(\zeta)}\,d\zeta \\ &+ (2\pi)^{-\frac{1}{2}} \int_{\Gamma} \frac{H(\zeta)e^{-i\zeta x}}{1 - \lambda(2\pi)^{\frac{1}{2}}K(\zeta)}\,d\zeta \end{aligned} \tag{2-15-12}$$

where the contour Γ is the boundary of the infinite strip $a < \eta < b$.

The third term on the right-hand side of equation (2-15-12) represents the solution of the homogeneous equation (2-15-9). Since the function $H(\zeta)$ is analytic in the strip $a < \eta < b$, this will be identically zero unless the function

$$1 - \lambda(2\pi)^{\frac{1}{2}}K(\zeta)$$

has zeros, or branch points, within the strip. If, for instance, this function has simple zeros $\xi_1, \xi_2, \ldots, \xi_n$ within the strip, and no branch points, then it follows from the calculus of residues that the solution $f_0(x)$ of the homogeneous equation (2-15-9) is given by the equation

$$f_0(x) = \sum_{r=1}^{n} A_r e^{-i\xi_r x} \tag{2-15-13}$$

where the constants A_r are given in terms of the analytic function $H(\zeta)$ by the formula

$$A_r = l(2\pi)^{\frac{1}{2}} \lim_{\zeta \to \zeta_r} \frac{(\zeta - \zeta_r)H(\zeta)}{1 - \lambda(2\pi)^{\frac{1}{2}}K(\zeta)}. \tag{2-15-14}$$

To illustrate the procedure we shall now consider the solution of the integral equation

$$f(x) = e^{\alpha|x|} + \lambda \int_{-\infty}^{\infty} e^{-|x-u|} f(u)\, du. \tag{2-15-15}$$

For this equation

$$k(x) = e^{-|x|}$$

so that

$$K(\zeta) = (2/\pi)^{\frac{1}{2}}(1 + \zeta^2)^{-1}$$

$K(\zeta)$ has simple poles at $\pm i$ and therefore is analytic in the strip $-1 < \eta < 1$.
Also it is easily shown that

$$G_+(\zeta) = i(2\pi)^{-\frac{1}{2}}(\zeta - i\alpha)^{-1}, \qquad G_-(\zeta) = -i(2\pi)^{-\frac{1}{2}}(\zeta + i\alpha)^{-1},$$

so that $G_+(\zeta)$ is analytic in the half-plane $\eta > \alpha$ and $G_-(\zeta)$ is analytic in the half-plane $\eta < -\alpha$. These half-planes overlap the strip in which $K(\zeta)$ is analytic—and hence the method of Fourier transforms can only be applied—if $\alpha < 1$. For this function $K(\zeta)$, we have

$$1 - \lambda(2\pi)^{\frac{1}{2}}K(\zeta) = \frac{\zeta^2 - \zeta_0^{\,2}}{\zeta^2 + 1}, \qquad \zeta_0^{\,2} = 2\lambda - 1.$$

The zeros of this function depend on the value of the parameter λ (which we shall assume to be real).

If $\lambda > \frac{1}{2}$, these poles are at the points $\pm\xi_0$ on the real axis, where $\xi_0 = (2\lambda - 1)^{\frac{1}{2}}$ and therefore contribute the terms

$$A_1 e^{i\xi_0 x} + A_2 e^{-i\xi_0 x}$$

to the solution. We observe that these are precisely the eigenfunctions of the equation (2-15-10) which we considered above.

On the other hand if $\lambda < -1$, the poles are at the points $\pm i\eta_0$, where $\eta_0 = (1 - 2\lambda)^{\frac{1}{2}}$. Since neither of these poles lie within the contour Γ it follows that the integral round Γ in equation (2-15-12) has the value zero.

We shall now write down the solution in the case $0 < \alpha < 1$, $\lambda > \frac{1}{2}$. From equation (2-15-12) we see that it can be written in the form

$$f(x) = -I_1 + I_2 + A_1 e^{i\xi_0 x} + A_2 e^{-i\xi_0 x} \tag{2-15-16}$$

where the integrals I_1 and I_2 are defined by the equations

$$I_1 = \frac{1}{2\pi i}\int_{ia-\infty}^{ia+\infty} \frac{(\zeta^2+1)e^{-i\zeta x}\,d\zeta}{(\zeta^2-\xi_0{}^2)(\zeta-i\alpha)}, \qquad (\alpha < a < 1)$$

$$I_2 = \frac{1}{2\pi i}\int_{ib-\infty}^{ib+\infty} \frac{(\zeta^2+1)e^{-i\zeta x}\,d\zeta}{(\zeta^2-\xi_0{}^2)(\zeta+i\alpha)}, \qquad (-1 < b < -\alpha)$$

and the constants A_1 and A_2 are arbitrary.

If $x > 0$, we add a semicircle in the *lower* half-plane to the path of each of these integrals and in that way obtain a closed contour along which to carry out the integrations by means of the calculus of residues. The integral over the semicircle is easily shown to tend to zero as its radius tends to infinity. The integrand of I_1 has simple poles at the points $i\alpha$, $\pm\xi_0$, all of which lie within the contour so that I_1 is equal to minus[1] the sum of the residues at these points, i.e.,

$$I_1 = \frac{1-\alpha^2}{\alpha^2+\xi_0{}^2}e^{\alpha x} - \frac{\xi_0{}^2+1}{2\xi_0}\left[\frac{e^{i\xi_0 x}}{\xi_0+i\alpha}+\frac{e^{-i\xi_0 x}}{\xi_0-i\alpha}\right]. \qquad (2\text{-}15\text{-}17)$$

The integrand of I_2 has no singularities within the contour so that, from Cauchy's theorem, we deduce that $I_2 = 0$.

If $x < 0$, we proceed in a similar way by adding a semicircle in the *upper* half-plane to the path of each integral. The integrand of I_1 has no singularities within the closed contour so formed and hence we deduce that $I_1 = 0$. On the other hand, the integrand of I_2 has simple poles at $-i\alpha$, $\pm\xi_0$, all lying within the contour. The value of I_2 is therefore given by the sum of the residues at these points, so that we find that

$$I_2 = -\frac{1-\alpha^2}{\alpha^2+\xi_0{}^2}e^{-\alpha x} + \frac{\xi_0{}^2+1}{2\xi_0}\left[\frac{e^{i\xi_0 x}}{\xi_0-i\alpha}+\frac{e^{-i\xi_0 x}}{\xi_0+i\alpha}\right]. \qquad (2\text{-}15\text{-}18)$$

In inserting the values of I_1 and I_2 given by (2-15-17) and (2-15-18) we note that we may omit the terms in the square brackets since they may be incorporated into the terms involving the arbitrary constants. We therefore find that the required solution is

$$f(x) = -\frac{1-\alpha^2}{\alpha^2+\xi_0{}^2}e^{\alpha|x|} + A_1 e^{i\xi_0 x} + A_2 e^{-i\xi_0 x}, \qquad (2\text{-}15\text{-}19)$$

where $\xi_0 = (2\lambda - 1)^{\frac{1}{2}}$, and A_1 and A_2 are arbitrary constants.

The method outlined here is called the *Wiener-Hopf technique*. It is used extensively in the solution of boundary value problems in mathematical physics (see Noble, 1958).

[1]The minus sign arises because the closed contour will be traversed in the negative direction.

2-16 THE SOLUTION OF PARTIAL DIFFERENTIAL EQUATIONS BY MEANS OF FOURIER TRANSFORMS

We shall now illustrate by means of some simple examples how Fourier transforms may be used in the solution of boundary value and initial value problems for linear partial differential equations.

2-16-1 LAPLACE'S EQUATION

In this subsection we shall consider some problems involving Laplace's equation

$$\Delta_n u(\mathbf{r}) = 0, \tag{2-16-1}$$

where n is the dimension of the space, u is a function of the position vector $\mathbf{r} = (x_1,x_2, \ldots ,x_n)$ and Δ_n denotes the Laplacian operator

$$\frac{\partial^2}{\partial {x_1}^2} + \frac{\partial^2}{\partial {x_2}^2} + \cdots + \frac{\partial^2}{\partial {x_n}^2}.$$

We begin with the problem already referred to in sec. 1-3.

(a) Laplace's equation in a half-plane It will be recalled—see equations (1-3-1) through (1-3-3)—that here we wish to determine a function $u(x,y)$ satisfying Laplace's equation

$$\Delta_2 u(x,y) = 0 \tag{2-16-2}$$

in the half-plane $y \geqslant 0$, the boundary condition

$$u(x,0) = f(x), \qquad -\infty < x < \infty \tag{2-16-3}$$

and the limiting condition

$$u(x,y) \to 0, \qquad \rho \to \infty \tag{2-16-4}$$

where $\rho = (x^2 + y^2)^{\frac{1}{2}}$.

If we introduce the Fourier transform

$$U(\xi,y) = \mathscr{F}[u(x,y);x \to \xi] \tag{2-16-5}$$

then it follows from equation (2-3-12) that Laplace's equation (2-16-2) is equivalent to the equation

$$\left(\frac{\partial^2}{\partial y^2} - \xi^2\right) U(\xi,y) = 0. \tag{2-16-6}$$

The boundary condition (2-16-3) is equivalent to the condition

$$U(\xi,0) = F(\xi)$$

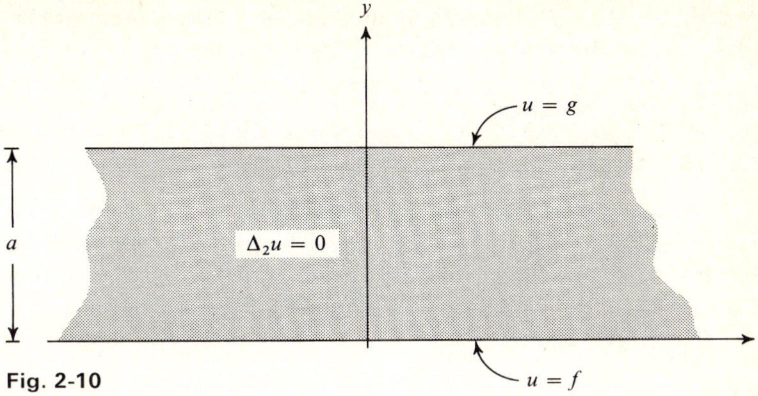

Fig. 2-10

where $F(\xi) = \mathscr{F}[f(x);\xi]$ and the limiting condition (2-16-4) implies the condition

$$U(\xi,y) \to 0, \qquad y \to \infty.$$

The condition $u(x,y) \to 0$ as $|x| \to \infty$ follows from the Riemann–Lebesgue lemma. We therefore obtain the expression

$$U(\xi,y) = F(\xi)e^{-|\xi|y} \tag{2-16-7}$$

for the Fourier transform of the function $u(x,y)$. Using the convolution theorem we see that

$$u(x,y) = (2\pi)^{-\frac{1}{2}} \int_{-\infty}^{\infty} f(t)g(x-t)\,dt,$$

where

$$g(x) = \mathscr{F}^{-1}[e^{-|\xi|y};\xi \to x] = \left(\frac{2}{\pi}\right)^{\frac{1}{2}} \frac{y}{x^2+y^2}.$$

Substituting this result into the last equation we obtain the expression

$$u(x,y) = \frac{y}{\pi}\int_{-\infty}^{\infty} \frac{f(t)\,dt}{(x-t)^2+y^2} \quad (y > 0), \tag{2-16-8}$$

for the required function.

(b) Laplace's equation in an infinite strip Suppose now that we wish to derive the solution of (2-16-2) in the strip $0 \leqslant y \leqslant a$ subject to the boundary conditions

$$u(x,0) = f(x), \qquad u(x,a) = g(x) \tag{2-16-9}$$

(cf. Fig. 2-10). This would give the steady-state distribution of temperature in a slab whose faces are maintained at prescribed temperatures. The transform $U(\xi,y)$ defined by equation (2-16-5) still satisfies the ordinary differential equation (2-16-6) but now, because of the relations (2-16-9) it satisfies the boundary conditions

$$U(\xi,0) = F(\xi), \qquad U(\xi,a) = G(\xi)$$

where $F(\xi)$ and $G(\xi)$ are the Fourier transforms of the prescribed functions $f(x)$ and $g(x)$ respectively. The solution of this two-point boundary value problem is easily shown to be

$$U(\xi,y) = F(\xi)\frac{\sinh \xi(a - y)}{\sinh \xi a} + G(\xi)\frac{\sinh \xi y}{\sinh \xi a}, \quad 0 < y < a. \qquad (2\text{-}16\text{-}10)$$

Using the convolution theorem (2-9-6) we see that the required solution may be written in the form

$$\begin{aligned} u(x,y) = {} & \frac{1}{2a}\int_{-\infty}^{\infty} f(t)K_1(x - t, a - y)\,dt \\ & + \frac{1}{2a}\int_{-\infty}^{\infty} g(t)K_1(x - t,y)\,dt, \qquad 0 < y < a \qquad (2\text{-}16\text{-}11) \end{aligned}$$

where the kernel $K_1(x,y)$ is defined by the equation

$$K_1(x,y) = \left(\frac{2}{\pi}\right)^{\frac{1}{2}} a\mathscr{F}^{-1}\left[\frac{\sinh(\xi y)}{\sinh(\xi a)};\xi \to x\right].$$

Since $\sinh(\xi y)/\sinh(\xi a)$ is an even function of ξ we see that this is equivalent to the relation

$$K_1(x,y) = \left(\frac{2}{\pi}\right)^{\frac{1}{2}} a\mathscr{F}_c\left[\frac{\sinh(\xi y)}{\sinh(\xi a)};\xi \to x\right].$$

This particular transform can be evaluated by means of the calculus of residues (see Problem 2-21) to give

$$K_1(x,y) = \frac{\sin(\pi y/a)}{\cosh(\pi x/a) + \cos(\pi y/a)}.$$

Substituting this expression into equation (2-16-11) we obtain finally the solution

$$\begin{aligned} u(x,y) = {} & \frac{1}{2a}\sin\left(\frac{\pi y}{a}\right)\int_{-\infty}^{\infty}\frac{f(t)\,dt}{\cosh \pi(x - t)/a - \cos \pi y/a} \\ & + \frac{1}{2a}\sin\left(\frac{\pi y}{a}\right)\int_{-\infty}^{\infty}\frac{g(t)\,dt}{\cosh \pi(x - t)/a + \cos \pi y/a}. \qquad (2\text{-}16\text{-}12) \end{aligned}$$

Another problem of this type is illustrated graphically in Fig. 2-11.

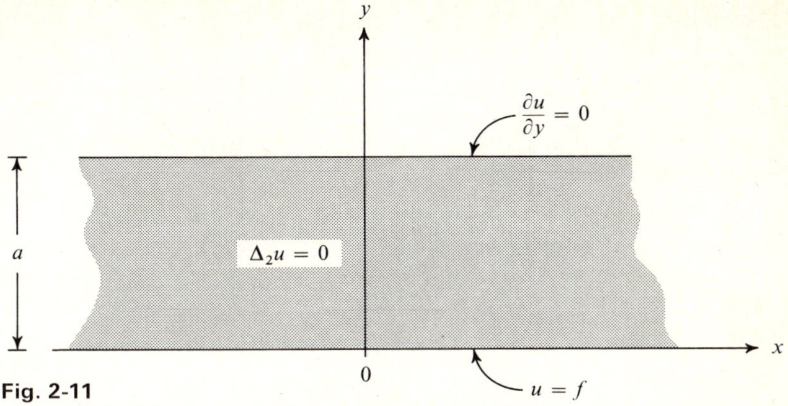

Fig. 2-11

This corresponds to the physical problem of determining the steady-state distribution of temperature in a thick slab when the temperature of one face of the slab is prescribed and the other face is insulated against the flow of heat. The conditions (2-16-9) are then replaced by the pair

$$u(x,0) = f(x), \qquad u_y(x,a) = 0$$

which in turn are transformed to

$$U(\xi,0) = F(\xi), \qquad U_y(\xi,0) = 0.$$

The appropriate solution of (2-16-6) is now

$$U(\xi,y) = F(\xi)\,\frac{\cosh \xi(a-y)}{\cosh \xi a}, \qquad (0 < y < a)$$

If we make use of the convolution theorem (2-9-6) we therefore find that

$$u(x,y) = \int_{-\infty}^{\infty} f(t) K_2(x-t, a-y)\,dt, \tag{2-16-13}$$

where the kernel $K_2(x,y)$ is defined by the equation

$$K_2(x,y) = (2\pi)^{-\frac{1}{2}} \mathscr{F}_c\left[\frac{\cosh(\xi y)}{\cosh(\xi a)}; \xi \to x\right].$$

The Fourier cosine transform involved in this definition is evaluated in Problem 2-22; it turns out that

$$K_2(x,y) = \frac{\cos\left(\dfrac{\pi y}{2a}\right)\cosh\left(\dfrac{\pi x}{2a}\right)}{\cosh\left(\dfrac{\pi x}{a}\right) + \cos\left(\dfrac{\pi y}{a}\right)},$$

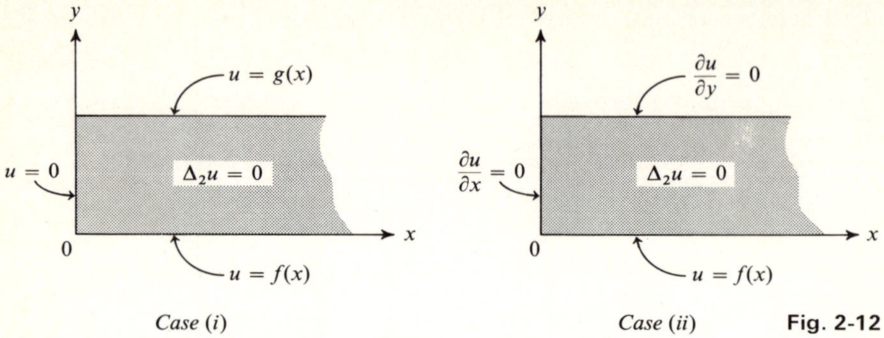

Fig. 2-12

from which we derive immediately the solution

$$u(x,y) = \sin\left(\frac{\pi y}{2a}\right) \int_{-\infty}^{\infty} \frac{f(t)\cosh\dfrac{\pi(x-t)}{2a}\,dt}{\cosh\dfrac{\pi(x-t)}{a} - \cos\dfrac{\pi y}{a}}. \tag{2-16-14}$$

(c) Laplace's equation in a semi-infinite strip To illustrate the method of solution of boundary value problems for a semi-infinite strip we consider the two problems illustrated in Fig. 2-12. In *case* (*i*), the boundary conditions are

$$u(0,y) = 0,\ (0 \leqslant y \leqslant a);\ u(x,0) = f(x),\ (x \geqslant 0);\ u(x,a) = g(x),\ (x \geqslant 0).$$

Physically, this corresponds to the situation in which the faces of the strip are maintained at prescribed temperatures—one face ($x = 0$) being at a constant temperature which is chosen as the reference temperature.

From equation (2-6-15) and the first of these conditions we see that in this case we should take a Fourier sine transform. We define

$$U_s(\xi,y) = \mathscr{F}_s[u(x,y); x \to \xi]$$

so that from equation (2-16-15) we see that

$$\left(\frac{\partial^2}{\partial y^2} - \xi^2\right) U_s(\xi,y) = 0$$

satisfying the conditions on $y = 0$ and $y = a$, we find that for $0 \leqslant y \leqslant a$,

$$U_s(\xi,y) = F_s(\xi)\frac{\sinh \xi(a-y)}{\sinh \xi a} + G_s(\xi)\frac{\sinh \xi y}{\sinh \xi a}.$$

In terms of the function defined in (*b*) above, we can write this equation in the form

$$U_s(\xi,y) = a^{-1}(\tfrac{1}{2}\pi)^{\frac{1}{2}}\{F_s(\xi)\mathscr{F}_c[K_1(x,a-y);x \to \xi] + G_s(\xi)\mathscr{F}_c[K_1(x,y);x \to \xi]\}.$$

Then, using the convolution theorem (2-10-8) we see that the solution is given by the equation

$$u(x,y) = \frac{1}{2a}\int_0^\infty f(t)\{K_1(|t-x|,a-y) - K_1(t+x,a-y)\}\,dt + \frac{1}{2a}\int_0^\infty g(t)\{K_1(|t-x|,y) - K_1(t+x,y)\}\,dt.$$

Similarly in case (*ii*), the boundary conditions are

$$u(x,0) = f(x),\ (x \geqslant 0);\ u_x(0,y) = 0,\ (0 \leqslant y \leqslant a);\ u_y(x,a) = 0,\ (x \geqslant 0).$$

The physical problem to which this corresponds is the one in which one of the infinite faces of the strip is subjected to a prescribed temperature distribution and the other two faces are insulated against the flow of heat.

An examination of equation (2-6-14) suggests that in this instance the appropriate transform to choose is the Fourier cosine transform

$$U_c(\xi,y) = \mathscr{F}_c[u(x,y);x \to \xi]$$

for the first of the above three boundary conditions then ensures that

$$\left(\frac{\partial^2}{\partial y^2} - \xi^2\right)U_c(\xi,y) = 0.$$

The remaining conditions provide the initial conditions

$$U_c(\xi,0) = F_c(\xi), \qquad \frac{\partial}{\partial y}U_c(\xi,a) = 0.$$

The appropriate form for $U_c(\xi,y)$ is therefore

$$U_c(\xi,y) = F_c(\xi)\frac{\cosh \xi(a-y)}{\cosh \xi a}$$

so that, in terms of the function $K_2(x,y)$ defined above, we have the expression

$$U_c(\xi,y) = (2\pi)^{\frac{1}{2}}F_c(\xi)\mathscr{F}_c[K_2(x,a-y);x \to \xi]$$

for the Fourier cosine transform (with respect to x) of the solution we are seeking. Making use of the convolution theorem (2-10-1) to invert this equation we see that the solution itself is given by the equation

$$u(x,y) = \int_0^\infty f(t)\{K_2(x+t,a-y) + K_2(|x-t|,a-y)\}\,dt$$

which may be rewritten in the form

$$u(x,y) = \sin\left(\frac{\pi y}{2a}\right)\int_0^\infty f(t)\left\{\frac{\cosh\dfrac{\pi(x+t)}{2a}}{\cosh\dfrac{\pi(x+t)}{a} - \cos\dfrac{\pi y}{a}} + \frac{\cosh\dfrac{\pi(x-t)}{2a}}{\cosh\dfrac{\pi(x-t)}{a} - \cos\dfrac{\pi y}{a}}\right\}dt$$

(d) The two-dimensional Laplace equation in orthogonal curvilinear coordinates The commonest use of Fourier transforms is in the discussion of boundary value problems relating to half-planes or to infinite or semi-infinite strips. They can, however, also be used in the study of boundary value problems in potential theory which are formulated in terms of orthogonal curvilinear coordinates (α,β) related to x and y through the equation

$$\alpha + i\beta = w(x + iy) \tag{2-16-15}$$

where w is a holomorphic function of the complex variable $x + iy$ and the region under consideration is specified by either $-\infty < \alpha < \infty$ or by $0 \leqslant \alpha < \infty$. The Laplacian $\Delta_2 u$ then transforms to

$$|w'|^2\left(\frac{\partial^2 u}{\partial\alpha^2} + \frac{\partial^2 u}{\partial\beta^2}\right)$$

so that Laplace's equation is equivalent to the equation

$$\frac{\partial^2 u}{\partial\alpha^2} + \frac{\partial^2 u}{\partial\beta^2} = 0. \tag{2-16-16}$$

To make the situation quite definite we shall consider the case in which the boundaries of the region under consideration are given by $\alpha = \pm\infty$, $\beta = \beta_1$, $\beta = \beta_2(<\beta_1)$. We shall assume that $u(\alpha,\beta) \to 0$ as $\alpha \to \pm\infty$, and that

$$u(\alpha,\beta_1) = f_1(\alpha), \qquad u(\alpha,\beta_2) = f_2(\alpha) \tag{2-16-17}$$

where the functions $f_1(\alpha)$ and $f_2(\alpha)$ are prescribed for all real values of α. If we introduce the Fourier transform

$$U(\xi,\beta) = \mathscr{F}[u(\alpha,\beta);\alpha \to \xi] \tag{2-16-18}$$

we see that it is determined by the solution of the two-point boundary value problem posed by the equations

$$\left(\frac{\partial^2}{\partial \beta^2} - \xi^2\right) U(\xi,\beta) = 0$$

$$U(\xi,\beta_1) = F_1(\xi), \qquad U(\xi,\beta_2) = F_2(\xi)$$

where, in the usual notation, F_1 and F_2 denote the Fourier transforms of f_1 and f_2.

This problem is easily seen to have the solution

$$U(\xi,\beta) = F_1(\xi)\frac{\sinh \xi(\beta - \beta_2)}{\sinh \xi(\beta_1 - \beta_2)} + F_2(\xi)\frac{\sinh \xi(\beta_1 - \beta)}{\sinh \xi(\beta_1 - \beta_2)}.$$

If we write

$$K(\alpha,\beta) = (2\pi)^{-\frac{1}{2}}\mathscr{F}^{-1}\left[\frac{\sinh \xi\beta}{\sinh \xi(\beta_1 - \beta_2)}; \xi \to \alpha\right]$$

we see, on making use of the convolution theorem, that we can write the required solution in the form

$$u(\alpha,\beta) = \int_{-\infty}^{\infty} f_1(t)K(\alpha - t,\beta - \beta_2)\,dt + \int_{-\infty}^{\infty} f_2(t)K(\alpha - t,\beta_1 - \beta)\,dt. \qquad (2\text{-}16\text{-}19)$$

In terms of the function $K_1(x,y)$ introduced above,

$$K(\alpha,\beta) = \tfrac{1}{2}(\beta_1 - \beta_2)^{-1}K_1(\alpha,\beta),$$

where, in the definition of K_1, a is taken to be $\beta_1 - \beta_2$. We therefore obtain the formula

$$K(\alpha,\beta) = \frac{1}{2(\beta_1 - \beta_2)} \cdot \frac{\sin\left(\dfrac{\pi\beta}{\beta_1 - \beta_2}\right)}{\cosh\left(\dfrac{\pi\alpha}{\beta_1 - \beta_2}\right) + \cos\left(\dfrac{\pi\beta}{\beta_1 - \beta_2}\right)}. \qquad (2\text{-}16\text{-}20)$$

In particular, in the symmetrical case in which $\beta_2 = -\beta_1$, and $f_1(\alpha) = f_2(\alpha) = f(\alpha)$, say, we find that

$$K(\alpha,\beta) = \frac{1}{4\beta_1} \cdot \frac{\sin\left(\dfrac{\pi\beta}{2\beta_1}\right)}{\cosh\left(\dfrac{\pi\alpha}{2\beta_1}\right) + \cos\left(\dfrac{\pi\beta}{2\beta_1}\right)}$$

and hence that the solution of the symmetrical problem is given by the formula

$$u(\alpha,\beta) = \frac{1}{2\beta_1}\cos\frac{\pi\beta}{2\beta_1}\int_{-\infty}^{\infty} f(t)M(\alpha - t,\beta,\beta_1)\,dt \qquad (2\text{-}16\text{-}21)$$

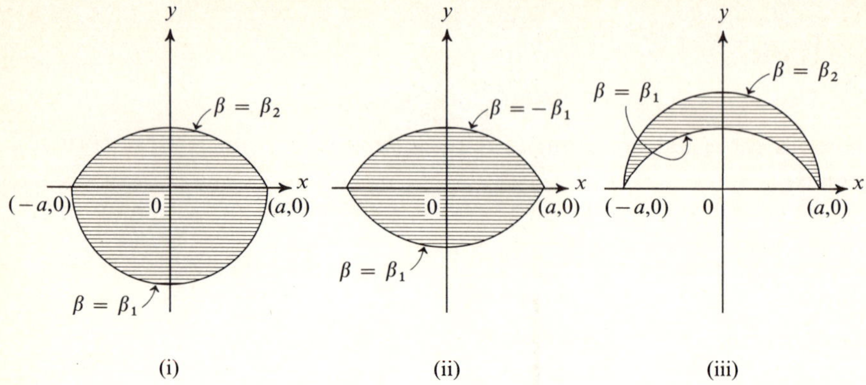

Fig. 2-13

where the function $M(\alpha,\beta,\beta_1)$ is defined by the equation

$$M(\alpha,\beta,\beta_1) = \frac{\cosh \dfrac{\pi\alpha}{2\beta_1}}{\cosh^2\left(\dfrac{\pi\alpha}{2\beta_1}\right) - \sin^2\left(\dfrac{\pi\beta}{2\beta_1}\right)}. \tag{2-16-22}$$

As an example of the use of this method we instance the solution of the Dirichlet problem for a two-dimensional region bounded by two circular arcs. If we introduce bipolar coordinates (α,β) in the plane through the equations

$$x = \frac{a \sinh \alpha}{\cosh \alpha + \cos \beta}, \qquad y = \frac{a \sin \beta}{\cosh \alpha + \cos \beta}$$

or by the single equation

$$x + iy = a \tanh\{\tfrac{1}{2}(\alpha + i\beta)\}$$

we find that $\beta = \beta_1$ is the circle with equation

$$x^2 + (y + a \cot \beta_1)^2 = a^2 \operatorname{cosec}^2 \beta_1,$$

and that, as $\alpha \to \pm\infty$, $(x,y) \to (\pm a,0)$ so that for $0 < \beta_1 < \frac{1}{2}\pi$, $-\frac{1}{2}\pi < \beta_2 < 0$ we get the region shown in (*i*) of Fig. 2-13. The solution of the Dirichlet problem for such a plane region is given by equations (2-16-19) and (2-16-20), or, in the symmetrical case shown in (*ii*) of Fig. 2-13 by equations (2-16-21) and (2-16-22). The solution for a crescent shaped domain of the type shown in Fig. 2-13 (*iii*) is also given by equations (2-16-19) and (2-16-20), the only difference being that, in this instance, β_1 and β_2 are of the same sign; in the case shown we have taken $-\frac{1}{2}\pi < \beta_2 < \beta_1 < 0$.

This method has been used extensively by Russian mathematicians in a wide variety of problems in elasticity; for an account of this work, the reader should consult secs. 7-22 of Uflyand (1963).

(e) The solution of $\Delta_3 u = 0$ for the half-space $z \geqslant 0$ As an example of the use of multiple Fourier transforms we consider the solution of the Dirichlet problem for a half-space. The problem is to determine a function $u(x,y,z)$ satisfying Laplace's equation

$$\Delta_3 u = 0 \tag{2-16-23}$$

in the half-space $z \geqslant 0$ and satisfying the boundary conditions

$$u(x,y,0) = f(x,y) \tag{2-16-24}$$

$$u(x,y,z) \to 0, \qquad \text{as} \qquad r \to \infty, (z \geqslant 0) \tag{2-16-25}$$

where, as usual, $r = (x^2 + y^2 + z^2)^{\frac{1}{2}}$.

To determine u we introduce its two-dimensional Fourier transform

$$U_{(2)}(\xi,\eta,z) = \mathscr{F}_{(2)}[u(x,y,z); x \to \xi, y \to \eta]. \tag{2-16-26}$$

Using equation (2-13-1) we see that equation (2-16-23) is equivalent to the ordinary differential equation

$$\frac{\partial^2 U_{(2)}}{\partial z^2} - \lambda^2 U_{(2)} = 0$$

where $\lambda^2 = \xi^2 + \eta^2$, and that the boundary conditions are equivalent to the relations

$$U_{(2)}(\xi,\eta,0) = F_2(\xi,\eta), \qquad U_{(2)}(\xi,\eta,z) \to 0 \text{ as } z \to \infty,$$

where $F_2(\xi,\eta)$ is the double Fourier transform of the prescribed function $f(x,y)$. This set of equations has the solution

$$U_{(2)}(\xi,\eta,z) = F_2(\xi,\eta)e^{-|\lambda|z}$$

so that the required solution is given by the convolution theorem for double Fourier transforms in the form

$$u(x,y,z) = \frac{1}{2\pi}\int_{-\infty}^{\infty}\int_{-\infty}^{\infty} f(x',y')g(x - x',y - y',z)\,dx'dy', \tag{2-16-27}$$

in which the function $g(x,y,z)$ is defined to be the inverse transform

$$g(x,y,z) = \mathscr{F}_{(2)}^{-1}[e^{-|\lambda|z}; \xi \to x, \eta \to y].$$

By differentiating both sides of equation (2-14-5) with respect to k, we see that

$$g(x,y,z) = z(x^2 + y^2 + z^2)^{-\frac{3}{2}}$$

and hence that the required solution is

$$u(x,y,z) = \frac{z}{2\pi}\int_{-\infty}^{\infty}\int_{-\infty}^{\infty}\frac{f(x',y')\,dx'\,dy'}{\{(x-x')^2+(y-y')^2+z^2\}^{\frac{3}{2}}}. \tag{2-16-28}$$

2-16-2 SOLUTIONS OF THE DIFFUSION EQUATION

(a) The linear diffusion equation on a semi-infinite line As an example of the use of the Fourier sine transform we consider the derivation of the solution of the linear diffusion equation

$$\frac{\partial^2 u}{\partial x^2} = \frac{1}{\kappa}\frac{\partial u}{\partial t} \tag{2-16-29}$$

in the semi-infinite line $x \geqslant 0$, satisfying the boundary conditions

$$u(0,t) = f(t), \qquad t \geqslant 0 \tag{2-16-30}$$

$$u(x,t) \to 0, \qquad \text{as } x \to \infty \tag{2-16-31}$$

and the initial condition

$$u(x,0) = 0. \tag{2-16-32}$$

If we introduce the Fourier sine transform

$$U_s(\xi,t) = \mathscr{F}_s[u(x,t);x \to \xi]$$

then, by equation (2-6-9), we see that

$$\mathscr{F}_s\left[\frac{\partial^2 u}{\partial x^2};x \to \xi\right] = -\xi^2 U_s(\xi,t) + (2/\pi)^{\frac{1}{2}}\xi f(t)$$

so that the partial differential equation (2-16-29) for the function $u(x,t)$ is equivalent to the ordinary differential equation

$$\frac{\partial U_s}{\partial t} + \kappa\xi^2 U_s = (2/\pi)^{\frac{1}{2}}\kappa\xi f(t)$$

for its Fourier sine transform $U_s(\xi,t)$. Further, the initial condition (2-16-32) is equivalent to the condition

$$U_s(\xi,0) = 0.$$

From these equations we deduce that

$$U_s(\xi,t) = \left(\frac{2}{\pi}\right)^{\frac{1}{2}}\kappa\xi\int_0^t f(\tau)e^{-\kappa(t-\tau)\xi^2}\,d\tau.$$

Applying the operator $\mathscr{F}_s$ to both sides of this equation we find that

$$u(x,t) = \left(\frac{2}{\pi}\right)^{\frac{1}{2}} \kappa \int_0^t f(\tau)\mathscr{F}_s[\xi e^{-\kappa(t-\tau)\xi^2}; \xi \to x]\, d\tau.$$

Now it is easily shown that

$$\mathscr{F}_s[\xi e^{-\kappa(t-\tau)\xi^2}; \xi \to x] = \frac{x}{2\sqrt{2}} \{\kappa(t - \tau)\}^{-\frac{3}{2}} \exp\{-x^2/4\kappa(t - \tau)\}$$

so that the required solution can be written in the form

$$u(x,t) = x(4\pi\kappa)^{-\frac{1}{2}} \int_0^t f(\tau) \exp\left[-\frac{x^2}{4\kappa(t - \tau)}\right]\frac{d\tau}{(t - \tau)^{\frac{3}{2}}} \tag{2-16-33}$$

If, instead of the boundary conditions (2-16-30), we have the condition

$$u_x(0,t) = f(t), \qquad (t \geqslant 0), \tag{2-16-34}$$

we must use the Fourier cosine transform

$$U_c(\xi,t) = \mathscr{F}_c[u(x,t); x \to \xi]$$

of the function $u(x,t)$ instead of its Fourier sine transform.

From equation (2-6-9) we find that

$$\mathscr{F}_c\left[\frac{\partial^2 u}{\partial x^2}; x \to \xi\right] = -\xi^2 U_c(\xi,t) - (2/\pi)^{\frac{1}{2}}f(t)$$

so that we see that $U_c(\xi,t)$ must be taken to be the solution of the initial value problem

$$\frac{\partial U_c}{\partial t} + \kappa\xi^2 U_c = -(2/\pi)^{\frac{1}{2}}\kappa f(t), \qquad U_c(\xi,0) = 0.$$

It is readily shown that

$$U_c(\xi,t) = -\left(\frac{2}{\pi}\right)^{\frac{1}{2}} \kappa \int_0^t f(\tau)e^{-\kappa(t-\tau)\xi^2}\, d\tau.$$

Applying the operator $\mathscr{F}_c$ to both sides of this equation we obtain the solution

$$u(x,t) = -\left(\frac{2}{\pi}\right)^{\frac{1}{2}} \kappa \int_0^t f(\tau)\mathscr{F}_c[e^{-\kappa(t-\tau)\xi^2}; \xi \to x]\, d\tau.$$

Since it is easily shown that

$$\mathscr{F}_c[e^{-\kappa(t-\tau)\xi^2}; \xi \to x] = \{2\kappa(t - \tau)\}^{-\frac{1}{2}}e^{-x^2/4\kappa(t-\tau)}$$

the required solution may be written in the form

$$u(x,t) = -\left(\frac{\kappa}{\pi}\right)^{\frac{1}{2}} \int_0^t f(\tau) \exp\left[-\frac{x^2}{4\kappa(t - \tau)}\right]\frac{d\tau}{\sqrt{(t - \tau)}}. \tag{2-16-35}$$

Another problem of the same kind is to determine the solution of the linear diffusion equation (2-16-29) satisfying the boundary conditions

$$u(0,t) = 0, \tag{2-16-36}$$

$$u(x,t) \to 0 \qquad \text{as } x \to \infty, \tag{2-16-37}$$

and the initial condition

$$u(x,0) = f(x). \tag{2-16-38}$$

Again we use the Fourier sine transform. Because of equation (2-16-36) we see that

$$\mathscr{F}_s\left[\frac{\partial^2 u}{\partial x^2}; x \to \xi\right] = -\xi^2 U_s(\xi,t),$$

and hence that equation (2-16-29) is equivalent to

$$\frac{\partial U_s}{\partial t} + \kappa\xi^2 U_s = 0.$$

Further, equation (2-16-38) is equivalent to the initial condition

$$U_s(\xi,0) = F_s(\xi),$$

where $F_s(\xi) = \mathscr{F}_s[f(t);\xi]$ is the Fourier sine transform of the prescribed function. The solution of this initial value problem is

$$U_s(\xi,t) = F_s(\xi)G_c(\xi,t)$$

where

$$G_c(\xi,t) = e^{-\kappa t\xi^2}.$$

From equation (2-10-8) we therefore deduce that

$$u(x,t) = (2\pi)^{-\frac{1}{2}} \int_0^\infty f(y)\{g(|x-y|,t) - g(x+y,t)\}\,dy$$

where

$$g(x,t) = \mathscr{F}_c[G_c(\xi,t);\xi \to x] = (2\kappa t)^{-\frac{1}{2}} e^{-x^2/4\kappa t}$$

In this way we obtain the solution

$$u(x,t) = (4\pi\kappa t)^{-\frac{1}{2}} \int_0^\infty f(y)\{e^{-(x-y)^2/4\kappa t} - e^{-(x+y)^2/4\kappa t}\}\,dy. \tag{2-16-39}$$

(b) The two-dimensional diffusion equation As a further example of the use of double Fourier transforms we derive the solution $u(x,y,t)$ of the diffusion equation

$$\Delta_2 u = \frac{1}{\kappa}\frac{\partial u}{\partial t}, \qquad (t > 0), \tag{2-16-40}$$

in the whole xy-plane when the value of u is prescribed initially:

$$u(x,y,0) = f(x,y).$$

We shall also assume that u and its partial derivatives with respect to x and y tend to zero as $(x^2 + y^2)^{\frac{1}{2}} \to \infty$.

We introduce the double transform

$$U_2(\xi,\eta,t) = \mathscr{F}_{(2)}[u(x,y,t); x \to \xi, y \to \eta].$$

From equation (2-13-12) we deduce that the transform $U_2(\xi,\eta,t)$ satisfies the first order initial value problem

$$\frac{\partial U_2}{\partial t} + \kappa\lambda^2 U_2 = 0, \qquad U_2(\xi,\eta,0) = F_2(\xi,\eta)$$

where $\lambda = (\xi^2 + \eta^2)^{\frac{1}{2}}$ and $F_2(\xi,\eta)$ is the double Fourier transform of the function $f(x,y)$. From the solution

$$U_2(\xi,\eta,t) = F_2(\xi,\eta)e^{-\kappa t\lambda^2}$$

we deduce that

$$u(x,y,t) = \frac{1}{2\pi}\int_{-\infty}^{\infty}\int_{-\infty}^{\infty} f(x',y')g(x - x',y - y',t)\,dx'dy' \qquad \text{(2-16-41)}$$

where

$$g(x,y,t) = \mathscr{F}_{(2)}^{-1}[e^{-\kappa t\lambda^2}; \xi \to x, \eta \to y].$$

Now, from equation (2-14-4) we know that

$$g(x,y,t) = \int_0^{\infty} \lambda e^{-\kappa t\lambda^2} J_0(\lambda r)\,d\lambda$$

where $r = (x^2 + y^2)^{\frac{1}{2}}$. Replacing the Bessel function by its series expansion and integrating term by term we find that the integral on the right-hand side of this equation is equal to

$$\sum_{n=0}^{\infty} \frac{(-\frac{1}{4}r^2)^n}{n!n!}\int_0^{\infty} \lambda^{2n+1}e^{-\kappa t\lambda^2}\,d\lambda$$

$$= (2\kappa t)^{-1}\sum_{n=0}^{\infty}\frac{(-r^2/4\kappa t)^n}{n!n!}\int_0^{\infty} z^n e^{-z}\,dz$$

$$= (2\kappa t)^{-1}e^{-r^2/4\kappa t}.$$

so that

$$g(x,y,t) = (2\kappa t)^{-1}\exp\left[-\frac{x^2 + y^2}{4\kappa t}\right],$$

and hence that we obtain the solution in the form

$$u(x,y,t) = \frac{1}{4\pi\kappa t}\int_{-\infty}^{\infty}\int_{-\infty}^{\infty} f(x',y') \times$$

$$\times \exp\left[-\frac{(x-x')^2+(y-y')^2}{4\kappa t}\right] dx'\,dy'. \qquad (2\text{-}16\text{-}42)$$

(c) The diffusion equation in an infinite region The three-dimensional problem of finding a function $u(\mathbf{r},t)$ satisfying

$$\Delta_3 u = \frac{1}{\kappa}\frac{\partial u}{\partial t} \qquad (2\text{-}16\text{-}43)$$

throughout the whole xyz-space, and the initial condition

$$u(\mathbf{r},0) = f(\mathbf{r})$$

can be solved by considering the triple Fourier transform

$$U_3(\xi,t) = \mathscr{F}_{(3)}[u(\mathbf{r},t);\mathbf{r}\to\boldsymbol{\xi}]$$

of the required function. From equation (2-12-12) we see that the initial value problem for $u(\mathbf{r},t)$ is equivalent to the initial value problem

$$\frac{\partial U_3}{\partial t} + \kappa\xi^2 U_3 = 0, \qquad U_3(\boldsymbol{\xi},0) = F_3(\boldsymbol{\xi})$$

for its Fourier transform. In the latter equation $F_3(\boldsymbol{\xi})$ denotes the three-dimensional Fourier transform of $f(\mathbf{r})$. This initial value problem has solution

$$U_3(\boldsymbol{\xi},t) = F_3(\boldsymbol{\xi})e^{-\kappa t\xi^2},$$

so that

$$u(\mathbf{r},t) = (2\pi)^{-\frac{3}{2}}\int_{E_3} f(\mathbf{y})g(\mathbf{r}-\mathbf{y},t)\,d\mathbf{y}, \qquad (2\text{-}16\text{-}44)$$

$$g(\mathbf{r},t) = \mathscr{F}_{(3)}^{-1}[e^{-\kappa t\xi^2};\boldsymbol{\xi}\to\mathbf{r}].$$

We could evaluate $g(\mathbf{r},t)$ by the same method as we used in the last subsection. Alternatively we could evaluate it by noticing that

$$\mathscr{F}_{(3)}^{-1}[e^{-\kappa t\xi^2};\boldsymbol{\xi}\to\mathbf{r}]$$
$$= \mathscr{F}[e^{-\kappa t\xi^2};\xi\to x]\mathscr{F}[e^{-\kappa t\eta^2};\eta\to y]\mathscr{F}[e^{-\kappa t\zeta^2};\zeta\to z],$$

and taking the value of each integral from equation (2-7-13) with a^2 replaced by $\frac{1}{4}\kappa t$. Whichever method we use we find that

$$g(\mathbf{r},t) = (2\kappa t)^{-\frac{3}{2}}e^{-r^2/4\kappa t}$$

so that we may write the solution (2-16-44) in the form

$$u(\mathbf{r},t) = (4\pi\kappa t)^{-\frac{3}{2}} \int_{E_3} f(\mathbf{y}) \exp\{-|\mathbf{r} - \mathbf{y}|^2/4\kappa t\}\, d\mathbf{y}. \tag{2-16-45}$$

2-16-3 VIBRATION PROBLEMS

(a) Free transverse vibrations of an infinite beam We consider a beam of uniform cross-sectional area S which, in its equilibrium position, lies along the x-axis. If we measure the transverse deflection $y(x,t)$—assumed to be small—of the beam at any point x, then it is readily shown[1] that

$$\frac{\partial^4 y}{\partial x^4} + \frac{1}{a^2}\frac{\partial^2 y}{\partial t^2} = \frac{P(x,t)}{EI}, \tag{2-16-46}$$

where E denotes the Young's modulus of the material forming the beam, I is the moment of inertia of the cross-section of the beam with respect to a line which is normal to both the x-axis and the y-axis and which passes through the centre of mass of S. The wave velocity a is defined by the equation

$$a^2 = \frac{EI}{\rho S}$$

where ρ is the mass of the beam per unit length and the function $P(x,t)$ is the transverse load per unit length.

It may also be shown that the bending moment $M(x,t)$ at the point x is given by the equation

$$M(x) = -EI\frac{\partial^2 y}{\partial x^2}$$

so that, if the end of the elastic beam is at the point $x = 0$ and if that end is freely hinged, then

$$y = \frac{\partial^2 y}{\partial x^2} = 0, \qquad \text{at } x = 0 \tag{2-16-47}$$

are the required boundary conditions at $x = 0$.

We consider first the free vibrations of a beam of infinite length produced by distorting the beam and then giving each point of it a prescribed transverse velocity. In other words, we assume that $P(x,t) \equiv 0$, and that, when $t = 0$,

$$y = f(x), \qquad \frac{\partial y}{\partial t} = ag''(x), \tag{2-16-48}$$

[1] See, for example, P. M. Morse (1948), p. 151.

where the functions $f(x)$ and $g(x)$ are prescribed for all real values of x.

To solve equation (2-16-46) we introduce the Fourier transform

$$Y(\xi,t) = \mathscr{F}[y(x,t); x \to \xi]$$

so that, by equation (2-3-12)

$$\mathscr{F}\left[\frac{\partial^4 y}{\partial x^4}; x \to \xi\right] = \xi^4 Y(\xi,t)$$

and hence the equation of motion is equivalent to the ordinary differential equation

$$\frac{\partial^2 Y}{\partial t^2} + a^2\xi^4 Y = 0 \tag{2-16-49}$$

for the Fourier transform of the transverse displacement. Further, the initial conditions (2-16-48) are equivalent to the initial conditions

$$Y(\xi,0) = F(\xi), \qquad Y_t(\xi,0) = -a\xi^2 G(\xi) \tag{2-16-50}$$

where $F(\xi)$ and $G(\xi)$ denote, respectively, the Fourier transforms of the functions $f(x)$ and $g(x)$. The initial value problem posed by equations (2-16-49) and (2-16-50) is easily shown to have the solution

$$Y(\xi,t) = F(\xi)\cos(a\xi^2 t) - G(\xi)\sin(a\xi^2 t). \tag{2-16-51}$$

Now, by equations (2-7-15) and (2-3-7) we see that

$$\mathscr{F}^{-1}[\cos(a\xi^2 t); \xi \to x] = \tfrac{1}{2}(at)^{-\frac{1}{2}}\left(\cos\frac{x^2}{4at} + \sin\frac{x^2}{4at}\right)$$

so that, by the convolution theorem

$$\mathscr{F}^{-1}[F(\xi)\cos(a\xi^2 t); \xi \to x]$$

$$= (8\pi at)^{-\frac{1}{2}} \int_{-\infty}^{\infty} f(x-u)\left(\cos\frac{u^2}{4at} + \sin\frac{u^2}{4at}\right) du$$

Changing the variable of integration from u to v, where $v^2 = u^2/(4at)$, we obtain the equation

$$\mathscr{F}^{-1}[F(\xi)\cos(at\xi^2); \xi \to x]$$

$$= (2\pi)^{-\frac{1}{2}} \int_{-\infty}^{\infty} f(x - 2va^{\frac{1}{2}}t^{\frac{1}{2}})(\cos v^2 + \sin v^2)\, dv$$

Similarly, from equation (2-7-16), we deduce that

$$\mathscr{F}^{-1}[G(\xi)\sin(at\xi^2); \xi \to x]$$

$$= (2\pi)^{-\frac{1}{2}} \int_{-\infty}^{\infty} g(x - 2va^{\frac{1}{2}}t^{\frac{1}{2}})(\cos v^2 - \sin v^2)\, dv,$$

so that from equation (2-16-51) we obtain the solution

$$y(x,t) = (2\pi)^{-\frac{1}{2}} \int_{-\infty}^{\infty} f(x - 2va^{\frac{1}{2}}t^{\frac{1}{2}})(\cos v^2 + \sin v^2)\, dv$$

$$+ (2\pi)^{-\frac{1}{2}} \int_{-\infty}^{\infty} g(x - 2va^{\frac{1}{2}}t^{\frac{1}{2}})(\cos v^2 - \sin v^2)\, dv, \qquad (2\text{-}16\text{-}52)$$

a result established by Boussinesq—see p. 282 of Vol. II, Part II of Todhunter and Pearson (1893).

In any given problem it may be simpler to use the solution in the form (2-16-51) instead of Boussinesq's solution (2-16-52). For example, if

$$f(x) = \varepsilon e^{-x^2/4\sigma^2}, \qquad g(x) = 0, \qquad (2\text{-}16\text{-}53)$$

we have

$$F(\xi) = 2^{\frac{1}{2}}\varepsilon\sigma e^{-\xi^2\sigma^2}, \qquad G(\xi) = 0$$

so that

$$Y(\xi,t) = 2^{\frac{1}{2}}\varepsilon\sigma e^{-\xi^2\sigma^2} \cos(at\xi^2) = 2^{-\frac{1}{2}}\varepsilon\sigma\{\exp(-Re^{-i\phi}) + \exp(-Re^{i\phi})\}$$

where we have written

$$\sigma^2 + iat = Re^{i\phi}.$$

From equation (2-7-13) we deduce that

$$y(x,t) = \varepsilon\sigma R^{-\frac{1}{2}} \exp\left(-\frac{x^2\cos\phi}{4R}\right)\cos\left(\frac{x^2\sin\phi}{4R} - \tfrac{1}{2}\phi\right).$$

Now by the definition of R and ϕ it follows immediately that

$$R\cos\phi = \sigma^2,\ R\sin\phi = at,\ R^2 = \sigma^4 + a^2t^2,\ \phi = \tan^{-1}(at/\sigma^2),$$

so that we may rewrite the solution in the form

$$y(x,t) = (1 + a^2t^2\sigma^{-4})^{-\frac{1}{4}} \exp\left[-\frac{\sigma^2x^2}{4(a^2t^2 + \sigma^4)}\right] \times$$

$$\times \cos\left[\frac{atx^2}{4(a^2t^2 + \sigma^2)} - \tfrac{1}{2}\tan^{-1}\frac{at}{\sigma^2}\right].$$

This result, which is in agreement with that found by Morse,[1] may also be obtained by substituting from equation (2-16-53) into the general solution (2-16-52).

[1] See equation (14.9) on p. 155 of P. M. Morse (1948).

(b) Motion of a semi-infinite beam when its end has a prescribed motion We shall now consider the motion of the semi-infinite elastic beam $x \geqslant 0$ when the end $x = 0$ is given a prescribed displacement. If we suppose that the beam is freely hinged but that the hinge is in motion, we have the boundary conditions

$$y(0,t) = f(t), \qquad y_{xx}(0,t) = 0, \qquad (t \geqslant 0) \tag{2-16-54}$$

and if we assume that the beam is originally at rest in its equilibrium position before the hinge is set in motion, we have the initial conditions

$$y(x,0) = y_t(x,0) = 0, \qquad (x \geqslant 0).$$

If we take the Fourier sine transform of both sides of equation (2-16-46) with respect to x and make use of equation (2-6-19) we find that the function

$$Y_s(\xi,t) = \mathscr{F}_s[y(x,t); x \to \xi]$$

satisfies the equation

$$\frac{\partial^2 Y_s}{\partial t^2} + a^2\xi^4 Y_s = \left(\frac{2}{\pi}\right)^{\frac{1}{2}} a^2\xi^3 f(t)$$

and the initial conditions

$$Y_s(\xi,0) = \frac{\partial Y_s(\xi,0)}{\partial t} = 0.$$

The solution of this initial value problem is known to be[1]

$$Y_s(\xi,t) = \left(\frac{2}{\pi}\right)^{\frac{1}{2}} a\xi \int_0^t f(t-u)\sin(au\xi^2)\,du.$$

If we differentiate both sides of equation (2-7-16) with respect to ξ we obtain the formula

$$\mathscr{F}_s[t\sin(\tfrac{1}{2}t^2);\xi] = 2^{-\frac{1}{2}}\xi[\sin(\tfrac{1}{2}\xi^2) + \cos(\tfrac{1}{2}\xi^2)]$$

from which, by using (2-5-7), we deduce that

$$\mathscr{F}_s[\xi\sin(au\xi^2);x] = \tfrac{1}{4}x(au)^{-\frac{3}{2}}\left[\sin\frac{x^2}{4au} + \cos\frac{x^2}{4au}\right],$$

so that

$$y(x,t) = (8\pi a)^{-\frac{1}{2}}x\int_0^t f(t-u)\left[\sin\frac{x^2}{4au} + \cos\frac{x^2}{4au}\right]\frac{du}{u^{\frac{3}{2}}}.$$

[1] This can be shown by elementary methods such as "variation of parameters," but there is a very simple proof of it obtained by using elementary properties of the Laplace transform—see sec. 3-14 below.

If we change the variable of integration from u to v where $v^2 = x^2/(2au)$, we obtain the solution

$$y(x,t) = \pi^{-\frac{1}{2}} \int_{(2at)^{-\frac{1}{2}}x}^{\infty} f\left(t - \frac{x^2}{2av^2}\right) \{\sin(\tfrac{1}{2}v^2) + \cos(\tfrac{1}{2}v^2)\}\, dv \quad (2\text{-}16\text{-}55)$$

originally due to Boussinesq—see p. 280 of Vol II, Part II of Todhunter and Pearson (1893).

(c) Free vibrations of a very large membrane We shall now consider some simple problems concerning the transverse vibrations of a membrane of uniform surface density, whose mass per unit area is σ, and which in the equilibrium position lies in the plane $z = 0$. We suppose that the membrane is stretched to a uniform tension T—in other words, that if a straight line of unit length is drawn in any direction on the surface of the membrane, the material on one side of the line exerts a force T on the material on the other side, the force acting in a direction perpendicular to the line itself.

If we denote the transverse displacement by $z(x,y,t)$ and the external force per unit area by $p(x,y,t)$ we obtain the equation of motion

$$\Delta_2 z = \frac{1}{c^2}\frac{\partial^2 z}{\partial t^2} - \frac{p(x,y,t)}{T}. \qquad (2\text{-}16\text{-}56)$$

If the vibrations are free we take $p \equiv 0$ so that the variation of z is governed by the wave equation

$$\Delta_2 z = \frac{1}{c^2}\frac{\partial^2 z}{\partial t^2}. \qquad (2\text{-}16\text{-}57)$$

We shall consider the free vibrations of a membrane of infinite extent, when its displacement and velocity are described initially:

$$z(x,y,0) = f(x,y), \qquad z_t(x,y,0) = g(x,y),$$

where the functions $f(x,y)$ and $g(x,y)$ are prescribed for all real values of x and y. To solve the wave equation (2-16-57) we introduce the double Fourier transform

$$Z(\xi,\eta,t) = \mathscr{F}_{(2)}[z(x,y,t); x \to \xi, y \to \eta].$$

Making use of equation (2-13-14) we see that the wave equation (2-16-57) is equivalent to the ordinary differential equation

$$\frac{\partial^2 Z}{\partial t^2} + \mu^2 Z = 0, \qquad \mu^2 = c^2(\xi^2 + \eta^2),$$

and the initial conditions on z to

$$Z(\xi,\eta,0) = F(\xi,\eta), \qquad Z_t(\xi,\eta,0) = G(\xi,\eta)$$

where $F(\xi,\eta)$ and $G(\xi,\eta)$ are respectively the double transforms of the prescribed functions $f(x,y)$ and $g(x,y)$.

The solution of this initial value problem is easily shown to be

$$Z(\xi,\eta,t) = F(\xi,\eta)\cos(\mu t) + \mu^{-1}G(\xi,\eta)\sin(\mu t). \tag{2-16-58}$$

In any particular problem the transforms $F(\xi,\eta)$ and $G(\xi,\eta)$ could be evaluated as a result of a double integration and the final form of $z(x,y,t)$ obtained from equation (2-16-58) by the Fourier inversion theorem for double transforms.

We can, however, obtain the inverse form of equation (2-16-58) by first noticing that it can be rewritten in the form

$$Z(\xi,\eta,t) = \left(F(\xi,\eta)\frac{\partial}{\partial t} + G(\xi,\eta)\right)\frac{\sin(\mu t)}{\mu}.$$

We then use the relation (2-14-4) to obtain the relation

$$\begin{aligned}\mathscr{F}_{(2)}^{-1}\left[\frac{\sin(\mu t)}{\mu};\xi \to x,\eta \to y\right] &= c^{-1}\mathscr{H}_0[\rho^{-1}\sin(ct\rho);\rho \to r] \\ &= c^{-1}(c^2t^2 - r^2)^{-\frac{1}{2}}H(ct - r),\end{aligned}$$

where it will be recalled $r = (x^2 + y^2)^{\frac{1}{2}}$. Making use of the convolution theorem for double Fourier transforms we see that the general solution may be written in the form

$$z(x,y,t) = \frac{1}{2\pi c}\iint_\Omega \left\{f(x',y')\frac{\partial}{\partial t} + g(x',y')\right\}\frac{1}{R}\,dx'dy'$$

where R is defined by the equation

$$R^2 = c^2t^2 - (x - x')^2 - (y - y')^2,$$

and Ω is the disk in the $x'y'$-plane defined by $R \geqslant 0$.

It will be observed that in all the cases we have considered in this section the problem of detérmining the solution of a boundary-value or initial-value problem for a partial differential equation has been reduced by means of a Fourier transform to the much simpler problem of deriving the solution of what is virtually an ordinary differential equation with data prescribed at one or two points. Once the solution of this simple problem has been found, the Fourier inversion theorem is used to derive the solution of the original more difficult problem. The process is described diagrammatically in Fig. 2-14 which is due to Dr. Murray Klampkin and expresses in a simple and direct way the whole philosophy of the use of transform methods in the solution of problems in applied mathematics. We shall find that the same procedure is followed in the application of other transforms to the solution of problems in physics and engineering. (The "path" from the formulation to

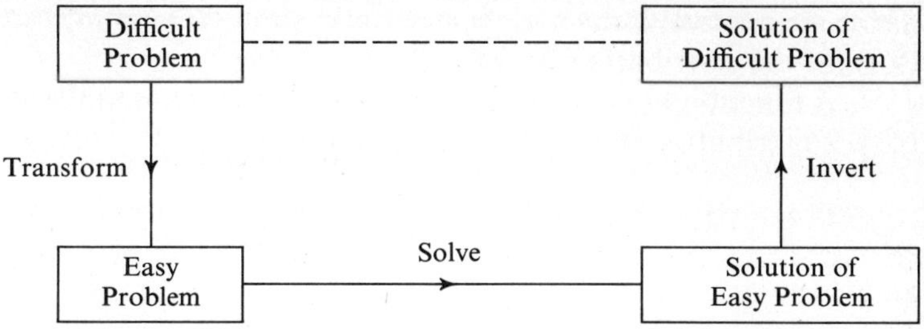

Fig. 2-14

the solution of a difficult problem is shown by a broken line because it can only seldom be realized—the usual path is along the full line!)

2-17 FOURIER TRANSFORMS IN STATISTICS

It should be noted that Fourier transforms arise in a natural way in statistics.[1]

If X is a random variable, then the *distribution function* $F(x)$ is a function defined by the relation that, for every real number x, $F(x)$ is the probability that $X \leqslant x$.

It is immediately obvious from this definition that the distribution function has the following properties:

(*a*) $\lim_{x \to -\infty} F(x) = 0,$

(*b*) $\lim_{x \to +\infty} F(x) = 1,$

(*c*) $F(x_1) \leqslant F(x_2)$ if $x_1 \leqslant x_2$.

When x is a continuous variable, it sometimes happens that there exists a function $p(x)$ such that

$$F(x) = \int_{-\infty}^{x} p(u)\,du \tag{2-17-1}$$

for all real values of x. When this *does* occur we say that the random variable X has an absolutely continuous distribution. The function $p(x)$ is then called the *probability density* or the *frequency function* of the distribution. When the random variable X has an absolutely continuous distribution, the *expectation* of X is given by the formula

$$EX = \int_{-\infty}^{\infty} xp(x)\,dx \tag{2-17-2}$$

provided, of course, that this integral exists. In a similar way, the equations

$$m^{(k)} = \int_{-\infty}^{\infty} x^k p(x)\,dx, \qquad \mu^{(k)} = \int_{-\infty}^{\infty} |x|^k p(x)\,dx \tag{2-17-3}$$

[1] See, for example, §**12** of Loève (1960).

define, respectively, the *k-th moment* and the *k-th absolute moment* (if they exist) of the distribution $F(x)$.

A function $F(x)$, defined on the whole real line, is said to be *compressed* of two distribution functions $F_1(x)$ and $F_2(x)$ if it can be expressed in the form

$$F(x) = \int_{-\infty}^{\infty} F_1(x-y)F_2'(y)\,dy, \tag{2-17-4}$$

and we write

$$F(x) = F_1 * F_2. \tag{2-17-5}$$

Comparing equations (2-17-4) and (2-9-1) we see that

$$F_1 * F_2 = (2\pi)^{\frac{1}{2}} F_1 \circ F_2'. \tag{2-17-6}$$

The *characteristic function f* of a distribution function F is defined for an absolutely continuous distribution by the equation

$$f(t) = \int_{-\infty}^{\infty} e^{itx} p(x)\,dx, \tag{2-17-7}$$

where $p(x)$ is related to $F(x)$ through the equation (2.17.1). We may rewrite equation (2-17-7) in the form

$$f(t) = (2\pi)^{\frac{1}{2}} \mathscr{F}[p(x);t]. \tag{2-17-8}$$

Writing equation (2-17-1) in the alternative form $F'(x) = p(x)$ and using equation (2-3-11) we see that equation (2-17-7) is equivalent to the relation

$$\mathscr{F}[F(x);t] = i(2\pi)^{-\frac{1}{2}} t^{-1} f(t). \tag{2-17-9}$$

Applying the convolution theorem to equations (2-17-5) and (2-17-6) we see that, for the composed distribution defined by equation (2-17-4)

$$\begin{aligned} i(2\pi)^{-\frac{1}{2}} t^{-1} f(t) &= (2\pi)^{\frac{1}{2}} \mathscr{F}[F_1(x);t] \cdot \mathscr{F}[p_2(x);t] \\ &= (p\pi)^{\frac{1}{2}} \cdot i(2\pi)^{-\frac{1}{2}} t^{-1} f_1(t) \cdot (2\pi)^{-\frac{1}{2}} f_2(t) \end{aligned}$$

where f_1 and f_2 are the characteristic functions of the two distributions. From this we see that equation (2-17-5) is equivalent to the simple relation

$$f(t) = f_1(t) f_2(t) \tag{2-17-10}$$

for the characteristic functions.

If the integral in the first of equations (2-17-3) exists then the k-th derivative of the characteristic function is given by

$$f^{(k)}(t) = i^k \int_{-\infty}^{\infty} x^k p(x) e^{ixt}\,dx$$

so that

$$m^{(k)} = (-i)^k f^{(k)}(0) \tag{2-7-11}$$

or

$$m^{(k)} = (-i)^k (2\pi)^{\frac{1}{2}} P^{(k)}(0) \tag{2-7-12}$$

where $P(t) = \mathscr{F}[p(x); x \to t]$ is the Fourier transform of the probability density.

Another way of putting these results is to say that $m^{(k)}$ is the coefficient of $(-it)^k/k!$ in the Maclaurin expansion of the characteristic function or $(2\pi)^{\frac{1}{2}}$ times the coefficient of the same power in the Maclaurin expansion of $P(t)$.

The *variance* of a distribution is defined by the equation

$$\sigma^2 = \int_{-\infty}^{\infty} (x - m^{(1)})^2 p(x)\, dx. \tag{2-17-13}$$

Expanding the square and using the definitions (2-17-3) we find that

$$\sigma^2 = m^{(2)} - \{m^{(1)}\}^2. \tag{2-17-14}$$

For example for the normal distribution

$$p(x) = \frac{1}{\sigma\sqrt{(2\pi)}}\, e^{-(x-m)^2/2\sigma^2} \tag{2-17-15}$$

we have the characteristic function

$$f(t) = \mathscr{F}^{-1}[\sigma^{-1} e^{-(x-m)^2/2\sigma^2}; t] = e^{itm}\mathscr{F}[e^{-\frac{1}{2}u^2}; \sigma t] = e^{itm - \frac{1}{2}\sigma^2 t^2}.$$

The k-th moment of the normal distribution is therefore the coefficient of $(it)^k/k!$ in the expansion

$$\sum_{r=0}^{\infty} \frac{(itm)^r}{r!} \sum_{s=0}^{\infty} \frac{(i\sigma t)^{2s}}{2^s s!}$$

so that

$$m^{(k)} = \sum_{s=0}^{K} \frac{m^{k-2s}\sigma^{2s} k!}{2^s s!(k - 2s)!}$$

where $K = \frac{1}{2}k$ if k is even and $\frac{1}{2}(k - 1)$ if k is odd. In this way we obtain the formulae

$$m^{(1)} = m, \qquad m^{(2)} = m^2 + \sigma^2, \qquad m^{(3)} = m(m^2 + 3\sigma^2),$$
$$m^{(4)} = m^4 + 6m^2\sigma^2 + 3\sigma^4$$

and so on. It should be observed that

$$m^{(2)} - \{m^{(1)}\}^2 = \sigma^2,$$

so that the σ^2 occurring in the probability density (2-17-14) of the normal distribution is, in fact, the variance.

These ideas can be extended to the discussion of multivariate distributions by the use of multiple Fourier transforms. For a full account of the use of characteristic functions—and hence of Fourier transforms!—in statistics and probability theory the reader is referred to Wintner (1947) and Lukacs (1960).

2-18 FOURIER TRANSFORMS IN QUANTUM MECHANICS

For a system containing a single electron it is well known[1] that the probability of finding the electron in a volume element of volume $d\mathbf{x} = dxdydz$ centered at the point with position vector $\mathbf{x} = (x,y,z)$ is $|\psi(\mathbf{x})|^2\, d\mathbf{x}$ where $\psi(\mathbf{x})$ is the relevant solution of the Schrödinger equation

$$\Delta_3\psi + 2(W - V)\psi = 0. \tag{2-18-1}$$

The function $V(\mathbf{x})$ is the potential energy of the field in which the electron is moving, W is the total energy of the electron and we have used *atomic* units, i.e., we have taken the charge of the electron, the mass of the electron and[2] $\hbar^2/me^2$ as the units of charge, mass, and length respectively.

In several situations—such as, for instance, determining the profile of the modified Compton line—it is desirable to know the *momentum wave function* $\chi(\mathbf{p})$ which has the interpretation that $|\chi(\mathbf{p})|^2\, d\mathbf{p}$ is the probability that the momentum of the electron lies in the range between $\mathbf{p}$ and $\mathbf{p} + d\mathbf{p}$.

The importance of Fourier transforms in quantum theory arises from the circumstance that, on the basis of the Dirac transformation theory, $\chi(\mathbf{p})$ is the three-dimensional Fourier transform of $\psi(\mathbf{x})$—provided of course that we are working in atomic units. We therefore have the relation

$$\chi(\mathbf{p}) = \mathscr{F}_{(3)}[\psi(\mathbf{x}); \mathbf{x} \to -\mathbf{p}]. \tag{2-18-2}$$

in atomic units, which is easily seen to be equivalent to the relation

$$\chi(\mathbf{p}) = \hbar^{-\frac{3}{2}}\mathscr{F}_{(3)}[\psi(\mathbf{x}); \mathbf{x} \to -\mathbf{p}/\hbar]. \tag{2-18-3}$$

It follows immediately from the Parseval relation (2-9-9) that if the spatial wave function $\psi(\mathbf{x})$ is normalized to unity, i.e., if

$$\int_{E_3} |\psi(\mathbf{x})|^2\, d\mathbf{x} = 1$$

then so is the momentum wave function, i.e.,

$$\int_{P_3} |\chi(\mathbf{p})|^2\, d\mathbf{p} = 1$$

where P_3 denotes the entire $p_1p_2p_3$-space.

[1] See, for instance, Chapter I of Mott and Sneddon (1948).

[2] As usual we denote by $\hbar$ the ratio $h/(2\pi)$, where h is Planck's constant.

It turns out that the quantity which enters into the calculation of the profile of the modified Compton line is the *momentum density function* $I(p)$ which is defined by the equation

$$I(p) = \int |\chi(\mathbf{p})|^2 p^2 \, d\omega, \tag{2-18-4}$$

where $d\omega$ is the element of solid angle for $\mathbf{p}$. Physically this obviously means that $I(p)\,dp$ is the probability that the magnitude of the momentum of the electron lies between p and $p + dp$. The Parseval relation therefore leads to the equation

$$\int_0^\infty I(p)\,dp = 1 \tag{2-18-5}$$

which is just what we would expect from the probabilistic interpretation of $I(p)$.

In the Schrödinger interpretation, the momentum $\mathbf{p}$ is represented by the differential operator $-i \cdot \text{grad}$ in the sense that the mean value of the x-component of $\mathbf{p}$ is given by the equation

$$\bar{p}_x = -i \int_{E_3} \psi^*(\mathbf{x}) \frac{\partial}{\partial x} \psi(\mathbf{x})\, d\mathbf{x}.$$

Now if $\chi(\mathbf{p})$ is related to $\psi(\mathbf{x})$ by the equation (2-18-2) it follows from equation (2-3-11) that

$$\mathscr{F}_{(3)}\left[-i\frac{\partial}{\partial x}; \mathbf{x} \to -\mathbf{p}\right] = p_x \chi(\mathbf{p}),$$

and hence from equation (2-9-8) that

$$\bar{p}_x = \int_{P_3} p_x |\chi(\mathbf{p})|^2 \, d\mathbf{p}$$

again in conformity with the probabilistic interpretation of the momentum wave function. It can similarly be shown that the mean momentum of the electron can be calculated from the formula

$$\bar{p} = \int_0^\infty p I(p)\,dp.$$

An especially simple situation arises if the wave function ψ is a function of $r = |\mathbf{x}|$ alone. From equation (2-14-16) we may then deduce that

$$\chi(\mathbf{p}) = p^{-1} X(p)$$

where

$$X(p) = \mathscr{F}_s[r\psi(r); r \to p].$$

Hence

$$I(p) = 4\pi|X(p)|^2.$$

In atomic units the ground state of the hydrogen atom may be described by the spatial wave function

$$\psi(\mathbf{x}) = \pi^{-\frac{1}{2}}e^{-r},$$

so that from equation (2-14-8) we deduce that

$$X(p) = \frac{2\sqrt{(2)}p}{\pi(1 + p^2)^2}$$

giving

$$\chi(p) = \frac{2\sqrt{2}}{\pi(1 + p^2)^2}$$

and

$$I(p) = \frac{32p^2}{\pi(1 + p^2)^4} \tag{2-18-6}$$

The mean momentum corresponding to this momentum density function is given by the formula

$$\bar{p} = \frac{32}{\pi}\int_0^\infty \frac{p^3\,dp}{(1 + p^2)^2}.$$

The integration is elementary and we find that (in atomic units)

$$\bar{p} = \frac{8}{3\pi}.$$

It will be recalled that—in conventional units—the unit of momentum is $mc/137$, where m is the mass of the electron and c is the velocity of light; the mean momentum therefore has the value

$$\frac{8mc}{411\pi}$$

In the nomenclature of spectroscopy the function $\pi^{-\frac{1}{2}}e^{-r}$ is the $1s$ wave function of a hydrogen atom; we have therefore shown that in a $1s$ orbit the mean momentum of the electron is $8mc/411\pi$ or $16me^2/(3h)$—since the fine-structure constant 137 is equal to $hc/(2\pi e^2)$.

The evaluation of the momentum wave function in the general case has been carried out by Podolsky and Pauling (1929). The calculation is cumbersome and will not be repeated here; the interested reader should consult the original paper or pp. 366–368 of Sneddon (1951).

In the more complicated case of an atomic or molecular system with n electrons we number the electrons $1, 2, \ldots, n$ and use $\mathbf{x}_1, \mathbf{x}_2, \ldots, \mathbf{x}_n$ for their spatial coordinates and $\mathbf{p}_1, \mathbf{p}_2, \ldots, \mathbf{p}_n$ for their momentum vectors. The momentum wave function $\chi(\mathbf{p}_1,\mathbf{p}_2, \ldots, \mathbf{p}_n)$ is obtained from the spatial wave function $\psi(\mathbf{x}_1,\mathbf{x}_2, \ldots, \mathbf{x}_n)$ by means of the $3n$-dimensional Fourier transform

$$\chi(\mathbf{p}_1,\mathbf{p}_2, \ldots, \mathbf{p}_n) = (2\pi)^{-3n/2} \int_{-\infty}^{\infty} \cdots \int_{-\infty}^{\infty} \psi(\mathbf{x}_1,\mathbf{x}_2, \ldots, \mathbf{x}_n) \exp\{i(\mathbf{x}_1 \cdot \mathbf{p}_1) + \cdots + i(\mathbf{x}_n \cdot \mathbf{p}_n)\}\, d\mathbf{x}_1 d\mathbf{x}_2 \ldots d\mathbf{x}_n$$

Calculations of the distribution of momenta in atoms based on this formula have been carried out by Coulson and Duncanson (1945,1948)—or see pp. 368–376 of Sneddon (1951). The determination of the distribution of momentum in more complex molecules has been considered by Coulson and Duncanson in a series of papers[1] to which the reader is referred for details.

PROBLEMS

2-1 If $f(t)$ is continuously differentiable in $(0, \infty)$ and the integral $\int_0^\infty f(t)\, dt$ and the series $\sum_{r=0} f(r\pi)$ are both absolutely convergent, prove that as $n \to \infty$,

$$\int_0^\infty f(t) \frac{\sin (2n+1)t}{\sin t}\, dt \to \tfrac{1}{2}\pi f(0) + \pi \sum_{r=1}^{\infty} f(r\pi)$$

$$\int_0^\infty f(t) \frac{\sin (2n+1)t}{\sin t} \cos t\, dt \to \tfrac{1}{2}\pi f(0) + \pi \sum_{r=1}^{\infty} (-1)^r f(r\pi).$$

2-2 Prove that if $\alpha > 0$,

$$\int_0^\infty e^{-2\alpha t} \frac{\sin (2n+1)t}{\sin t}\, dt \to \tfrac{1}{2}\pi \coth (\pi\alpha)$$

$$\int_0^\infty e^{-2\alpha t} \frac{\sin (2n+1)t}{\sin t} \cos t\, dt \to \tfrac{1}{2}\pi \tanh (\pi\alpha)$$

as $n \to \infty$.

2-3 If $f(t)$ is continuously differentiable in $(0, a)$, $a > 0$, and if $0 < s < 1$, prove that as $n \to \infty$,

$$n^s \int_0^a t^{s-1} f(t) \sin (nt)\, dt \to f(0) \int_0^\infty x^{s-1} \sin x\, dx$$

$$n^s \int_0^a t^{s-1} f(t) \cos (nt)\, dt \to f(0) \int_0^\infty x^{s-1} \cos x\, dx.$$

[1] Coulson (1941); Coulson and Duncanson (1941,1942); Duncanson (1941).

2-4 If χ is the characteristic function of a finite interval on the real line, i.e., if

$$\chi(x) = \begin{cases} 1, & x \in (a,b), \quad -\infty < a < b < \infty \\ 0, & x < a, \quad x > b, \end{cases}$$

calculate its Fourier transform X and show that

$$\|X\| = \|\chi\|.$$

2-5 If χ_1 and χ_2 are the characteristic functions of two finite non-overlapping intervals, L_1 and L_2, show that

$$(\chi_1, \chi_2) = 0, \qquad (X_1, X_2) = 0$$

where

$$(f,g) = \int_{-\infty}^{\infty} f(x)\bar{g}(x)\,dx.$$

2-6 If χ is the characteristic function of a finite interval, and X is its Fourier transform, prove that

$$\frac{1}{\sqrt{(2\pi)}}\int_{-N}^{N} e^{-i\xi x}X(\xi)\,d\xi \to \chi(x)$$

as $N \to \infty$.

2-7 If $s(x)$ is a step function

$$s(x) = \sum_{r=1}^{n} c_r\chi_r(x)$$

where $\chi_r(x)$ is the characteristic function of the interval $L_r = \{x : a_r < x < a_{r+1}\}$ and the intervals L_r are disjoint and their union is bounded, prove that

$$\|S\| = \|s\|,$$

where S is the Fourier transform of s.

Show also that $F(\xi) \to 0$ as $|\xi| \to \infty$.

2-8 If $s(x)$ is a step-function and if $K(x)$ satisfies the conditions:

(*a*) $K(x)$ is integrable and bounded in the interval $(0, \infty)$,

(*b*) $\int_0^T K(x)\,dx = o(T)$,

prove that

$$\lim_{\xi\to\infty}\int_0^{\infty} s(x)K(\xi x)\,dx = 0.$$

Using the fact that the set of all step functions is everywhere dense[1] in the space of functions of class L, prove that if $f(x) \in L(0, \infty)$,

$$\lim_{\xi \to \infty} \int_0^\infty f(x) K(\xi x)\, dx = 0.$$

2-9 Putting $f(x) = x^{-\frac{1}{2}}$ in the sine and cosine forms of Fourier's integral theorem show that

$$\int_0^\infty \frac{\sin x}{\sqrt{x}}\, dx = \int_0^\infty \frac{\cos x}{\sqrt{x}}\, dx = \sqrt{(\tfrac{1}{2}\pi)}$$

and deduce that

$$\mathscr{F}_s[t^{-\frac{1}{2}}; \xi] = \mathscr{F}_c[t^{-\frac{1}{2}}; \xi] = \xi^{-\frac{1}{2}}.$$

2-10 Prove that if $a > 0$, $n > 0$,

$$\mathscr{F}_c[e^{-ax}x^{n-1}; \xi] = \sqrt{(2/\pi)}\Gamma(n) r^{-n} \cos(n\theta)$$
$$\mathscr{F}_s(e^{-ax}x^{n-1}; \xi] = \sqrt{(2/\pi)}\Gamma(n) r^{-n} \sin(n\theta)$$

where $r = (a^2 + \xi^2)^{\frac{1}{2}}$, $\theta = \tan^{-1}(\xi/a)$, and deduce that, if $n > 1$ and $x \geqslant 0$,

$$\int_0^{\frac{1}{2}\pi} \cos(n\theta) \cos^{n-2}\theta \cos(x \tan\theta)\, d\theta = \frac{\pi}{2\Gamma(n)} e^{-x} x^{n-1}$$

$$\int_0^{\frac{1}{2}\pi} \sin(n\theta) \cos^{n-2}\theta \sin(x \tan\theta)\, d\theta = \frac{\pi}{2\Gamma(n)} e^{-x} x^{n-1}.$$

Multiplying both sides of these last two equations by $e^{-x}x^{m-1}$ and integrating with respect to x from 0 to ∞ and adding, show that

$$\int_0^{\frac{1}{2}\pi} \cos(m - n)\theta \cos^{m+n-2}\theta\, d\theta$$

$$= \frac{\pi}{2^{m+n-1}} \cdot \frac{\Gamma(m + n - 1)}{\Gamma(m)\Gamma(n)}, \qquad (m > 0, n > 0)$$

Deduce that, if $0 < z < 1$,

$$\Gamma(z)\Gamma(1 - z) = \pi \operatorname{cosec}(\pi z).$$

2-11 Find the Fourier cosine transform of $e^{-at}\cos(at)$ and $e^{-at}\sin(at)$ and deduce the Fourier cosine transforms of $(t^4 + k^4)^{-1}$ and $t^2(t^4 + k^4)^{-1}$, $(k > 0)$.

[1]This means that given a function $f(x) \in L(0, \infty)$, there exists a step-function $s_\varepsilon(x)$ with the property that

$$\int_0^\infty |f(x) - s_\varepsilon(x)|\, dx \leqslant \varepsilon.$$

2-12 From the Fourier sine transform of $e^{-at}\sin(at)$ and $e^{-at}\cos(at)$ deduce the Fourier sine transforms of the functions: (*a*) $t(t^4+k^4)^{-1}$, (*b*) $t(t^4+k^4)^{-2}$, (*c*) $t^3(t^4+k^4)^{-1}$, (*d*) $t^3(t^4+k^4)^{-2}$, where k is a positive constant.

2-13 Show that if $a > 0$, $b > 0$,

$$\mathscr{F}[(1-t)^{a-1}(1+t)^{b-1}H(1-|t|);\xi] = 2^{a+b-\frac{3}{2}}\xi^{-\frac{1}{2}}B(a,b)\,{}_1F_1(b;a+b;2i\xi)$$

where ${}_1F_1(p;q;x)$ denotes the confluent hypergeometric function defined by the equation

$${}_1F_1(p;q;x) = 1 + \frac{p}{q}x + \frac{p(p+1)}{q(q+1)}\frac{x^2}{2!} + \cdots$$

2-14 If $P_n(t)$ is the Legendre polynomial defined by Rodrigues' formula

$$P_n(t) = \frac{1}{2^n n!}\frac{d^n}{dt^n}(t^2-1)^n$$

and H denotes the Heaviside unit function, prove that

$$\mathscr{F}[P_n(t)H(1-|t|);\xi] = i^n\xi^{-\frac{1}{2}}J_{n+\frac{1}{2}}(\xi)$$

and deduce that

$$\mathscr{F}[t^{-\frac{1}{2}}J_{n+\frac{1}{2}}(t);\xi] = i^nP_n(\xi)H(1-|\xi|).$$

2-15 If the function $f(x)$ is continuous and possesses a Fourier cosine transform $F_c(\xi)$, show that[1]

$$\sqrt{\xi}\left[\tfrac{1}{2}F_c(0) + \sum_{n=1}^{\infty} F_c(n\xi)\right] = \sqrt{x}\left[\tfrac{1}{2}f(0) + \sum_{n=1}^{\infty} f(nx)\right]$$

where $\xi x = 2\pi$, $\xi > 0$.

Deduce that

(*a*) $$\sqrt{\xi}\left[\tfrac{1}{2} + \sum_{n=1}^{\infty} e^{-\frac{1}{2}n^2\xi^2}\right] = \sqrt{x}\left[\tfrac{1}{2} + \sum_{n=1}^{\infty} e^{-\frac{1}{2}n^2x^2}\right];$$

(*b*) $$\sqrt{(\tfrac{1}{2}\xi)}\left[\tfrac{1}{2} + \sum_{n=1}^{\infty}\cos(\tfrac{1}{2}n^2\xi^2) + \sum_{n=1}^{\infty}\sin(\tfrac{1}{2}n^2\xi^2)\right] = \sqrt{x}\left[\tfrac{1}{2} + \sum_{n=1}^{\infty}\cos(\tfrac{1}{2}n^2x^2)\right];$$

(*c*) $$\tfrac{1}{2} + \sum_{n=1}^{\infty} e^{-n^2\pi}\cosh(2\pi n/p) = e^{\pi/p^2}\left[\tfrac{1}{2} + \sum_{n=1}^{\infty} e^{-n^2\pi}\cos(2n\pi/p)\right], \qquad p > 0.$$

[1] This is known as *Poisson's formula*.

2-16 If $\nu > \frac{1}{2}$, $\xi > 0$, $\xi x = 2\pi$, show that

$$(\tfrac{1}{2}\xi)^{1-\nu}\left[\frac{\xi^{\nu}}{2^{\nu+1}\Gamma(\nu+1)} + \sum_{n=1}^{\infty} n^{-\nu}\mathrm{J}_{\nu}(n\xi)\right] = \frac{\Gamma(\frac{1}{2})}{\Gamma(\nu+\frac{1}{2})}\left[\frac{1}{2} + \sum_{n=1}^{p}(1-n^2x^2)^{\nu-\frac{1}{2}}\right]$$

where $p = [x^{-1}]$, the integral part of x^{-1}.

Deduce that if m is a positive integer,

$$\tfrac{1}{2}m\pi + \sum_{n=1}^{\infty} n^{-1}\mathrm{J}_1(2mn\pi) = 1 + 2\sum_{n=1}^{m-1}(1 - n^2/m^2)^{\frac{1}{2}}.$$

2-17 Use the method of sec. 2-8 to evaluate $\mathscr{F}[(1+t^4)^{-1};\xi]$ and $\mathscr{F}[(1+t^6)^{-1};\xi]$.

2-18 Show that if b and ξ are real,

$$\mathscr{F}_s\left[\frac{t^2-b^2}{t(t^2+b^2)};\xi\right] = \sqrt{(2\pi)}\{e^{-|b\xi|} - \tfrac{1}{2}\}\operatorname{sgn}(\xi).$$

2-19 C is the closed contour formed by the segment of the imaginary axis between the points $\pm i$ and that half of the unit circle lying to the right of the imaginary axis, indented as necessary. By considering

$$\int_C (z + z^{-1})^{a-2} z^{t-1}\, dz,$$

where a is real and $1 < a < 2$, and then inverting the Fourier transform so obtained, prove that

$$\mathscr{F}\left[\frac{1}{\Gamma(\frac{1}{2}a - \frac{1}{2}t)\Gamma(\frac{1}{2}a + \frac{1}{2}t)};\xi\right] = \frac{2^{a-\frac{3}{2}}\cos^{a-2}\xi}{\Gamma(\frac{1}{2})\Gamma(a-1)} H(\tfrac{1}{2}\pi - |\xi|).$$

2-20 By integrating the function $e^{i\xi z}\operatorname{cosech} z$ round the rectangle formed by the lines $y = 0$, $y = 2\pi$, $x = \pm R$ indented at the points $z = 0, 2\pi i$ and letting $R \to \infty$ prove that

$$\mathscr{F}_s[\operatorname{cosech} x;\xi] = \sqrt{(\tfrac{1}{2}\pi)}\tanh(\tfrac{1}{2}\pi\xi).$$

Deduce that

$$\mathscr{F}_c[x\operatorname{cosech} x;\xi] = \sqrt{(2\pi^3)}e^{\pi\xi}(e^{\pi\xi}+1)^{-2}.$$

2-21 By integrating the functions

$$\frac{\sinh(kz)}{\sinh z}e^{i\xi z}, \qquad \frac{\cosh(kz)}{\sinh z}e^{i\xi z}, \qquad (0 < k < 1)$$

round the rectangle formed by the lines $x = \pm R$, $y = 0$, $y = 2\pi$, suitably indented, find

$$\mathscr{F}_c\left[\frac{\sinh(kx)}{\sinh x};\xi\right] \quad \mathscr{F}_s\left[\frac{\cosh(kx)}{\sinh x};\xi\right]$$

and deduce that if $0 < \alpha < \beta$,

$$\mathscr{F}_c\left[\frac{\sinh(\alpha x)}{\sinh(\beta x)};\xi\right] = \sqrt{(\tfrac{1}{2}\pi)}\beta^{-1}\frac{\sin(\pi\alpha/\beta)}{\cosh(\pi\xi/\beta) + \cos(\pi\alpha/\beta)},$$

$$\mathscr{F}_s\left[\frac{\cosh(\alpha x)}{\sinh(\beta x)};\xi\right] = \sqrt{(\tfrac{1}{2}\pi)}\beta^{-1}\frac{\sinh(\pi\xi/\beta)}{\cosh(\pi\xi/\beta) + \cos(\pi\alpha/\beta)}.$$

2-22 Using a method similar to that of the previous problem, show that

$$\mathscr{F}_c\left[\frac{\cosh(\alpha x)}{\cosh(\beta x)};\xi\right] = \sqrt{(2\pi)}\beta^{-1}\frac{\cos(\pi\alpha/2\beta)\cosh(\pi\xi/2\beta)}{\cos(\pi\alpha/\beta) + \cosh(\pi\xi/\beta)}$$

$$\mathscr{F}_s\left[\frac{\sinh(\alpha x)}{\cosh(\beta x)};\xi\right] = \sqrt{(2\pi)}\beta^{-1}\frac{\sin(\pi\alpha/2\beta)\sinh(\pi\xi/\beta)}{\cos(\pi\alpha/\beta) + \cosh(\pi\xi/\beta)}.$$

2-23 Show that if $\xi > 0$,

$$\frac{2}{\pi}\int_0^\infty \frac{\sin x\cos\xi x}{x}\,dx = H(1-\xi).$$

2-24 If $f(t) = (1 - |t|)H(1 - |t|)$, find $F(\xi)$ and deduce that

(a) $\displaystyle\int_0^\infty \left(\frac{\sin x}{x}\right)^2 dx = \tfrac{1}{2}\pi$

(b) $\displaystyle\int_0^\infty \left(\frac{\sin x}{x}\right)^3 dx = \tfrac{3}{8}\pi$

(c) $\displaystyle\int_0^\infty \left(\frac{\sin x}{x}\right)^4 dx = \tfrac{1}{3}\pi.$

2-25 Show that the Fourier transform of the function

$$\frac{3}{4a^3}(a^2 - t^2)H(1 - |t|/a), \qquad a > 0,$$

is $(2\pi)^{-\frac{1}{2}}\phi(\xi a)$, where

$$\phi(z) = 3\frac{\sin z}{z^3} - \frac{\cos z}{z^2}.$$

2-26 If $a > 0$, $b > 0$, show that

(*a*) $\mathscr{F}_s[t^{-1}e^{-at};\xi] = (2/\pi)^{\frac{1}{2}}\tan^{-1}(\xi/a)$,

(*b*) $\mathscr{F}_c[t^{-1}(e^{-bt} - e^{-at});\xi] = (2\pi)^{-\frac{1}{2}}\log\dfrac{a^2+\xi^2}{b^2+\xi^2}$, $\quad (a > b)$

(*c*) $\mathscr{F}_s[t^{-1}(e^{-bt} - e^{-at});\xi] = (2/\pi)^{\frac{1}{2}}\tan^{-1}\left[\dfrac{(a-b)\xi}{ab+\xi^2}\right]$, $\quad (a > b)$.

2-27 If $F(\xi) = \mathscr{F}_c[e^{-\frac{1}{2}t^2};\xi]$, show that

$$\frac{dF}{d\xi} = -\xi F, \qquad F(0) = 1$$

and *hence* deduce that $F(\xi) = e^{-\frac{1}{2}\xi^2}$.

2-28 Show that

$$\mathscr{F}_c[\cos(\tfrac{1}{2}t^2);\xi] = \frac{2}{\sqrt{(2\pi)}}\int_{-\infty}^{\infty}\cos(\tfrac{1}{2}t^2 - \xi t)\,dt$$

and deduce that

$$\mathscr{F}_c[\cos(\tfrac{1}{2}t^2);\xi] = a\cos(\tfrac{1}{2}\xi^2) + b\sin(\tfrac{1}{2}\xi^2)$$

where a and b are constants. Show also that

$$\mathscr{F}_c[\sin(\tfrac{1}{2}t^2);\xi] = b\cos(\tfrac{1}{2}\xi^2) - a\sin(\tfrac{1}{2}\xi^2).$$

Deduce that

$$\mathscr{F}_c[\cos(\tfrac{1}{2}t^2);\xi] = \frac{1}{\sqrt{2}}[\cos(\tfrac{1}{2}\xi^2) + \sin(\tfrac{1}{2}\xi^2)];$$

$$\mathscr{F}_c[\sin(\tfrac{1}{2}t^2);\xi] = \frac{1}{\sqrt{2}}[\cos(\tfrac{1}{2}\xi^2) - \sin(\tfrac{1}{2}\xi^2)].$$

2-29 Show that the function $\cos(\frac{1}{2}t^2 - \frac{1}{8}\pi)$ is self-reciprocal under the Fourier cosine transform.

2-30 Show that if $a > 0$,

$$\mathscr{F}[(a^2 - t^2)^{-\frac{1}{2}}H(a - |t|);\xi] = \sqrt{(\tfrac{1}{2}\pi)}\mathrm{J}_0(a\xi)$$

and deduce that if $0 < a < b$,

$$\int_0^{\infty}\mathrm{J}_0(ax)\mathrm{J}_0(bx)\,dx = \frac{2}{\pi b}K(a/b)$$

where $K(k)$ denotes the complete elliptic integral

$$\int_0^{\frac{1}{2}\pi}(1 - k^2\sin^2\theta)^{-\frac{1}{2}}\,d\theta.$$

2-31 Using Parseval's theorem, show that if a and b are positive constants

$$\int_0^\infty \sin(ax)\sin(bx)\frac{dx}{x^2} = \tfrac{1}{2}\pi \min(a,b).$$

2-32 Making use of the Fourier cosine transform (2-7-21) and the Parseval relation (2-10-2), prove that

$$\int_0^\infty \frac{x^{-s}\,dx}{1+x^2} = \tfrac{1}{2}\pi \sec(\tfrac{1}{2}\pi s), \qquad 0 < s < 1.$$

2-33 Using the Parseval relation (2-10-2), show that

$$\mathscr{F}_s[t^{-1}J_0(at);\xi] = \begin{cases} \sqrt{(\tfrac{1}{2}\pi)}, & 0 < a < \xi, \\ \sqrt{(2/\pi)}\sin^{-1}(\xi/a), & 0 < \xi < a. \end{cases}$$

2-34 Using the Parseval relation (2-10-1), show that if a and k are positive

$$\mathscr{F}_c[(k^2+t^2)^{-1}J_0(at);\xi] = \sqrt{(2/\pi)}e^{-k\xi}\int_0^a \frac{\cosh(ku)\,du}{(a^2-u^2)^{\frac{1}{2}}}$$

and deduce that

$$\mathscr{F}_c[(k^2+t^2)^{-1}J_0(at);\xi] = \sqrt{(\tfrac{1}{2}\pi)}k^{-1}e^{-k\xi}I_0(ak), \qquad (\xi > a)$$

where I_0 denotes the modified Bessel function of zero order.

By a similar method show that

$$\mathscr{F}_s[t(k^2+t^2)^{-1}J_0(at);\xi] = \sqrt{(\tfrac{1}{2}\pi)}e^{-k\xi}I_0(ak), \qquad (\xi > a).$$

2-35 Making use of the Fourier cosine transform (2-7-19) and the Parseval relation (2-10-2) deduce that if $0 < a < b$,

$$\int_0^\infty x^{-\mu-\nu}J_\mu(ax)J_\nu(bx)\,dx$$
$$= \frac{a^{2\mu}b^{2\nu-1}\Gamma(\tfrac{1}{2})}{2^{\mu+\nu}\Gamma(\nu+\tfrac{1}{2})\Gamma(\mu+1)}\,{}_2F_1(\tfrac{1}{2}-\nu,\tfrac{1}{2};\mu+1;a^2/b^2).$$

In particular, deduce the result of Problem 2-30 and the result

$$\int_0^\infty x^{-1}J_0(ax)J_1(bx)\,dx = (2b/\pi)E(a/b), \qquad (0 < a < b)$$

where $E(k)$ denotes the complete elliptic integral

$$E(k) = \int_0^{\frac{1}{2}\pi}(1-k^2\sin^2\theta)^{\frac{1}{2}}\,d\theta, \qquad (|k| < 1).$$

Show also that if $\mu + \nu > 0$,

$$\int_0^\infty x^{-\mu-\nu}J_\mu(x)J_\nu(x)\,dx = \frac{2^{-\mu-\nu}\Gamma(\tfrac{1}{2})\Gamma(\mu+\nu)}{\Gamma(\mu+\tfrac{1}{2})\Gamma(\nu+\tfrac{1}{2})\Gamma(\mu+\nu+\tfrac{1}{2})}.$$

2-36 Making use of the result proved in Problem 2-14, prove that if m and n are positive integers,

$$\int_{-\infty}^{\infty} x^{-1} \mathrm{J}_{m+\frac{1}{2}}(x) \mathrm{J}_{n+\frac{1}{2}}(x)\, dx = (n + \tfrac{1}{2})^{-1} \delta_{m,n}$$

where $\delta_{m,n}$ denotes the Kronecker delta.

2-37 The Bessel function of the second kind $Y_\nu(x)$ is defined by the equation

$$Y_\nu(x) = \operatorname{cosec}(\nu\pi)[\cos(\nu\pi) \mathrm{J}_\nu(x) - \mathrm{J}_{-\nu}(x)].$$

Prove that

$$(a)\ \mathscr{M}[Y_\nu(x); s] = -2^{s-1}\pi^{-1}\Gamma(\tfrac{1}{2}s + \tfrac{1}{2}\nu)\Gamma(\tfrac{1}{2}s - \tfrac{1}{2}\nu)\cos(\tfrac{1}{2}s - \tfrac{1}{2}\nu)\pi$$

$$(b)\ \mathscr{M}[x^{\frac{1}{2}} Y_\nu(x); s] = 2^{s-\frac{1}{2}} \frac{\Gamma(\frac{1}{4} + \frac{1}{2}s + \frac{1}{2}\nu)}{\Gamma(\frac{3}{4} - \frac{1}{2}s + \frac{1}{2}\nu)} \cot(\tfrac{3}{4}\pi - \tfrac{1}{2}s\pi + \tfrac{1}{2}\nu\pi).$$

2-38 Struve's function of order ν is defined by the equation

$$\mathbf{H}_\nu(x) = \sum_{r=0}^{\infty} \frac{(-1)^r (\frac{1}{2}x)^{\nu+2r+1}}{\Gamma(r + \frac{3}{2})\Gamma(r + \nu + \frac{3}{2})}, \qquad (\nu > -\tfrac{3}{2}).$$

Prove that

$$\mathscr{F}_s[t^{-\nu}\mathbf{H}_\nu(t); \xi] = \frac{2^{\frac{1}{2}-\nu}}{\Gamma(\nu + \frac{1}{2})}(1 - \xi^2)^{\nu-\frac{1}{2}} H(1 - \xi)$$

and deduce that

$$(a)\ \mathscr{M}[\mathbf{H}_\nu(x); s] = \frac{2^{s-1}\tan(\frac{1}{2}\pi s + \frac{1}{2}\pi\nu)\Gamma(\frac{1}{2}s + \frac{1}{2}\nu)}{\Gamma(\frac{1}{2}\nu - \frac{1}{2}s + 1)}$$

$$(b)\ \mathscr{M}[x^{\frac{1}{2}}\mathbf{H}_\nu(x); s] = \frac{2^{s-\frac{1}{2}}\tan(\frac{1}{4}\pi + \frac{1}{2}\pi s + \frac{1}{2}\pi\nu)\Gamma(\frac{1}{4} + \frac{1}{2}s + \frac{1}{2}\nu)}{\Gamma(\frac{1}{2}\nu - \frac{1}{2}s + \frac{3}{4})}$$

2-39 Prove that if $x = (x_1{}^2 + x_2{}^2)^{\frac{1}{2}}$, $\xi = (\xi_1{}^2 + \xi_1{}^2)^{\frac{1}{2}}$

$$(a)\ \mathscr{F}_{(2)}[x_1 f(x); x_1 \to \xi_1, x_2 \to \xi_2] = i\xi^{-1}\xi_1 \mathscr{H}_1[xf(x); \xi]$$

$$(b)\ \mathscr{F}_{(2)}[x_1 x_2 f(x); x_1 \to \xi_1, x_2 \to \xi_2]$$
$$= \xi^{-2}\xi_1\xi_2 \mathscr{H}_0[x^2 f(x); \xi] - 2\xi^{-3}\xi_1\xi_2\mathscr{H}_1[xf(x); \xi]$$

$$(c)\ \mathscr{F}_{(2)}[x^{-2}x_1 e^{-kx}; x_1 \to \xi_1, x_2 \to \xi_2]$$
$$= i\xi^{-2}\xi_1\{1 - k(k^2 + \xi^2)^{-\frac{1}{2}}\}.$$

2-40 Show that if $x = (x_1{}^2 + x_2{}^2)^{\frac{1}{2}}$,

$$\mathscr{F}_{(2)}[e^{-x^2/a^2}; x_1 \to \xi_1, x_2 \to \xi_2] = \tfrac{1}{2}a^2 e^{-\frac{1}{4}\xi^2 a^2}$$

where $\xi = (\xi_1{}^2 + \xi_2{}^2)^{\frac{1}{2}}$.

2-41 If

$$f(x_1,x_2) = \mathscr{F}_{(2)}^{-1}[F(\xi_1,\xi_2);\xi_1 \to x_1,\xi_2 \to x_2],$$

and $\xi = (\xi_1^2 + \xi_2^2)^{\frac{1}{2}}$, show that

$$\mathscr{F}_{(2)}^{-1}[\xi^{-1}e^{-k\xi}F(\xi_1,\xi_2);\xi_1 \to x_1,\xi_2 \to x_2]$$
$$= \frac{1}{2\pi}\int_{-\infty}^{\infty}\int_{-\infty}^{\infty} \frac{f(u_1,u_2)\,du_1du_2}{[(x_1-u_1)^2+(x_2-u_2)^2+k^2]^{\frac{1}{2}}}.$$

2-42 Show that the integral equation

$$\tfrac{1}{2}a\int_{-\infty}^{\infty} f(t)e^{-a|x-t-b|}\,dt = g(x)$$

in which a and b are positive constants and the function $g(x)$ possesses derivatives of all orders, has the solution

$$f(x) = g(x) + bg'(x) + \sum_{n=2}^{\infty} c_n g^{(n)}(x)$$

in which the coefficients c_n are given by the equation

$$c_n = \frac{b^n}{n!} - \frac{b^{n-2}}{a^2(n-2)!}, \qquad (n \geqslant 2).$$

2-43 A function $y(x,t)$ satisfies the diffusion equation

$$\kappa\frac{\partial^2 y}{\partial x^2} = \frac{\partial y}{\partial t}$$

on the half-line $x \geqslant 0$ for $t > 0$, the initial condition

$$y(x,0) = cxe^{-\frac{1}{4}x^2/a^2}$$

c and a being constants, and the boundary conditions $y(0,t) = 0$ and $y(x,t) \to 0$ as $x \to \infty$. Prove that

$$y(x,t) = cx(1+\kappa t/a^2)^{-\frac{3}{2}}\exp\left[-\frac{x^2}{4(a^2+\kappa t)}\right].$$

2-44 The function $u(x,y,z)$ is harmonic in the half-space $z \geqslant 0$ and tends to zero as $r \to \infty$. If, also,

$$\left[\frac{\partial u}{\partial z}\right]_{z=0} = -f(x,y)$$

show that

$$u(x,y,z) = \frac{1}{2\pi}\int_{-\infty}^{\infty}\int_{-\infty}^{\infty} R^{-1}f(s,t)\,dsdt,$$

where $R^2 = (x-s)^2 + (y-t)^2 + z^2$.

2-45 The function $u(x, y)$ is harmonic in the positive quadrant and satisfies the boundary conditions $u_x(0, y) = f(y)$, $y \geqslant 0$; $u(x, 0) = 0$, $x \geqslant 0$.

Show that

$$u_y(x, 0) = -\frac{2}{\pi}\int_0^\infty \frac{tf(t)\,dt}{x^2 + t^2}.$$

Find $u_y(x, 0)$ if $f(y) = qH(b - y)$ where q and b are positive constants.

2-46 A function $\theta(x, t)$ satisfies the diffusion equation

$$\frac{\partial^2\theta}{\partial x^2} = \frac{\partial\theta}{\partial t}$$

for $x \geqslant 0$, $t \geqslant 0$, the initial condition

$$\theta(x, 0) = f(x),$$

in which the function f is prescribed, and the boundary condition

$$\frac{\partial\theta}{\partial x} - h\theta = 0, \qquad x = 0.$$

Show that

$$\theta(x, t) = (2\pi)^{-\frac{1}{2}}\int_0^\infty w(u, t)\left[f(|x - u|) + f(x + u) + h\int_{|x-u|}^{x+u} f(v)\,dv\right]du$$

where

$$w(x, t) = (2t)^{-\frac{1}{2}}e^{-x^2/4t} - (\tfrac{1}{2}\pi)^{\frac{1}{2}}he^{hx + h^2t}\operatorname{Erfc}(\tfrac{1}{2}xt^{-\frac{1}{2}} + ht^{-\frac{1}{2}}).$$

2-47 The axisymmetric function $\theta(\rho, z)$ satisfies Laplace's equation in the semi-infinite cylinder $0 \leqslant \rho \leqslant a$, $z \geqslant 0$, and the boundary conditions

$$\theta(\rho, 0) = 1, \ (0 \leqslant \rho \leqslant a); \qquad \theta(a, z) = 0, \ (z \geqslant 0).$$

Show that

$$\theta(\rho, z) = \mathscr{F}_s[\xi^{-1}\{1 - I_0(\xi\rho)/I_0(\xi a)\}; \xi \to z].$$

2-48 The axisymmetric function $\theta(\rho, z)$ satisfies Laplace's equation in the semi-infinite cylinder $0 \leqslant \rho \leqslant a$, $z \geqslant 0$, and the boundary conditions

$$\left[\frac{\partial\theta}{\partial z} - h\theta\right]_{z=0} = 0, \qquad 0 \leqslant \rho \leqslant a;$$

$$\theta(a, z) = f(z), \qquad z \geqslant 0.$$

Show that

$$\theta(\rho, z) = \mathscr{R}^{-1}[\bar{f}(\xi)I_0(\xi\rho)/I_0(\xi a); \xi \to z]$$

where

$$\bar{f}(\xi) = \mathscr{R}[f(z); \xi].$$

2-49 In the two-dimensional irrotational flow of a perfect fluid the components of the velocity $\{u(x,y),\ v(x,y)\}$, in the x- and y-directions respectively, satisfy the first-order equations

$$\frac{\partial u}{\partial y} - \frac{\partial v}{\partial x} = 0, \qquad \frac{\partial u}{\partial x} + \frac{\partial v}{\partial y} = 0.$$

If the fluid is introduced to the half-plane $y \geqslant 0$ through the slit $|x| \leqslant a$ in the rigid plane $y = 0$ in such a way that

$$v(x,0) = f(x)H(a - |x|),$$

and if u and v both tend to zero as $(x^2 + y^2)^{\frac{1}{2}} \to \infty$, show that

$$u(x,y) = \frac{1}{\pi}\int_{-a}^{a} \frac{(x-t)f(t)}{(x-t)^2 + y^2}\,dt, \qquad v(x,y) = \frac{y}{\pi}\int_{-a}^{a} \frac{f(t)\,dt}{(x-t)^2 + y^2}.$$

Calculate u and v in the special cases:

(*a*) $f(x) = V_0$, constant;

(*b*) $f(x) = c(a^2 - x^2)^{-\frac{1}{2}}$, where c is a constant.

Verify that in case (*b*), $u(x,0) = 0$, $|x| < a$.

2-50 If, in the last problem, it happens that there exists a velocity potential $\phi(x,y)$ such that

$$u = -\frac{\partial \phi}{\partial x}, \qquad v = -\frac{\partial \phi}{\partial x}$$

show that

$$\phi(x,y) = \mathscr{F}^{-1}[|\xi|^{-1}F(\xi)e^{-|\xi|y};\xi \to x]$$

where $F(\xi) = \mathscr{F}[f(x)H(a - |x|);\xi]$.

Show also that if a stream function $\psi(x,y)$ exists with the property

$$u = -\frac{\partial \psi}{\partial y}, \qquad v = \frac{\partial \psi}{\partial x},$$

then

$$\psi(x,y) = -i\mathscr{F}^{-1}[\xi^{-1}F(\xi)e^{-|\xi|y};\xi \to x].$$

2-51 The diffusion of heat in a solid medium which is generating heat is governed by the differential equation

$$\frac{\partial u}{\partial t} = \kappa\Delta_3 u + f(\mathbf{x},t)$$

and the initial condition

$$u(\mathbf{x},0) = g(\mathbf{x})$$

in which the functions f and g are prescribed.

Prove that

$$u(\mathbf{x},t) = (4\pi\kappa t)^{-\frac{3}{2}} \int_{E_3} g(\mathbf{y}) \exp(-|\mathbf{x} - \mathbf{y}|^2/4\kappa t)\, d\mathbf{y}$$
$$+ (4\pi\kappa)^{-\frac{1}{2}} \int_0^t (t - \tau)^{-\frac{3}{2}}\, d\tau \int_{E_3} f(\mathbf{y},\tau) \exp\{-|\mathbf{x} - \mathbf{y}|^2/4\kappa(t - \tau)\}\, d\mathbf{y}.$$

2-52 If the scalar $\phi(\mathbf{x})$ and the vector $\mathbf{u}(\mathbf{x})$ both belong to $\mathscr{C}^1(E_3)$, show that

$$\mathscr{F}_{(3)}[\operatorname{grad}\phi(\mathbf{x});\mathbf{x} \to \xi] = -i\boldsymbol{\xi}\Phi(\boldsymbol{\xi}),$$
$$\mathscr{F}_{(3)}[\operatorname{div}\mathbf{u}(\mathbf{x});\mathbf{x} \to \xi] = -i\boldsymbol{\xi}\cdot\mathbf{U}(\boldsymbol{\xi}),$$
$$\mathscr{F}_{(3)}[\operatorname{curl}\mathbf{u}(\mathbf{x});\mathbf{x} \to \xi] = -i\boldsymbol{\xi} \times \mathbf{U}(\boldsymbol{\xi}),$$

where $\Phi(\boldsymbol{\xi}) = \mathscr{F}_{(3)}[\phi(\mathbf{x});\mathbf{x} \to \boldsymbol{\xi}]$, $\mathbf{U}(\boldsymbol{\xi}) = \mathscr{F}_{(3)}[\mathbf{u}(\mathbf{x});\mathbf{x} \to \boldsymbol{\xi}]$.

2-53 In the three-dimensional irrotational flow of a perfect fluid the velocity vector $\mathbf{u}(\mathbf{x})$ satisfies the equations

$$\operatorname{div}\mathbf{u} = 0, \qquad \operatorname{curl}\mathbf{u} = 0.$$

If the fluid is introduced to the half-space $z \geqslant 0$ through the aperture $\{(x,y):(x,y,0) \in \Omega\}$ in the rigid boundary $z = 0$ in such a way that $u_3(x,y,0) = f(x,y)$, $(x,y) \in \Omega$, and if $|\mathbf{u}| \to 0$ as $r = |\mathbf{x}| \to \infty$, show that

$$\mathbf{u}(\mathbf{x}) = (2\pi)^{-1} \iint_\Omega \frac{(x - x',y - y',z)f(x',y')\, dx'dy'}{[(x - x')^2 + (y - y')^2 + z^2]^{\frac{3}{2}}}.$$

2-54 The components of stress $(\sigma_{xx},\sigma_{xy},\sigma_{yy})$ in an infinite two-dimensional elastic medium satisfy the equations of equilibrium

$$\frac{\partial\sigma_{xx}}{\partial x} + \frac{\partial\sigma_{xy}}{\partial y} + \sigma f(x,y) = 0, \qquad \frac{\partial\sigma_{xy}}{\partial x} + \frac{\partial\sigma_{yy}}{\partial y} + \sigma g(x,y) = 0$$

where $f(x,y)$ and $g(x,y)$ are the prescribed components of the body force per unit mass and σ is the density. They also satisfy the compatibility condition

$$\frac{\partial^2\sigma_{xx}}{\partial y^2} + \frac{\partial^2\sigma_{yy}}{\partial x^2} = 2\frac{\partial^2\sigma_{xy}}{\partial x\partial y}.$$

Show that if f and g possess two-dimensional Fourier transforms, then

$$\sigma_{xx} = -\frac{\sigma}{4\pi}\int_{-\infty}^{\infty}\int_{-\infty}^{\infty} A(x - y', y - y')\{(x - x')f(x', y')$$
$$- (y - y')g(x', y')\}\, dx'dy'$$

where the kernel $A(x,y)$ is defined by the equation

$$A(x,y) = \frac{3x^2 + y^2}{(x^2 + y^2)^2}.$$

Determine the corresponding expressions for σ_{xy} and σ_{yy}.

2-55 The function $u(x_1, x_2, \ldots, x_n)$ satisfies $\Delta_n u = 0$ in the region $x_n \geqslant 0$ and tends to zero as $(x_1{}^2 + x_2{}^2 + \cdots + x_n{}^2)^{\frac{1}{2}} \to \infty$. If also,

$$u(x_1, x_2, \ldots, x_{n-1}, 0) = f(x_1, \ldots, x_{n-1})$$

show that

$$u(x_n, \ldots, x_n)$$
$$= \frac{2^{n-1}\Gamma(\frac{1}{2}n)x_n}{\pi^{\frac{1}{2}n}} \int_{E_{n-1}} \frac{f(t_1, \ldots, t_{n-1})\, dt_1 \ldots dt_{n-1}}{[(x_1{}^2 - t_1{}^2) + \cdots + (x_{n-1}{}^2 - t_{n-1}{}^2) + x_n{}^2]^{\frac{1}{2}n}}$$

2-56 The function $u(x,y,z)$ satisfies the biharmonic equation $\Delta_3\Delta_3 u = 0$ in the half-space $z \geqslant 0$ and tends to zero as $(x^2 + y^2 + z^2)^{\frac{1}{2}} \to \infty$. On the surface $z = 0$ the function u satisfies the conditions $u = f(x,y)$, $\partial u/\partial z = hu$. Prove that

$$u(x,y,z) = \frac{1}{2\pi}\int_{-\infty}^{\infty}\int_{-\infty}^{\infty} f(s,t)K(x - s, y - t, z)\, dsdt$$

where

$$K(x,y,z) = z^2[h(x^2 + y^2 + z^2) + 3z](x^2 + y^2 + z^2)^{-\frac{3}{2}}.$$

2-57 Find the characteristic function of the *rectangular distribution* whose probability density is

$$p(x) = \begin{cases} (2\sigma)^{-1}, & |x - m| \leqslant \sigma \\ 0, & |x - m| > \sigma \end{cases}$$

and deduce that the moments are given by the formula

$$m^{(k)} = k!\sum_{r=0}^{K} \frac{\sigma^{2r} m^{k-2r}}{(2r + 1)!(k - 2r)!},$$

where K is $\frac{1}{2}k$ if k is even and $\frac{1}{2}(k - 1)$ if k is odd.

2-58 Find the characteristic function of the *Cauchy distribution* whose probability density is $p(x) = \theta[\pi^2\theta^2 + (x - m)^2]^{-1}$.

2-59 Show that the *Laplace distribution* defined by the probability density

$$p(x) = \frac{1}{2\sigma} \exp\left[-\frac{|x - m|}{\sigma}\right]$$

has characteristic function

$$f(t) = (1 + \sigma^2 t^2)^{-1} e^{imt}$$

and find the k-th moment of the distribution.

2-60 The probability density of the *gamma distribution* is

$$p(x) = \frac{\lambda}{\Gamma(n)} e^{-\lambda x}(\lambda x)^{n-1} H(x),$$

where $n > 0$ and $H(x)$ is the Heaviside unit function.

Find the characteristic function of the distribution.

2-61 Show that the probability density corresponding to the one-dimensional spatial wave function

$$\psi(x) = \pi^{-\frac{1}{4}} a^{-\frac{1}{2}} e^{-\frac{1}{2}x^2/a^2}$$

is a normal distribution with mean zero and variance $\sigma_x = a/\sqrt{2}$. Find the mean and variance σ_p of the probability density corresponding to the momentum wave function of the particle and show that[1]

$$\sigma_x \sigma_p = \tfrac{1}{2}.$$

2-62 Show that Schrödinger's equation (2-18-1) is equivalent to the integral equation

$$(\tfrac{1}{2}\mathbf{p}^2 - W)\chi(\mathbf{p}) + (2\pi)^{-\frac{3}{2}} \int_{-\infty}^{\infty}\int_{-\infty}^{\infty}\int_{-\infty}^{\infty} v(\mathbf{p} - \mathbf{p}')\chi(\mathbf{p}')\, d\mathbf{p}' = 0$$

for the momentum wave function $\chi(\mathbf{p})$, where $v(\mathbf{p}) = \mathscr{F}_{(3)}[V(\mathbf{r}); \mathbf{r} \to \mathbf{p}]$. Assuming that W is negative show that this can be reduced to the form[2]

$$\phi(\mathbf{p}) = \int_{-\infty}^{\infty}\int_{-\infty}^{\infty}\int_{-\infty}^{\infty} K(\mathbf{p},\mathbf{p}')\phi(\mathbf{p}')\, d\mathbf{p}'$$

Show that in the one-dimensional case in which

$$V(x) = V_0 e^{-a|x|}, \qquad (a > 0),$$

[1] This is a form of Heisenberg's uncertainty relation—See de Bruyn (1968).

[2] Cf. Svartholm (1945,1947).

the corresponding integral equation for the momentum wave function $\chi(p)$ is

$$(p^2 + p_0{}^2)\chi(p) = \frac{\lambda}{\pi}\int_{-\infty}^{\infty} \frac{a^3\chi(p')\,dp'}{a^2 + (p - p')^2}$$

where $p_0{}^2 = -2W$, $\lambda = -2V_0/a^2$.

Verify that this equation has a solution of the form

$$\chi(p) = \sum_{m=0}^{\infty} \frac{c_m p}{p^2 + (am + p_0)^2}$$

provided that the coefficients c_m satisfy the equation

$$\sum_{m=0}^{\infty} pc_m\left[1 - \frac{am(am + 2p_0)}{p^2 + (am + p_0)^2}\right] = a^2 \sum_{m=1}^{\infty} c_{m-1}\frac{p}{p^2 + (am + p_0)^2}$$

and hence the relations

$$mc_m(m + 2p_0) + \lambda ac_{m-1} = 0, \qquad \sum_{m=0}^{\infty} c_m = 0.$$

Deduce that the possible values of W are given by the relation

$$e^{-p_0/a}J_{2p_0/a}(2\lambda^{\frac{1}{2}}) = 0.$$

3

The Laplace Transform

3-1 THE LAPLACE TRANSFORM

We saw in the proof of Fourier's inversion theorem that, if the integral

$$\int_{-\infty}^{\infty} |f(t)|\,dt \tag{3-1-1}$$

is not convergent, then the Fourier transform $F(\xi)$ of the function $f(t)$ need not exist for real values of ξ. This situation does in fact arise in many cases of interest. For instance a function which arises frequently in mathematical physics is $f(t) = \sin(\omega t)$, with ω real, and this function is such that the integral (3-1-1) diverges. In many physical problems, especially those in which transient effects are of interest, we have a function $f(t)$ of this type but which, in addition, may be assumed to be identically equal to zero when $t < 0$. Since $f(t)$ does not make the integral (3-1-1) convergent, we consider instead the function

$$f_1(t) = e^{-\gamma t} f(t) \mathrm{H}(t) \tag{3-1-2}$$

where γ is a positive constant of such a nature that the integral

$$\int_0^\infty e^{-\gamma t} f(t)\, dt$$

does converge. Since $f_1(t)$ now satisfies the conditions of Fourier's theorem and is identically zero if $t < 0$, it follows that

$$f_1(t) = \frac{1}{2\pi} \int_{-\infty}^{\infty} e^{-i\xi t}\, d\xi \int_0^\infty f_1(u) e^{i\xi u}\, du.$$

Substituting from equation (3-1-2) into this equation, we obtain the result

$$f(t) = \frac{e^{\gamma t}}{2\pi} \int_{-\infty}^{\infty} e^{-i\xi t}\, d\xi \int_0^\infty f(u) e^{-(\gamma - i\xi)u}\, du.$$

If we now write $p = \gamma - i\xi$ and define $\bar{f}(p)$ by the equation

$$\bar{f}(p) = \int_0^\infty f(t) e^{-pt}\, dt \tag{3-1-3}$$

we see that we can rewrite this last equation in the form

$$f(t) = \frac{1}{2\pi i} \int_{\gamma - i\infty}^{\gamma + i\infty} \bar{f}(p) e^{pt}\, dp. \tag{3-1-4}$$

As we observed in Chapter 1, the function $\bar{f}(p)$ defined by equation (3-1-3) is called the *Laplace transform* of the function $f(t)$. We use the notation

$$\bar{f}(p) = \mathscr{L}[f(t); p] \tag{3-1-5}$$

for functions of a single variable t. In a similar manner we shall write the relation between the functions $f(t)$ and $\bar{f}(p)$ in the inverse form

$$f(t) = \mathscr{L}^{-1}[\bar{f}(p); t].$$

In the case of functions of several variables $t_1, \ldots, t_n$ we use the notation

$$\bar{f}(t_1, \ldots, p, \ldots, t_n) = \mathscr{L}[f(t_1, \ldots, t_r, \ldots, t_n); t_r \to p]$$

to denote the integral

$$\int_0^\infty f(t_1, \ldots, t_r, \ldots, t_n) e^{-pt_r}\, dt_r.$$

We shall also use the inverse notation

$$f(t_1, \ldots, t_r, \ldots, t_n) = \mathscr{L}^{-1}[\bar{f}(t_1, \ldots, p, \ldots, t_n); p \to t_r].$$

The above heuristic argument suggests that the *inversion theorem* for the Laplace transform is of the form (3-1-4). We shall return to this topic a little later—in sec. 3-10 below. For the moment we shall consider some of the more elementary properties of the Laplace transform.

An integral of the form (3-1-3) in which the function $f(t)$ is integrable over the integral $(0,a)$ for any positive a, is called a *Laplace integral.* The integral does not exist for every function $f(t)$. For instance, if $f(t) = e^{t^2}$, then the integral is divergent for all values of p. For some functions $f(t)$, the integral exists for some values of p but not for others. For example if $f(t) = e^t$, the integral exists if $\operatorname{Re} p > 1$, but not if $\operatorname{Re} p \leqslant 1$. We therefore begin by considering sufficient conditions for the Laplace integral of a function $f(t)$ to exist.

Lemma 1 *If*

(*a*) *$f(t)$ is integrable over any finite interval $[a,b]$, $0 < a < b$,*

(*b*) *there exists a real number c such that for any arbitrary $b > 0$,*

$$\int_b^\lambda e^{-ct} f(t)\, dt$$

tends to a finite limit as $\lambda \to \infty$,

(*c*) *for arbitrary $a > 0$,*

$$\int_\varepsilon^a |f(t)|\, dt$$

tends to a finite limit as $\varepsilon \to 0+$,

then $\mathscr{L}[f(t);p]$ exists for $\operatorname{Re} p \geqslant c$.

The proof is based on elementary properties of integrals. If a and ε are arbitrary ($\varepsilon < a$) and $c > 0$,

$$\left|\int_\varepsilon^a e^{-pt} f(t)\, dt\right| \leqslant \int_\varepsilon^a e^{-ct}|f(t)|\, dt \leqslant \int_\varepsilon^a |f(t)|\, dt$$

while for $c < 0$

$$\left|\int_\varepsilon^a e^{-pt} f(t)\, dt\right| \leqslant \int_\varepsilon^a e^{-ct}|f(t)|\, dt \leqslant e^{-ca} \int_\varepsilon^a |f(t)|\, dt.$$

In both cases it follows from condition (c) that the integral

$$\int_0^a e^{-pt} f(t)\, dt$$

exists for arbitrary positive values of a. The result then follows from the identity

$$\int_0^\lambda e^{-pt} f(t)\,dt = \left(\int_0^a + \int_a^b + \int_b^\lambda\right) e^{-pt} f(t)\,dt,$$

by letting $\lambda \to \infty$ and using conditions (*a*) and (*b*).

Lemma 2 *If the Laplace integral for a given function converges for $p = c$, a real number, it converges in the half-plane* $\mathrm{Re}\,p \geqslant c$, *the convergence being uniform in the angle* $|\arg(p - c)| < \frac{1}{2}\pi$.

This follows from the inequality

$$|e^{-pt} f(t)| = e^{-t\mathrm{Re}(p-c)} |e^{-ct} f(t)| \leqslant |e^{-ct} f(t)| \qquad (t > 0).$$

In the discussion of the convergence of Laplace integrals it is useful to have at our disposal:

Lemma 3 *If*

(*a*) *$f(t)$ is integrable over any finite interval (a,b), $0 < a < b$,*

(*b*) *for arbitrary positive a, the integral*

$$\int_\varepsilon^a |f(t)|\,dt$$

tends to a finite limit as $\varepsilon \to 0+$,

(*c*) *$f(t) = \mathrm{O}(e^{ct})$ as $t \to \infty$,*

then the integral defining $\mathscr{L}[f(t);p]$ converges absolutely if $\mathrm{Re}\,p > c$.

If, instead of (*c*), *we have the condition* (*c*)* *$f(t) = \mathrm{O}(t^\gamma)$ as $t \to \infty$, then the Laplace integral converges absolutely for* $\mathrm{Re}\,p > 0$.

The proof of this is trivial and is left as an exercise to the reader.

Theorem 1 *If $f(t)$ satisfies the conditions of Lemma* 1, *its Laplace transform $\bar{f}(p) = \mathscr{L}[f(t);p]$ is an analytic function of p in the half-plane* $\mathrm{Re}\,p > c$.

To prove this result we consider the function

$$\bar{f}(\lambda,p) = \int_0^\lambda e^{-pt} f(t)\,dt, \qquad \mathrm{Re}\,p > c \tag{3-1-6}$$

If we apply the rule for integration by parts to the integral

$$(p - p_1) \int_0^\infty e^{-(p-p_1)t} \bar{f}(t,p_1)\,dt$$

($\mathrm{Re}\,p > \mathrm{Re}\,p_1 > c$) we find that it has the value

$$\left[-e^{-(p-p_1)t}\bar{f}(t,p_1)\right]_0^\infty + \int_0^\infty e^{-(p-p_1)t}e^{-p_1t}f(t)\,dt$$

which is simply the Laplace transform $\bar{f}(p)$ of $f(t)$; i.e., we have shown that

$$\bar{f}(p) = (p - p_1)\int_0^\infty e^{-(p-p_1)t}\bar{f}(t,p_1)\,dt, \qquad (\text{Re}\,p > \text{Re}\,p_1 > c), \tag{3-1-7}$$

where the function $\bar{f}(t,p_1)$ is defined by equation (3-1-6).

We now consider the function

$$\Delta_f(p,h) = \frac{\bar{f}(p+h) - \bar{f}(p)}{h} - \phi(p) \tag{3-1-8}$$

where the function $\phi(p)$ is defined by the equation

$$\phi(p) = \int_0^\infty e^{-(p-p_1)t}\{1 - (p - p_1)t\}\bar{f}(t,p_1)\,dt. \tag{3-1-9}$$

Using (3-1-7) and (3-1-9) it is a simple matter to show that

$$\Delta_f(p,h) = \int_0^\infty e^{-(p-p_1)t}(e^{-ht} - 1)\bar{f}(t,p_1)\,dt$$
$$+ (p - p_1)\int_0^\infty e^{-(p-p_1)t}\{t - h^{-1}(1 - e^{-ht})\}\bar{f}(t,p_1)\,dt.$$

By condition (*c*) of Lemma 1, the function $\bar{f}(t,p_1)$ is bounded and continuous for $t > 0$ and $\text{Re}\,p_1 > c$. Hence there exists a positive number $M(p_1)$ such that for $t > 0$,

$$|\bar{f}(t,p_1)| < M(p_1).$$

Until now, we have taken p_1 to be arbitrary except in so far as $\text{Re}\,p_1 > c$. We now define a positive number η by the relation

$$\eta = \tfrac{1}{3}\text{Re}\,(p - c)$$

and choose

$$p_1 = c + \eta.$$

We then find that

$$|\Delta_f(p,h)| \leqslant M(p_1)\int_0^\infty e^{-2\eta t}|e^{-ht} - 1|\,dt$$
$$+ 2\eta M(p_1)\int_0^\infty e^{-2\eta t}|t - h^{-1}(1 - e^{-ht})|\,dt.$$

Now

$$|e^{-ht} - 1| < |h|te^{\eta t}$$

and

$$|t - h^{-1}(1 - e^{-ht})| < |h|t^2e^{\eta t}$$

so that

$$|\Delta_f(p,h)| < |h|M(p_1)\int_0^\infty te^{-\eta t}(1 + 2\eta t)\,dt$$

from which it follows immediately that $\Delta_f(p,h) \to 0$ as $h \to 0$. In other words the derivative of $\bar{f}(p)$ exists if $\operatorname{Re} p > c$, and is equal to

$$\phi(p) = -\int_0^\infty tf(t)e^{-pt}\,dt.$$

This also shows that if $\mathscr{L}[f(t);p]$ exists for $\operatorname{Re} p > c$, $\mathscr{L}[tf(t);p]$ exists for $\operatorname{Re} p > c$, and

$$\mathscr{L}[tf(t);p] = -\frac{d\bar{f}}{dp}. \tag{3-1-10}$$

By induction over n we can show that all the derivatives of $\bar{f}(p)$ exist for $\operatorname{Re} p > c$, and hence that $\bar{f}(p)$ is an analytical function of the complex variable p in the domain of convergence.

In an intuitive way we see that as $\operatorname{Re} p \to \infty$, the Laplace transform of $f(t)$ will tend to zero. We state without proof the appropriate lemma:

Lemma 4 *If $f(t)$ satisfies the conditions of lemma 1, then $\bar{f}(p) \to 0$, as $|p| \to \infty$, $|\arg(p - c)| \leqslant \theta < \frac{1}{2}\pi$.*

This lemma and Theorem 1 enable us to settle an important theoretical point. It is conceivable that to any arbitrary function $\bar{f}(p)$ there corresponds a function $f(t)$ such that $\mathscr{L}[f(t);p] = \bar{f}(p)$. By these results we see that this is *not* so: if $\bar{f}(p)$ is not analytic in some half-plane $\operatorname{Re} p > c$, (where c is finite), or if $\bar{f}(p)$ does not tend to zero as $|p| \to \infty$ in a sector $|\arg(p - c)| \leqslant \theta < \frac{1}{2}\pi$, then $\bar{f}(p)$ cannot be the Laplace transform of a function $f(t)$ which satisfies the conditions of lemma 1.

It should be emphasized that though the conditions (a)–(c) of lemma 1 on $f(t)$ are *sufficient* for the existence of $\bar{f}(p)$, they are not *necessary*.

Another theoretical point of some importance is to settle the question as to whether it is possible for two different functions $f_1(t)$ and $f_2(t)$ to have the same Laplace transform $\bar{f}(p)$. To state the answer to this question concisely we introduce the idea of *null function*. A null function is a function $N(t)$ which is zero everywhere except possibly on a set of measure zero (Cf. 2-11); it therefore has the property that

$$\int_0^t N(u)\,du = 0 \tag{3-1-11}$$

for arbitrary positive values of t. It should be observed that if, in addition, it is postulated that $N(t)$ be continuous, then $N(t) \equiv 0$, for all positive values of t.

We have the result

Theorem 2 (Lerch's theorem) *If $f_1(t)$ and $f_2(t)$ both satisfy conditions (a)–(c) of lemma 1 and if $\mathscr{L}[f_1(t);p]$ exists for $\operatorname{Re} p > c_1$, $\mathscr{L}[f_2(t);p]$ exists for $\operatorname{Re} p > c_2$, and if*

$$\mathscr{L}[f_2(t);p] = \mathscr{L}[f_1(t);p],$$

for $\operatorname{Re} p > c = \max(c_1,c_2)$, then $f_2(t) - f_1(t)$ is a null function.

If, in addition, $f_1(t)$ and $f_2(t)$ are continuous on the whole real line, then $f_1(t) \equiv f_2(t)$, $t > 0$.

We can state the theorem in another way by saying that $f_1(t) = f_2(t)$ (a.e.). We shall not prove this theorem now but shall treat it as a corollary to the inversion theorem for Laplace transforms when we come to discuss it (in sec. 3-10 below). Before doing that we shall discuss some of the simpler properties of the Laplace transform.

3-2 CALCULATION OF THE LAPLACE TRANSFORMS OF SOME ELEMENTARY FUNCTIONS

We begin by calculating the Laplace transforms of some elementary functions. From the definition, we have

$$\mathscr{L}[1;p] = \int_0^\infty e^{-pt}\,dt = [-p^{-1}e^{-pt}]_0^\infty = p^{-1}. \tag{3-2-1}$$

Similarly, integrating by parts, we find that

$$\mathscr{L}[t;p] = \int_0^\infty te^{-pt}\,dt = [-p^{-1}e^{-pt}t]_0^\infty + p^{-1}\int_0^\infty e^{-pt}\,dt$$

which shows that

$$\mathscr{L}[t;p] = p^{-2}. \tag{3-2-2}$$

These are special cases of a general formula. If v is a constant such that $\operatorname{Re} v > -1$, then

$$\mathscr{L}[t^v;p] = \int_0^\infty t^v e^{-pt}\,dt = p^{-v-1}\int_0^\infty u^v e^{-u}\,du$$

where we have made the substitution $u = pt$. The integral is Euler's integral for the gamma function $\Gamma(v+1)$ so that we deduce the formula

$$\mathscr{L}[t^v;p] = p^{-v-1}\Gamma(v+1). \tag{3-2-3}$$

We have already calculated the Laplace transforms of the functions $\sin t$ and $\cos t$ in sec. 2-7. If, in equations (2-7-1) and (2-7-2) we replace a by p and then put $\xi = 1$, we obtain the formulae

$$\mathscr{L}[\sin t; p] = (p^2 + 1)^{-1}, \tag{3-2-4}$$

$$\mathscr{L}[\cos t; p] = p(p^2 + 1)^{-1}. \tag{3-2-5}$$

One of the simplest of all Laplace transforms is that of e^{at}, viz.,

$$\mathscr{L}[e^{at}; p] = (p - a)^{-1}, \qquad (\operatorname{Re} p > \operatorname{Re} a). \tag{3-2-6}$$

Using the fact that $\mathscr{L}$ is a linear operator, we deduce that if a is real and positive,

$$\mathscr{L}[\cosh(at); p] = \tfrac{1}{2}\mathscr{L}[e^{at}; p] + \tfrac{1}{2}\mathscr{L}[e^{-at}; p] = \tfrac{1}{2}(p - a)^{-1} + \tfrac{1}{2}(p + a)^{-1}$$

if $\operatorname{Re} p > a$. Hence we have the formula

$$\mathscr{L}[\cosh(at); p] = p(p^2 - a^2)^{-1}, \qquad (\operatorname{Re} p > a > 0). \tag{3-2-7}$$

In a similar way, we can show that

$$\mathscr{L}[\sinh(at); p] = a(p^2 - a^2)^{-1}, \qquad (\operatorname{Re} p > a > 0). \tag{3-2-8}$$

To derive some of the Laplace transforms we require later we now consider the evaluation of the integral

$$I(a,b) = \int_0^\infty e^{-a^2u^2 - b^2u^{-2}}\, du \tag{3-2-9}$$

which has the properties

$$I(a,0) = \int_0^\infty e^{-a^2u^2}\, du = \tfrac{1}{2}\pi^{\frac{1}{2}}a^{-1} \tag{3-2-10}$$

$$\frac{\partial I}{\partial b} = -2b\int_0^\infty u^{-2}e^{-a^2u^2 - b^2u^{-2}}\, du. \tag{3-2-11}$$

If we make the substitution $v = au$ in equation (3-2-9) we find that

$$I(a,b) = a^{-1}\int_0^\infty e^{-v^2 - a^2b^2v^{-2}}\, dv = a^{-1}I(1,ab).$$

Similarly if we make the substitution $v = bu^{-1}$ in equation (3-2-11) we find that

$$\frac{\partial I(a,b)}{\partial b} = -2I(1,ab)$$

Eliminating $I(1,ab)$ from these last two equations we find that $I(a,b)$ satisfies the simple differential equation

$$\frac{\partial I}{\partial b} = -2aI.$$

whose solution

$$I(a,b) = I(a,0)e^{-2ab}$$

is easily derived. If we make use of equation (3-2-10) we therefore obtain the equation

$$\int_0^\infty e^{-a^2u^2-b^2u^{-2}}\,du = \frac{\sqrt{\pi}}{2a}e^{-2ab}. \tag{3-2-12}$$

If, for instance, we put $t = u^2$ in the definition

$$\mathscr{L}[t^{-\frac{1}{2}}e^{-c/t};p] = \int_0^\infty t^{-\frac{1}{2}}e^{-c/t-pt}\,dt$$

we find that the integral on the right becomes

$$\int_0^\infty e^{-pu^2-cu^{-2}}\,du$$

and it follows from equation (3-2-12) that

$$\mathscr{L}[t^{-\frac{1}{2}}e^{-c/t};p] = \sqrt{\frac{\pi}{p}}\,e^{-2\sqrt{(cp)}}. \tag{3-2-13}$$

If we now differentiate both sides of this equation with respect to c we obtain the formula

$$\mathscr{L}[t^{-\frac{3}{2}}e^{-c/t};p] = \sqrt{\frac{\pi}{c}}\,e^{-2\sqrt{(cp)}}. \tag{3-2-14}$$

3-3 RULES OF MANIPULATION OF THE LAPLACE TRANSFORM

We shall now develop some general rules for the manipulation of the Laplace transform by means of which we may extend the list of transforms we have already calculated.

First of all we have the *shift theorem* which enables us to write down the Laplace transform of $e^{-at}f(t)$ when that of $f(t)$ is known. From the definition of the Laplace transform we have

$$\mathscr{L}[e^{-at}f(t);p] = \int_0^\infty f(t)e^{-(p+a)t}\,dt$$

from which we deduce immediately that

$$\mathscr{L}[e^{-at}f(t);p] = \mathscr{L}[f(t);p+a]. \tag{3-3-1}$$

For example, using equation (3-2-3) and this result, we see that

$$\mathscr{L}[t^{\nu}e^{-at};p] = (p + a)^{-\nu-1}\Gamma(\nu + 1), \qquad (\operatorname{Re}\nu > -1, \operatorname{Re}p > -a). \tag{3-3-2}$$

Similarly the Laplace transform of $f(bt)$, $(b > 0)$, is defined to be

$$\int_0^\infty f(bt)e^{-pt}\,dt.$$

Changing the variable of integration from t to u where $u = bt$, we see that this integral is equal to

$$b^{-1}\int_0^\infty f(u)e^{-pu/b}\,du$$

from which we deduce the rule

$$\mathscr{L}[f(bt);p] = b^{-1}\mathscr{L}[f(t);p/b], \qquad (b > 0). \tag{3-3-3}$$

For instance, using this result in conjunction with equations (3-2-4) and (3-2-5) we see that

$$\mathscr{L}[\sin(bt);p] = b(p^2 + b^2)^{-1} \tag{3-3-4}$$

$$\mathscr{L}[\cos(bt);p] = p(p^2 + b^2)^{-1} \tag{3-3-5}$$

Combining equations (3-3-1) and (3-3-3) we obtain the result

$$\mathscr{L}[f(bt)e^{-at};p] = b^{-1}\mathscr{L}[f(t);(p + a)/b], \qquad (\operatorname{Re}p > a, b > 0), \tag{3-3-6}$$

and, as special cases,

$$\mathscr{L}[e^{-at}\sin(bt);p] = b\{(p + a)^2 + b^2\}^{-1} \tag{3-3-7}$$

$$\mathscr{L}[e^{-at}\cos(bt);p] = (p + a)\{(p + a)^2 + b^2\}^{-1}. \tag{3-3-8}$$

We often use these results in the inverse form

$$\mathscr{L}^{-1}\left[\frac{1}{(p + a)^2 + b^2};t\right] = b^{-1}e^{-at}\sin(bt) \tag{3-3-9}$$

$$\mathscr{L}^{-1}\left[\frac{p}{(p + a)^2 + b^2};t\right] = e^{-at}\left[\cos(bt) - \frac{a}{b}\sin(bt)\right]. \tag{3-3-10}$$

As an example of the use of these formulae we shall calculate

$$\mathscr{L}^{-1}\left[\frac{p + 4}{p^2 + 2p + 5};t\right]$$

Since

$$p^2 + 2p + 5 = (p + 1)^2 + 2^2, \qquad p + 4 = (p + 1) + 3$$

we have

$$\mathscr{L}^{-1}\left[\frac{p+4}{(p+1)^2+4};t\right] = \mathscr{L}^{-1}\left[\frac{p+1}{(p+1)^2+4};t\right]$$

$$+ \mathscr{L}^{-1}\left[\frac{3}{(p+1)^2+4};t\right]$$

$$= e^{-t}\{\cos(2t) + \tfrac{3}{2}\sin(2t)\}.$$

Although in any particular case it is usually simplest to calculate the inverse Laplace transform in this way, we can easily develop a general formula by means of which to calculate

$$\mathscr{L}^{-1}\left[\frac{f_1(p)}{f_2(p)};t\right]$$

where $f_2(p)$ is a polynomial of degree n, say, in p and $f_1(p)$ is a polynomial of degree less than n.

Suppose, for instance, that $f_2(p)$ has n simple zeros $a_1, a_2, \ldots, a_n$. Then if we choose the polynomials in such a way that the coefficient of p^n in $f_2(p)$ is unity, we can write

$$\bar{f}(p) = \sum_{r=1}^{n} A_r(p - a_r)^{-1}$$

where

$$\sum_{r=1}^{n} A_r(p - a_1) \ldots (p - a_{r-1})(p - a_{r+1}) \ldots (p - a_n) \equiv f_1(p)$$

so that—as we see by putting $p = a_r$ in this identity—

$$A_r = \frac{f_1(a_r)}{(a_r - a_1) \ldots (a_r - a_{r-1})(a_r - a_{r+1}) \ldots (a_r - a_n)}.$$

Hence with this value for A_r, we find that

$$\mathscr{L}^{-1}\left[\frac{f_1(p)}{f_2(p)};t\right] = \sum_{r=1}^{n} A_r e^{a_r t}.$$

For example, in the case considered above $f_1(p) = p + 4$, $f_2(p) = (p+1)^2 + 4$, so that $a_1 = -1 + 2i$, $a_2 = -1 - 2i$, $A_1 = (3 + 2i)/4i$, and so

$$\mathscr{L}^{-1}\left[\frac{p+4}{p^2+2p+5};t\right] = \frac{1}{4i}\{(3+2i)e^{-t+2it} - (3-2i)e^{-t-2it}\}$$

in agreement with the result above.

A similar result can be deduced in the case in which the polynomial $f_2(p)$ has multiple zeros; this is left as an exercise to the reader (see Problem 3-16).

We sometimes encounter the problem of finding the Laplace transform of $f(t)\mathrm{H}(t-a)$, $(a > 0)$, where H denotes the Heaviside unit function. We have

$$\mathscr{L}[f(t)\mathrm{H}(t-a);p] = \int_a^\infty e^{-pt}f(t)\,dt = \int_0^\infty e^{-p(u+a)}f(u+a)\,du$$

from which we deduce that

$$\mathscr{L}[f(t)\mathrm{H}(t-a);p] = e^{-pa}\mathscr{L}[f(t+a);p]. \tag{3-3-11}$$

For example, if $f(t) = \cos t$, then $f(t+a) = \cos a \cos t - \sin a \sin t$, so that

$$\mathscr{L}[\cos t\mathrm{H}(t-a);p] = e^{-pa}\left\{\cos a\,\frac{p}{p^2+1} - \sin a\,\frac{1}{p^2+1}\right\}$$

$$= \frac{e^{-pa}}{p^2+1}(p\cos a - a\sin a)$$

We have already proved in sec. 3-1 that

$$\mathscr{L}[tf(t);p] = -\frac{d}{dp}\bar{f}(p). \tag{3-3-12}$$

Repeating the process we find that, if n is a positive integer,

$$\mathscr{L}[t^n f(t);p] = (-1)^n\frac{d^n}{dp^n}\bar{f}(p). \tag{3-3-13}$$

For example, from equation (3-3-5) we deduce that

$$\mathscr{L}[t\cos(at);p] = (p^2-a^2)(p^2+a^2)^{-2}. \tag{3-3-14}$$

Similarly, from equation (3-3-4) we deduce that

$$\mathscr{L}[t\sin(at);p] = 2ap(p^2+a^2)^{-2}. \tag{3-3-15}$$

Differentiating these last two equations once more with respect to p, we have the formulae

$$\mathscr{L}[t^2\cos(at);p] = 2p(p^2-3a^2)(p^2+a^2)^{-3} \tag{3-3-16}$$

$$\mathscr{L}[t^2\sin(at);p] = 2a(3p^2-a^2)(p^2+a^2)^{-3} \tag{3-3-17}$$

Results of this kind can also be obtained by equating real and imaginary parts in the formula

$$\mathscr{L}[t^n e^{iat};p] = \frac{n!}{(p-ia)^{n+1}} = n!(p+ia)^{n+1}(p^2+a^2)^{-n-1} \tag{3-3-18}$$

where n is a positive integer.

Since $\mathscr{L}$ is a linear operator it follows immediately from (3-3-13) that

$$\mathscr{L}[(c_0 t^n + c_1 t^{n-1} + \cdots + c_n)f(t);p]$$
$$= c_0 \mathscr{L}[t^n f(t);p] + \cdots + c_n \mathscr{L}[f(t);p]$$
$$= \left\{c_0\left(-\frac{d}{dp}\right)^n + c_1\left(-\frac{d}{dp}\right)^{n-1} + \cdots + c_n\right\}\bar{f}(p)$$

showing that if $p_n(t)$ is a polynomial of degree n in t,

$$\mathscr{L}[p_n(t)f(t);p] = p_n(-\mathrm{D})\bar{f}(p) \tag{3-3-19}$$

where D denotes the differential operator d/dp.

If we integrate both sides of the equation

$$\mathscr{L}[f(t);p] = \bar{f}(p)$$

with respect to p from p to ∞, we find that

$$\mathscr{L}[t^{-1}f(t);p] = \int_p^\infty f(u_1)\,du_1 \tag{3-3-20}$$

provided, of course, that both integrals exist. Repeating the process we find that, if n is a positive integer,

$$\mathscr{L}[t^{-n}f(t);p] = \int_p^\infty du_n \ldots \int_{u_3}^\infty du_2 \int_{u_2}^\infty f(u_1)\,du_1. \tag{3-3-21}$$

As an example of this result we see that

$$\mathscr{L}\left[\frac{\sin(at)}{t};p\right] = \int_p^\infty \frac{a}{p^2+a^2}\,dp = \left[\tan^{-1}\frac{p}{a}\right]_p^\infty = \tfrac{1}{2}\pi - \tan^{-1}\left(\frac{p}{a}\right),$$

so that

$$\mathscr{L}\left[\frac{\sin(at)}{t};p\right] = \cot^{-1}\left(\frac{p}{a}\right). \tag{3-3-22}$$

3-4 LAPLACE TRANSFORMS OF DERIVATIVES

In applications of the theory of the Laplace transform to the solution of ordinary differential equations we require the formula for the Laplace transform of the derivative of a function. Integrating the defining integral by parts we find that

$$\mathscr{L}[f'(t);p] = [f(t)e^{-pt}]_0^\infty + p\int_0^\infty f(t)e^{-pt}\,dt.$$

A necessary condition for the existence of the Laplace transform of $f(t)$ is that

$$\lim_{t\to\infty} f(t)e^{-pt} = 0$$

so that if we assume that $f(t)$ is continuous for $t > 0$, this formula is equivalent to

$$\mathscr{L}[f'(t);p] = p\bar{f}(p) - f(0). \tag{3-4-1}$$

Repeating this process we see that if $f \in \mathscr{C}^1\ (0,\infty)$

$$\mathscr{L}[f''(t);p] = p^2 f(p) - pf(0) - f'(0) \tag{3-4-2}$$

and it is easily shown by induction that if n is a positive integer and $f(t) \in \mathscr{C}^{(n-1)}(0,\infty)$,

$$\mathscr{L}[f^{(n)}(t);p] = p^n\bar{f}(p) - p^{n-1}f(0) - p^{n-2}f'(0) - \cdots - f^{(n-1)}(0). \tag{3-4-3}$$

In particular if

$$f^{(r)}(0) = 0, \qquad (r = 0,1,2, \ldots, n-1),$$

then

$$\mathscr{L}[f^{(n)}(t);p] = p^n\bar{f}(p), \tag{3-4-4}$$

so that (in this sense) the operator d^n/dt^n is transformed under the Laplace transform to p^n. It is this simple result which provides the basis of operational calculus.

In applications to partial differential equations we have to consider the Laplace transform with respect to t, say, of a function $f(x_1, \ldots, x_n,t)$ of several variables. It is then easily shown from equation (3-4-4) that, if $\partial^r f/\partial t^r$ is continuous for all positive values of t, and for $r = 0,1,2, \ldots, n$, then

$$\mathscr{L}\left[\frac{\partial^n f}{\partial t^n}; t \to p\right] = p^n\bar{f}(x_1, \ldots, x_n,p) - \sum_{r=0}^{n} p^{n-1-r}f_r(x_1, \ldots, x_n), \tag{3-4-5}$$

where

$$\bar{f}(x_1, \ldots, x_n,p) = \mathscr{L}[f(x_1, \ldots, x_n,t); t \to p]$$

and

$$f_r(x_1, \ldots, x_n) = \left[\frac{\partial^r f}{\partial t^r}\right]_{t=0}$$

For any partial derivative not involving differentiation with respect to t we immediately have formulae of the type

$$\mathscr{L}\left[\frac{\partial^m f}{\partial x_1{}^q \ldots \partial x_n{}^r}; t \to p\right] = \frac{\partial^m}{\partial x_1{}^q \ldots \partial x_n{}^r} \bar{f}(x_1, \ldots, x_n, p), \tag{3-4-6}$$

(with $m = q + \cdots + r$). For mixed derivatives involving t we have formulae of the type

$$\mathscr{L}\left[\frac{\partial^2 f}{\partial x_1 \partial t}; t \to p\right] = \frac{\partial}{\partial x_1} \mathscr{L}\left[\frac{\partial f}{\partial t}; t \to p\right] = p\frac{\partial \bar{f}}{\partial x_1} - \frac{\partial f_0}{\partial x_1} \tag{3-4-7}$$

and so on.

These results have, of course, to be modified if, instead of being continuous, f is only piecewise-continuous on the real line. For example, suppose that $f(t)$ has a finite discontinuity at $t = a$ but is otherwise continuous in $(0,\infty)$. Then, proceeding from the definition, we see that

$$\begin{aligned}\mathscr{L}[f'(t);p] &= \int_0^{a-} f'(t)e^{-pt}\,dt + \int_{a+}^{\infty} f'(t)e^{-pt}\,dt \\ &= [f(t)e^{-pt}]_0^{a-} + p\int_0^{a-} f(t)e^{-pt}\,dt + [f(t)e^{-pt}]_{a+}^{\infty} \\ &\quad + p\int_{a+}^{\infty} f(t)e^{-pt}\,dt\end{aligned}$$

and, as a consequence, we deduce that equation (3-4-1) has to be replaced by the formula

$$\mathscr{L}[f'(t);p] = p\bar{f}(p) - f(0) - e^{-pa}[f]_a. \tag{3-4-8}$$

The case in which $f'(t)$ has a finite number of points of finite discontinuity is obviously covered by the formula

$$\mathscr{L}[f'(t);p] = p\bar{f}(p) - f(0) - \sum_{r=1}^{n} e^{-pa_r}[f]_{a_r} \tag{3-4-9}$$

where $a_1, \ldots, a_n$ are the points of discontinuity of f.

If, in addition, $f'(t)$ has points of discontinuity $b_1, \ldots, b_m$,

$$\begin{aligned}\mathscr{L}[f''(t);p] &= p^2\bar{f}(p) - pf(0) - f'(0) - p\sum_{r=1}^{n} e^{-pa_r}[f]_{a_r} \\ &\quad - \sum_{s=1}^{m} e^{-pb_s}[f']_{b_s}\end{aligned} \tag{3-4-10}$$

There are corresponding results for functions of more than one variable. For instance, suppose that $f(x,t)$ is a function of two independent variables x and t, where $x \in D$, an interval on the real line and $t > 0$, and

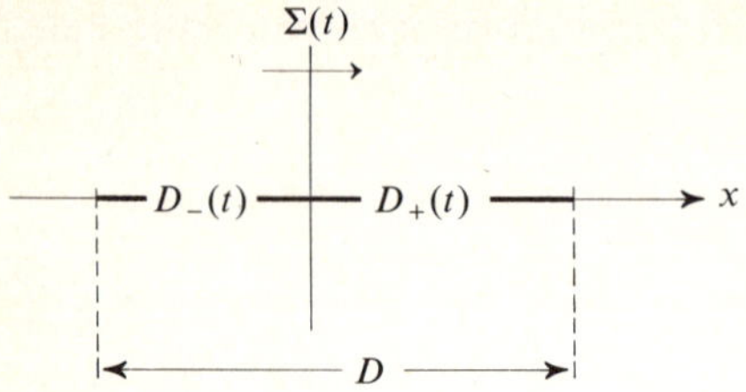

Fig. 3-1

that $\Sigma(t)$ is a curve in the xt-plane with equation

$$\Phi(x,t) = 0$$

where

$$\Phi(x,t) = \phi(x) - t.$$

We shall also assume that $\Sigma(t)$ propagates along the x-axis in the manner shown in Fig. 3-1, and divides the interval D into two subintervals $D_-(t)$ and $D_+(t)$ on which $\Phi < 0$ and $\Phi > 0$ respectively. We now define the jump across Σ by the relation

$$[f]_\Sigma = f_+(x,t) - f_-(x,t).$$

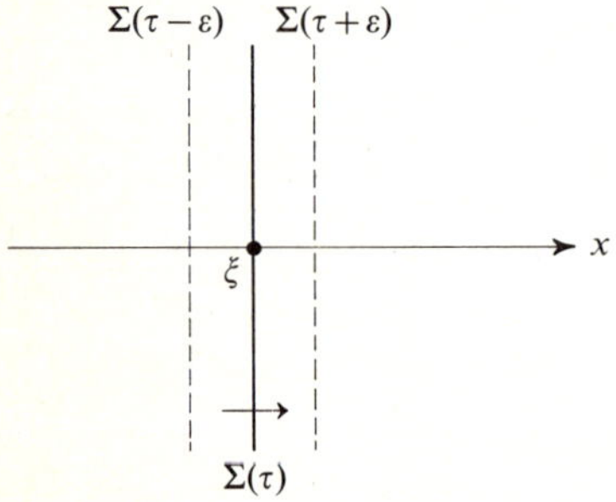

Fig. 3-2

From the consideration of a simple diagram such as Fig. 3-2 we see that if the jump comes at an instant τ then

$$[f]_\tau = \lim_{\varepsilon \to 0} \{f(\xi,\tau + \varepsilon) - f(\xi,\tau + \varepsilon)\}$$

is given in terms of the jump across Σ by the simple relation

$$[f]_\tau = -[f]_\Sigma$$

so that, in this notation, equation (3-4-8) is equivalent to the relation

$$\mathscr{L}\left[\frac{\partial f}{\partial t}; t \to p\right] = p\bar{f}(x,p) - f(x,0) + [f]_\Sigma e^{-p\phi(x)} \tag{3-4-11}$$

where

$$\bar{f}(x,p) = \mathscr{L}[f(x,t);t \to p]. \tag{3-4-12}$$

To find the Laplace transform of $\partial f/\partial x$, we write

$$\bar{f}(x,p) = \int_0^{\phi(x)-} f(x,t)e^{-pt}\,dt + \int_{\phi(x)+}^{\infty} f(x,t)e^{-pt}\,dt$$

and differentiate with respect to x to obtain

$$\frac{\partial}{\partial x}\bar{f}(x,p) = [f\{x,\phi(x)-\} - f\{x,\phi(x)+\}]\phi'(x)e^{-p\phi(x)} + \int_0^{\infty} \frac{\partial f}{\partial x} e^{-pt}\,dt$$

and hence to deduce the relation

$$\mathscr{L}\left[\frac{\partial f}{\partial x};t \to p\right] = \frac{\partial}{\partial x}\bar{f}(x,p) - [f]_\Sigma \phi'(x)e^{-p\phi(x)}. \tag{3-4-13}$$

If we apply equation (3-4-11) twice in succession we obtain the formula

$$\mathscr{L}\left[\frac{\partial^2 f}{\partial t^2};t \to p\right] = p^2\bar{f}(x,p) - pf(x,0) - f_t(x,0) + \left\{p[f]_\Sigma + \left[\frac{\partial f}{\partial t}\right]_\Sigma\right\} e^{-p\phi(x)} \tag{3-4-14}$$

and if we apply equation (3-4-13) twice in succession and make use of the relation

$$\frac{\partial [f]_\Sigma}{\partial x} = \left[\frac{\partial f}{\partial x}\right]_\Sigma + \phi'(x)\left[\frac{\partial f}{\partial t}\right]_\Sigma \tag{3-4-15}$$

we obtain the formula

$$\mathscr{L}\left[\frac{\partial^2 f}{\partial x^2};t \to p\right] = \frac{\partial^2 \bar{f}}{\partial x^2} - \left\{(\phi''(x) - p\{\phi'(x)\}^2)[f]_\Sigma + 2\phi'(x)\left[\frac{\partial f}{\partial x}\right]_\Sigma + \{\phi'(x)\}^2\left[\frac{\partial f}{\partial t}\right]_\Sigma\right\} e^{-p\phi(x)}. \tag{3-4-16}$$

Similar results hold when $f(\mathbf{r},t)$ is a function of a vector $\mathbf{r} \in D$ and a scalar variable $t \geqslant 0$. $\Sigma(t)$ is now a surface with equation

$$\phi(\mathbf{r}) - t = 0, \tag{3-4-17}$$

dividing D into two sub-regions $D_+(t)$, $D_-(t)$. (Cf. Fig. 3-3.) The analysis goes through in much the same way as when $\mathbf{r}$ is a scalar. In terms of the Laplace transform

$$\bar{f}(\mathbf{r},p) = \mathscr{L}[f(\mathbf{r},t);t \to p]$$

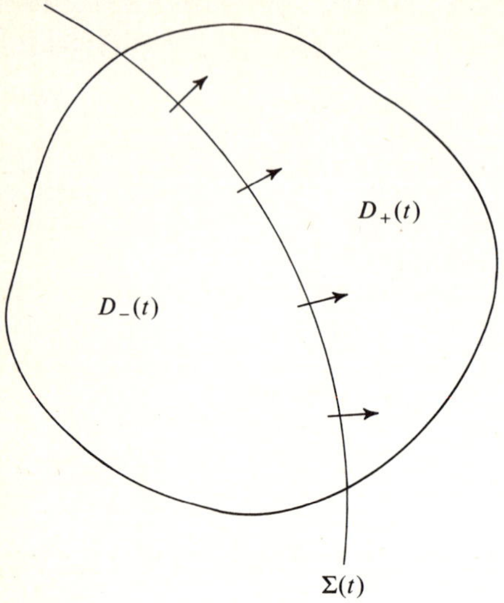

Fig. 3-3

we find that the analogues of equations (3-4-11) and (3-4-12) are respectively

$$\mathscr{L}\left[\frac{\partial f}{\partial t}; t \to p\right] = p\bar{f}(\mathbf{r},p) - f(\mathbf{r},0) + [f]_\Sigma e^{-p\phi(\mathbf{r})} \tag{3-4-18}$$

$$\mathscr{L}[\operatorname{grad} f; t \to p] = \operatorname{grad} \bar{f}(\mathbf{r},p) - [f]_\Sigma e^{-p\phi(\mathbf{r})} \operatorname{grad} \phi(\mathbf{r}). \tag{3-4-19}$$

These formulae are due to Chadwick and Powdrill (1964).

3-5 RELATIONS INVOLVING INTEGRALS

If we consider the function

$$F(t) = \int_0^t f(u)\,du \tag{3-5-1}$$

then

$$F'(t) = f(t), \qquad F(0) = 0, \tag{3-5-2}$$

and from equation (3-4-1) we have that

$$\mathscr{L}[f(t);p] = p\mathscr{L}[F(t);p]$$

a result which can be written in the form

$$\mathscr{L}\left[\int_0^t f(u)\,du; p\right] = p^{-1}\bar{f}(p) \tag{3-5-3}$$

where $\bar{f}(p)$ is the Laplace transform of $f(t)$.

For example, if Si(t) denotes the sine integral

$$\mathrm{Si}(t) = \int_0^t \frac{\sin u}{u}\, du \tag{3-5-4}$$

we find from equation (3-3-20) that

$$\mathscr{L}[\mathrm{Si}(t);p] = p^{-1}\cot^{-1} p. \tag{3-5-5}$$

The result (3-5-3) can be generalized to the form

$$\mathscr{L}\left[\int_0^t \cdots \int_0^{u_2} f(u_1)\, du_1 \ldots du_n; t \to p\right] = p^{-n}\bar{f}(p) \tag{3-5-6}$$

If we combine (3-3-20) with (3-5-3) we obtain the relation

$$\mathscr{L}\left[\int_0^t u^{-1}f(u)\, du;p\right] = p^{-1}\int_p^\infty \bar{f}(\zeta)\, d\zeta. \tag{3-5-7}$$

On the other hand if we interchange the order of the integrations on the right-hand side of the equation

$$\mathscr{L}\left[\int_t^\infty u^{-1}f(u)\, du;p\right] = \int_0^\infty e^{-pt}\, dt \int_t^\infty u^{-1}f(u)\, du$$

we find that it becomes

$$\int_0^\infty u^{-1}f(u)\, du \int_0^u e^{-pt}\, dt = p^{-1}\int_0^\infty u^{-1}(1 - e^{-pu})f(u)\, du.$$

Similarly, if we integrate both sides of the equation

$$\bar{f}(p) = \int_0^\infty f(u)e^{-pu}\, du$$

with respect to p from 0 to p we find that

$$\int_0^p \bar{f}(\zeta)\, d\zeta = \int_0^\infty u^{-1}(1 - e^{-pu})f(u)\, du$$

and hence that we have the result

$$\mathscr{L}\left[\int_t^\infty u^{-1}f(u)\, du;p\right] = p^{-1}\int_0^p \bar{f}(\zeta)\, d\zeta. \tag{3-5-8}$$

Adding (3-5-7) and (3-5-8) we find that

$$\mathscr{L}\left[\int_0^\infty u^{-1}f(u)\, du;p\right] = p^{-1}\int_0^\infty \bar{f}(\zeta)\, d\zeta$$

which is equivalent to the relation

$$\int_0^\infty u^{-1}f(u)\, du = \int_0^\infty \bar{f}(\zeta)\, d\zeta. \tag{3-5-9}$$

To illustrate the use of equation (3-5-8), we put $f(u) = \cos u$, $\bar{f}(\zeta) = \zeta(1 + \zeta^2)^{-1}$ to find

$$\mathscr{L}[\mathrm{Ci}(t);p] = -\frac{1}{2p}\log(1 + p^2) \tag{3-5-10}$$

where $\mathrm{Ci}(t)$ denotes the cosine integral defined by the equation

$$\mathrm{Ci}(t) = -\int_t^\infty \frac{\cos u}{u}\,du. \tag{3-5-11}$$

As an example of the use of equation (3-5-9) we choose $f(u) = \sin u$, $\bar{f}(\zeta) = (1 + \zeta^2)^{-1}$ in which case (3-5-9) gives the well-known result

$$\int_0^\infty \frac{\sin u}{u}\,du = \tfrac{1}{2}\pi.$$

Finally we shall consider a group of formulae by means of which it is possible to evaluate the Laplace transforms of definite integrals involving certain functions in terms of the Laplace transforms of the functions themselves. We suppose that

$$\mathscr{L}[k(t,u);t \to p] = a(p)e^{-ub(p)} \tag{3-5-12}$$

then, multiplying both sides of this equation by $f(u)$ and integrating with respect to u from 0 to ∞, we find that

$$\mathscr{L}\left[\int_0^\infty k(t,u)f(u)\,du;p\right] = a(p)\int_0^\infty f(u)e^{-ub(p)}\,du.$$

The integral on the right-hand side of this equation is the Laplace transform of the function $f(t)$ with argument $b(p)$ so that we have the equation

$$\mathscr{L}\left[\int_0^\infty k(t,u)f(u)\,du;p\right] = a(p)\bar{f}\{b(p)\} \tag{3-5-13}$$

where, as usual, $\bar{f}(p) = \mathscr{L}[f(t);p]$.

For instance, if we take

$$k(t,u) = (\pi t)^{-\frac{1}{2}}e^{-u^2/4t}$$

it follows from (3-2-13) that

$$\mathscr{L}[k(t,u);t \to p] = p^{-\frac{1}{2}}e^{-u\sqrt{p}},$$

so that $a(p) = p^{-\frac{1}{2}}$, $b(p) = p^{\frac{1}{2}}$, and we deduce from (3-5-13) that

$$\mathscr{L}\left[(\pi t)^{-\frac{1}{2}}\int_0^\infty f(u)e^{-u^2/4t}\,du;p\right] = p^{-\frac{1}{2}}\bar{f}(p^{\frac{1}{2}}) \tag{3-5-14}$$

where $\bar{f}(p)$ is the Laplace transform of the function $f(t)$. Similarly, if we take

$$k(t,u) = (4\pi t^3)^{-\frac{1}{2}} u e^{-u^2/4t}$$

we see from equation (3-2-14) that, in this instance, $a(p) = 1$, $b(p) = p^{\frac{1}{2}}$, and that equation (3-5-13) leads to

$$\mathscr{L}\left[(4\pi t^3)^{-\frac{1}{2}} \int_0^\infty u f(u) e^{-u^2/4t}\, du; p\right] = \bar{f}(p^{\frac{1}{2}}). \tag{3-5-15}$$

3-6 THE ERROR FUNCTION

In many applications of mathematical techniques to the solution of physical problems, we encounter the function Erf(z) defined by the equation

$$\operatorname{Erf}(z) = \frac{2}{\sqrt{\pi}} \int_0^z e^{-t^2}\, dt. \tag{3-6-1}$$

Because of its occurrence in the analysis of errors of observations it is called the *error function*. It is also useful to have at our disposal its 'complement' the function Erfc(z) defined by the equation

$$\operatorname{Erfc}(z) = 1 - \operatorname{Erf}(z). \tag{3-6-2}$$

Since

$$\frac{2}{\sqrt{\pi}} \int_0^\infty e^{-t^2}\, dt = 1$$

this definition can be written in the alternative form

$$\operatorname{Erfc}(z) = \frac{2}{\sqrt{\pi}} \int_z^\infty e^{-t^2}\, dt. \tag{3-6-3}$$

By simple changes of variable in these results we can derive the formulae

$$\int_0^x e^{-au^2}\, du = \tfrac{1}{2}(\pi/a)^{\frac{1}{2}} \operatorname{Erf}(xa^{\frac{1}{2}}) \tag{3-6-4}$$

$$\int_x^\infty e^{-au^2}\, du = \tfrac{1}{2}(\pi/a)^{\frac{1}{2}} \operatorname{Erfc}(xa^{\frac{1}{2}}). \tag{3-6-5}$$

The relation of the functions Erf(x) and Erfc(x) to the Gaussian distribution $2\pi^{-\frac{1}{2}}e^{-t^2}$ is shown in Fig. 3-4 and the variation of these functions with x is shown in Figs. 3-5 and 3-6 respectively.

From the fundamental relation (3-6-2) we see that

$$\mathscr{L}[f(t)\operatorname{Erfc}\{g(t)\};p] = \mathscr{L}[f(t);p] - \mathscr{L}[f(t)\operatorname{Erf}\{g(t)\};p]. \tag{3-6-6}$$

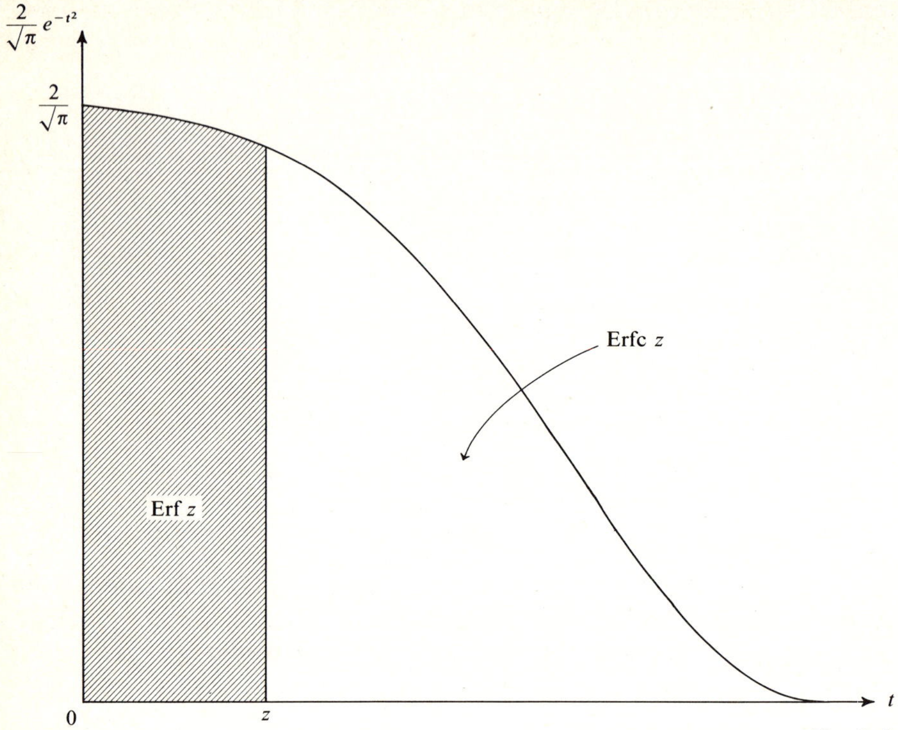

Fig. 3-4

In particular

$$\mathscr{L}[\mathrm{Erfc}\{g(t)\};p] = \frac{1}{p} - \mathscr{L}[\mathrm{Erf}\{g(t)\};p].$$

Since

$$\tfrac{1}{4}t^2 + pt = (\tfrac{1}{2}t + p)^2 - p^2$$

we may write

$$\mathscr{L}[e^{-\frac{1}{4}t^2};p] = e^{p^2} \int_0^\infty e^{-(p+\frac{1}{2}t)^2}\,dt.$$

Changing the variable of integration from t to u where $u = p + \frac{1}{2}t$, we find that the integral on the right takes the value

$$2\int_p^\infty e^{-u^2}\,du = \pi^{\frac{1}{2}}\mathrm{Erfc}(p)$$

so that

$$\mathscr{L}[e^{-\frac{1}{4}t^2};p] = \pi^{\frac{1}{2}} e^{p^2}\mathrm{Erfc}(p). \tag{3-6-7}$$

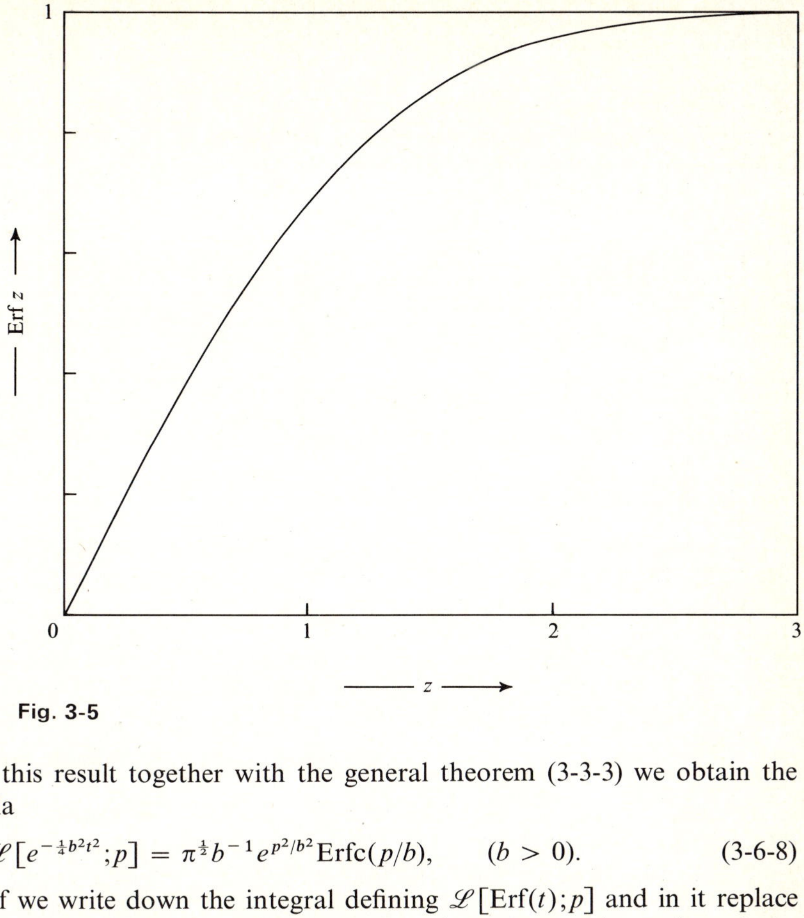

Fig. 3-5

Using this result together with the general theorem (3-3-3) we obtain the formula

$$\mathscr{L}[e^{-\frac{1}{4}b^2t^2};p] = \pi^{\frac{1}{2}}b^{-1}e^{p^2/b^2}\operatorname{Erfc}(p/b), \qquad (b > 0). \tag{3-6-8}$$

If we write down the integral defining $\mathscr{L}[\operatorname{Erf}(t);p]$ and in it replace $\operatorname{Erf}(t)$ by its definition (3-6-1) and then interchange the order of the integrations (Cf. Fig. 3-7) we find that

$$\mathscr{L}[\operatorname{Erf} t;p] = \int_0^\infty e^{-pt}\,dt\,\frac{2}{\sqrt{\pi}}\int_0^t e^{-u^2}\,du = \frac{2}{\sqrt{\pi}}\int_0^\infty e^{-u^2}\,du\int_u^\infty e^{-pt}\,dt$$

$$= \frac{2}{p\sqrt{\pi}}\int_0^\infty e^{-u^2-pu}\,du.$$

Now

$$u^2 + pu = (u + \tfrac{1}{2}p)^2 - \tfrac{1}{4}p^2$$

so that the integral on the right of this last equation is

$$e^{\frac{1}{4}p^2}\int_0^\infty e^{-(u+\frac{1}{2}p)^2}\,du = e^{\frac{1}{4}p^2}\int_{\frac{1}{2}p}^\infty e^{-u^2}\,du$$

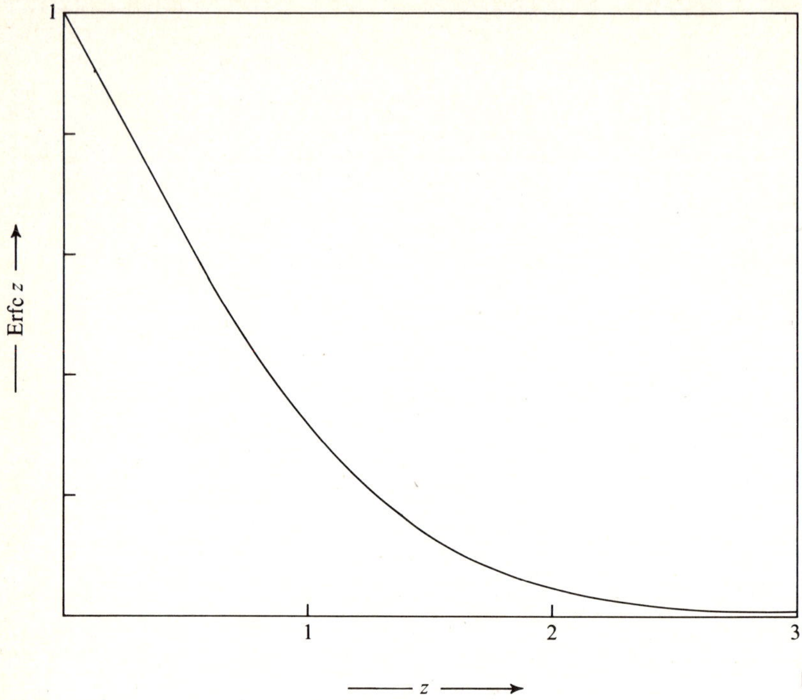

Fig. 3-6

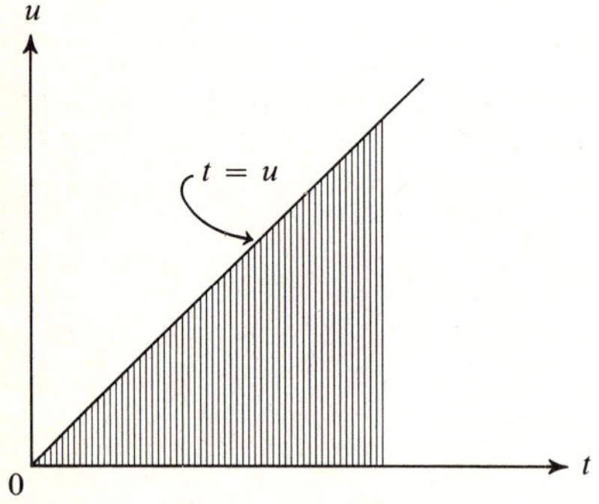

Fig. 3-7

and so we have shown that

$$\mathscr{L}[\mathrm{Erf}\,t;p] = p^{-1}e^{\frac{1}{4}p^2}\mathrm{Erfc}(\tfrac{1}{2}p). \tag{3-6-9}$$

Applying the theorem (3-3-3) to this result we find that

$$\mathscr{L}[\mathrm{Erf}(bt);p] = p^{-1}e^{\frac{1}{4}p^2/b^2}\mathrm{Erfc}(\tfrac{1}{2}p/b)$$

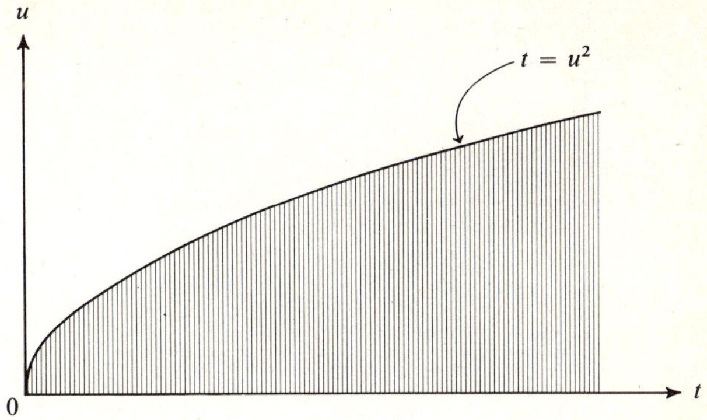

Fig. 3-8

or, replacing b by $1/(2a)$, that

$$\mathscr{L}\left[\operatorname{Erf}\left(\frac{t}{2a}\right);p\right] = p^{-1}e^{a^2p^2}\operatorname{Erfc}(ap). \tag{3-6-10}$$

From equation (3-6-6) we then deduce the complementary equation

$$\mathscr{L}\left[\operatorname{Erfc}\left(\frac{t}{2a}\right);p\right] = \frac{1}{p}\left\{1 - e^{a^2p^2}\operatorname{Erfc}(ap)\right\}. \tag{3-6-11}$$

In a similar way (except that, now, the field of integration is that shown in Fig. 3-8) we see that

$$\mathscr{L}[\operatorname{Erf}(t^{\frac{1}{2}});p] = \int_0^\infty e^{-pt}\,dt\,\frac{2}{\sqrt{\pi}}\int_0^{\sqrt{t}} e^{-u^2}\,du$$

$$= \frac{2}{\sqrt{\pi}}\int_0^\infty e^{-u^2}\,du\int_{u^2}^\infty e^{-pt}\,dt = \frac{2}{p\sqrt{\pi}}\int_0^\infty e^{-(1+p)u^2}\,du$$

from which we deduce immediately that

$$\mathscr{L}[\operatorname{Erf}(t^{\frac{1}{2}});p] = p^{-1}(p+1)^{-\frac{1}{2}}.$$

Applying the theorem (3-3-3) to this result we find that

$$\mathscr{L}[\operatorname{Erf}(b^{\frac{1}{2}}t^{\frac{1}{2}});p] = b^{\frac{1}{2}}p^{-1}(p+b)^{-\frac{1}{2}} \tag{3-6-12}$$

and then making use of the rule (3-3-1) that

$$\mathscr{L}[e^{bt}\operatorname{Erf}(b^{\frac{1}{2}}t^{\frac{1}{2}});p] = b^{\frac{1}{2}}p^{-\frac{1}{2}}(p-b)^{-1}. \tag{3-6-13}$$

These results can be written in the alternative forms

$$\mathscr{L}^{-1}[p^{-1}(p+b)^{-\frac{1}{2}};t] = b^{-\frac{1}{2}}\operatorname{Erf}(b^{\frac{1}{2}}t^{\frac{1}{2}}) \tag{3-6-14}$$

$$\mathscr{L}^{-1}[p^{-\frac{1}{2}}(p-b)^{-1};t] = b^{-\frac{1}{2}}e^{bt}\operatorname{Erf}(b^{\frac{1}{2}}t^{\frac{1}{2}}). \tag{3-6-15}$$

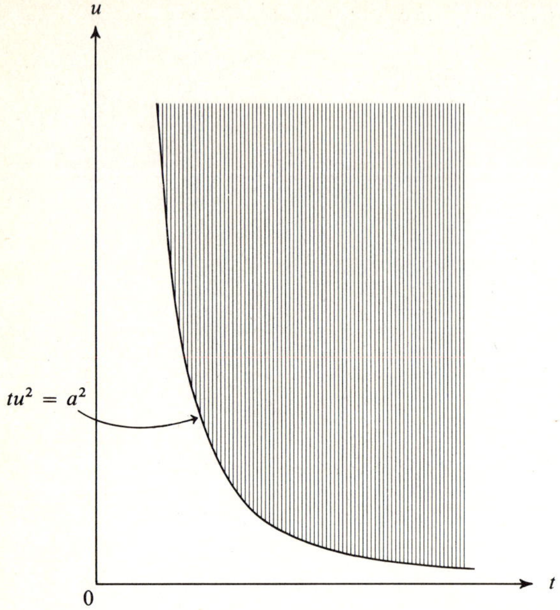

Fig. 3-9

From equations (3-6-13) and (3-6-6) we deduce that

$$\mathscr{L}[e^{bt}\operatorname{Erfc}(b^{\frac{1}{2}}t^{\frac{1}{2}});p] = (p-b)^{-1}(1-bp^{-\frac{1}{2}}) = p^{-\frac{1}{2}}(p^{\frac{1}{2}}+b^{\frac{1}{2}})^{-1}. \tag{3-6-16}$$

By converting the repeated integral into a double integral over the region shaded in Fig. 3-9 and then carrying out the integrations in the reverse order we find that

$$\begin{aligned}\mathscr{L}[\operatorname{Erfc}(at^{-\frac{1}{2}});p] &= 2\pi^{-\frac{1}{2}}\int_0^\infty e^{-pt}\,dt\int_{at^{-\frac{1}{2}}}^\infty e^{-u^2}\,du\\ &= 2\pi^{-\frac{1}{2}}\int_0^\infty e^{-u^2}\,du\int_{a^2u^{-2}}^\infty e^{-pt}\,dt\\ &= 2p^{-1}\pi^{-\frac{1}{2}}\int_0^\infty e^{-u^2-pa^2u^{-2}}\,du.\end{aligned}$$

Evaluating this last integral by means of equation (3-2-12) we obtain the formula

$$\mathscr{L}[\operatorname{Erfc}(at^{-\frac{1}{2}});p] = p^{-1}e^{-2a\sqrt{p}}. \tag{3-6-17}$$

3-7 LAPLACE TRANSFORMS OF BESSEL FUNCTIONS

We now consider the Laplace transforms of some functions involving the Bessel function of the first kind of order v defined by the equation

$$J_v(t) = \sum_{r=0}^{\infty} \frac{(-1)^r(\frac{1}{2}t)^{v+2r}}{r!\Gamma(r+v+1)}. \qquad (3\text{-}7\text{-}1)$$

It follows directly from this definition that

$$\mathscr{L}[t^v J_v(t);p] = \sum_{r=0}^{\infty} \frac{(-1)^r 2^{-v-2r}}{r!\Gamma(r+v+1)} \mathscr{L}[t^{2v+2r};p].$$

Using the result (3-2-3) and then the duplication formula for the gamma function, we have the relation

$$\frac{1}{\Gamma(r+v+1)} \mathscr{L}[t^{2v+2r};p] = \frac{\Gamma(2v+2r+1)}{\Gamma(r+v+1)} p^{-2v-2r-1}$$

$$= \pi^{-\frac{1}{2}} 2^{2v+2r} \Gamma(v+r+\tfrac{1}{2}) p^{-2v-2r-1}$$

($\operatorname{Re} p > 0$, $v > -\frac{1}{2}$), so we deduce that

$$\mathscr{L}[t^v J_v(t);p] = 2^v \pi^{-\frac{1}{2}} p^{-2v-1} \sum_{r=0}^{\infty} \frac{\Gamma(v+\frac{1}{2}+r)}{r!} (-p^{-2})^r.$$

If we make use of the binomial theorem

$$(1-x)^{-\alpha} = \sum_{r=0}^{\infty} \frac{\Gamma(\alpha+r)}{r!\Gamma(\alpha)} x^r, \qquad |x| < 1$$

we find that if $v > -\frac{1}{2}$, $|p| > 1$, the series on the right converges to

$$\Gamma(v+\tfrac{1}{2})(1+p^{-2})^{-v-\frac{1}{2}}$$

so that we obtain the formula

$$\mathscr{L}[t^v J_v(t);p] = \frac{2^v \Gamma(v+\frac{1}{2})}{\Gamma(\frac{1}{2})} (p^2+1)^{-v-\frac{1}{2}}, \qquad \operatorname{Re} p > 1. \qquad (3\text{-}7\text{-}1)$$

In particular

$$\mathscr{L}[J_0(t);p] = (p^2+1)^{-\frac{1}{2}}, \qquad \operatorname{Re} p > 1. \qquad (3\text{-}7\text{-}2)$$

Using the theorem (3-3-3) we therefore have that, if $\operatorname{Re} p > a > 0$,

$$\mathscr{L}[t^\nu \mathrm{J}_\nu(at);p] = \frac{(2a)^\nu \Gamma(\nu + \frac{1}{2})}{\sqrt{\pi}(p^2 + a^2)^{\nu+\frac{1}{2}}}, \qquad \nu > -\tfrac{1}{2} \tag{3-7-3}$$

and that

$$\mathscr{L}[\mathrm{J}_0(at);p] = (p^2 + a^2)^{-\frac{1}{2}}. \tag{3-7-4}$$

The formula (3-7-3) has, as yet, been proved only when $\operatorname{Re} p > a > 0$, but as long as $\operatorname{Re}(p + ia) > 0$, $\operatorname{Re}(p - ia) > 0$ both sides of (3-7-3) are analytic functions of a so that, by the principle of analytic continuation, the result is true for this wider range of values of a. In particular if a is real, it will be true for $\operatorname{Re} p > 0$.

The result can be established in another way. For simplicity we shall illustrate the method by establishing (3-7-4) for positive real values of a. From Poisson's integral

$$\mathrm{J}_0(at) = \frac{2}{\pi}\int_0^{\frac{1}{2}\pi} \cos(at\cos\theta)\, d\theta$$

we have

$$\mathscr{L}[\mathrm{J}_0(at);p] = \frac{2}{\pi}\int_0^{\frac{1}{2}\pi} d\theta\, \mathscr{L}[\cos(at\cos\theta);p] = \frac{2}{\pi}\int_0^{\frac{1}{2}\pi} \frac{p\,d\theta}{p^2 + a^2\cos^2\theta}.$$

The integral can be written

$$\frac{2p}{\pi}\int_0^{\frac{1}{2}\pi} \frac{\sec^2\theta\, d\theta}{p^2 + a^2 + p^2\tan^2\theta} = \frac{2p}{\pi}\int_0^{\infty} \frac{dx}{p^2 + a^2 + p^2x^2}$$

which is easily shown to have the value $(p^2 + a^2)^{-\frac{1}{2}}$.

In a similar way we have the result

$$\mathscr{L}[t^{\nu+1}\mathrm{J}_\nu(at);p] = \sum_{r=0}^{\infty} \frac{(-1)^r 2^{-\nu-2r}\Gamma(2\nu + 2r + 2)}{r!\,\Gamma(\nu + r + 1)p^{2\nu+2r+2}}.$$

Using the duplication formula for the gamma function we can show that

$$\frac{2^{-\nu-2r}\Gamma(2\nu + 2r + 2)}{\Gamma(\nu + r + 1)} = 2^{\nu+1}\pi^{-\frac{1}{2}}\Gamma(\nu + r + \tfrac{3}{2}),$$

so that if $\operatorname{Re} p > 1$,

$$\begin{aligned}\mathscr{L}[t^{\nu+1}\mathrm{J}_\nu(t);p] &= 2^{\nu+1}\pi^{-\frac{1}{2}}p^{-2\nu-2}\sum_{r=0} \frac{(-p^{-2})^r\Gamma(\nu + r + \frac{3}{2})}{r!} \\ &= 2^{\nu+1}\pi^{-\frac{1}{2}}p^{-2\nu-2}(1 + p^{-2})^{-\nu-\frac{3}{2}}\Gamma(\nu + \tfrac{3}{2})\end{aligned}$$

showing that

$$\mathscr{L}[t^{\nu+1}J_\nu(t);p] = \frac{2^{\nu+1}p\Gamma(\nu+\frac{3}{2})}{\sqrt{\pi}(p^2+1)^{\nu+\frac{3}{2}}}, \qquad (\text{Re}\,p > 1). \qquad (3\text{-}7\text{-}5)$$

Using the general result (3-3-3) we find that

$$\mathscr{L}[t^{\nu+1}J_\nu(at);p] = \frac{2^{\nu+1}a^\nu p\Gamma(\nu+\frac{3}{2})}{\sqrt{\pi}(p^2+a^2)^{\nu+\frac{3}{2}}}, \qquad (\text{Re}\,p > a > 0) \qquad (3\text{-}7\text{-}6)$$

and in particular that

$$\mathscr{L}[tJ_0(at);p] = p(p^2+a^2)^{-\frac{3}{2}}, \qquad (\text{Re}\,p > a > 0). \qquad (3\text{-}7\text{-}7)$$

Again the principle of analytic continuation can be used to extend the results to a more extensive range of values of a and p than is permissible by the argument we have used.

From the definition of the Bessel function of the first kind we have the relation

$$t^{\frac{1}{2}\nu}J_\nu(2t^{\frac{1}{2}}) = \sum_{r=0}^{\infty} \frac{(-1)^r t^{\nu+r}}{r!\,\Gamma(\nu+r+1)}$$

so that, if $\text{Re}\,p > 0$,

$$\mathscr{L}[t^{\frac{1}{2}\nu}J_\nu(2t^{\frac{1}{2}});p] = \sum_{r=0}^{\infty} \frac{(-1)^r}{r!\,\Gamma(r+\nu+1)} \mathscr{L}[t^{\nu+r};p]$$

$$= p^{-\nu-1} \sum_{r=0}^{\infty} \frac{(-1/p)^r}{r!}$$

from which we deduce that, if $\text{Re}\,p > 0$,

$$\mathscr{L}[t^{\frac{1}{2}\nu}J_\nu(2t^{\frac{1}{2}});p] = p^{-\nu-1}e^{-1/p}$$

and hence that if $a > 0$, $\text{Re}\,p > 0$,

$$\mathscr{L}[t^{\frac{1}{2}\nu}J_\nu(2\sqrt{at});p] = a^{\frac{1}{2}\nu}p^{-\nu-1}e^{-a/p}. \qquad (3\text{-}7\text{-}8)$$

In particular

$$\mathscr{L}[J_0(2\sqrt{at});p] = \frac{1}{p}e^{-a/p}, \qquad (a > 0,\ \text{Re}\,p > 0). \qquad (3\text{-}7\text{-}9)$$

Similar results hold for the modified Bessel function of the first kind defined by the equation

$$\mathrm{I}_\nu(t) = \sum_{r=0}^{\infty} \frac{(\frac{1}{2}t)^{\nu+2r}}{r!\Gamma(r+\nu+1)}.$$

For instance, by exactly the same procedure as before

$$\mathscr{L}[t^\nu \mathrm{I}_\nu(t);p] = \pi^{-\frac{1}{2}} 2^\nu p^{-2\nu-1} \sum_{r=0}^{\infty} \frac{\Gamma(\nu + r + \frac{1}{2})}{r!} p^{-2r}$$

showing that if $\mathrm{Re}\,p > 1$, $\nu > -\frac{1}{2}$.

$$\mathscr{L}[t^\nu \mathrm{I}_\nu(t);p] = \pi^{-\frac{1}{2}} 2^\nu \Gamma(\nu + \tfrac{1}{2})(p^2 - 1)^{-\nu-\frac{1}{2}}. \tag{3-7-10}$$

Hence, using (3-3-3) we have that if $\nu > -\frac{1}{2}$

$$\mathscr{L}[t^\nu \mathrm{I}_\nu(at);p] = \frac{(2a)^\nu \Gamma(\nu + \frac{1}{2})}{\sqrt{\pi}(p^2 - a^2)^{\nu+\frac{1}{2}}}, \qquad (\mathrm{Re}\,p > a > 0). \tag{3-7-11}$$

In a similar way we can show that if $\nu > -\frac{1}{2}$

$$\mathscr{L}[t^{\nu+1} \mathrm{I}_\nu(at);p] = \frac{2^{\nu+1} a^\nu p \Gamma(\nu + \frac{3}{2})}{\sqrt{\pi}(p^2 - a^2)^{\nu+\frac{3}{2}}}, \qquad (\mathrm{Re}\,p > a > 0). \tag{3-7-12}$$

If we repeat the argument leading to (3-7-8) we obtain the result

$$\mathscr{L}[t^{\frac{1}{2}\nu} \mathrm{I}_\nu(2\sqrt{at});p] = a^{\frac{1}{2}\nu} p^{-\nu-1} e^{a/p}, \qquad (\mathrm{Re}\,p > 0, a > 0, \nu > -\tfrac{1}{2}). \tag{3-7-13}$$

The special cases corresponding to $\nu = 0$ are of special interest; they are

$$\mathscr{L}[\mathrm{I}_0(at);p] = (p^2 - a^2)^{-\frac{1}{2}}, \qquad (\mathrm{Re}\,p > a > 0), \tag{3-7-14}$$

$$\mathscr{L}[t\mathrm{I}_0(at);p] = p(p^2 - a^2)^{-\frac{3}{2}}, \qquad (\mathrm{Re}\,p > a > 0), \tag{3-7-15}$$

$$\mathscr{L}[\mathrm{I}_0(2\sqrt{at});p] = p^{-1} e^{a/p}, \qquad (\mathrm{Re}\,p > 0; a > 0). \tag{3-7-16}$$

We can derive an interesting integral theorem from equation (3-7-8). From the definition of the Laplace transform

$$\mathscr{L}\left[t^{\frac{1}{2}\nu} \int_0^\infty x^{-\frac{1}{2}\nu} \mathrm{J}_\nu(2\sqrt{xt}) f(x)\,dx;p\right]$$

$$= \int_0^\infty x^{-\frac{1}{2}\nu} f(x) \mathscr{L}[t^{\frac{1}{2}\nu} \mathrm{J}_\nu(2\sqrt{xt});p]\,dx$$

$$= \int_0^\infty x^{-\frac{1}{2}\nu} f(x) p^{-\nu-1} x^{\frac{1}{2}\nu} e^{-x/p}\,dx$$

from equation (3-7-8). Hence we have proved that

$$\mathscr{L}\left[t^{\frac{1}{2}\nu}\int_0^\infty x^{-\frac{1}{2}\nu}J_\nu(2\sqrt{xt})f(x)\,dx;p\right] = p^{-\nu-1}\bar{f}(p^{-1}) \tag{3-7-17}$$

where $\bar{f}(p)$ denotes the Laplace transform of $f(t)$. This result is known as *Tricomi's theorem* (Cf. Tricomi (1935)).

From this result we deduce immediately that if

$$p^{\nu+1}\bar{f}(p) = \bar{f}(p^{-1}) \tag{3-7-18}$$

then

$$\int_0^\infty x^{-\frac{1}{2}\nu}J_\nu(2\sqrt{xt})f(x)\,dx = t^{-\frac{1}{2}\nu}f(t)$$

Hence if $f(t)$ satisfies the equation (3-7-18) and if

$$g(t) = t^{-\nu}f(\tfrac{1}{2}t^2) \tag{3-7-19}$$

then

$$\int_0^\infty xg(x)J_\nu(xt)\,dx = g(t).$$

Recalling the definition of the Hankel transform (Cf. sec. 2-14 above) we see that if the function $g(t)$ is defined by the pair of equations (3-7-18), (3-7-19), then

$$\mathscr{H}_\nu[g(x);t] = g(t)$$

i.e., the function $g(t)$ is self-reciprocal under the Hankel transform of order ν.

We shall now derive some Laplace transforms involving the generalized Bessel function $J_\lambda{}^\mu(z)$ introduced by E. M. Wright (1935) and defined by the equation

$$J_\lambda{}^\mu(z) = \sum_{r=0}^\infty \frac{(-z)^r}{r!\Gamma(1+\lambda+\mu r)}, \qquad (\mu > 0); \tag{3-7-21}$$

it is related to the Bessel function of the first kind through the equation

$$J_\nu(2z^{\frac{1}{2}}) = z^{\frac{1}{2}\nu}J_\nu{}^1(z). \tag{3-7-22}$$

If $\operatorname{Re} p > a > 0$,

$$\mathscr{L}[t^\lambda e^{at}J_\lambda{}^\mu(t^\mu x);p] = \int_0^\infty t^\lambda e^{at}J_\lambda{}^\mu(t^\mu x)e^{-pt}\,dt$$

$$= \sum_{r=0}^\infty \frac{(-x)^r}{r!\Gamma(1+\lambda+\mu r)}\mathscr{L}[t^{\lambda+\mu r};p-a]$$

$$= \sum_{r=0}^{\infty} \frac{(-x)^r}{r!\Gamma(1+\lambda+\mu r)} \frac{\Gamma(\lambda+\mu r+1)}{(p-a)^{\lambda+\mu r+1}}$$

$$= (p-a)^{-\lambda-1} \exp\{-x(p-a)^{-\mu}\}. \qquad (3\text{-}7\text{-}23)$$

From this we deduce that

$$\mathscr{L}\left[t^{\lambda}e^{at}\int_0^{\infty} e^{-bx}J_{\lambda}^{\mu}(t^{\mu}x)f(x)\,dx;p\right]$$

$$= \int_0^{\infty} e^{-bx}f(x)\mathscr{L}[t^{\lambda}e^{at}J_{\lambda}^{\mu}(t^{\mu}x);p]\,dx$$

$$= (p-a)^{-\lambda-1}\int_0^{\infty} f(x)\exp\{-[b+(p-a)^{-\mu}]x\}\,dx$$

showing that, if $\bar{f}(p)$ is the Laplace transform of the function $f(t)$,

$$\mathscr{L}\left[t^{\lambda}e^{at}\int_0^{\infty} e^{-bx}J_{\lambda}^{\mu}(t^{\mu}x)f(x)\,dx;p\right] = (p-a)^{-\lambda-1}\bar{f}\{b+(p-a)^{-\mu}\} \qquad (3\text{-}7\text{-}24)$$

a result which is due to Kumar (1952).

Putting $a = b$, $\lambda = \nu$, $\mu = 1$ in this result and making use of (3-7-22) we see that

$$\mathscr{L}\left[t^{\frac{1}{2}\nu}e^{at}\int_0^{\infty} e^{-ax}x^{-\frac{1}{2}\nu}J_{\nu}(2\sqrt{xt})f(x)\,dx;p\right]$$

$$= (p-a)^{-\nu-1}\bar{f}\{a+(p-a)^{-1}\}. \qquad (3\text{-}7\text{-}25)$$

3-8 PERIODIC FUNCTIONS

We now consider the Laplace transform of a function $f(t)$ which is periodic with period τ and which is bounded so that its Laplace transform exists for $\operatorname{Re} p > 0$. By splitting up the infinite range of integration in the integral defining the Laplace transform we have for a function of this kind

$$\mathscr{L}[f(t);p] = \sum_{r=0}^{\infty}\int_{r\tau}^{(r+1)\tau} f(t)e^{-pt}\,dt$$

$$= \sum_{r=0}^{\infty}\int_0^{\tau} f(u+r\tau)e^{-p(u+r\tau)}\,du$$

$$= \sum_{r=0}^{\infty} e^{-pr\tau}\int_0^{\tau} f(u)e^{-pu}\,du,$$

since, because of the periodic nature of the function f,

$$f(u+r\tau) = f(u), \quad (r = 1,2,\ldots).$$

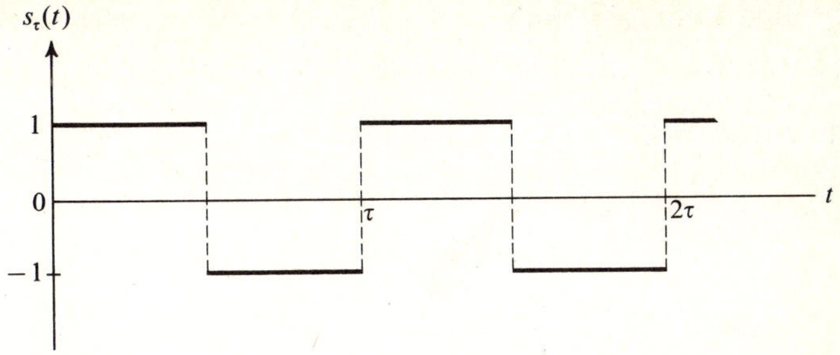

Fig. 3-10

Now

$$\sum_{r=0}^{\infty} e^{-pr\tau} = (1 - e^{-p\tau})^{-1}$$

so that for a periodic function we have the formula

$$\mathscr{L}[f(t);p] = (1 - e^{-p\tau})^{-1} \int_0^\tau f(u)e^{-pu}\,du. \qquad (3\text{-}8\text{-}1)$$

In particular we have the relation

$$\mathscr{L}[f(t);p] = (1 - e^{-2p\pi})^{-1} \int_0^{2\pi} f(u)e^{-pu}\,du, \qquad (3\text{-}8\text{-}2)$$

for a function of period 2π.

For example, if

$$f(t) = |\sin t|,$$

we take $\tau = \pi$ in the formula (3-8-1) and use the fact that

$$(1 + p^2) \int_0^\pi \sin te^{-pt}\,dt = 1 + e^{-p\pi}$$

to derive the result

$$\mathscr{L}[|\sin t|;p] = (1 + p^2)^{-1} \coth(\tfrac{1}{2}p\pi), \qquad (\text{Re}\,p > 0). \qquad (3\text{-}8\text{-}3)$$

Similarly, if we define the function $S_\tau(t)$ by the equations

$$S_\tau(t) = \text{H}(t) - 2\text{H}(t - \tfrac{1}{2}\tau), \qquad 0 < t < \tau$$
$$S_\tau(t + r\tau) = S_\tau(t), \qquad (r = 1,2, \ldots)$$

(Cf. Fig. 3-10) and use the result

$$\int_0^\tau S_\tau(t)e^{-pt}\,dt = \int_0^{\frac{1}{2}\tau} e^{-pt}\,dt - \int_{\frac{1}{2}\tau}^{\tau} e^{-pt}\,dt = \frac{1}{p}(1 - 2e^{-\frac{1}{2}p\tau} + e^{-p\tau})$$
$$= \frac{1}{p}(1 - e^{-\frac{1}{2}p\tau})^2,$$

we find from (3-8-1) that

$$\mathscr{L}[S_\tau(t);p] = p^{-1}\tanh(\tfrac{1}{2}p\tau), \qquad (\text{Re}\,p > 0). \tag{3-8-4}$$

In particular

$$\mathscr{L}[S_{2\pi}(t);p] = p^{-1}\tanh(p\pi), \qquad (\text{Re}\,p > 0). \tag{3-8-5}$$

If $f(t)$ is such that it has a Fourier series

$$f(t) = \tfrac{1}{2}a_0 + \sum_{n=1}^{\infty}\{a_n\cos(2n\pi t/\tau) + b_n\sin(2n\pi t/\tau)\}$$

its Fourier coefficients are given by the equations

$$a_n = \frac{2}{\tau}\int_0^\tau f(u)\cos(2n\pi u/\tau)\,du, \qquad b_n = \frac{2}{\tau}\int_0^\tau f(u)\sin(2n\pi u/\tau)\,du. \tag{3-8-6}$$

Also

$$\mathscr{L}[f(t);p] = \tfrac{1}{2}a_0\mathscr{L}[1;p] + \sum_{n=1}^{\infty}\{a_n\mathscr{L}[\cos(2n\pi t/\tau);p] + b_n\mathscr{L}[\sin(2n\pi t/\tau);p]\},$$

so that if we make use of equations (3-3-4) and (3-3-5) we find that

$$\mathscr{L}[f(t);p] = \frac{1}{2p}a_0 + \tau\sum_{n=1}^{\infty}\frac{p\tau a_n + 2\pi n b_n}{p^2\tau^2 + 4\pi^2 n^2}. \tag{3-8-7}$$

In particular, if $\tau = 2\pi$ we obtain the relation

$$\mathscr{L}[f(t);p] = \frac{1}{2p}a_0 + \sum_{n=1}^{\infty}\frac{pa_n + nb_n}{p^2 + n^2} \tag{3-8-8}$$

for the Laplace transform of a periodic function $f(t)$ of period 2π with Fourier coefficients

$$a_n = \frac{1}{\pi}\int_0^{2\pi} f(u)\cos(nu)\,du, \qquad b_n = \frac{1}{\pi}\int_0^{2\pi} f(u)\sin(nu)\,du. \tag{3-8-9}$$

We now have two different expressions—(3-8-1) and (3-8-7)— for the Laplace transform of a periodic function of period τ. If we equate these two expressions we obtain the identity

$$\frac{1}{2p}a_0 + \tau\sum_{n=1}\frac{p\tau a_n + 2\pi n b_n}{p^2\tau^2 + 4\pi^2 n^2} = \frac{1}{1 - e^{-p\tau}}\int_0^\tau f(u)e^{-pu}\,du \tag{3-8-10}$$

where the constants a_n, b_n are given by equations (3-8-6). In the special case in which $\tau = 2\pi$ this reduces to the simpler form

$$\frac{1}{2p}a_0 + \sum_{n=1}^{\infty} \frac{pa_n + nb_n}{p^2 + n^2} = \frac{1}{1 - e^{-2\pi p}} \int_0^{2\pi} f(u)e^{-pu}\, du \qquad \text{(3-8-11)}$$

where, now, a_n and b_n are given by equations (3-8-9).

To illustrate the use of these formulae we consider the case in which the function $f(t)$ is of period 2π and

$$f(t) = \begin{cases} 1 & 0 < t < \pi \\ 0 & \pi < t < 2\pi. \end{cases}$$

It follows from equations (3-8-9) that for this function

$$a_0 = 1, \qquad a_n = 0, (n = 1,2, \ldots)$$

and

$$b_{2r} = 0, \qquad b_{2r-1} = (r - \tfrac{1}{2})^{-1}\pi, (r = 1,2, \ldots).$$

Also we have

$$\int_0^{2\pi} f(u)e^{-pu}\, du = \int_0^{\pi} e^{-pu}\, du = p^{-1}(1 - e^{-p\pi}).$$

Substituting these values into equation (3-8-11) we find that

$$\frac{1}{2} + \frac{2p}{\pi} \sum_{r=1}^{\infty} \frac{1}{p^2 + (2r - 1)^2} = \frac{1}{1 + e^{-\pi p}} \qquad \text{(3-8-12)}$$

a result which is known as the *Mittag–Leffler identity*.

3-9 THE CONVOLUTION OF TWO FUNCTIONS

We now introduce the idea of the convolution $(f{*}g)$ of two functions f and g which is similar to the convolution $f \circ g$ introduced in sec. 2-9, but more appropriate to the Laplace transform. We define $f{*}g$ by the equation

$$f{*}g(t) = \int_0^t f(t - \tau)g(\tau)\, d\tau. \qquad \text{(3-9-1)}$$

Using the theorem

$$\int_0^t \psi(\tau)\, d\tau = \int_0^t \psi(t - \tau)\, d\tau$$

of elementary calculus, we see that the integral on the right-hand side of

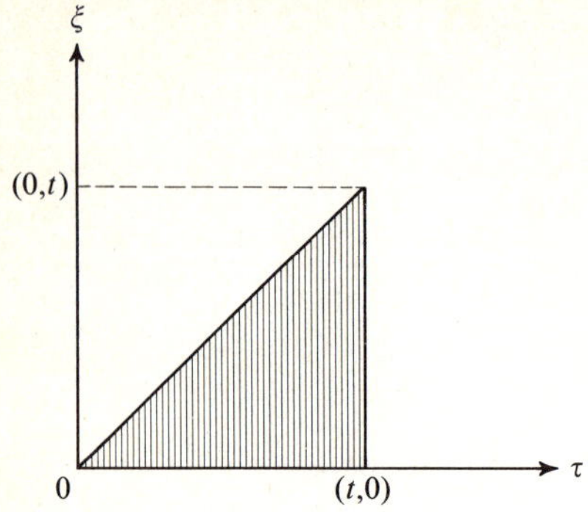

Fig. 3-11

(3-9-1) can be written as

$$\int_0^t f(\tau)g(t-\tau)\,d\tau$$

from which it follows that

$$f*g = g*f. \tag{3-9-2}$$

It is obvious from the definition that if λ is a real number

$$(\lambda f)*g = f*(\lambda g) = \lambda(f*g) \tag{3-9-3}$$

and also that the distributive law

$$f*(g+h) = f*g + f*h \tag{3-9-4}$$

holds. It is a little bit more difficult to prove that the associative law

$$f*(g*h) = (f*g)*h \tag{3-9-5}$$

is valid. From the definition we have that

$$(f*g)*h = \int_0^t h(t-\tau)\,d\tau \int_0^\tau f(\xi)g(\tau-\xi)\,d\xi.$$

Changing the order of the integrations (Cf. Fig. 3-11) we find that the right-hand side of this last equation can be written as

$$\int_0^t f(\xi)\,d\xi \int_\xi^t g(\tau-\xi)h(t-\tau)\,d\tau.$$

If we change the variable of integration in the inner integral from τ to η where $\tau - \xi = \eta$ we see that the double integral is

$$\int_0^t f(\xi)\,d\xi \int_0^{t-\xi} g(\eta)h(t-\xi-\eta)\,d\eta$$

which is simply $f*(g*h)$.

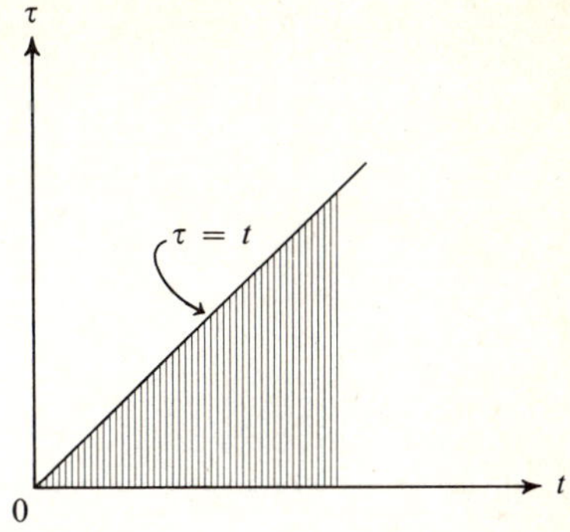

Fig. 3-12

We now calculate the Laplace transform of the convolution $f*g$ in terms of the Laplace transforms $\bar{f}(p)$, $\bar{g}(p)$ of the functions $f(t)$, $g(t)$. By the definitions of the Laplace transform and of $f*g$ we find that

$$\mathscr{L}[f*g;p] = \int_0^\infty e^{-pt}\,dt \int_0^t f(\tau)g(t-\tau)\,d\tau.$$

Changing the order of the integrations (cf. Fig. 3-12) and then changing the variable of integration in the inner integral from t to $\eta = t - \tau$ we find that

$$\mathscr{L}[f*g;p] = \int_0^\infty f(\tau)\,d\tau \int_\tau^\infty e^{-pt}g(t-\tau)\,dt$$
$$= \int_0^\infty f(\tau)e^{-p\tau}\,d\tau \int_0^\infty e^{-p\eta}g(\eta)\,d\eta$$

showing that

$$\mathscr{L}[f*g;p] = \bar{f}(p)\bar{g}(p). \tag{3-9-6}$$

This may also be written in the inverse form

$$\mathscr{L}^{-1}[\bar{f}(p)\bar{g}(p);t] = (f*g)(t). \tag{3-9-7}$$

In particular, if $g \equiv f$ we have the equations

$$\mathscr{L}[f*f;p] = \{\bar{f}(p)\}^2 \tag{3-9-8}$$
$$\mathscr{L}^{-1}[\{\bar{f}(p)\}^2;t] = f*f(t)$$

To illustrate the use of these formulae we first of all consider the pair of functions

$$f(t) = t^{a-1},\ (a > 0), \qquad g(t) = t^{b-1},\ (b > 0)$$

which have Laplace transforms

$$\bar{f}(p) = \Gamma(a)p^{-a}, \qquad \bar{g}(p) = \Gamma(b)p^{-b}.$$

Inserting these forms into (3-9-7) we find that if $a > 0$, $b > 0$,

$$\int_0^t \tau^{a-1}(t-\tau)^{b-1}\,d\tau = \Gamma(a)\Gamma(b)\mathscr{L}^{-1}[p^{-a-b};t] = \frac{\Gamma(a)\Gamma(b)}{\Gamma(a+b)}\,t^{a+b}.$$

If we now let $t = 1$ we find that

$$\int_0^1 \tau^{a-1}(1-\tau)^{b-1}\,d\tau = \frac{\Gamma(a)\Gamma(b)}{\Gamma(a+b)}, \qquad (a > 0, b > 0). \tag{3-9-10}$$

Denoting the integral on the left by the beta function $\mathrm{B}(a,b)$ this is the well-known relation between the beta and the gamma function.

To illustrate the use of equation (3-9-9) we take

$$f(t) = \mathrm{J}_0(t), \qquad \bar{f}(p) = (1+p^2)^{-\frac{1}{2}},$$

the form of $\bar{f}(p)$ being derived from (3-7-2). From (3-9-9) we see that

$$\int_0^t \mathrm{J}_0(\tau)\mathrm{J}_0(t-\tau)\,d\tau = \mathscr{L}^{-1}[(1+p^2)^{-1};t]$$

so that, by using (3-2-4), we find that

$$\int_0^t \mathrm{J}_0(\tau)\mathrm{J}_0(t-\tau)\,d\tau = \sin t. \tag{3-9-11}$$

The definition (3-9-1) can be written in an alternative way in the special case in which $f(0) = 0$. When this happens we can write

$$f(t-x_2) = \int_0^{t-x_2} f'(x_1)\,dx_1$$

and write the right-hand side of (3-9-1) as the repeated integral

$$\int_0^t g(x_2)\,dx_2 \int_0^{t-x_2} f'(x_1)\,dx_1$$

which can in turn be represented as the double integral

$$\iint_{T_2} f'(x_1)g(x_2)\,dx_1 dx_2$$

where T_2 is the region $x_1 + x_2 \leqslant t$, $x_1 \geqslant 0$, $x_2 \geqslant 0$, it being assumed that $t > 0$ (Cf. Fig. 3-13). Hence if we replace the functions $f'(x_1)$ and $g(x_2)$ by $f_1(x_1)$ and $f_2(x_2)$, and hence $\bar{f}(p)$ and $\bar{g}(p)$ by $p^{-1}\bar{f}_1(p)$ and $\bar{f}_2(p)$, we see that the convolution theorem (3-9-6) can be written in the alternative form

$$\mathscr{L}\left[\iint_{T_2} f_1(x_1)f_2(x_2)\,dx_1 dx_2; t \to p\right] = p^{-1}\bar{f}_1(p)\bar{f}_2(p) \tag{3-9-12}$$

Applying the convolution theorem (3-9-6) twice in succession we find

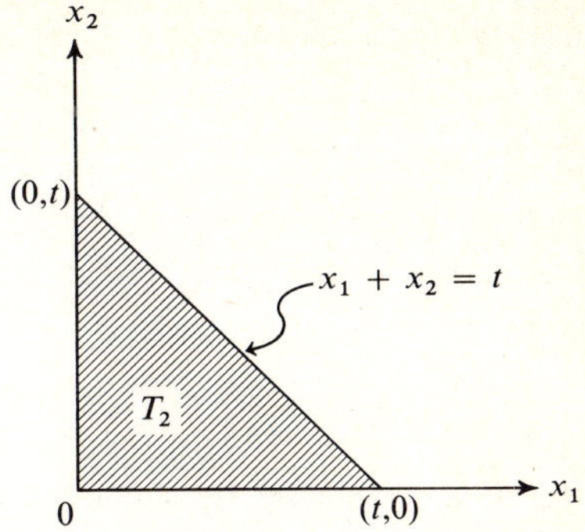

Fig. 3-13

that

$$\mathscr{L}[f_1*(f_2*g);p] = \bar{f}_1(p)\mathscr{L}[f_2*g;p] = \bar{f}_1(p)\bar{f}_2(p)\bar{g}(p). \tag{3-9-13}$$

Now if $g(0) = 0$,

$$f_2*g(t - x_1) = \int_0^{t-x_1} f_2(x_2)g(t - x_1 - x_2)\, dx_2$$
$$= \int_0^{t-x_1} f_2(x_2)\, dx_2 \int_0^{t-x_1-x_2} f_3(x_3)\, dx_3$$

where we have written $f_3(x_3) = g'(x_3)$, so that

$$f_1*(f_2*g) = \int_0^t f_1(x_1)\, dx_1 \int_0^{t-x_1} f_2(x_2)\, dx_2 \int_0^{t-x_1-x_2} f_3(x_3)\, dx_3$$

and it is easily shown that the integral on the right can be written

$$\iiint_{T_3} f_1(x_1)f_2(x_2)f_3(x_3)\, dx_1dx_2dx_3$$

where T_3 is the tetrahedron $x_1 + x_2 + x_3 \leqslant t$, $x_1 \geqslant 0$, $x_2 \geqslant 0$, $x_3 \geqslant 0$ with $t > 0$ (cf. Fig. 3-14). Replacing $g(p)$ by $p^{-1}\bar{f}_3(p)$ in (3-9-13) we find that that equation can be written in the form

$$\left[\iiint_{T_3} f_1(x_1)f_2(x_2)f_3(x_3)\, dx_1dx_2dx_3; t \to p\right] = p^{-1}\bar{f}_1(p)\bar{f}_2(p)\bar{f}_3(p). \tag{3-9-14}$$

This result can be generalized in an obvious way. If T_n is the domain defined by the inequations

$$x_1 + x_2 + \cdots + x_n \leqslant t, \qquad x_i \geqslant 0, (i = 1,2, \ldots, n), (t > 0)$$

then

$$\mathscr{L}\left[\int \cdots \int_{T_n} f_1(x_1) \ldots f_n(x_n)\, dx_1 \ldots dx_n; t \to p\right]$$
$$= p^{-1}\bar{f}_1(p) \ldots \bar{f}_n(p). \tag{3-9-15}$$

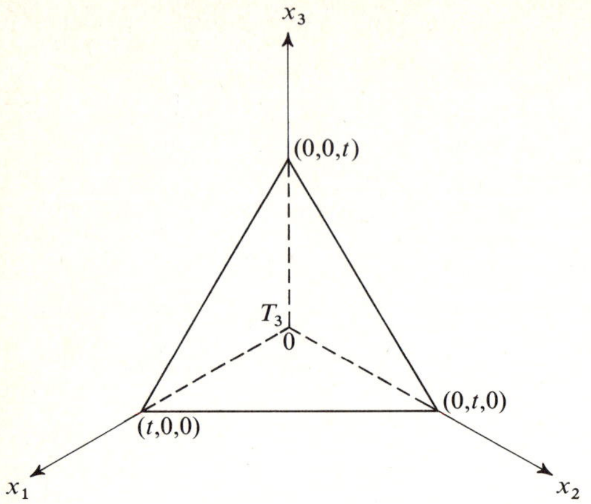

Fig. 3-14

3-10 THE INVERSION FORMULA FOR THE LAPLACE TRANSFORM

In the preceding sections we have derived certain simple results concerning the Laplace transform $\bar{f}(p)$ of a function $f(x)$. Almost all of these results give information about the transform $\bar{f}(p)$ when the function $f(x)$ is prescribed. In this section we shall consider the converse problem—that of deriving information about the original function $f(x)$ when we have some knowledge of its Laplace transform. When $\bar{f}(p)$ is prescribed any formula enabling us to derive the form of the original function $f(x)$ is called an *inversion formula* for the Laplace transform. In this section we shall consider the simplest type of inversion formula; in sec. 3-12 we shall discuss other types of inversion formula.

We saw in sec. 3-1—by a heuristic argument—that $f(x)$ would appear to be given in terms of the transform $\bar{f}(p)$ by the complex integral

$$f(x) = \frac{1}{2\pi i} \int_{c-i\infty}^{c+i\infty} \bar{f}(p) e^{px} \, dp. \tag{3-10-1}$$

An *outline* of the proof of this formula is as follows: If $f(x)$ has a continuous derivative, and if $f(x) = 0(e^{\gamma x})$ for large positive values of x, where $\gamma > 0$, then the Laplace integral

$$\bar{f}(p) = \int_0^\infty f(u) e^{-pu} \, du, \qquad (p = \sigma + i\tau)$$

converges absolutely and uniformly with respect to σ and τ in the half-plane $\mathrm{Re}\, p > \gamma$ and $\bar{f}(p)$ is an analytic function of p in the same half-plane.

Substituting the Laplace integral expression for $\bar{f}(p)$ into the integral

$$\frac{1}{2\pi i}\int_{c-iw}^{c+iw} \bar{f}(p)e^{px}\,dp, \qquad (c > \gamma) \tag{3-10-2}$$

we find that the integral has the value

$$\frac{1}{2\pi i}\int_{c-iw}^{c+iw} e^{px}\,dp \int_0^\infty f(u)e^{-pu}\,du.$$

Because of the conditions on the function $f(x)$ we may invert the order of the two integrations to obtain the expression

$$\frac{1}{2\pi i}\int_0^\infty f(u)\,du \int_{c-iw}^{c+iw} e^{p(x-u)}\,dp.$$

Carrying out the inner integration

$$\frac{1}{2i}\int_{c-iw}^{c+iw} e^{p(x-u)}\,dp = \frac{\sin\{w(x-u)\}}{x-u}\,e^{(x-u)c}$$

we have the expression

$$\frac{1}{\pi}\int_0^\infty f(u)e^{(x-u)c}\,\frac{\sin\{w(x-u)\}}{x-u}\,du$$

for the integral (3-10-2). By a simple change of variable we may rewrite this integral in the form

$$\frac{1}{\pi}\int_{-x}^\infty g(t)\,\frac{\sin(wt)}{t}\,dt = \frac{1}{\pi}\int_0^\infty g(t)\,\frac{\sin(wt)}{t}\,dt + \frac{1}{\pi}\int_0^\infty g(-t)\,\frac{\sin(wt)}{t}\,dt$$

where $g(t)$ denotes the function $e^{-ct}f(x+t)$. Making use of Dirichlet's integral formula (Corollary 1 to the localization lemma of sec. 2-2) we see that, provided $c > \gamma$, the integrals on the right-hand side of this equation reduce to $\frac{1}{2}f(x+)$ and $\frac{1}{2}f(x-)$ respectively. Since $f(x)$ has been assumed to be a continuous function of x, each of these terms is equal to $f(x)$ and their sum is consequently equal to $f(x)$. We therefore have the inversion formula

$$\frac{1}{2\pi i}\lim_{w\to\infty}\int_{c-iw}^{c+iw} \bar{f}(p)e^{px}\,dp = f(x). \tag{3-10-3}$$

The following theorem gives conditions on the function $\bar{f}(p)$ sufficient to ensure the validity of the inversion theorem (3-10-3). The conditions imposed on the function $f(x)$ as a result of those on $\bar{f}(p)$ should also be noted.

Theorem 3 *If $\bar{f}(p)$ is an analytic function of the complex variable p and is of order $0(p^{-k})$ in some half-plane* $\operatorname{Re} p \geqslant \gamma$ *where γ and k are real*

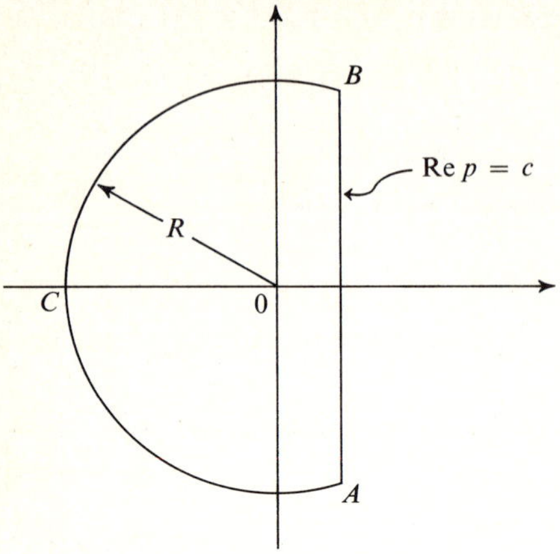

Fig. 3-15

constants and $k > 1$, *then the integral*

$$\frac{1}{2\pi i} \lim_{w\to\infty} \int_{c-iw}^{c+iw} e^{px}\bar{f}(p)\, dp$$

along any line $\operatorname{Re} p = c > \gamma$ *converges to a function* $f(x)$ *which is independent of* c *and whose Laplace transform is* $\bar{f}(p)$, $\operatorname{Re} p > \gamma$.

Furthermore, the function $f(x)$ *is continuous for each* $x \geqslant 0$ *and is* $0(e^{cx})$ *as* $x \to \infty$.

3-11 THE CALCULATION OF INVERSE TRANSFORMS

The inversion theorem for the Laplace transform can be used for the calculation of inverse transforms. If $\bar{f}(p)$ is analytic in the complex p-plane except for a finite number of isolated singularities at $p_1, p_2, \ldots, p_n$, we evaluate

$$\mathscr{L}^{-1}[\bar{f}(p); t]$$

by integrating the function

$$\bar{g}(p) = \bar{f}(p)e^{pt}$$

along the path $ABCA$ shown in Fig. 3-15. In this path AB is the line $\operatorname{Re} p = c$, where c has the interpretation given to it in Theorem 3, so that AB lies to the *right* of all the singularities of $\bar{f}(p)$, and the radius R of the circular arc BCA is chosen so large that all the singularities of $\bar{f}(p)$ are enclosed by the contour. By the inversion theorem

$$\mathscr{L}^{-1}[\bar{f}(p); t] = \lim_{R\to\infty} \frac{1}{2\pi i} \int_{AB} \bar{g}(p)\, dp.$$

To evaluate

$$\lim_{R\to\infty} \frac{1}{2\pi i}\int_{BCA} \bar{g}(p)\,dp \tag{3-11-2}$$

we make use of Lemma 2 of sec. 2-7. By Cauchy's theorem

$$\frac{1}{2\pi i}\int_{AB} \bar{g}(p)\,dp + \frac{1}{2\pi i}\int_{BCA} \bar{g}(p)\,dp = \sum_{r=1}^{n} \operatorname{res} \bar{g}(p_r) \tag{3-11-3}$$

so that in cases in which the limit (3-11-2) has the value zero, we see by letting $R \to \infty$ in equation (3-11-3) and using (3-11-1) that

$$\mathscr{L}^{-1}[\bar{f}(p);t] = \sum_{r=1}^{n} \operatorname{res} \bar{g}(p_r). \tag{3-11-4}$$

The result (3-11-4) may be extended to the case in which $\bar{f}(p)$ has an infinite number of singularities. We choose a radius R_n for BCA such that $R_n \neq |p_n|$ and $R_n \to \infty$ as $n \to \infty$. If the limit (3-11-2) is again zero and the series on the right-hand side of equation (3-11-4) converges as $n \to \infty$, we find that

$$\mathscr{L}^{-1}[\bar{f}(p);t] = \sum_{r=1}^{\infty} \operatorname{res} \bar{g}(p_r). \tag{3-11-5}$$

As a simple illustration of the use of the formula (3-11-4) we shall calculate $\mathscr{L}^{-1}[(p^2 + \omega^2)^{-1};t]$ where ω is real and positive. The function

$$\bar{g}(p) = (p^2 + \omega^2)^{-1}e^{pt}, \qquad (t > 0)$$

has simple poles at $p = \pm i\omega$ with residues $(\pm 2i\omega)^{-1}e^{\pm i\omega t}$ respectively. Since

$$\lim_{|p|\to\infty} p\bar{g}(p) = 0$$

if Re $p < 0$, it follows that the integral round the circular arc tends to zero as $R \to \infty$, so that equation (3-11-4) gives

$$\mathscr{L}^{-1}[\bar{f}(p);t] = \frac{1}{2i\omega}(e^{i\omega t} - e^{-i\omega t}) = \frac{\sin(\omega t)}{\omega}.$$

To show the use of the method in a more complicated situation we shall determine the function whose Laplace transform is

$$\bar{f}(p) = \frac{I_0\{\rho(p/\kappa)^{\frac{1}{2}}\}}{I_0\{a(p/\kappa)^{\frac{1}{2}}\}}, \qquad 0 < \rho < a.$$

This function has poles at $p_r = -\kappa\xi_r^2$, $(r = 1,2, \ldots)$, where $\xi_1, \xi_2, \ldots$ are the positive zeros of $J_0(a\xi)$. From the theory of Bessel functions we know that all these zeros are real and simple. If we take the radius of the circular arc ABC to be $(n + \frac{1}{2})^2\kappa\pi^2/a^2$, there will be no poles of the function $\bar{f}(p)$ on this arc, and from the asymptotic expansion of the modified Bessel function we can readily show that if $0 < \rho < a$, the integral round the circular arc tends to zero as $n \to \infty$.

Writing

$$\bar{g}(p) = \frac{I_0\{\rho(p/\kappa)^{\frac{1}{2}}\}}{I_0\{a(p/\kappa)^{\frac{1}{2}}\}} e^{pt}$$

we see that

$$\operatorname{res} \bar{g}(p_r) = \frac{I_0\{\rho(p_r/\kappa)^{\frac{1}{2}}\}}{[d/dp\, I_0\{a(p/\kappa)^{\frac{1}{2}}\}]_{p=p_r}} e^{p_r t}$$

Since $I_0'(x) = I_1(x)$, we have the relation

$$\left[\frac{d}{dp} I_0\{a(p/\kappa)^{\frac{1}{2}}\}\right]_{p=p_r} = \tfrac{1}{2}a(\kappa p_r)^{-\frac{1}{2}} I_1\{a(p_r/\kappa)^{\frac{1}{2}}\}$$

and using the relation between the modified Bessel function I_1 with imaginary argument and the Bessel function J_1 of the first kind we see that the right-hand side of this equation is equal to

$$\frac{a}{2\kappa\xi_r} J_1(a\xi_r).$$

We therefore have the relation

$$\operatorname{res} \bar{g}(p_r) = \frac{2\kappa}{a} \frac{\xi_r J_0(\rho\xi_r)}{J_1(a\xi_r)} e^{-\kappa\xi_r^2 t}$$

and hence from (3-11-5) that

$$\mathscr{L}^{-1}\left[\frac{I_0\{\rho(p/\kappa)^{\frac{1}{2}}\}}{I_0\{a(p/\kappa)^{\frac{1}{2}}\}}; p \to t\right] = \frac{2\kappa}{a} \sum_{r=1}^{\infty} \frac{\xi_r J_0(\rho\xi_r)}{J_1(a\xi_r)} e^{-\kappa\xi_r^2 t} \tag{3-11-6}$$

the sum being taken over the positive zeros of the function $J_0(a\xi)$.

The method has to be modified if the function $\bar{f}(p)$ has branch points. This is done by deforming the contour of Fig. 3-15 to exclude the branch point, and then—by using Lemma 1 of sec. 2-7—to evaluate the integral along the small path encircling the branch point.

To illustrate the method we shall calculate the inverse Laplace transform of the function

$$\bar{f}(p) = e^{-a\sqrt{p}}.$$

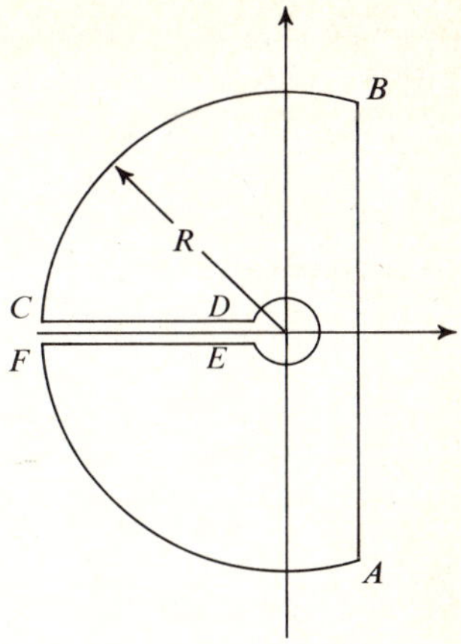

Fig. 3-16

This function has a branch point at the origin $p = 0$ so we consider the integral of the function

$$\bar{g}(p) = e^{pt - a\sqrt{p}}$$

round the contour shown in Fig. 3-16, in which the arcs BC, CA are arcs of a circle with centre O and radius R and the small circle surrounding the origin has radius ε.

On CD we may write $p = x^2 e^{i\pi}$ where x is real and positive so that

$$\frac{1}{2\pi i}\int_{CD} \bar{g}(p)\, dp = \frac{1}{\pi i}\int_{\varepsilon}^{R} x e^{-x^2 t + iax}\, dx$$

and on EF we may write $p = x^2 e^{-i\pi}$ where x is real and positive so that

$$\frac{1}{2\pi i}\int_{EF} \bar{g}(p)\, dp = \frac{1}{\pi i}\int_{\varepsilon}^{R} x e^{-x^2 t + iax}\, dx$$

and hence as $R \to \infty$ and $\varepsilon \to 0$,

$$\frac{1}{2\pi i}\left(\int_{CD} + \int_{EF}\right)\bar{g}(p)\, dp \to -\frac{2}{\pi}\int_{0}^{\infty} x e^{-x^2 t}\sin(ax)\, dx$$

$$= -(2/\pi)^{\frac{1}{2}}\mathscr{F}_s[xe^{-x^2 t}; x \to a].$$

Since $p\bar{g}(p) \to 0$ as $|p| \to \infty$, the integrals round BC, FA tend to zero as $R \to \infty$, and since

$$\lim_{|p| \to 0} p\bar{g}(p) = 0$$

it follows that the integral of $\bar{g}(p)$ round the small circle enclosing the origin tends to zero as $\varepsilon \to 0$. Finally, we see that as $R \to \infty$ the integral along AB tends to $\mathscr{L}^{-1}[\bar{f}(p); t]$.

Since the function $\bar{g}(p)$ has no singularities within the contour, it follows that as $R \to \infty$, $\varepsilon \to 0$ the sum of these integrals tends to zero. In this way we find that

$$\mathscr{L}^{-1}[e^{-a\sqrt{p}};t] = (2/\pi)^{\frac{1}{2}}\mathscr{F}_s[xe^{-x^2t};x \to a]$$

and from equation (2-7-14) we deduce that

$$\mathscr{L}^{-1}[e^{-a\sqrt{p}};t] = a(4\pi t^3)^{-\frac{1}{2}}e^{-a^2/4t}. \tag{3-11-7}$$

It should be noted that, if we integrate both sides of this equation with respect to a from a to ∞, we obtain the formula

$$\mathscr{L}^{-1}[p^{-\frac{1}{2}}e^{-a\sqrt{p}};p \to t] = (\pi t)^{-\frac{1}{2}}e^{-a^2/4t}. \tag{3-11-8}$$

Repeating this process we obtain the formula

$$\mathscr{L}^{-1}[p^{-1}e^{-a\sqrt{p}};p \to t] = \text{Erfc}(\tfrac{1}{2}at^{-\frac{1}{2}}). \tag{3-11-9}$$

3-12 OTHER FORMS OF INVERSION FORMULA

Useful inversion formulae for the calculation of the original function $f(t)$ have been derived in the case in which its Laplace transform $\bar{f}(p)$ can be expressed in the form of an infinite series

$$\bar{f}(p) = \sum_{n=0}^{\infty} a_n\bar{\phi}_n(p) \tag{3-12-1}$$

where the a_n are constants and the functions $\phi_n(p)$ are respectively the Laplace transforms of the functions $\phi_n(t)$, $(n = 0,1,2, \ldots)$. Applying the operator $\mathscr{L}^{-1}$ to both sides of this equation and invoking Lerch's theorem we obtain the inversion formula

$$f(t) = \sum_{n=0}^{\infty} a_n\phi_n(t). \tag{3-12-2}$$

This method was introduced by Tricomi (1935) for the special case in which the Laplace transform can be expressed in the form

$$\bar{f}(p) = \frac{1}{p+b}\sum_{n=0}^{\infty} a_n\left(\frac{p-b}{p+b}\right)^n \tag{3-12-3}$$

with b a positive constant.

If we define the Laguerre polynomial[1] $L_n(t)$ by the equation

$$\text{L}_n(t) = \sum_{r=0}^{n}\binom{n}{r}\frac{(-t)^r}{r!}$$

[1] For the properties of these polynomials see Chapter V of Sneddon (1961*b*).

where $\binom{n}{r}$ denotes the binomial coefficient, then by equation (3-2-3) we deduce that

$$\mathscr{L}[\mathrm{L}_n(t);p] = \sum_{r=0}^{n} \binom{n}{r} (-1)^r p^{-r-1} = \frac{1}{p}\left(\frac{p-1}{p}\right)^n.$$

Making use of the rule of manipulation (3-3-6) we obtain the equation

$$\mathscr{L}[e^{-bt}\mathrm{L}_n(2bt);p] = \frac{1}{p+b}\left(\frac{p-b}{p+b}\right)^n.$$

If, therefore, we take

$$\phi_n(t) = e^{-bt}\mathrm{L}_n(2bt), \qquad \bar{\phi}_n(p) = \frac{1}{p+b}\left(\frac{p-b}{p+b}\right)^n$$

in equations (3-12-1) and (3-12-2) we find that the function

$$\bar{f}(p) = \frac{1}{p+b}\sum_{n=0}^{\infty} a_n \left(\frac{p-b}{p+b}\right)^n \tag{3-12-4}$$

is the Laplace transform of the function

$$f(t) = e^{-bt}\sum_{n=0}^{\infty} a_n \mathrm{L}_n(2bt) \tag{3-12-5}$$

Tricomi's method of determining $\mathscr{L}^{-1}[\bar{f}(p);t]$ consists of the following steps:

(*a*) Given the function $\bar{f}(p)$, we make the substitution

$$p = b\left(\frac{1+\sigma}{1-\sigma}\right).$$

(*b*) We manipulate the resulting function of σ to give the expansion

$$\bar{f}(p) = \frac{1-\sigma}{2b}\sum_{n=0}^{\infty} a_n \sigma^n.$$

(*c*) We interpret the original function $f(t) = \mathscr{L}^{-1}[\bar{f}(p);t]$ as the series

$$f(t) = \sum_{n=0}^{\infty} a_n B_n(bt)$$

where the functions $B_n(x)$ are defined by the equation

$$B_n(x) = e^{-x}\mathrm{L}_n(2x). \tag{3-12-6}$$

A special case considered by Ward (1954) is that in which

$$\bar{f}(p) = \frac{1}{(p+1)(p+2)}.$$

If we make the substitution

$$p = \frac{1+\sigma}{1-\sigma}$$

in this expression we find that

$$\bar{f}(p) = \frac{(1-\sigma)^2}{2(3-\sigma)} = \tfrac{1}{2}(1-\sigma)\cdot\tfrac{1}{3}(1-\tfrac{1}{3}\sigma)^{-1}(1-\sigma).$$

Since

$$(1-\sigma)(1-\tfrac{1}{3}\sigma)^{-1} = 1 - 2\sum_{n=1}^{\infty}\left(\frac{\sigma}{3}\right)^n$$

it follows that

$$\bar{f}(p) = \frac{1-\sigma}{2}\left[\frac{1}{3} - \frac{2}{3}\sum_{n=1}^{\infty}\left(\frac{\sigma}{3}\right)^n\right]$$

Table 3-1 The accuracy of Tricomi's method for $\bar{f}(p) = \{(p+1)(p+2)\}^{-1}$. The function $f_7(t)$ is the sum of the first seven terms of the series (3-12-7).

t	$f(t)$	$f_7(t)$
0.0	0.00000	0.00112
0.05	0.04639	0.04655
0.1	0.08611	0.08608
0.2	0.14841	0.14824
0.3	0.19201	0.19185
0.5	0.23865	0.23866
0.6	0.24762	0.24770
0.75	0.24924	0.24936
0.85	0.24473	0.24486
1.0	0.23254	0.23263
1.5	0.17334	0.17326
2.0	0.11702	0.11695
2.5	0.07535	0.07538
3.0	0.04731	0.04739
4.0	0.01798	0.01797

and hence that

$$f(t) = \tfrac{1}{3}B_0(t) - \tfrac{2}{3}\sum_{n=1}^{\infty} 3^{-n}B_n(t). \tag{3-12-7}$$

The function $f(t)$ is known exactly in this case. Since

$$\bar{f}(p) = (p+1)^{-1} - (p+2)^{-1}$$

it follows that

$$f(t) = e^{-t} - e^{-2t}.$$

An estimate of the accuracy of Tricomi's method in numerical calculations can be made by means of Table 3-1 which is taken from Ward (1954) and which compares $f(t)$ with $f_7(t)$ which is the sum of the first seven terms of the series on the right-hand side of equation (3-12-7). From this we notice that (provided numerical tables of the Laguerre functions are available) Tricomi's method gives a high degree of accuracy. This conclusion reached from an inspection of Table 3-1 is confirmed by a consideration of other numerical examples discussed by Ward; an interested reader is referred to Ward's paper.

A similar result can be derived involving Legendre polynomials. If $f(t)$ is defined at each point of the positive real line, then the function

$$e^{-(h-1)t}f(t), \qquad (h > 1)$$

may be expanded as a Fourier–Legendre series

$$\sum_{n=0}^{\infty} a_n \mathrm{P}_n(x), \qquad (x = 1 - 2e^{-t}) \tag{3-12-8}$$

where the coefficients are given by the formula

$$\begin{aligned} a_n &= (n + \tfrac{1}{2}) \int_{-1}^{1} f(t)e^{-(h-1)t}\mathrm{P}_n(x)\,dx \\ &= (2n+1)\int_0^{\infty} f(t)e^{-ht}\mathrm{P}_n(1 - 2e^{-t})\,dt. \end{aligned}$$

Making use of Murphy's formula[1]

$$\mathrm{P}_n(x) = \sum_{r=0}^{n} \frac{(-n)_r(n+1)_r}{(r!)^2}\left(\frac{1-x}{2}\right)^r$$

we see that

$$a_n = (2n+1)\sum_{r=0}^{n} \frac{(-n)_r(n+1)_r}{(r!)^2}\bar{f}(h+r). \tag{3-12-9}$$

[1] See, for example, sec. 15 of Sneddon (1961b).

Hence we have the inversion theorem

$$f(t) = e^{(h-1)t} \sum_{n=0}^{\infty} a_n \mathrm{P}_n(1 - 2e^{-t}) \tag{3-12-10}$$

with the coefficients a_n given by the formula (3-12-9). This formula is useful when the transform $\bar{f}$ is known at a discrete set of points.

An inversion formula of an entirely different kind has been established by Post (1930). We first define an operator $\mathrm{L}_{k,x}$ by the equation

$$\mathrm{L}_{k,x}\bar{f}(p) = (-1)^k \bar{f}^{(k)}(k/x)(k/x)^{k+1}. \tag{3-12-11}$$

For all positive values of x and for positive integral values of k, Post's inversion theorem states that if $f \in \mathrm{L}(0,R)$, for every $R > 0$, and if the integral defining $\bar{f}(p)$ converges for some p, then for $x > 0$,

$$f(x) = \lim_{k \to \infty} \mathrm{L}_{k,x}\bar{f}(p). \tag{3-12-12}$$

For the proof of this theorem the reader is referred to Post's original memoir or to Chapter VII of Widder (1941).

Post's inversion formula is only applicable if we have an explicit expression for the k-th derivative of $\bar{f}(p)$. It is analogous to the Maclaurin formula $a_n = f^{(n)}(0)/n!$ for the coefficients of the power series expansion $\sum a_n x^n$ of the function $f(x)$, and is of great significance in theoretical investigations on the Laplace transform.

3-13 TAUBERIAN THEOREMS

We saw in sec. 2-7 that the behavior of a function $f(t)$ in the neighborhood of the origin is, in some sense, reflected in the behavior of its Fourier transform $F(\xi)$ as $|\xi| \to \infty$. That there is some such relation between a function and its Laplace transform—at least for a restricted class of functions—is obvious from an examination of the behavior of the Laplace transform of the polynomial

$$f_n(t) = \sum_{r=0}^{n} c_r \frac{t^r}{r!}. \tag{3-13-1}$$

From equation (3-3-2) we see that

$$\bar{f}_n(p) = \mathscr{L}[f_n(t);p] = \sum_{r=0}^{n} c_r p^{-r-1}. \tag{3-13-2}$$

From these equations we deduce immediately that

$$\lim_{p \to \infty} p\bar{f}_n(p) = \lim_{t \to 0+} f_n(t) \tag{3-13-3}$$

(since they are both equal to the constant c_0), and that

$$\lim_{p\to 0} p^{n+1}\bar{f}_n(p) = n! \lim_{t\to\infty} t^{-n} f_n(t) \tag{3-13-4}$$

(since they are both equal to the constant c_n).

That the relation (3-13-3) holds for functions other than polynomials is obvious from an inspection of the transform pair

$$f(t) = \cos t, \qquad \bar{f}(p) = \frac{p}{p^2 + 1},$$

which obviously satisfy the relation

$$\lim_{p\to\infty} pf(p) = f(0+). \tag{3-13-5}$$

We notice also that the transform pair

$$f(t) = 1 - e^{-t}, \qquad \bar{f}(p) = \frac{p}{p(p+1)}$$

satisfy the relation

$$\lim_{p\to 0+} p\bar{f}(p) = f(\infty) \tag{3-13-6}$$

analogous to (3-13-4).

Sufficient conditions for the validity of equation (3-13-5) are given by:

Theorem 4 *If the function $f(t)$ satisfies the inequality $|f(t)| < Me^{\alpha t}$, for all $t > 0$, M being a positive constant,*

$$\lim_{p\to\infty} p\bar{f}(p) = f(0+).$$

To prove this theorem, we suppose that p is real and that $p > \alpha + \delta$, $\delta > 0$. Then by a simple change of variable in the defining integral for $\bar{f}(p)$, we find that

$$p\bar{f}(p) = \int_0^\infty f(x/p)e^{-x}\,dx$$

so that we may write

$$p\bar{f}(p) - f(0) = \int_0^\infty \{f(x/p) - f(0)\}e^{-x}\,dx$$

and then put it in the alternative form

$$p\bar{f}(p) - f(0) = I_1 + I_2,$$

where

$$I_1 = \int_0^R \{f(x/p) - f(0)\}e^{-x}\,dx, \qquad I_2 = \int_R^\infty \{f(x/p) - f(0)\}e^{-x}\,dx$$

with R a positive number whose value is not yet prescribed.

By the hypotheses on $f(t)$ and p we have that

$$|f(x/p)| < M\exp\left(-\frac{\alpha x}{\alpha+\delta}\right)$$

so that

$$\left|\int_R^\infty f(x/p)e^{-x}\,dx\right| < M\int_R^\infty \exp\left(-\frac{\alpha x}{\alpha+\delta}\right)dx$$
$$= \left(1+\frac{\alpha}{\delta}\right)M\exp\left(-\frac{\alpha R}{\alpha+\delta}\right)$$

from which we deduce that

$$\left|\int_R^\infty f(x/p)e^{-x}\,dx\right| < \left(1+\frac{\alpha}{\delta}\right)Me^{-R}$$

and hence that

$$|I_2| < \left\{\left(1+\frac{\alpha}{\delta}\right)M + |f(0)|\right\}e^{-R}.$$

From this inequality it is obvious that for any prescribed positive constant ε we can find R_1 such that

$$|I_2| < \tfrac{1}{2}\varepsilon, \qquad R > R_1.$$

Having fixed R_1 we now choose $p_1(\varepsilon)$ so large that

$$|f(x/p) - f(0)| < \tfrac{1}{2}\varepsilon, \qquad p > p_1, \qquad 0 < x < R_1$$

and hence that

$$|I_1| < \tfrac{1}{2}\varepsilon\int_0^R e^{-x}\,dx < \tfrac{1}{2}\varepsilon.$$

Consequently for $p > p_1(\varepsilon)$ we have

$$|p\bar{f}(p) - f(0)| < \varepsilon$$

which establishes the theorem.

Similarly, sufficient conditions for the validity of equation (3-13-6) are given by:

Theorem 5 *If the function $f(t)$ is bounded for all $t > 0$:*

$$|f(t)| < M,$$

and the limit

$$\lim_{t \to \infty} f(t) = f(\infty)$$

exists, then

$$\lim_{p \to 0+} p\bar{f}(p) = f(\infty).$$

To prove this result we note that we may write

$$p\bar{f}(p) - f(\infty) = I_1 + I_2$$

where

$$I_1 = \int_0^R \{f(x/p) - f(\infty)\}e^{-x}\,dx, \qquad I_2 = \int_R^\infty \{f(x/p) - f(\infty)\}e^{-x}\,dx.$$

By the hypotheses we have

$$|I_1| < 2M \int_0^R e^{-x}\,dx = 2M(1 - e^{-R})$$

so that we choose a value R_1 of R such that

$$|I_1| < \tfrac{1}{2}\varepsilon, \qquad R = R_1.$$

For this value of R choose p_1 to be such that

$$|f(x/p) - f(\infty)| < \tfrac{1}{2}\varepsilon, \qquad 0 < p < p_1, \qquad x > R_1$$

and hence

$$|I_2| < \tfrac{1}{2}\varepsilon \int_{R_1}^\infty e^{-x}\,dx < \tfrac{1}{2}\varepsilon.$$

From these inequalities it follows that for $0 < p < p_1$,

$$|p\bar{f}(p) - f(\infty)| < \varepsilon$$

thus establishing the theorem.

As a corollary to this theorem we deduce that if $f(t) \to 0$ as $t \to \infty$, then

$$\lim_{p \to 0+} p\bar{f}(p) = 0.$$

Similarly, if $f'(t)$ also tends to zero as $t \to \infty$ and it is known that $f(0)$ is finite, then

$$\lim_{p \to 0+} p\mathscr{L}[f'(t);p] = 0$$

which, because of the assumption on $f(0)$ is equivalent to

$$\lim_{p \to 0+} p^2 \bar{f}(p) = 0.$$

We can easily generalize this to the case in which $f^{(n)}(t)$ tends to zero as $t \to \infty$ and $f(0), f'(0), \ldots, f^{(n-1)}(0)$ are all finite. We then have

$$\lim_{p \to 0+} p^{n+1} \bar{f}(p) = 0. \tag{3-13-7}$$

This is the generalization of the simple result (3-13-4) established for polynomials.

By methods similar to those used in the proofs of these last two theorems we can derive asymptotic expansions for the Laplace transform $\bar{f}(p)$ of a function $f(t)$ which has a Maclaurin expansion in a neighborhood of $t = 0$.

It will be recalled that a series

$$\sum_{r=0}^{\infty} a_r x^{-r}$$

is said to be *asymptotic* to a function $\phi(x)$ if

$$\phi(x) - \sum_{r=0}^{n-1} a_r x^{-r} = O(x^{-n})$$

as $x \to \infty$ for every positive integer n. We write

$$\phi(x) \sim \sum_{r=0}^{\infty} a_r x^{-r}.$$

The series itself may be either convergent or divergent.

The basic result in the asymptotic expansion of the Laplace transform of a function is contained in:

Theorem 6 (Watson's lemma) *If the function $f(t)$ satisfies the inequality*

$$|f(t)| < Me^{\alpha t},$$

and if, in some neighborhood of $t = 0$, $f(t)$ has a Maclaurin expansion

$$\sum_{r=0}^{\infty} c_r t^r / r!$$

then the Laplace transform of $f(t)$ has the asymptotic expansion

$$\bar{f}(p) \sim \sum_{r=0}^{\infty} c_r p^{-r-1}$$

To prove this result we note that if $f(t)$ has a Maclaurin expansion in a neighborhood of the origin, then for τ sufficiently small we can write

$$f(t) = S_n(t) + R_n(t), \qquad 0 < t < \tau$$

where

$$S_n(t) = \sum_{r=0}^{n-1} c_r \frac{t^r}{r!}, \qquad c_r = f^{(r)}(0)$$

and there is a positive constant A such that

$$|R_n(t)| < At^n/n!$$

Then

$$\left|\int_0^\tau \{f(t) - S_n(t)\}e^{-pt}\,dt\right| < A\int_0^\tau \frac{t^n}{n!}e^{-pt}\,dt < \frac{A}{n!}\int_0^\infty t^n e^{-pt}\,dt$$
$$= Ap^{-n-1}.$$

Also by the hypothesis on $f(t)$

$$\left|\int_\tau^\infty f(t)e^{-pt}\,dt\right| \leqslant M\int_\tau^\infty e^{-(p-\alpha)t}\,dt = \frac{M}{p-\alpha}e^{-(p-\alpha)\tau}$$

and

$$\int_\tau^\infty S_n(t)e^{-pt}\,dt = e^{-p\tau}\int_0^\infty S_n(t+\tau)e^{-pt}\,dt.$$

Now

$$\int_0^\infty S_n(t+\tau)e^{-pt}\,dt = \sum_{r=0}^{n-1}\sum_{s=0}^{r} \frac{c_r}{s!(r-s)!}\tau^{r-s}\int_0^\infty t^s e^{-pt}\,dt$$

$$= \sum_{r=0}^{n-1}\sum_{s=0}^{r} \frac{c_r}{(r-s)!}\frac{\tau^{r-s}}{p^{s+1}}$$

so that

$$\int_\tau^\infty S_n(t)e^{-pt}\,dt = O(p^{-1}e^{-p}).$$

Hence, writing

$$\bar{f}(p) - \bar{S}_n(p) = \int_0^\tau \{f(t) - S_n(t)\}e^{-pt}\,dt + \int_\tau^\infty f(t)e^{-pt}\,dt$$
$$- \int_\tau^\infty S_n(t)e^{-pt}\,dt$$

we see that

$$\bar{f}(p) - \bar{S}_n(p) = A(p^{-n-1}) + O(p^{-1}e^{-p}).$$

Since

$$\bar{S}_n(p) = \sum_{r=0}^{n-1} c_r p^{-r-1}$$

the theorem follows.

As an illustration of the use of this theorem we shall derive the asymptotic expansion of $\operatorname{Erfc}(x)$.

If we take

$$f(t) = \operatorname{Erf}(t)$$

then it is easily shown that

$$S_n(t) = 2\pi^{-\frac{1}{2}} \sum_{r=0}^{n-1} \frac{(-1)^r}{r!} \cdot \frac{t^{2r+1}}{2r+1}$$

and hence that

$$\bar{S}_n(p) = 2\pi^{-\frac{1}{2}} \sum_{r=0}^{n-1} \frac{(-1)^r}{r!} \cdot \frac{(2r)!}{p^{2r+2}}.$$

Using the duplication formula for the gamma function we see that we can write this in the form

$$\bar{S}_n(p) = \frac{1}{\pi p} \sum_{r=0}^{n-1} \frac{(-1)^r \Gamma(r+\frac{1}{2})}{(\frac{1}{2}p)^{2r+1}}.$$

From equation (3-6-9) we see that, in this case,

$$\bar{f}(p) = p^{-1} e^{\frac{1}{4}p^2} \operatorname{Erfc}(\tfrac{1}{2}p).$$

Writing $p = 2x$, we obtain the asymptotic expansion

$$\operatorname{Erfc}(x) \sim \frac{e^{-x^2}}{x\sqrt{\pi}} \sum_{r=1}^{\infty} \frac{(-1)^{r-1}(\frac{1}{2})_r}{x^{2r}}. \tag{3-13-8}$$

3-14 ORDINARY DIFFERENTIAL EQUATIONS

In this section we shall illustrate briefly how the properties of the Laplace transform may be exploited to obtain the solution of certain simple types of ordinary differential equations.

3-14-1 INITIAL VALUE PROBLEM FOR A LINEAR EQUATION WITH CONSTANT COEFFICIENTS

We begin by considering the solution of the differential equation

$$\mathrm{L}y = f(x), \qquad x > 0 \tag{3-14-1}$$

in which L denotes the linear differential operator with constant coefficients defined by the equation

$$\mathrm{L}y = a_n \frac{d^n y}{dx^n} + a_{n-1} \frac{d^{n-1} y}{dx^{n-1}} + \cdots + a_1 \frac{dy}{dx} + a_0 y, \qquad (a_n \neq 0). \tag{3-14-2}$$

We shall assume that the coefficients $a_0, a_1, \ldots, a_n$ are all real.

First of all we consider the solution of the equation (3-14-1) satisfying the initial conditions

$$y^{(r)}(0) = 0, \qquad (r = 0,1, \ldots, n-1). \tag{3-14-3}$$

We see from equation (3-4-4) that for any function satisfying the initial conditions (3-14-3)

$$\mathscr{L}\left[\frac{d^r y}{dx^r}; p\right] = p^r \bar{y}(p), \qquad (r = 0,1, \ldots, n)$$

where $\bar{y}(p) = \mathscr{L}[y(x);p]$, so that taking the Laplace transform of both sides of equation (3-14-1) we find that

$$\Lambda(p)\bar{y}(p) = \bar{f}(p)$$

where $\Lambda(p)$ denotes the polynomial[1] defined by the equation

$$\Lambda(p) = \sum_{r=0}^{n} a_r p^r. \tag{3-14-4}$$

Solving this equation for $\bar{y}(p)$ and making use of the convolution theorem we see that the solution of this particular initial value problem may be written in the form

$$y(x) = \int_0^x f(u)G(x-u)\,du \tag{3-14-5}$$

where $G(x)$ is defined by the equation

$$G(x) = \mathscr{L}^{-1}\left[\frac{1}{\Lambda(p)}; p \to x\right]. \tag{3-14-6}$$

[1]The polynomial $\Lambda(p)$ is sometimes referred to as the *characteristic polynomial* of the linear differential operator L.

The function $G(x)$, defined by equations (3-14-6) and (3-14-4) is called the *Green's function* of the initial value problem of the equation (3-14-1). It can be thought of in a slightly different way. Suppose that $y = g(x)$ is the solution of the differential equation

$$\mathrm{L}g = H(x), \qquad g(0) = g'(0) = \cdots = g^{(n-1)}(0) \tag{3-14-7}$$

in which $H(x)$ denotes the Heaviside unit function, then its Laplace transform is given by the equation

$$\Lambda(p)\bar{g}(p) = p^{-1}$$

which shows that

$$\bar{G}(p) = \frac{1}{\Lambda(p)} = p\bar{g}(p)$$

and hence that

$$G(x) = g'(x). \tag{3-14-8}$$

For example, the solution of the equation

$$\frac{d^2y}{dx^2} + \omega^2 y = f(x), \qquad y(0) = y'(0) = 0 \tag{3-14-9}$$

is given by equation (3-14-5) with

$$G(x) = \mathscr{L}^{-1}[(p^2 + \omega^2)^{-1}; x] = \omega^{-1} \sin(\omega x)$$

that is, is given by the formula

$$y(x) = \frac{1}{\omega}\int_0^x f(t) \sin\{\omega(x - t)\}\, dt. \tag{3-14-10}$$

The same result is obtained by observing that the initial value problem

$$\frac{d^2g}{dx^2} + \omega^2 g = H(x), \qquad g(0) = g'(0) = 0$$

has the solution

$$g(x) = \mathscr{L}^{-1}[p^{-1}(p^2 + \omega^2)^{-1}; x].$$

Since

$$\frac{1}{p(p^2 + \omega^2)} = \frac{1}{\omega^2}\left[\frac{1}{p} - \frac{p}{p^2 + \omega^2}\right]$$

it follows that

$$g(x) = \omega^{-2}\{1 - \cos(\omega x)\},$$

and hence from (3-14-8) that

$$G(x) = \omega^{-1} \sin(\omega x)$$

as before.

We complete the solution of the differential equation (3-14-1) in the general case by considering the homogeneous n-th order equation

$$\mathrm{L}y = 0, \tag{3-14-11}$$

in which the operator L is again defined by (3-14-2), and the function $y(x)$ satisfies the initial conditions

$$y^{(m)}(0) = y_m, \qquad (m = 0,1, \ldots, n-1) \tag{3-14-12}$$

$y_0, y_1, \ldots, y_{n-1}$ denoting arbitrary constants. From (3-4-4) we find that

$$\mathscr{L}\left[\frac{d^r y}{dx^r}; p\right] = p^r \bar{y}(p) - \sum_{m=0}^{r-1} y_m p^{r-m-1}.$$

Hence we find that

$$\mathscr{L}[\mathrm{L}y; p] = \Lambda(p)\bar{y}(p) - \phi(p)$$

where $\phi(p)$ is defined by the equation

$$\phi(p) = \sum_{r=0}^{n} \sum_{m=0}^{r-1} y_m p^{r-m-1}$$

and is obviously a polynomial of degree $n - 1$ in p. The transform of equation (3-14-11) is therefore

$$\bar{y}(p) = \frac{\phi(p)}{\Lambda(p)}. \tag{3-14-13}$$

For example, if $\Lambda(p)$ has n simple zeros $p_1, p_2, \ldots, p_n$, so that

$$\Lambda(p) = a_0(p - p_1)(p - p_2) \ldots (p - p_n)$$

in which $p_1, p_2, \ldots, p_n$ are all distinct, then (3-14-13) can be rewritten in the form

$$\bar{y}(p) = \sum_{r=1}^{n} \frac{c_r}{p - p_r}$$

where $c_1, c_2, \ldots, c_n$ are real constants. In this case the solution of equation (3-14-11) is of the form

$$y(x) = \sum_{r=1}^{n} c_r e^{p_r x}$$

showing that to each simple factor $p - p_r$ in $\Lambda(p)$ there corresponds a term

$$c_r e^{p_r x}$$

in the general solution.

If there is a multiple factor $(p - p_r)^q$ in $\Lambda(p)$, then in the partial fraction expansion of $\bar{y}(p)$ we get terms of the form

$$\gamma_q(p - p_r)^{-q} + \gamma_{q-1}(p - p_r)^{-q+1} + \cdots + \gamma_1(p - p_r)^{-1}$$

which, in turn, lead to the terms

$$(\Gamma_1 + \Gamma_2 x + \cdots + \Gamma_q x^{q-1})e^{p_r x}$$

in the general solution. The constants γ_s are related to the Γ_s by the equation

$$\Gamma_s = \frac{\gamma_s}{(s - 1)!}.$$

In general the coefficients of the linear differential operator are real, so that if the zeros $p_1, p_2, \ldots$ are all real then the constants c_r (or Γ_s) in the above solutions will also be real. If the zero p_r of $\Lambda(p)$ is complex it can be written in the form

$$p_r = \lambda_r + i\omega_r,$$

then, since the coefficients a_r of the polynomial $\Lambda(p)$ are assumed to be real, there will also be a zero $\lambda_r - i\omega_r$. In the partial fraction expansion for $\bar{y}(p)$ these two zeros will lead to a pair of terms

$$\frac{\alpha_r}{p - \lambda_r - i\omega_r} + \frac{\beta_r}{p - \lambda_r + i\omega_r}$$

which can be written in the form

$$\frac{(p - \lambda_r)A_r}{(p - \lambda_r)^2 + \omega_r^2} + \frac{\omega_r B_r}{(p - \lambda_r)^2 + \omega_r^2}$$

and hence lead to a pair of terms

$$\{A_r \cos(\omega_r x) + B_r \sin(\omega_r x)\}e^{\lambda_r x}.$$

Similarly a factor in $\Lambda(p)$ of the form

$$[(p - \lambda_r)^2 + \omega_r^2]^q$$

leads to terms of the form

$$(A_{r,0} + \cdots + A_{r,q-1}x^{q-1})e^{\lambda_r x}\cos(\omega_r x) + (B_{r,0} + B_{r,1}x + \cdots + B_{r,q-1}x^{q-1})e^{\lambda_r x}\sin(\omega_r x).$$

If, therefore, we wish to find the general solution of a homogeneous equation with constant coefficients we need only examine the zeros of the polynomial $\Lambda(p)$.

As examples, we have:

Example *Find the general solution of*

$$\frac{d^3y}{dx^3} - 2\frac{d^2y}{dx^2} - \frac{dy}{dx} + 2y = 0.$$

Here $\Lambda(p) = p^3 - 2p^2 - p + 2 = (p+1)(p-1)(p-2)$ so that the zeros of $\Lambda(p)$ are $-1, 1, 2$ and the general solution is

$$y(x) = c_1e^{-x} + c_2e^x + c_3e^{2x}.$$

Example *Find the general solution of*

$$\frac{d^4y}{dx^4} - 2\frac{d^2y}{dx^2} + y = 0.$$

Here $\Lambda(p) = p^4 - 2p^2 + 1 = (p-1)^2(p+1)^2$ so that the general solution is

$$y(x) = (c_1 + c_2x)e^x + (c_3 + c_4x)e^{-x}.$$

Example *Find the general solution of the equation*

$$\frac{d^4y}{dx^4} - 4\frac{d^3y}{dx^3} + 8\frac{d^2y}{dx^2} - 8\frac{dy}{dx} + 4y = 0.$$

Here $\Lambda(p) = \{(p-1)^2 + 1\}^2$ so that the required general solution is

$$y(x) = \{(c_1 + c_2x)\cos x + (c_3 + c_4x)\sin x\}e^x.$$

On the other hand if we wish to find the solution of an actual initial value problem it is usually simpler to use the *method* of the Laplace transform rather than to remember formulae of the above type and then to determine the values of the constants involved. As an illustration we have:

Example *Find the solution of the differential equation*

$$\frac{d^2y}{dx^2} + 6\frac{dy}{dx} + 9y = \sin x$$

which satisfies the initial conditions $y(0) = 1$, $y'(0) = 0$.

Since in this case

$$\mathscr{L}\left[\frac{d^2y}{dx^2}; p\right] = p^2\bar{y}(p) - p, \qquad \mathscr{L}\left[\frac{dy}{dx}; p\right] = p\bar{y}(p) - 1$$

we see that the Laplace transform $\bar{y}(p)$ of the required solution satisfies the algebraic equation

$$p^2\bar{y}(p) - p + 6\{p\bar{y}(p) - 1\} + 9\bar{y}(p) = (p^2 + 1)^{-1}$$

which has solution

$$\bar{y}(p) = \frac{p + 6}{(p + 3)^2} + \frac{1}{(p^2 + 1)(p + 3)^2}.$$

Now it is easily shown that

$$\frac{p + 6}{(p + 3)^2} = \frac{1}{p + 3} + \frac{3}{(p + 3)^2}$$

so that

$$\frac{1}{(p^2 + 1)(p + 3)^2} = \frac{3}{50}\frac{1}{p + 3} + \frac{1}{10}\frac{1}{(p + 3)^2} - \frac{1}{50}\frac{3p - 4}{p^2 + 1}$$

so that

$$\bar{y}(p) = \frac{53}{50}\frac{1}{p + 3} + \frac{31}{10}\frac{1}{(p + 3)^2} - \frac{1}{50}\frac{3p - 4}{p^2 + 1}.$$

Inverting this formula we find that the required solution is given by the equation

$$y(x) = \left[\frac{53}{50} + \frac{31}{10}x\right]e^{-3x} - \frac{1}{50}(3\cos x - 4\sin x).$$

3-14-2 TWO-POINT BOUNDARY VALUE PROBLEM FOR A LINEAR EQUATION WITH CONSTANT COEFFICIENTS

In a two-point boundary value problem we have to find the solution of a differential equation of the n-th order satisfying r conditions at the point $x = 0$, say, and the remaining $n - r$ necessary conditions at the second point $x = a$, say. For instance, we might wish to find the solution of the equation

$$\frac{d^3y}{dx^3} - 2\frac{d^2y}{dx^2} - \frac{dy}{dx} + 2y = 0 \qquad (3\text{-}14\text{-}14)$$

satisfying the conditions

$$y(0) = 0, \qquad y'(0) = 1, \qquad y(1) = 0. \qquad (3\text{-}14\text{-}15)$$

We use the Laplace transform method by assigning arbitrary values to the $n - r$ derivatives not specified at $x = 0$ and then, from the general solution so obtained, determine the values of these constants by fitting the conditions at the point $x = a$. To illustrate the method we shall derive the solution of the two-point boundary value problem described by equations (3-14-14) and (3-14-15).

In this case we assume that $y''(0) = c$ so that

$$\mathscr{L}\left[\frac{d^3y}{dx^3};p\right] = p^3\bar{y}(p) - c - p,$$

$$\mathscr{L}\left[\frac{d^2y}{dx^2};p\right] = p^2\bar{y}(p) - 1, \qquad \mathscr{L}\left[\frac{dy}{dx};p\right] = p\bar{y}(p)$$

and hence we find that $\bar{y}(p)$, the Laplace transform of the solution, satisfies the algebraic equation

$$(p^3 - 2p^2 - p + 2)\bar{y}(p) = p + c + 1.$$

We therefore have the equation

$$\bar{y}(p) = \frac{p + c + 1}{(p + 1)(p - 1)(p - 2)}.$$

This can be written in the alternative form

$$\bar{y}(p) = \tfrac{1}{6}c(p + 1)^{-1} - (1 + \tfrac{1}{2}c)(p - 1)^{-1} + (1 + \tfrac{1}{3}c)(p - 2)^{-1}$$

which is equivalent to the solution

$$y(x) = \tfrac{1}{6}ce^{-x} - (1 + \tfrac{1}{2}c)e^{x} + (1 + \tfrac{1}{3}c)e^{2x}. \tag{3-14-16}$$

To satisfy the condition $y(1) = 0$ we must therefore choose the constant c to be such that

$$\tfrac{1}{6}ce^{-1} - (1 + \tfrac{1}{2}c)e + (1 + \tfrac{1}{3}c)e^2 = 0$$

i.e., we must choose c to be given by the equation

$$c = -\frac{6e^2}{(e - 1)(2e + 1)}. \tag{3-14-17}$$

The required solution is therefore given by equations (3-14-16) and (3-14-17).

The method can be applied when, instead of knowing the value of $y(x)$ or of some of its derivatives at a second point $x = a$, we know its behavior as $x \to \infty$. We may illustrate the method by:

Example *Find the solution of the differential equation*

$$\frac{d^2y}{dx^2} + \frac{dy}{dx} - 2y = 0$$

satisfying the initial condition $y(0) = 1$ *and the limiting condition* $y(x) \to 0$ *as* $x \to \infty$.

In this case we do not know the value of $y'(0)$ so we ascribe to it the value c. The differential equation is then transformed to the algebraic equation

$$(p^2 + p - 2)\bar{y}(p) = p + c + 1$$

showing that

$$y(p) = \frac{p + c + 1}{(p - 1)(p + 2)}.$$

When we expand this in partial fractions we cannot have a term in $(p - 1)^{-1}$ because this would lead to a term proportional to e^x in the solution and this would cause $y(x)$ to tend to infinity as $x \to \infty$. Hence the numerator must vanish when $p = 1$, i.e., we must have $c + 2 = 0$, or $c = -2$. Hence, we have

$$\bar{y}(p) = \frac{1}{p + 2}$$

so that the required solution is

$$y(x) = e^{-2x}.$$

3-14-3 LINEAR DIFFERENTIAL EQUATIONS WITH VARIABLE COEFFICIENTS

The method, outlined above, of solving ordinary linear differential equations with constant coefficients by means of the Laplace transform may be modified to provide the solution of certain linear equations with variable coefficients. We shall illustrate the procedure by finding the solution of the equation

$$x\frac{d^2y}{dx^2} - \frac{dy}{dx} - xy = 0$$

satisfying the initial condition $y(0) = 0$.

In this case

$$\mathscr{L}\left[\frac{d^2y}{dx^2}; p\right] = p^2\bar{y}(p) - y'(0)$$

so that

$$\mathscr{L}\left[x\frac{d^2y}{dx^2}; p\right] = -\frac{d}{dp}\{p^2\bar{y}(p)\}.$$

Also

$$\mathscr{L}\left[\frac{dy}{dx};p\right] = p\bar{y}(p), \qquad \mathscr{L}[xy;p] = -\frac{d\bar{y}}{dp}$$

so that the Laplace transform $\bar{y}(p)$ of the required solution satisfies the first-order equation

$$-\frac{d}{dp}\{p^2\bar{y}(p)\} - p\bar{y}(p) + \frac{d\bar{y}}{dp} = 0$$

$$(p^2 - 1)\frac{d\bar{y}}{dp} + 3p\bar{y}(p) = 0.$$

It is easily shown that the solution of this equation is

$$\bar{y}(p) = c(p^2 - 1)^{-\frac{3}{2}}$$

where c denotes an arbitrary constant. Hence the required solution is

$$y(x) = c\mathscr{L}^{-1}[(p^2 - 1)^{-\frac{3}{2}};x] = cxI_1(x)$$

where $I_1(x)$ is the modified Bessel function of the first kind of order unity—compare equation (3-7-11) above.

3-14-4 SIMULTANEOUS DIFFERENTIAL EQUATIONS WITH CONSTANT COEFFICIENTS

The method of the Laplace transform can also be applied to simultaneous differential equations with constant coefficients. We illustrate the procedure by an example:

Example *Solve the differential equations*

$$\ddot{x} - x + \dot{y} - y = 0, \qquad -2\dot{x} - 2x + \ddot{y} - y = e^{-t}$$

subject to the initial conditions

$$x(0) = 0, \qquad \dot{x}(0) = -1, \qquad y(0) = 1, \qquad \dot{y}(0) = 1.$$

(Here we are using the notation $\dot{x} = dx/dt$, etc.)

Taking the Laplace transform of both sides of the equation and writing $\bar{x}(p)$ and $\bar{y}(p)$ for the Laplace transforms of $x(t)$ and $y(t)$ respectively, we find that they transform to

$$p^2\bar{x}(p) + 1 - \bar{x} + p\bar{y} - 1 - \bar{y} = 0$$

$$-2p\bar{x} - 2\bar{x} + p^2\bar{y} - p - 1 - \bar{y} = (p + 1)^{-1}$$

which may be simplified to

$$\bar{y}(p) = -(1 + p)\bar{x}(p)$$

$$(p - 1)\bar{y}(p) - 2\bar{x}(p) = 1 + (p + 1)^{-2}$$

with solution

$$\bar{x}(p) = -\frac{1}{p^2 + 1} - \frac{1}{(p+1)^2(p^2+1)}$$

$$\bar{y}(p) = \frac{p+1}{p^2+1} + \frac{1}{(p+1)(p^2+1)}.$$

The right-hand sides of these equations may be expanded in partial fractions to yield the expressions

$$\bar{x}(p) = \tfrac{1}{2}\frac{p}{p^2+1} - \tfrac{1}{2}\frac{1}{p+1} - \tfrac{1}{2}\frac{1}{(p+1)^2} - \frac{1}{p^2+1}$$

$$\bar{y}(p) = \tfrac{1}{2}\frac{p}{p^2+1} + \tfrac{3}{2}\frac{1}{p^2+1} + \tfrac{1}{2}\frac{1}{p+1}$$

from which we deduce the solution

$$x(t) = \tfrac{1}{2}\cos t - \sin t - \tfrac{1}{2}(1+t)e^{-t}$$
$$y(t) = \tfrac{1}{2}\cos t + \tfrac{3}{2}\sin t + \tfrac{1}{2}e^{-t}.$$

3-14-5 AUTONOMOUS LINEAR SYSTEMS

An autonomous linear system of ordinary differential coefficients may be written in the matrix form

$$\frac{dy}{dx} = Ay + \eta, \tag{3-14-19}$$

where y and η are column vectors

$$y = \begin{bmatrix} y_1 \\ y_2 \\ \vdots \\ y_n \end{bmatrix} \qquad \eta = \begin{bmatrix} \eta_1 \\ \eta_2 \\ \vdots \\ \eta_n \end{bmatrix}$$

$\eta_1, \eta_2, \ldots, \eta_n$ being constants and A is an $n \times n$ matrix (a_{ij}). If we write

$$\bar{y}(p) = \begin{bmatrix} \bar{y}_1(p) \\ y_2(p) \\ \vdots \\ y_n(p) \end{bmatrix} \qquad y(0) = \begin{bmatrix} y_1(0) \\ y_2(0) \\ \vdots \\ y_n(0) \end{bmatrix}$$

then the Laplace transform of the vector equation (3-14-19) is

$$p\bar{y}(p) = A\bar{y}(p) + p^{-1}\eta + y(0)$$

with solution

$$\bar{y}(p) = (pI - A)^{-1}\{p^{-1}\eta + y(0)\}. \tag{3-14-20}$$

The solution to the original system is then found by an application of the inversion theorem for the Laplace transform.

As an example we shall consider the solution of the set of simultaneous ordinary differential equations

$$\frac{dy_1}{dx} = 2y_1 - 2y_2 + 3y_3$$

$$\frac{dy_2}{dx} = y_1 + y_2 + y_3$$

$$\frac{dy_3}{dx} = y_1 + 3y_2 - y_3$$

satisfying the initial conditions $y_1(0) = 1$, $y_2(0) = y_3(0) = 0$. In this case $\eta = 0$ and

$$pI - A = \begin{bmatrix} p-2 & 2 & -3 \\ -1 & p-1 & -1 \\ -1 & -3 & p+1 \end{bmatrix}, \qquad y(0) = \begin{bmatrix} 1 \\ 0 \\ 0 \end{bmatrix}.$$

The determinant of the matrix $pI - A$ is easily shown to be

$$D = (p+2)(p-1)(p-3)$$

and

$$(pI - A)^{-1} = D^{-1}\begin{bmatrix} p^2-4 & -2p+7 & 3p-5 \\ p+2 & p^2-p-5 & p+1 \\ p+2 & 3p-8 & p^2-3p+4 \end{bmatrix}$$

so that

$$\bar{y}(p) = (pI - \mathrm{A})^{-1}y(0)$$

$$= D^{-1}\begin{bmatrix} p^2-4 \\ p+2 \\ p+2 \end{bmatrix} = \begin{bmatrix} (p-2)(p-1)^{-1}(p-3)^{-1} \\ (p-1)^{-1}(p-3)^{-1} \\ (p-1)^{-1}(p-3)^{-1} \end{bmatrix}$$

Using the Laplace inversion theorem we obtain the solution

$$y(x) = \tfrac{1}{2}\begin{bmatrix} e^{3x} + e^x \\ e^{3x} - e^x \\ e^{3x} - e^x \end{bmatrix}.$$

3-15 PARTIAL DIFFERENTIAL EQUATIONS

We illustrate the application of the Laplace transform to the solution of partial differential equations by considering the solution of the diffusion equation in certain circumstances.

3-15-1 THE DIFFUSION EQUATION IN A SEMI-INFINITE LINE

We begin by considering the problems discussed already in sec. 2-16. In that section we transformed the equation

$$\frac{\partial^2\theta}{\partial x^2} = \frac{1}{\kappa}\frac{\partial\theta}{\partial t} \tag{3-15-1}$$

by integrating both sides of this equation with respect to the space-variable x; in this section we take the Laplace transform with respect to the time t. If we write

$$\bar{\theta}(x,p) = \mathscr{L}[\theta(x,t);t \to p]$$

then, if we assume that $\theta(x,0) = 0$, we find from equation (3-4-4) that

$$\mathscr{L}\left[\frac{\partial\theta}{\partial t};t \to p\right] = p\bar{\theta}(x,p).$$

It follows that the transform of (3-5-1) is

$$\frac{\partial^2\bar{\theta}}{\partial x^2} = \frac{p}{\kappa}\bar{\theta}. \tag{3-15-2}$$

If we wish to find the solution in the semi-infinite line $x > 0$ satisfying the boundary condition

$$\theta(0,t) = f(t), \qquad (t > 0) \tag{3-15-3}$$

and the limiting condition

$$\theta(x,t) \to 0, \qquad x \to \infty$$

we note that the transform $\bar{\theta}(x,p)$ must satisfy the conditions

$$\bar{\theta}(0,p) = \bar{f}(p), \qquad \bar{\theta}(x,p) \to 0, \qquad \text{as } x \to \infty,$$

so that the appropriate solution of (3-15-2) is

$$\bar{\theta}(x,p) = \bar{f}(p)e^{-x\sqrt{(p/\kappa)}}. \tag{3-15-4}$$

Making use of the convolution theorem we see that

$$\theta(x,t) = \int_0^t f(u)g(x,t-u)\,du \tag{3-15-5}$$

where

$$g(x,t) = \mathscr{L}^{-1}[e^{-x\sqrt{(p/\kappa)}}; p \to t].$$

From equation (3-2-14) we find that

$$g(x,t) = \frac{1}{\sqrt{(4\pi\kappa t^3)}} e^{-x^2/4\kappa t}$$

so that the required solution is

$$\theta(x,t) = \frac{1}{\sqrt{(4\pi\kappa)}} \int_0^t f(u) \exp\left[-\frac{x^2}{4\kappa(t-u)}\right] \cdot \frac{du}{(t-u)^{\frac{3}{2}}}. \tag{3-15-6}$$

It will be noted that this is in agreement with equation (2-16-33) obtained by taking the Fourier sine transform of equation (3-15-1).

For the calculation of $\theta(x,t)$ corresponding to particular forms of $f(t)$ it is often easier to make use of equation (3-15-4) than of equation (3-15-6). To illustrate this point we consider two special cases:

Case (i) $\theta(0,t) = \theta_0(T/t)^{\frac{1}{2}}$, *with* θ_0 *and* T *positive constants.*

In this case

$$\bar{f}(p) = \theta_0 T^{\frac{1}{2}} \mathscr{L}[t^{-\frac{1}{2}}; p] = \theta_0(\pi T/p)^{\frac{1}{2}},$$

so that equation (3-15-4) gives the expression

$$\bar{\theta}(x,p) = \theta_0(\pi T)^{\frac{1}{2}} p^{-\frac{1}{2}} e^{-x\sqrt{(p/\kappa)}}$$

for the Laplace transform of the solution $\theta(x,t)$. If we make use of equation (3-2-13) we obtain the solution

$$\theta(x,t) = \theta_0(T/t)^{\frac{1}{2}} e^{-x^2/4\kappa t}. \tag{3-15-7}$$

Case (ii) $\theta(0,t) = \theta_0$, *a constant.*

In this case

$$\bar{f}(p) = \theta_0 p^{-1}$$

so that

$$\bar{\theta}(x,p) = \theta_0 p^{-1} e^{-x\sqrt{(p/\kappa)}}.$$

From equation (3-6-17) we then deduce the solution

$$\theta(x,t) = \theta_0 \operatorname{Erfc}(\tfrac{1}{2}x\kappa^{-\frac{1}{2}}t^{-\frac{1}{2}}).$$

To return to the general case: we now consider the case in which $\theta(x,t)$ satisfies the same conditions as before except that equation (3-15-3) is replaced by the condition

$$\frac{\partial}{\partial x}\theta(0,t) = f(t). \tag{3-15-8}$$

The problem, then, is to solve equation (3-15-2) subject to the boundary conditions

$$\frac{\partial}{\partial x}\bar{\theta}(0,p) = \bar{f}(p), \qquad \bar{\theta}(x,p) \to 0, \qquad x \to \infty.$$

We therefore obtain the expression

$$\bar{\theta}(x,p) = -\kappa^{\frac{1}{2}}\bar{f}(p)p^{-\frac{1}{2}}e^{-x\sqrt{(p/\kappa)}}$$

which leads to the solution (3-15-5) with

$$g(x,t) = -\kappa^{\frac{1}{2}}\mathscr{L}^{-1}[p^{-\frac{1}{2}}e^{-x\sqrt{(p/\kappa)}};p \to t].$$

From (3-15-7) we deduce that

$$g(x,t) = -(\kappa/\pi t)^{\frac{1}{2}}e^{-x^2/4\kappa t}. \tag{3-15-9}$$

The required solution is given by (3-15-5) with the function $g(x,t)$ given by (3-15-9)—in agreement with equation (2-16-35).

3-15-2 DIFFUSION IN SLABS

We now consider the solution of the diffusion equation (3-15-1) in the slab $0 \leqslant x \leqslant a$. We shall consider the solution $\theta(x,t)$ satisfying the boundary conditions

$$\theta(0,t) = f(t), \qquad \theta(a,t) = 0, \qquad (t > 0) \tag{3-15-10}$$

and the initial condition

$$\theta(x,0) = 0, \qquad (0 \leqslant x \leqslant a). \tag{3-15-11}$$

Because of (3-15-11) the Laplace transform of the linear diffusion equation again takes the form (3-15-2). The transformed boundary conditions are

$$\bar{\theta}(x,p) = \bar{f}(p), \qquad \bar{\theta}(a,p) = 0. \tag{3-15-12}$$

The solution of the two-point boundary value problem posed by equations (3-15-2) and (3-15-12) is easily seen to be

$$\bar{\theta}(x,p) = \bar{f}(p)\frac{\sinh\{(a-x)\sqrt{(p/\kappa)}\}}{\sinh\{a\sqrt{(p/\kappa)}\}}$$

The solution of the problem is therefore given by equation (3-15-5) but now

$$g(x,t) = \mathscr{L}^{-1}\left[\frac{\sinh\{(a-x)\sqrt{(p/\kappa)}\}}{\sinh\{a\sqrt{(p/\kappa)}\}};p \to t\right], \qquad (0 \leqslant x \leqslant a)$$

Regarded as a function of the complex variable p, the function $\bar{g}(x,p)$ is a single-valued function with poles at $p = p_n = -\kappa n^2\pi^2/a^2$, $(n = 1,2,\ldots)$.

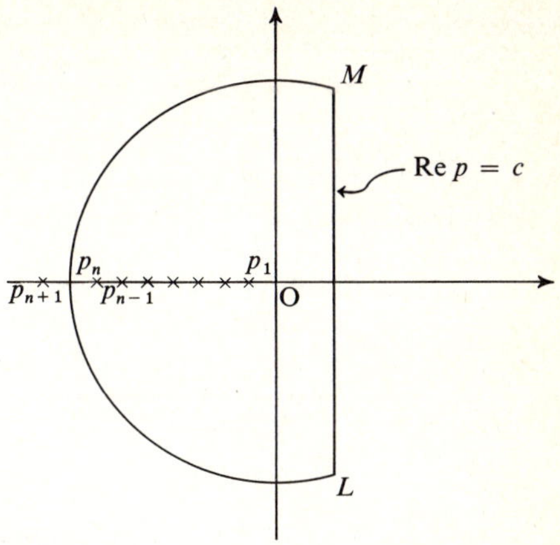

Fig. 3-17

We consider the integral of $\bar{g}(x,p)e^{pt}$ round the closed contour of Fig. 3-17, where LM is parallel to the imaginary axis in the p-plane and is at a distance c from it, and the radius of the circle MNL is equal to

$$\kappa(n + \tfrac{1}{2})^2\pi^2/a^2.$$

Letting $n \to \infty$, we can easily show that the integral round the circular arc tends to zero so that $g(x,t)$ is equal to the sum of the residues of the integrand at its poles. The residue at the pole $p = p_n$ is

$$\frac{\sinh\{(a - x)\sqrt{(p_n/\kappa)}\}}{[d/dp \sinh\{a\sqrt{(p/\kappa)}\}]_{p=p_n}} e^{p_n t}$$

and it is easily shown that this residue is

$$\frac{2\kappa\pi}{a^2} n \sin\left(\frac{n\pi x}{a}\right) e^{-n^2\pi^2\kappa t/a^2}$$

so that we have the equation

$$g(x,t) = \frac{2\kappa\pi}{a^2} \sum_{n=1}^{\infty} n \sin\left(\frac{n\pi x}{a}\right) e^{-n^2\pi^2\kappa t/a^2}.$$

If we substitute this expression for $g(x,t)$ into equation (3-15-5) we obtain the solution

$$\theta(x,t) = \frac{2\pi\kappa}{a^2} \sum_{n=1}^{\infty} b_n(t) \sin\left(\frac{n\pi x}{a}\right) \qquad (3\text{-}15\text{-}13)$$

with

$$b_n(t) = n \int_0^t f(u) \exp\left[-\frac{n^2\pi^2\kappa}{a^2}(t-u) \right] du. \tag{3-15-14}$$

For example, if $\theta(0,t) = \theta_0$, a constant, the corresponding functions $b_n(t)$ are given by the formula

$$b_n(t) = \frac{a^2\theta_0}{\pi^2\kappa n}(1 - e^{-n^2\pi^2\kappa t/a^2}),$$

so that

$$\theta(x,t) = \frac{2}{\pi}\theta_0 \sum_{n=1}^{\infty} \frac{1}{n}(1 - e^{-n^2\pi^2\kappa t/a^2}) \sin\left(\frac{n\pi x}{a}\right). \tag{3-15-15}$$

3-15-3 DIFFUSION IN CIRCULAR CYLINDERS

We now consider a simple problem on the axisymmetric diffusion of heat in a cylinder of radius a. The method of solution is similar to that discussed above in the case of diffusion in a finite slab, but is included here to afford a comparison with the solution of the same problem derived by means of the theory of finite Hankel transforms (see sec. 8-3 below).

If $\theta(r,t)$ is the temperature at a point distant r from the axis of the cylinder, at a time t, and if κ is the diffusivity of the material of the cylinder, then we have to solve the partial differential equation

$$\frac{\partial^2\theta}{\partial r^2} + \frac{1}{r}\frac{\partial\theta}{\partial r} = \frac{1}{\kappa}\frac{\partial\theta}{\partial t}, \qquad (t \geqslant 0,\ 0 \leqslant r \leqslant a). \tag{3-15-16}$$

We shall suppose that the surface temperature is prescribed:

$$\theta(a,t) = f(t), \qquad (t \geqslant 0), \tag{3-15-17}$$

and that, initially, the temperature of every point of the cylinder is zero

$$\theta(r,0) = 0, \qquad (0 \leqslant r \leqslant a). \tag{3-15-18}$$

From equations (3-4-4) and (3-5-18), we see that

$$\mathscr{L}\left[\frac{\partial\theta}{\partial t}; t \to p\right] = p\bar{\theta}(r,p),$$

where

$$\bar{\theta}(r,p) = \mathscr{L}[\theta(r,t); t \to p].$$

Hence the Laplace transform of the temperature satisfies the simple boundary value problem

$$\frac{\partial^2\bar{\theta}}{\partial r^2} + \frac{1}{r}\frac{\partial\bar{\theta}}{\partial r} - \frac{p}{\kappa}\bar{\theta} = 0$$

with

$$\bar{\theta}(a,p) = \bar{f}(p)$$

and $\bar{\theta}(r,p)$ finite when $r = 0$†. The solution of this problem is easily seen to be

$$\bar{\theta}(r,p) = \bar{f}(p)\frac{\mathrm{I}_0(r\sqrt{p/\kappa})}{\mathrm{I}_0(a\sqrt{p/\kappa})}, \qquad 0 \leqslant r \leqslant a.$$

Using the convolution theorem for the Laplace transform we obtain the solution

$$\theta(r,t) = \int_0^t f(u)g(r,t-u)\,du \tag{3-15-19}$$

where

$$g(r,t) = \mathscr{L}^{-1}\left[\frac{\mathrm{I}_0(r\sqrt{p/\kappa})}{\mathrm{I}_0(a\sqrt{p/\kappa})};p \to t\right].$$

From equation (3-11-6) we deduce that

$$g(r,t) = \frac{2\kappa}{a}\sum_{n=1}^{\infty} \xi_n \frac{\mathrm{J}_0(r\xi_n)}{\mathrm{J}_1(a\xi_n)} e^{-\kappa t \xi_n^2}$$

so that, if ξ_n are the positive zeros of $\mathrm{J}_0(a\xi)$,

$$\theta(r,t) = \frac{2\kappa}{a}\sum_{n=1}^{\infty} c_n(t)\xi_n \frac{\mathrm{J}_0(r\xi_n)}{\mathrm{J}_1(a\xi_n)} \tag{3-15-20}$$

where the coefficients $c_n(t)$ are defined by the equation

$$c_n(t) = \int_0^t f(u)e^{-\kappa\xi_n^2(t-u)}\,du \tag{3-15-21}$$

3-16 INTEGRAL EQUATIONS

The Laplace transform provides a useful tool for the solution of integral equations of Volterra type in which the kernel $K(x,t)$ is a function of $t - x$ alone.

We begin by considering equations of the first kind

$$\int_0^t f(x)K(t-x)\,dx = g(t), \qquad (t > 0) \tag{3-16-1}$$

with $g(0) = 0$.

†Here we make use of the physical condition that $\theta(r,t)$—and hence $\bar{\theta}(r,p)$—cannot be infinite along the axis, $r = 0$, of the cylinder.

Taking the Laplace transform of both sides and using the convolution theorem we have the relation

$$\bar{f}(p)\bar{K}(p) = \bar{g}(p)$$

which may be solved to give the equation $\bar{f}(p) = \bar{g}(p)/\bar{K}(p)$. It usually happens that $\mathscr{L}^{-1}[1/\bar{K}(p); x]$ does *not* exist. When this occurs but $\mathscr{L}^{-1}[1/p\bar{K}(p); x]$ exists we write

$$\bar{f}(p) = p\bar{g}(p)\bar{L}(p) \tag{3-16-2}$$

for the determination of the Laplace transform of the unknown function $f(x)$ in terms of $\bar{g}(p)$ and

$$\bar{L}(p) = [p\bar{K}(p)]^{-1}$$

(Cf. the similar procedure in sec. 2-15). If the inverse transform

$$L(x) = \mathscr{L}^{-1}[\{p\bar{K}(p)\}^{-1}; x] \tag{3-16-3}$$

exists then we can apply the convolution theorem for the Laplace transform to equation (3-16-2) to obtain the solution of the integral equation (3-16-1) either in the form

$$f(x) = \frac{d}{dx}\int_0^x g(t)L(x - t)\,dt \tag{3-16-4}$$

or in the form

$$f(x) = \int_0^x g'(t)L(x - t)\,dt. \tag{3-16-5}$$

For instance, if

$$K(x) = x^{-\alpha}, \qquad 0 < \alpha < 1$$

$$\bar{K}(p) = \Gamma(1 - \alpha)p^{\alpha - 1}$$

and

$$\bar{L}(p) = \frac{p^{-\alpha}}{\Gamma(1 - \alpha)},$$

so that

$$L(x) = \frac{x^{\alpha - 1}}{\Gamma(1 - \alpha)\Gamma(\alpha)} = \frac{\sin(\alpha\pi)}{\pi}x^{\alpha - 1}.$$

Thus we have shown that the solution of the integral equation

$$\int_0^t \frac{f(x)\,dx}{(t - x)^\alpha} = g(t), \qquad t > 0, \qquad 0 < \alpha < 1, \qquad g(0) = 0 \tag{3-16-6}$$

can be written either as

$$f(x) = \pi^{-1} \sin(\pi\alpha) \frac{d}{dx} \int_0^x \frac{g(t)\,dt}{(x-t)^{1-\alpha}} \tag{3-16-7}$$

or as

$$f(x) = \pi^{-1} \sin(\pi\alpha) \int_0^x \frac{g'(t)\,dt}{(x-t)^{1-\alpha}}. \tag{3-16-8}$$

The solution of the equation

$$\int_a^t \frac{f(x)\,dx}{(t-x)^\alpha} = g(t), \qquad t > a, \qquad 0 < \alpha < 1, \qquad g(a) = 0 \tag{3-16-9}$$

is easily deduced by writing it in the form

$$\int_0^{t+a} \frac{f(u+a)\,du}{(t+a-x)^\alpha} = g(t+a)$$

which has solution

$$f(u+a) = \pi^{-1} \sin(\pi\alpha) \frac{d}{du} \int_0^u \frac{g(t+a)\,dt}{(u-t)^{1-\alpha}}.$$

This, in turn, can be reduced to the equivalent form

$$f(x) = \pi^{-1} \sin(\pi\alpha) \frac{d}{dx} \int_a^x \frac{g(t)\,dt}{(x-t)^{1-\alpha}}. \tag{3-16-10}$$

By a similar process we can show that the equation

$$\int_t^b \frac{f(x)\,dx}{(x-t)^\alpha} = g(t), \qquad 0 < t < b, \qquad 0 < \alpha < 1, \qquad g(b) = 0 \tag{3-16-11}$$

has solution

$$f(x) = -\pi^{-1} \sin(\pi\alpha) \frac{d}{dx} \int_x^b \frac{g(t)\,dt}{(t-x)^{1-\alpha}}. \tag{3-16-12}$$

The solutions of the equations

$$\int_a^t \frac{f(x)\,dx}{\{\phi(t) - \phi(x)\}^\alpha} = g(t), \qquad a < t < b, \qquad 0 < \alpha < 1 \tag{3-16-13}$$

$$\int_t^b \frac{f(x)\,dx}{\{\phi(x) - \phi(t)\}^\alpha} = g(t), \qquad a < t < b, \qquad 0 < \alpha < 1 \tag{3-16-14}$$

in which the prescribed function $\phi(t)$ is monotonic strictly increasing and differentiable for $a < t < b$, and $\phi'(t) \neq 0$ in this interval, can be derived from equations (3-16-10) and (3-16-12) by a simple change of variables.

If we put $\phi(t) = \tau$, $\phi(x) = \xi$, $\phi(a) = \beta$, $\phi(b) = \gamma \ldots f(x)/\phi'(x) = h(\xi)$, $g(t) = G(\tau)$, we see that (3-16-13) reduces to

$$\int_\beta^\tau \frac{h(\xi)\,d\xi}{(\tau - \xi)^\alpha} = G(\tau), \qquad \beta < \tau < \gamma, \qquad 0 < \alpha < 1$$

where the conditions on $\phi(t)$ ensure the existence of the inverse function $t = \phi^{-1}(\tau)$ and the fact that $\beta < \xi < \gamma$. The solution of this equation is

$$h(\xi) = \pi^{-1} \sin(\pi\alpha) \frac{d}{d\xi} \int_\beta^\xi \frac{G(\tau)\,d\tau}{(\xi - \tau)^{1-\alpha}}.$$

Returning to the original variables we then see that the solution of the integral equation (3-16-13) is

$$f(x) = \pi^{-1} \sin(\pi\alpha) \frac{d}{dx} \int_a^x \frac{\phi'(t)g(t)\,dt}{\{\phi(x) - \phi(t)\}^{1-\alpha}}. \tag{3-16-15}$$

The same substitutions reduce equation (3-16-14) to the form (3-16-11) whence it is easily shown that its solution is

$$f(x) = -\pi^{-1} \sin(\pi\alpha) \frac{d}{dx} \int_x^b \frac{\phi'(t)g(t)\,dt}{\{\phi(t) - \phi(x)\}^{1-\alpha}}. \tag{3-16-16}$$

For instance, if we take $a = 0$, $b = \pi$, $\phi(t) = -\cos t$, we find that the integral equation

$$\int_0^t \frac{f(x)\,dx}{(\cos x - \cos t)^{\frac{1}{2}}} = g(t), \qquad 0 < t < \pi \tag{3-16-17}$$

has solution

$$f(x) = \frac{1}{\pi}\frac{d}{dx} \int_0^x \frac{\sin(t)g(t)\,dt}{(\cos t - \cos x)^{\frac{1}{2}}}, \qquad 0 < x < \pi \tag{3-16-18}$$

and that the integral equation

$$\int_t^\pi \frac{f(x)\,dx}{(\cos t - \cos x)^{\frac{1}{2}}} = g(t), \qquad 0 < t < \pi \tag{3-16-19}$$

has solution

$$f(x) = -\frac{1}{\pi}\frac{d}{dx} \int_x^\pi \frac{\sin(t)g(t)\,dt}{(\cos x - \cos t)^{\frac{1}{2}}}, \qquad 0 < x < \pi. \tag{3-16-20}$$

Similarly, if we take $a = 0$, $b = \infty$, $\phi(t) = t^2$, we find that the integral equation

$$\int_0^t \frac{f(x)\,dx}{(t^2 - x^2)^\alpha} = g(t), \qquad t > 0, \qquad 0 < \alpha < 1 \tag{3-16-21}$$

has solution

$$f(x) = 2\pi^{-1}\sin(\alpha\pi)\frac{d}{dx}\int_0^x \frac{tg(t)\,dt}{(x^2 - t^2)^{1-\alpha}}, \qquad x > 0, \qquad 0 < \alpha < 1 \tag{3-16-22}$$

and that the equation

$$\int_t^\infty \frac{f(x)\,dx}{(x^2 - t^2)^\alpha} = g(t), \qquad t > 0, \qquad 0 < \alpha < 1 \tag{3-16-23}$$

has solution

$$f(x) = -2\pi^{-1}\sin(\pi\alpha)\frac{d}{dx}\int_x^\infty \frac{tg(t)\,dt}{(t^2 - x^2)^{1-\alpha}}, \qquad x > 0, \qquad 0 < \alpha < 1. \tag{3-16-24}$$

It can, of course, happen that for a given kernel $K(x)$ the inverse transform

$$\mathscr{L}^{-1}[\{p\bar{K}(p)\}^{-1}; x]$$

does not exist. For example it does not exist if $K(x) = x^{\frac{1}{2}}$. In these circumstances it may turn out that although this inverse transform does not exist, it is possible to find a positive integer m such that the function

$$\bar{M}(p) = p^{-m}\{\bar{K}(p)\}^{-1}$$

does have an inverse. If it is known, in addition, that

$$g^{(r)}(0) = 0, \qquad (r = 0, 1, \ldots, m - 1), \tag{3-16-25}$$

then we can solve the equation (3-16-1), by writing

$$\bar{f}(p) = p^m \bar{g}(p)\bar{M}(p)$$

and noting that, because of the conditions (3-16-25),

$$\mathscr{L}[g^{(m)}(t); p] = p^m \bar{g}(p)$$

so that we deduce immediately from the convolution theorem that

$$f(x) = \int_0^x g^{(m)}(t) M(x - t)\,dt \tag{3-16-26}$$

where the kernel $M(x)$ is defined by the equation

$$M(x) = \mathscr{L}^{-1}[p^{-m}\{\bar{K}(p)\}^{-1}; x]. \tag{3-16-27}$$

For example, if we are considering the integral equation

$$\int_0^t (t - x)^{m-1-\alpha} f(x)\,dx = g(t), \qquad (t > 0, 0 < \alpha < 1) \tag{3-16-28}$$

in which m is a positive integer and $g(x)$ is a prescribed function satisfying the initial conditions (3-16-25) then

$$K(x) = x^{m-\alpha-1}$$

so that

$$\bar{K}(p) = p^{\alpha-m}\Gamma(m-\alpha)$$

and hence

$$M(x) = \mathscr{L}^{-1}\left[\frac{p^{-\alpha}}{\Gamma(m-\alpha)};x\right] = \frac{x^{\alpha-1}}{\Gamma(\alpha)\Gamma(m-\alpha)}.$$

The solution of the equation (3-16-28) can therefore be written in the form

$$f(x) = \frac{1}{\Gamma(\alpha)\Gamma(m-\alpha)}\int_0^x \frac{g^{(m)}(t)\,dt}{(x-t)^{1-\alpha}}. \tag{3-16-29}$$

In circumstances in which the inverse transform of $\bar{M}(p)$ does not exist although the conditions (3-16-25) are satisfied it is possible to modify this method of solution. The modification depends on the fact that if the conditions (3-16-25) are satisfied,

$$\mathscr{L}[\phi_m(\mathrm{D}_t)g(t);p] = \phi_m(p)\bar{g}(p)$$

where $\phi_m(p)$ is a polynomial whose degree is equal to (or less than) m, and D_t denotes the differential operator d/dt. We can therefore rewrite the basic relation (3-16-2) in the alternative form

$$\bar{f}(p) = \phi_m(p)\bar{g}(p)\cdot\bar{N}(p)$$

where

$$\bar{N}(p) = \frac{1}{\phi_m(p)\bar{K}(p)}.$$

Making use of the convolution theorem we therefore obtain the solution of the integral equation (3-16-1), when the function $g(x)$ satisfies the initial conditions (3-16-25), in the form

$$f(x) = \int_0^x N(x-t)\phi_m(\mathrm{D}_t)g(t)\,dt \tag{3-16-30}$$

where the kernel $N(x)$ is defined by the equation

$$N(x) = \mathscr{L}^{-1}\left[\frac{1}{\phi_m(p)\bar{K}(p)};x\right]. \tag{3-16-31}$$

For example to solve the integral equation

$$\int_0^t \mathrm{J}_0(t-x)f(x)\,dx = g(t), \qquad g(0) = g'(0) = 0 \tag{3-16-32}$$

we note that

$$\bar{K}(p) = (p^2 + 1)^{-\frac{1}{2}}$$

so that $\mathscr{L}^{-1}[\{\bar{K}(p)\}^{-1};x]$ does not exist. On the other hand

$$\mathscr{L}^{-1}\left[\frac{1}{(p^2+1)\bar{K}(p)};x\right] = \mathscr{L}^{-1}[(p^2+1)^{-\frac{1}{2}};x] = J_0(x)$$

so that we may take $\phi(p) = (p^2 + 1)$ to obtain the solution

$$f(x) = \int_0^x J_0(x-t)(D_t^2 + 1)g(t)\, dt. \tag{3-16-33}$$

It may be possible to find a solution by this method even when the conditions (3-16-25) are not satisfied. For example, suppose that we wish to find the solution of the equation (3-16-32) when $g'(0)$ is non-zero. Proceeding as before we can write the solution in the form

$$\begin{aligned}\bar{f}(p) &= (p^2+1)^{-\frac{1}{2}}\cdot(p^2+1)\bar{g}(p)\\ &= g'(0)(p^2+1)^{-\frac{1}{2}} + (p^2+1)^{-\frac{1}{2}}\mathscr{L}[(D_t^2+1)g(t);p].\end{aligned}$$

Applying the operator $\mathscr{L}^{-1}$ to both sides of this last equation we obtain the desired solution in the form

$$f(x) = g'(0)J_0(x) + \int_0^x J_0(x-t)(D_t^2+1)g(t)\, dt.$$

Finally we consider the solution of the integral equation

$$f(t) + \int_0^t f(x)K(t-x)\, dx = g(t) \tag{3-16-34}$$

where, for simplicity, we assume that $g(0) = 0$ and $f(0) = 0$. If we take the Laplace transform of both sides of this equation we obtain the relation

$$\bar{f}(p)\{1 + \bar{K}(p)\} = g(p).$$

As in the above discussion we can interpret this basic equation in a variety of ways. Writing it in the form

$$\bar{f}(p) = p\bar{g}(p)\bar{L}(p), \qquad \bar{L}(p) = \frac{1}{p\{1+\bar{K}(p)\}}$$

and recalling that $g(0) = 0$ we see that we can write the solution in the form

$$f(x) = \int_0^x L(x-t)g'(t)\, dt$$

provided that the inverse transform

$$L(x) = \mathscr{L}^{-1}\left[\frac{1}{p\{1 + K(p)\}}; x\right]$$

exists.

In the special case in which the function $g(x)$ satisfies the initial conditions (3-16-25) we can write the solution in the form

$$f(x) = \int_0^x M(x - t)\phi_m(\mathrm{D}_t)g(t)\, dt$$

provided that we can find a polynomial ϕ_m such that the inverse transform

$$M(x) = \mathscr{L}^{-1}[[\phi_m(p)\{1 + K(p)\}]^{-1}; x]$$

exists.

Alternatively, we could write the solution of the basic equation in the form

$$\bar{f}(p) = \bar{g}(p) - \bar{N}(p)\bar{g}(p), \qquad \bar{N}(p) = \frac{\bar{K}(p)}{1 + \bar{K}(p)}$$

and hence the solution of the integral equation (3-16-34) in the form

$$f(x) = g(x) - \int_0^x N(x - t)g(t)\, dt, \qquad N(x) = \mathscr{L}^{-1}\left[\frac{\bar{K}(p)}{1 + \bar{K}(p)}; x\right].$$

For example, for the equation

$$f(x) + \int_0^x f(t) \sin\{\omega(x - t)\}\, dt = g(x), \qquad (x > 0)$$

we have

$$\bar{K}(p) = \frac{\omega}{p^2 + \omega^2}, \qquad \bar{N}(p) = \frac{\omega}{p^2 + \omega(1 + \omega)}$$

so that the solution may be written in the form

$$f(x) = g(x) - \frac{\omega}{\Omega}\int_0^x g(t) \sin\{\Omega(x - t)\}\, dt, \qquad \Omega^2 = \omega(1 + \omega).$$

3-17 DIFFERENCE EQUATIONS

Difference equations arise in a natural way in physics and engineering. As an example we consider the ladder network shown in Fig. 3-18. Suppose that we denote the current in the n-th loop by i_n, then applying Ohm's law to the $(n + 1)$-th loop we obtain the relation

$$R_1 i_{n+1} + R_2(i_{n+1} - i_n) + R_2(i_{n+1} - i_{n+2}) = 0, \qquad 1 \leqslant n \leqslant N - 2.$$

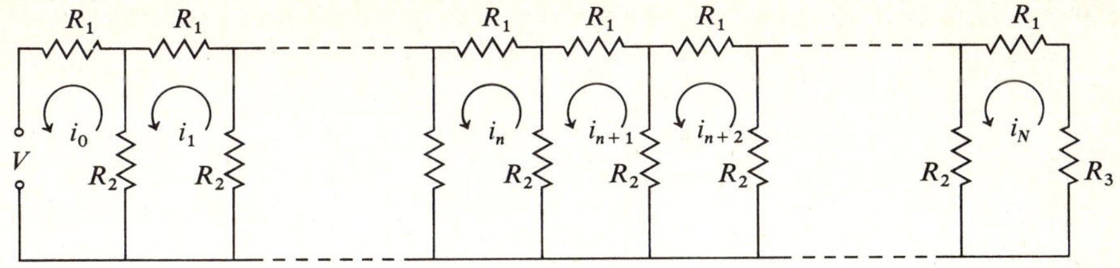

Fig. 3-18

This may be written in the simplified form

$$i_{n+2} - 2\cosh\alpha i_{n+1} - i_n = 0 \tag{3-17-1}$$

where α is defined through the equation

$$\cosh\alpha = 1 + \frac{R_1}{2R_2}. \tag{3-17-2}$$

The equations for the end loops are

$$R_1 i_0 + R_2(i_1 - i_0) = V, \qquad R_2(i_N - i_{N-1}) + (R_1 + R_3)i_N = 0$$

and these may be written in the forms

$$(1 - 2\cosh\alpha)i_0 + i_1 = V/R_2 \tag{3-17-3}$$

$$(R_1 + R_2 + R_3)i_N = R_2 i_{N-1}. \tag{3-17-4}$$

The problem is, of course, to derive a formula for the solution of the difference equation (3-17-1) satisfying the end conditions (3-17-3) and (3-17-4).

The technique for solving equations of this type by means of the Laplace transform depends on the following method of representing a sequence $\{a_n\}$ by a function $a(t)$ defined by the equation

$$a(t) = \sum_{n=0}^{\infty} a_n S_n(t) \tag{3-17-5}$$

where the functions $S_0, S_1, S_2, \ldots$ are defined by the equation

$$S_n(t) = \mathrm{H}(t - n) - \mathrm{H}(t - n - 1). \tag{3-17-6}$$

From this definition we see that

$$a(t) = a_n, \qquad n < t < (n + 1)$$

so that the graph of $a(t)$ is of the form shown in Fig. 3-19.

Also, from the definition (3-17-5) we see that

$$a(t + 1) = \sum_{n=0}^{\infty} a_n\{\mathrm{H}(t + 1 - n) - \mathrm{H}(t - n)\} = \sum_{n=1}^{\infty} a_n S_{n-1}(t)$$

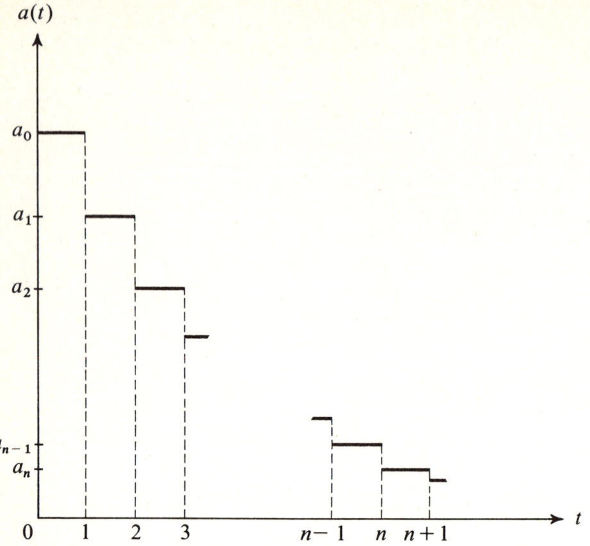

Fig. 3-19

so that we obtain the relation

$$a(t+1) = \sum_{n=0}^{\infty} a_{n+1} S_n(t). \tag{3-17-7}$$

In a similar way we may derive the relation

$$a(t+2) = \sum_{n=0}^{\infty} a_{n+2} S_n(t). \tag{3-17-8}$$

Since

$$\mathscr{L}[S_n(t);p] = p^{-1}(1 \quad e^{-p})e^{-np}$$

we deduce immediately from the definition (3-17-5) of the function $a(t)$ that its Laplace transform is given by the equation

$$\bar{a}(p) = p^{-1}(1 - e^{-p})\zeta(p) \tag{3-17-9}$$

where the function $\zeta(p)$ is given in terms of the sequence $\{a_n\}$ by the relation

$$\zeta(p) = \sum_{n=0}^{\infty} a_n e^{-np} \tag{3-17-10}$$

It is convenient also to use the notation

$$\zeta(p) = \mathscr{D}[\{a_n\};p]. \tag{3-17-11}$$

From the identities

$$\int_k^\infty a(t)e^{-pt}\,dt = \bar{a}(p) - \int_0^k a(t)e^{-pt}\,dt$$

$$\int_0^k a(t)e^{-pt}\,dt = p^{-1}(1 - e^{-p})\sum_{r=0}^{k-1} a_r e^{-pr}$$

(in which k is a positive integer) we deduce that

$$\mathscr{L}[a(t+k);p] = p^{-1}(1 - e^{-p})\left[\zeta(p) - \sum_{r=0}^{k-1} a_r e^{-pr}\right]e^{pk} \tag{3-17-12}$$

and, in particular, that

$$\mathscr{L}[a(t+1);p] = p^{-1}(1 - e^{-p})e^{p}[\zeta(p) - a_0] \tag{3-17-13}$$

$$\mathscr{L}[a(t+2);p] = p^{-1}(1 - e^{-p})e^{2p}[\zeta(p) - a_0 - a_1 e^{-p}]. \tag{3-17-14}$$

From equations (3-17-7) and (3-17-13) and the definition (3-17-11) of the operator $\mathscr{D}$ we see that

$$\mathscr{D}[\{a_{n+1}\};p] = e^{p}[\zeta(p) - a_0]. \tag{3-17-15}$$

Similarly, we may rewrite equations (3-17-14) and (3-17-12) in the forms

$$\mathscr{D}[\{a_{n+2}\};p] = e^{2p}[\zeta(p) - a_0 - a_1 e^{-p}] \tag{3-17-16}$$

$$\mathscr{D}[\{a_{n+k}\};p] = e^{kp}\left[\zeta(p) - \sum_{r=0}^{k-1} a_r e^{-pr}\right]. \tag{3-17-17}$$

If α is a constant, it follows immediately from the definition (3-17-10) that

$$\mathscr{D}[\{\alpha^n\};p] = e^{p}(e^{p} - \alpha)^{-1}. \tag{3-17-18}$$

Similarly, if $\{y_n\}$ is the sequence defined by the rule

$$y_n = \begin{cases} \alpha^{n-k}, & n \geqslant k \\ 0, & n < k \end{cases} \tag{3-17-19a}$$

we find that

$$\mathscr{D}[\{y_n\};p] = e^{-(k-1)p}(e^{p} - \alpha)^{-1}. \tag{3-17-19b}$$

We also observe that

$$\mathscr{D}[\{(n+1)\alpha^n\};p] = \sum_{n=0}^{\infty} (n+1)\alpha^n e^{-np} = (1 - \alpha e^{-p})^{-2}$$

and hence that

$$\mathscr{D}[\{(n+1)\alpha^n\};p] = e^{2p}(e^{p} - \alpha)^{-2}. \tag{3-17-20}$$

From this result and the relation

$$\mathscr{D}[\{n\alpha^n\};p] = \alpha e^{-p} \sum_{n=0}^{\infty} (n + 1)\alpha^n e^{-np}$$

we deduce that

$$\mathscr{D}[\{n\alpha^n\};p] = \alpha e^p(e^p - \alpha)^{-2}. \tag{3-17-21}$$

If we define a sequence $\{y_n\}$ by the rule

$$y_n = \begin{cases} \binom{n}{k}\alpha^n, & n \geqslant k, \\ 0, & n < k, \end{cases} \tag{3-17-22a}$$

then we have

$$\mathscr{D}[\{y_n\};p] = \sum_{n=k}^{\infty} \binom{n}{k} e^{-np}\alpha^n = \sum_{r=0}^{\infty} \frac{(k + 1)_r}{r!} e^{-(k+r)p}\alpha^{k+r}$$

$$= \alpha^k e^{-kp}(1 - \alpha e^{-p})^{-k-1}$$

from which we deduce that

$$\mathscr{D}[\{y_n\};p] = \alpha^k e^p(e^p - \alpha)^{-k-1}. \tag{3-17-22b}$$

Also from the definition (3-17-10) we deduce that for a general sequence

$$\mathscr{D}[\{na_n\};p] = -\frac{d\zeta}{dp}. \tag{3-17-23}$$

The operator $\mathscr{D}$ is closely related to the operator of the Z-transform. This latter transform is defined through the equation

$$\mathscr{Z}[\{a_n\};z] = \sum_{n=0}^{\infty} a_n z^{-n} \tag{3-17-24}$$

so that we have the relation

$$\mathscr{D}[\{a_n\};p] = \mathscr{Z}[\{a_n\};e^p]. \tag{3-17-25}$$

Since the Z-transform is related to the Laplace transform through this equation we see that the results obtained by the use of the Z-transform can just as easily be derived by means of the Laplace transform as outlined here. Also, by means of this formula, we may make use of tables of the Z-transform (see, for instance, Appendix I of Jury, 1964) to calculate $\mathscr{D}$-transforms.

We shall now use these results to derive the solutions of some simple difference equations. We begin with the equation

$$a_{n+1} = \alpha a_n + \beta \tag{3-17-26}$$

in which α and β denote constants. If we write

$$\zeta(p) = \mathscr{D}[\{a_n\};p]$$

it follows from equations (3-17-15) and (3-17-18) that

$$e^p[\zeta(p) - a_0] = \alpha\zeta(p) + \beta e^p(e^p - 1)^{-1}.$$

Solving this equation for $\zeta(p)$ we find that

$$\zeta(p) = a_0 \frac{e^p}{e^p - \alpha} + \frac{\beta e^p}{\alpha - 1}\left[\frac{1}{e^p - \alpha} - \frac{1}{e^p - 1}\right].$$

From equation (3-17-8) we deduce that

$$a_n = a_0\alpha^n + \frac{\frac{2}{3}\beta}{\alpha - 1}(\alpha^n - 1). \tag{3-17-27}$$

As an example of the solution of a homogeneous difference equation of the second degree we consider the equation

$$a_{n+2} - (b + c)a_{n+1} + bca_n = 0 \tag{3-17-28}$$

where b and c denote unequal real numbers. If we apply the $\mathscr{D}$-operator to both sides of this equation and make use of equations (3-17-15) and (3-17-16) we obtain the relation

$$[\zeta(p) - a_0 - a_1e^{-p}]e^{2p} - (b + c)[\zeta(p) - a_0]e^p + bc\zeta(p) = 0$$

which may be solved to give the expression

$$\zeta(p) = \frac{a_0e^{2p} + a_1e^p - (b + c)a_0e^p}{e^{2p} - (b + c)e^p + bc}.$$

Resolving this expression into partial fractions we find that

$$\zeta(p) = \frac{a_1 - ca_0}{b - c}\cdot\frac{e^p}{e^p - b} - \frac{a_1 - ba_0}{b - c}\cdot\frac{e^p}{e^p - c}.$$

Identifying the separate terms by means of equation (3-17-18) we obtain the solution

$$a_n = \frac{a_1 - ca_0}{b - c}b^n - \frac{a_1 - na_0}{b - c}c^n. \tag{3-17-29}$$

For instance, by taking $b = e^\alpha$, $c = e^{-\alpha}$, we find that the solution of the difference equation (3-17-1) is

$$i_n = \frac{\sinh(n\alpha)}{\sinh\alpha}i_1 - \frac{\sinh(n - 1)\alpha}{\sinh\alpha}i_0 \tag{3-17-30}$$

where, because of equations (3-17-3) and (3-17-4), i_0 and i_1 are determined by the linear equations

$$(R_1 - R_2)i_0 + R_2 i_1 = V,$$

$$\left[(R_1 + R_2 + R_3)\frac{\sinh(N\alpha)}{\sinh\alpha} - R_2\frac{\sinh(N-1)\alpha}{\sinh\alpha}\right]i_1$$

$$= \left[(R_1 + R_2 + R_3)\frac{\sinh(N-1)\alpha}{\sinh\alpha} - R_2\frac{\sinh(N-2)\alpha}{\sinh\alpha}\right]i_0.$$

The solution has to be modified in an obvious way in the case when $b = c$. The equation then becomes

$$a_{n+2} - 2ca_{n+1} + c^2 a_n = 0 \tag{3-17-31}$$

so that

$$\{a_n\} = \mathscr{D}^{-1}[\zeta(p)]$$

where

$$\zeta(p) = \frac{a_0 e^{2p} + a_1 e^p - 2ca_0 e^p}{(e^p - c)^2}$$

Since the right-hand side of this equation can be resolved into partial fractions

$$a_0 \frac{e^p}{e^p - c} + (a_1 - ca_0)\frac{e^p}{(e^p - c)^2}$$

it follows from equations (3-17-18) and (3-17-21) that the required solution of equation (3-17-31) is

$$a_n = a_0 c^n + (a_1 - ca_0)nc^{n-1}. \tag{3-17-32}$$

The general solution of the inhomogeneous equation

$$a_{n+2} - (b + c)a_{n+1} + bca_n = x^n \tag{3-17-33}$$

in which no two of the numbers b,c,x are equal, may be obtained by adding to the right-hand side of the solution (3-17-29) the solution a_n^* of equation (3-17-33) corresponding to $a_0 = a_1 = 0$.

Applying the $\mathscr{D}$-operator to both sides of equation (3-17-33) and making use of equations (3-17-15) and (3-17-16) with $a_0 = a_1 = 0$ and equation (3-17-18) with $\alpha = x$, we find that

$$[e^{2p} - (b + c)e^p + bc]\zeta*(p) = e^p(e^p - x)^{-1}$$

from which we deduce that

$$\zeta*(p) = \frac{1}{(b-c)(b-x)(c-x)} \left[(c-x)\frac{e^p}{e^p-b} + (x-b)\frac{e^p}{e^p-c} + (b-c)\frac{e^p}{e^p-x}\right]$$

and hence that

$$a_n^* = \frac{(c-x)b^n + (x-b)c^n + (b-c)x^n}{(b-c)(b-x)(c-x)}. \tag{3-17-34}$$

3-18 THE DOUBLE LAPLACE TRANSFORM

The *double Laplace transform* of a function of two variables $f(x,y)$ defined in the positive quadrant of the xy-plane is defined by the equation

$$\bar{f}(p,q) = \mathscr{L}_2[f(x,y);(p,q)] = \mathscr{L}[\mathscr{L}\{f(x,y);x \to p\};y \to q] \tag{3-18-1}$$

and hence by the double integral

$$\bar{f}(p,q) = \int_0^\infty \int_0^\infty f(x,y)e^{-(px+qy)}\,dxdy \tag{3-18-2}$$

whenever that integral exists.

From this definition we deduce immediately that

$$\mathscr{L}_2[f(ax,by);(p,q)] = (ab)^{-1}\bar{f}(p/a,q/b), \qquad (a > 0, b > 0) \tag{3-18-3}$$

and that

$$\mathscr{L}_2[f(ax)g(by);(p,q)] = (ab)^{-1}\bar{f}(p/a)\bar{g}(q/b), \qquad (a > 0, b > 0) \tag{3-18-4}$$

where $\bar{f}(p) = \mathscr{L}[f(x);p]$ and $\bar{g}(q) = \mathscr{L}[g(y);q]$.

In particular, we have the formula

$$\mathscr{L}_2[f(x);(p,q)] = q^{-1}\bar{f}(p). \tag{3-18-5}$$

To evaluate the double transform of a function $f(x + y)$ which depends on x and y only in the combination $x + y$ we change the independent variables from (x,y) to (ξ,η) where

$$x = \tfrac{1}{2}(\xi - \eta), \qquad y = \tfrac{1}{2}(\xi + \eta)$$

(Cf. Fig. 3-20). Then since $dxdy = \frac{1}{2}d\xi\, d\eta$ we have

$$\mathscr{L}_2[f(x+y);(p,q)] = \tfrac{1}{2}\int_0^\infty f(\xi)\,d\xi \int_{-\xi}^{\xi} e^{-\frac{1}{2}(p+q)\xi - \frac{1}{2}(p-q)\eta}\,d\eta.$$

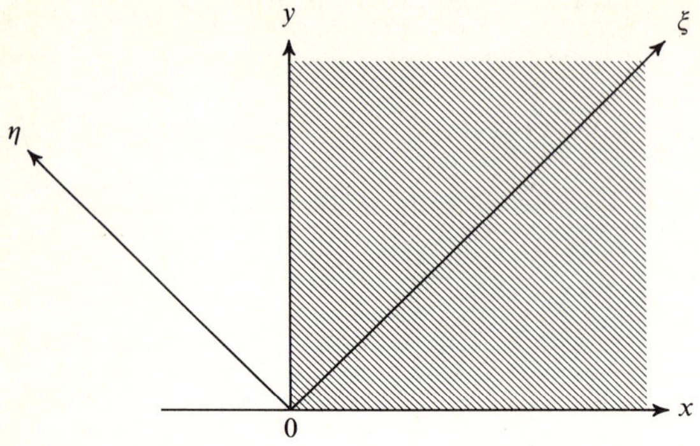

Fig. 3-20

The value of the inner integral is easily shown to be

$$\frac{2}{p-q}(e^{-q\xi} - e^{-p\xi})$$

so that

$$\mathscr{L}_2[f(x+y);(p,q)] = \frac{1}{p-q}[\bar{f}(q) - \bar{f}(p)] \tag{3-18-6}$$

where

$$\bar{f}(p) = \mathscr{L}[f(t);p]$$

There is no such simple formula for a general function of the form $f(x-y)$. If, however, $f(t)$ is an *even* function of t, we have

$$\mathscr{L}_2[f(x-y);(p,q)] = \iint_{Q_1} f(x-y)e^{-px-qy}\,dxdy + \iint_{Q_2} f(\gamma - x)e^{-px-qy}\,dxdy \tag{3-18-7}$$

while if $f(t)$ is an *odd* function of t, we have

$$\mathscr{L}_2[f(x-y);(p,q)] = \iint_{Q_1} f(x-y)e^{-px-qy}\,dxdy - \iint_{Q_2} f(y-x)e^{-px-qy}\,dxdy \tag{3-18-8}$$

where Q_1 and Q_2 are the infinite wedges shown in Fig. 3-21.

If we make the substitutions

$$x = \tfrac{1}{2}(\xi + \eta), \qquad y = \tfrac{1}{2}(\eta - \xi)$$

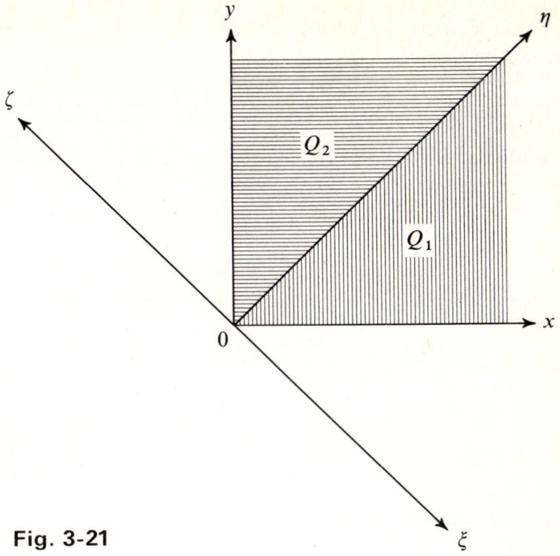

Fig. 3-21

(again see Fig. 3-21), we find that

$$\iint_{Q_1} f(x-y)e^{-px-qy}\,dxdy = \tfrac{1}{2}\int_0^\infty f(\xi)\,d\xi \int_\xi^\infty e^{-\frac{1}{2}(p-q)\xi-\frac{1}{2}(p+q)\eta}\,d\eta$$

$$= \frac{1}{p+q}\int_0^\infty f(\xi)e^{-p\xi}\,d\xi$$

$$= \frac{\bar{f}(p)}{p+q}.$$

On the other hand if we make the substitutions

$$x = \tfrac{1}{2}(\eta - \zeta), \qquad y = \tfrac{1}{2}(\eta + \zeta)$$

we find that

$$\iint_{Q_2} f(y-x)e^{-px-qy}\,dxdy = \tfrac{1}{2}\int_0^\infty f(\zeta)\,d\zeta \int_\zeta^\infty e^{-\frac{1}{2}(q-p)\zeta-\frac{1}{2}(p+q)\eta}\,d\eta$$

$$= \frac{1}{p+q}\int_0^\infty f(\zeta)e^{-q\zeta}\,d\zeta$$

$$= \frac{\bar{f}(q)}{p+q}$$

From equations (3-18-7) and (3-18-8) respectively we see that if $f(u)$ is an *even* function of u,

$$\mathscr{L}_2[f(x-y);(p,q)] = \frac{1}{p+q}[\bar{f}(p) + \bar{f}(q)] \tag{3-18-9}$$

while if $f(u)$ is an *odd* function of u,

$$\mathscr{L}_2[f(x-y);(p,q)] = \frac{1}{p+q}[\bar{f}(p) - \bar{f}(q)]. \tag{3-18-10}$$

Another formula of some interest is

$$\mathscr{L}_2[f(x)\mathrm{H}(x-y);(p,q)] = \int_0^\infty f(x)e^{-px}\,dx \int_0^x e^{-qy}\,dy$$

$$= \frac{1}{q}\int_0^\infty f(x)\{e^{-px} - e^{-(p+q)x}\}\,dx$$

which can be written in the form

$$\mathscr{L}_2[f(x)\mathrm{H}(x-y);(p,q)] = \frac{1}{q}\{\bar{f}(p) - \bar{f}(p+q)\}. \tag{3-18-11}$$

Similarly

$$\mathscr{L}_2[f(x)\mathrm{H}(y-x);(p,q)] = \int_0^\infty f(x)e^{-px}\,dx \int_x^\infty e^{-qy}\,dy$$

$$= \frac{1}{q}\int_0^\infty f(x)e^{-(p+q)x}\,dx$$

which can be expressed as the formula

$$\mathscr{L}_2[f(x)\mathrm{H}(y-x);(p,q)] = \frac{1}{q}\bar{f}(p+q). \tag{3-18-12}$$

The similar formula

$$\mathscr{L}_2[f(x)\mathrm{H}(x+y);(p,q)] = \frac{1}{q}\bar{f}(p)$$

is trivial—it is merely a restatement of equation (3-18-5).

As a special case of (3-18-11) we take $f(x) = 1, \bar{f}(p) = p^{-1}$, to obtain

$$\mathscr{L}_2[\mathrm{H}(x-y);(p,q)] = \frac{1}{p(p+q)}. \tag{3-18-13}$$

There are formulae for the double transforms of partial derivatives of a function of two variables which are straightforward analogues of the results of sec. 3-4. For example, from the definition of $\mathscr{L}_2$ we have

$$\mathscr{L}_2\left[\frac{\partial u}{\partial x};(p,q)\right] = \int_0^\infty e^{-qy}\,dy \int_0^\infty \frac{\partial u}{\partial x} e^{-px}\,dx$$

$$= \int_0^\infty e^{-qy}\,dy \left\{p \int_0^\infty u(x,y)e^{-px}\,dx - u(0,y)\right\}$$

so that we have the formula

$$\mathscr{L}_2\left[\frac{\partial u}{\partial x};(p,q)\right] = p\bar{u}(p,q) - \bar{u}_1(q) \tag{3-18-14}$$

where

$$\bar{u}(p,q) = \mathscr{L}_2[u(x,y);(p,q)], \qquad \bar{u}_1(q) = \mathscr{L}[u(0,y);q].$$

Similarly we can derive the relation

$$\mathscr{L}_2\left[\frac{\partial u}{\partial y};(p,q)\right] = q\bar{u}(p,q) - \bar{u}_2(p) \tag{3-18-15}$$

where $u(p,q)$ has the same interpretation as before and

$$\bar{u}_2(p) = \mathscr{L}[u(x,0);p].$$

Higher derivatives can be dealt with in a precisely similar way. For example, it is easily shown that

$$\mathscr{L}_2\left[\frac{\partial^2 u}{\partial x \partial y};(p,q)\right] = \mathscr{L}\left[\mathscr{L}\left\{\frac{\partial^2 u}{\partial x \partial y};y \to q\right\};x \to p\right]$$

$$= \mathscr{L}\left[q\mathscr{L}\left\{\frac{\partial u}{\partial x};y \to q\right\} - u_x(x,0);x \to p\right]$$

$$= q\mathscr{L}_2\left[\frac{\partial u}{\partial x};(p,q)\right] - pu_2(p) + u(0,0)$$

since

$$\mathscr{L}\left[\frac{\partial u(x,0)}{\partial x};p\right] = p\bar{u}_2(p) - u(0,0).$$

Substituting into this last equation from equation (3-18-14) we obtain the result

$$\mathscr{L}\left[\frac{\partial^2 u}{\partial x \partial y};(p,q)\right] = pq\bar{u}(p,q) - q\bar{u}_1(q) - p\bar{u}_2(p) + u(0,0). \tag{3-18-16}$$

In an exactly similar way we can show that

$$\mathscr{L}_2\left[\frac{\partial^2 u}{\partial x^2};(p,q)\right] = p^2\bar{u}(p,q) - p\bar{u}_1(q) - \bar{u}_3(q) \tag{3-18-17}$$

$$\mathscr{L}_2\left[\frac{\partial^2 u}{\partial y^2};(p,q)\right] = q^2\bar{u}(p,q) - q\bar{u}_2(p) - \bar{u}_4(p) \tag{3-18-18}$$

where $\bar{u}_1(q)$ and $\bar{u}_2(p)$ are defined above and $\bar{u}_3(q)$ and $\bar{u}_4(p)$ are defined by the equations

$$\bar{u}_3(q) = \mathscr{L}[u_x(0,y);q], \qquad \bar{u}_4(p) = \mathscr{L}[u_y(x,0);p].$$

There is also an obvious generalization of the convolution theorem. We define the two-dimensional convolution by

$$f**g = \int_0^x ds \int_0^y f(s,t)g(x - s,y - t)\,dt \tag{3-18-19}$$

and it is readily shown that

$$\mathscr{L}_2[f**g;(p,q)] = \bar{f}(p,q)\bar{g}(p,q). \tag{3-18-20}$$

Tables of double Laplace transforms can be constructed from standard tables of Laplace transforms by means of equation (3-18-1), or directly by double integrations. Extensive tables are given in Ditkin and Prudnikov (1958) but the reader should note that these authors list the form of $F(p,q) = pq\bar{f}(p,q)$—not that of $\bar{f}(p,q)$.

A result which is used in the application of the double Laplace transform to the solution of partial differential equations is that if the function

$$u(x,y) = \mathscr{L}_2^{-1}[\bar{u}(p,q);(x,y)]$$

is to be analytic, the image function $\bar{u}(p,q)$ must be an analytic function for all values of p and q in the region defined by the inequalities $\operatorname{Re} p > \alpha$, $\operatorname{Re} q > \beta$, where α and β are suitably chosen (real) constants. We use this condition to remove a difficulty which is peculiar to this method of solving partial differential equations. To derive expressions for the double Laplace transforms of some of the terms of the differential equation it is sometimes necessary to introduce more boundary (or initial) conditions than the problem demands. We use the condition that the function $\bar{f}(p,q)$ must be analytic to derive "conditions of consistency" by means of which we may determine the forms of the extra functions we have introduced.

To illustrate the procedure we consider the problem of determining, in the positive quadrant, the solution of the partial differential equation

$$\frac{\partial u}{\partial x} = \frac{\partial u}{\partial y} \tag{3-18-21}$$

satisfying the boundary conditions

$$u(x,0) = a(x), \qquad x > 0.$$

We know from the theory of first-order partial differential equations that these conditions are sufficient to determine the unique solution, but to use the method of double Laplace transforms we need to know the value of $u(0,y)$, $(y > 0)$, in order to calculate the double transform of $u_x(x,y)$ by means of equation (3-18-14). We *assume* that $u(0,y) = b(y)$. We then have

$$\mathscr{L}_2\left[\frac{\partial u}{\partial x};(p,q)\right] = p\bar{u}(p,q) - \bar{b}(q)$$

$$\mathscr{L}_2\left[\frac{\partial u}{\partial y};(p,q)\right] = q\bar{u}(p,q) - \bar{a}(p)$$

where

$$\bar{a}(p) = \mathscr{L}[a(x);p], \qquad \bar{b}(q) = \mathscr{L}[b(y);q].$$

The equation (3-18-21) is therefore equivalent to the equation

$$\bar{u}(p,q) = \frac{\bar{b}(q) - \bar{a}(p)}{p - q}$$

for the double Laplace transform of the desired solution $u(x,y)$.

Since $\bar{u}(p,q)$ must be an analytic function for all values of p and q in some region $\operatorname{Re} p > \alpha$, $\operatorname{Re} q > \beta$, the numerator must vanish when $p = q$. In other words we must have

$$\bar{b}(p) = \bar{a}(p)$$

so that, by Lerch's theorem, we deduce that

$$b(y) = a(y), \qquad (y > 0).$$

The required solution is therefore given by the equation

$$u(x,y) = \mathscr{L}_2^{-1}\left[\frac{\bar{a}(q) - \bar{a}(p)}{p - q};(x,y)\right]$$

Another complex of problems in whose solution the double Laplace transform is used is that of functional equations. As an example, we consider *Cauchy's functional equation*

$$f(x + y) = f(x) + f(y). \tag{3-18-22}$$

Making use of equations (3-18-5) and (3-18-6) we see that this equation is equivalent to the relation

$$\frac{1}{p - q}\{\bar{f}(q) - \bar{f}(p)\} = \frac{1}{q}\bar{f}(p) + \frac{1}{p}\bar{f}(q)$$

which can be rewritten in the separable form

$$p^2\bar{f}(p) = q^2\bar{f}(q).$$

Equating each side to the arbitrary constant k, we see that

$$\bar{f}(p) = kp^{-2},$$

showing that the solution of Cauchy's functional equation (3-18-22) is

$$f(x) = kx,$$

where k is an arbitrary constant.

3-19 THE ITERATED LAPLACE TRANSFORM

The special case of the double Laplace transform when the variables p and q are equal has been studied in some detail by Bartels and Churchill (1942), who call such a double transform an *iterated Laplace transform*. Thus, the iterated transform of the function $f(x,y)$ is defined by the equation

$$\mathscr{I}[f(x,y);p] = \mathscr{L}_2[f(x,y);(p,p)] = \int_0^\infty \int_0^\infty f(x,y)e^{-p(x+y)}\,dxdy. \tag{3-19-1}$$

It is obvious that the principal properties of such a transform can be derived from those of the last section by writing $q = p$. There is, however, one interesting result due to Bartels and Churchill which does not emerge from the properties of the double Laplace transform. This concerns the *generalized convolution* $f^{(2)}(x)$ of a function $f(x,y)$ defined by the equation

$$f^{(2)}(x) = \int_0^x f(x-y,y)\,dy. \tag{3-19-2}$$

The Laplace transform of $f^{(2)}(x)$ is defined by the integral

$$\mathscr{L}[f^{(2)}(x);p] = \int_0^\infty e^{-px}\,dx \int_0^x f(x-y,y)\,dy.$$

If we make the transformation

$$x = u + v, \qquad y = v \tag{3-19-3}$$

with Jacobian

$$\frac{\partial(x,y)}{\partial(u,v)} = 1,$$

we see that

$$\iint_A f(x-y,y)e^{-px}\,dxdy = \iint_{A'} f(u,v)e^{-p(u+v)}\,dudv$$

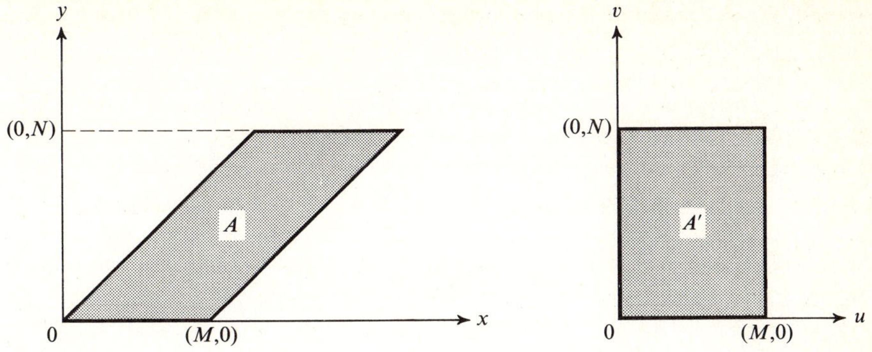

Fig. 3-22

where A is the parallelogram in the xy-plane shown in Fig. 3-22, and A' is the rectangle in the uv-plane onto which it is mapped by the transformation (3-19-3). If we now let $M \to \infty$, $N \to \infty$ we see that

$$\mathscr{L}[f^{(2)}(x);p] = \mathscr{I}[f(x,y);p] \tag{3-19-4}$$

that is, that the Laplace transform of the generalized convolution $f^{(2)}(x)$ of the function $f(x,y)$ is identical with the iterated Laplace transform of the function $f(x,y)$.

3-20 THE STIELTJES TRANSFORM

If we take the Laplace transform with respect to p of the function $\bar{f}(p)$ which is itself the Laplace transform of the function $f(x)$, we have the relation

$$\mathscr{L}[\bar{f}(p);y] = \int_0^\infty e^{-py}\,dp \int_0^\infty f(x)e^{-px}\,dx.$$

Changing the order in which we perform the integrations and making use of the elementary result

$$\int_0^\infty e^{-px-py}\,dp = \frac{1}{x+y}$$

we see that

$$\mathscr{L}[\mathscr{L}\{f(x);p\};y] = \int_0^\infty \frac{f(x)}{x+y}\,dx.$$

If a function is such that the integral on the right-hand side of this last equation converges for (complex) values of y belonging to a set Ω, the equation

$$\bar{f}_2(y) \equiv \mathscr{S}[f(x);y] = \int_0^\infty \frac{f(x)\,dx}{x+y} \tag{3-20-1}$$

is said to define the Stieltjes transform of $f(x)$ for $y \in \Omega$.

The Stieltjes transform arose in connection with Stieltjes' moment problem for the semi-infinite interval (see Shohat and Tamarkin, 1943), and hence is related to certain continued fractions. From our (purely formal!) approach it is obvious that the properties of the Stieltjes transform will be closely related to those of the Laplace transform; for the details the reader is referred to Widder (1937) or to Chapter VIII of Widder (1941). These works are concerned principally with the conditions to be satisfied by the function f and the set Ω for the Stieltjes transform $\bar{f}_2(y)$ to exist. In particular it can be shown that if the integral (3-20-1) converges, the function $\bar{f}_2(y)$ is analytic in the complex plane cut along the negative y-axis. We shall consider here those formal properties of the transform that can be established by elementary means.

For example, the simple identity

$$\frac{x}{x+y} = 1 - \frac{y}{x+y}$$

leads to the property

$$\mathscr{S}[xf(x);y] = \int_0^\infty f(x)\,dx - y\bar{f}_2(y) \tag{3-20-2}$$

and the identity

$$\frac{1}{(x+a)(x+y)} = \frac{1}{a-y}\left[\frac{1}{x+y} - \frac{1}{x+a}\right]$$

leads to the relation

$$\mathscr{S}\left[\frac{f(x)}{x+a};y\right] = \frac{1}{a-y}[\bar{f}_2(y) - \bar{f}_2(a)]. \tag{3-20-3}$$

By changing the variable of integration in the defining integral from x to t where $t = ax$, we deduce that

$$\mathscr{S}[f(ax);y] = \bar{f}_2(ay), \qquad (a > 0). \tag{3-20-4}$$

Similarly, since by substituting $x = u^{-1}$ we can show that

$$\int_0^\infty \frac{x^{-1}f(x^{-1})\,dx}{x+y} = \int_0^\infty \frac{f(u)\,du}{1+yu}$$

we deduce that

$$\mathcal{S}[x^{-1}f(x^{-1});y] = y^{-1}\bar{f}_2(y^{-1}). \tag{3-20-5}$$

Combining (3-20-4) and (3-20-5) we obtain the formula

$$\mathcal{S}[x^{-1}f(a/x);y] = y^{-1}\bar{f}_2(a/y). \tag{3-20-6}$$

The substitution $x = u^2$ shows that

$$\mathcal{S}[f(x^{\frac{1}{2}});y] = 2\int_0^\infty \frac{uf(u)\,du}{u^2 + y}.$$

Using the partial fraction expansion

$$\frac{2u}{u^2 + y} = \frac{1}{u + iy^{\frac{1}{2}}} + \frac{1}{u - iy^{\frac{1}{2}}}$$

we obtain the equation

$$\mathcal{S}[f(x^{\frac{1}{2}});y] = \bar{f}_2(iy^{\frac{1}{2}}) + \bar{f}_2(-iy^{\frac{1}{2}}). \tag{3-20-7}$$

Applying the rule for integrating by parts, we see that

$$\int_0^\infty \frac{f'(x)\,dx}{x + y} = \left[\frac{f(x)}{x + y}\right]_0^\infty + \int_0^\infty \frac{f(x)\,dx}{(x + y)^2}$$

which can be written as the formula

$$\mathcal{S}[f'(x);y] = -y^{-1}f(0) - \frac{d\bar{f}_2}{dy} \tag{3-20-8}$$

for the Stieltjes transform of the derivative of a function.

In some cases it is a simple matter to calculate the Stieltjes transform directly from the definition (3-20-1). For example, if $|\arg a| < \pi$, $|\arg y| < \pi$, we have

$$\mathcal{S}[(x + a)^{-1};y] = \int_0^\infty \frac{dx}{(x + a)(x + y)} = \frac{1}{a - y}\log\left|\frac{a}{y}\right|. \tag{3-20-9}$$

On the other hand it is often easier to calculate the Stieltjes transform as a repeated Laplace transform. For instance

$$\mathcal{L}[x^{-\frac{1}{2}}e^{-ax};p] = \pi^{\frac{1}{2}}(p + a)^{-\frac{1}{2}},$$
$$\mathcal{L}[(p + a)^{-\frac{1}{2}};y] = \pi^{\frac{1}{2}}y^{-\frac{1}{2}}e^{ay}\operatorname{Erfc}(a^{\frac{1}{2}}y^{\frac{1}{2}})$$

so that we immediately have the formula

$$\mathcal{S}[x^{-\frac{1}{2}}e^{-ax};y] = \pi y^{-\frac{1}{2}}e^{ay}\operatorname{Erfc}(a^{\frac{1}{2}}y^{\frac{1}{2}}). \tag{3-20-10}$$

Extensive tables of the Stieltjes transform are given on pp. 216–232 of Vol. 2 of Erdelyi *et al.* (1954).

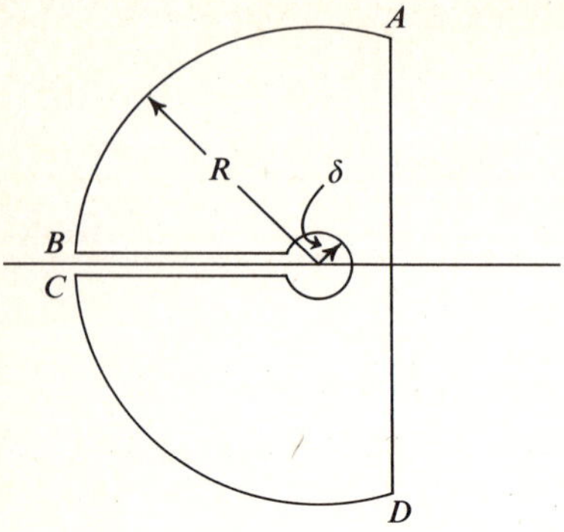

Fig. 3-23

We can derive the inversion formula for the Stieltjes transform in a simple way by considering the integral of the function $g(z)e^{pz}$ round the contour C in the z-plane shown in Fig. 3-23, it being assumed that $g(z)$ is analytic in the z-plane cut along the negative real axis, and satisfies the growth condition

$$\lim_{|z|\to\infty} g(z)e^{pz} = 0, \qquad \operatorname{Re} z \leqslant c$$

for large $|z|$. By Cauchy's theorem

$$\int_C g(z)e^{pz}\,dz = 0$$

so that

$$\int_{c-iR}^{c+iR} g(z)e^{pz}\,dz = \int_R^\delta g(xe^{-i\pi})e^{-px}\,dx - \int_\delta^R g(xe^{i\pi})e^{-px}\,dx$$
$$+ \left(\int_\Gamma + \int_{AB} + \int_{CD}\right) g(z)e^{pz}\,dz$$

where Γ denotes the circle $|z| = \delta$. If $g(z)$ is such that $zg(z) \to 0$ as $|z| \to 0$, then

$$\lim_{\delta\to 0} \int_\Gamma g(z)e^{pz}\,dz = 0$$

and from the growth condition imposed on $g(z)$ we see that

$$\lim_{R\to\infty} \int_{AB} g(z)e^{pz}\,dz = 0, \qquad \lim_{R\to\infty} \int_{CD} g(z)e^{pz}\,dz = 0$$

so that letting $\varepsilon \to 0$, $R \to \infty$, we obtain the result

$$2\pi i \mathscr{L}^{-1}[g(z);p] = \mathscr{L}[g(xe^{-i\pi}) - g(xe^{i\pi});p]$$

Applying the operator $\mathscr{L}$ to both sides of this equation we find that

$$\mathscr{L}\{[\mathscr{L}(2\pi i)^{-1}\{g(xe^{-i\pi}) - g(xe^{i\pi})\};p];z\} = g(z).$$

Comparing this equation with the relation

$$\mathscr{L}\{\mathscr{L}[f(x);p];y\} = \mathscr{S}[f(x);y]$$

we see that

$$\mathscr{S}^{-1}[g(y);x] = (2\pi i)^{-1}\{g(xe^{-i\pi}) - g(xe^{i\pi})\}. \tag{3-20-11}$$

For a rigorous proof of this inversion theorem the reader is referred to pp. 340–341 of Widder (1941) where the theorem is proved in the form

Theorem 7 *If $f(x)$ is absolutely integrable in $[0,R]$ for every positive R and is such that the integral*

$$\bar{f}_2(y) = \int_0^\infty \frac{f(x)\,dx}{x + y}$$

converges, then for every positive x at which $f(x+)$ and $f(x-)$ exist,

$$\lim_{\varepsilon\to 0+} \frac{\bar{f}_2(-x - i\varepsilon) - \bar{f}_2(-x + i\varepsilon)}{2\pi i} = \tfrac{1}{2}\{f(x+) + f(x-)\}.$$

3-21 THE HILBERT TRANSFORM

Closely related to the Stieltjes transform is the *Hilbert transform* defined by the equation

$$\bar{f}_H(x) \equiv \mathscr{H}[f(t);x] = \frac{1}{\pi}\int_{-\infty}^{\infty} \frac{f(t)\,dt}{t - x} \tag{3-21-1}$$

where x is real and the integral is a Cauchy principal value.[1] We sometimes say that $\bar{f}_H(x)$ is *conjugate* to $f(x)$.

If $x > 0$, we may write

$$\bar{f}_H(x) = \frac{1}{\pi}\int_0^\infty \frac{f(t)\,dt}{t - x} + \frac{1}{\pi}\int_{-\infty}^{0} \frac{f(t)\,dt}{t - x}$$

$$= \frac{1}{2\pi}\int_0^\infty \frac{f(t)\,dt}{t + xe^{i\pi}} + \frac{1}{2\pi}\int_0^\infty \frac{f(t)\,dt}{t + xe^{-i}} - \frac{1}{\pi}\int_0^\infty \frac{f(-t)\,dt}{t - x}$$

[1] I.e. the integral is interpreted in the sense

$$\lim_{\varepsilon\to 0}\left(\int_{-\infty}^{x-\varepsilon} + \int_{x+\varepsilon}^{\infty}\right)\frac{f(t)\,dt}{t - x}.$$

a relationship which may be interpreted in the form

$$\mathscr{H}[f(t);x] = (2\pi)^{-1}\mathscr{S}[f(t);xe^{i\pi}] + (2\pi)^{-1}\mathscr{S}[f(t);xe^{-i\pi}] - \pi^{-1}\mathscr{S}[f(-t);x] \qquad (x > 0) \tag{3-21-2}$$

In a similar way we can show that

$$\mathscr{H}[f(t);x] = \pi^{-1}\mathscr{S}[f(t);-x] - (2\pi)^{-1}\mathscr{S}[f(-t);|x|e^{i\pi}] - (2\pi)^{-1}\mathscr{S}[f(-t);|x|e^{-i\pi}] \qquad (x < 0). \tag{3-21-3}$$

The Hilbert transform occurs occasionally in mathematical physics so it is useful to have its inversion formula available. Formally we may write (3-21-1) in the form

$$\bar{f}_H(x) = (2\pi)^{-\frac{1}{2}} \int_{-\infty}^{\infty} f(t)g(x-t)\,dt$$

where

$$g(x) = -(2/\pi)^{\frac{1}{2}} x^{-1}$$

so that its Fourier transform is

$$G(\xi) = -i\,\mathrm{sgn}\,(\xi).$$

Applying the convolution theorem we obtain the relation

$$\bar{F}_H(\xi) = -i\,\mathrm{sgn}\,(\xi)F(\xi) \tag{3-21-4}$$

between $\bar{F}_H(\xi)$, the Fourier transform of $\bar{f}_H(x)$, and $F(\xi)$, the Fourier transform of $f(x)$. Writing this relation in the form

$$F(\xi) = \mathrm{i}\,\mathrm{sgn}\,(\xi)\bar{F}_H(\xi)$$

and reversing the steps of the above argument we see that

$$f(x) = -\frac{1}{\pi}\int_{-\infty}^{\infty} \frac{\bar{f}_H(t)\,dt}{t-x} \tag{3-21-5}$$

that is, that the relationship between $f(x)$ and $\bar{f}_H(x)$ is skew-symmetrical in the sense that if $\bar{f}_H(x)$ is conjugate to $f(x)$, then $-f(x)$ is conjugate to $\bar{f}_H(x)$.

For a rigorous proof that this is indeed so the reader is referred to pp. 121–125 of Titchmarsh (1948) where the result is discussed in the form of:

Theorem 8 *If $f(x)$ is square integrable over $(-\infty,\infty)$, then the formula*

$$\bar{f}_H(x) = -\frac{1}{\pi}\frac{d}{dx}\int_{-\infty}^{\infty} f(t)\log\left|1-\frac{x}{t}\right|dt \tag{3-21-6}$$

defines almost everywhere a function $\bar{f}_H(x)$ which is also square-integrable. The reciprocal formula

$$f(t) = \frac{1}{\pi}\frac{d}{dt}\int_{-\infty}^{\infty} \bar{f}_H(x)\log\left|1 - \frac{t}{x}\right| dx \tag{3-21-7}$$

holds almost everywhere and

$$\|f\| = \|\bar{f}_H\|. \tag{3-21-8}$$

We note that the inversion formula can be written in the operator form

$$\mathscr{H}^{-1} = -\mathscr{H} \tag{3-21-9}$$

The elementary properties of the Hilbert transform:

$$\mathscr{H}[f(a + t); x] = \mathscr{H}[f(t); x + a] \tag{3-21-10}$$

$$\mathscr{H}[f(at); x] = \mathscr{H}[f(t); ax], \qquad (a > 0) \tag{3-21-11}$$

$$\mathscr{H}[f(-at); x] = -\mathscr{H}[f(t); -ax], \qquad (a > 0) \tag{3-21-12}$$

$$\mathscr{H}[f'(t); x] = \frac{d}{dx}\bar{f}_H(x) \tag{3-21-13}$$

$$\mathscr{H}[tf(t); x] = x\bar{f}_H(x) + \pi^{-1}\int_{-\infty}^{\infty} f(t)\, dt \tag{3-21-14}$$

follow directly from the definition (3-21-1) and may readily be established by the reader.

The calculation of the Hilbert transform of any prescribed function may be carried out either directly from the definition (3-21-1) or by use of the formulae (3-21-2) and (3-21-3). Extensive tables are given on pp. 244–262 of Vol. 2 of Erdelyi *et al.* (1954).

As an example of the use of both methods we consider the Hilbert transform of the function $t(t^2 + a^2)^{-1}$ where, for convenience, we take a to be real and positive. Integrating both sides of the identity

$$\frac{t}{(t^2 + a^2)(t - x)} = \frac{1}{x^2 + a^2}\left[\frac{a^2}{t^2 + a^2} + \frac{x}{t - x} - \frac{xt}{t^2 + a^2}\right]$$

with respect to t over the whole real line we obtain the formula

$$\mathscr{H}\left[\frac{t}{t^2 + a^2}; x\right] = \frac{a}{x^2 + a^2}. \tag{3-21-15}$$

On the other hand, if we use the formula

$$\mathscr{S}\left[\frac{t}{t^2 + a^2}; x\right] = \frac{1}{x^2 + a^2}\left[\tfrac{1}{2}\pi a + y\log(y/a)\right]$$

in equation (3-21-2) we find that for $x > 0$

$$\mathscr{H}\left[\frac{t}{t^2 + a^2}; x\right] = \frac{1}{x^2 + a^2}\left[\frac{1}{4}a - \frac{x}{2\pi}\log(xe^{i\pi}/a)\right.$$

$$\left. + \frac{1}{4}a - \frac{x}{2\pi}\log(xe^{-i\pi}/a) + \frac{1}{2}a + \frac{x}{\pi}\log(x/a)\right]$$

which gives the result (3-21-15) in the case $x > 0$. In a similar fashion formula (3-21-3) gives the same result in the case $x < 0$.

An alternative method, due to Titchmarsh, is available for the generation of conjugate pairs. It also shows clearly the relation between the Hilbert transform and the Fourier transform.

If we write

$$a(t) = \frac{1}{\pi}\int_{-\infty}^{\infty} f(u)\cos(tu)\,du, \qquad b(t) = \frac{1}{\pi}\int_{-\infty}^{\infty} f(u)\sin(tu)\,du \tag{3-21-16}$$

we may express Fourier's integral in the form

$$f(x) = \tfrac{1}{2}\int_{-\infty}^{\infty} \{a(t) + ib(t)\}e^{-ixt}\,dt$$

i.e., in the form

$$f(x) = \int_0^{\infty} \{a(t)\cos(xt) + b(t)\sin(xt)\}\,dt.$$

If we now define a function

$$\Phi(z) = \int_0^{\infty} \{a(t) - ib(t)\}e^{izt}\,dt, \qquad (z = x + iy) \tag{3-21-17}$$

and write

$$\Phi(z) = U(x,y) + iV(x,y) \tag{3-21-18}$$

we see that

$$\lim_{y\to 0} U(x,y) = f(x).$$

Also we see that

$$-\lim_{y\to 0} V(x,y) = \int_0^{\infty} \{b(t)\cos(xt) - a(t)\sin(xt)\}\,dt$$

so that substituting for $a(t)$ and $b(t)$ from equations (3-21-16) we find that

$$-\lim_{y\to 0} V(x,y) = \frac{1}{\pi}\int_0^\infty dt \int_{-\infty}^{\infty} \sin\{(u-x)t\}f(u)\,du$$

$$= \lim_{\lambda\to\infty} \frac{1}{\pi}\int_{-\infty}^{\infty} \frac{1-\cos\{\lambda(u-x)\}}{u-x} f(u)\,du$$

$$= \lim_{\lambda\to\infty} \frac{1}{\pi}\int_0^\infty \frac{1-\cos(\lambda t)}{t}\{f(x+t)-f(x-t)\}\,dt.$$

Now if $f(t)$ is sufficiently smooth it follows from the Riemann–Lebesgue lemma that

$$\lim_{\lambda\to\infty} \int_0^\infty \cos(\lambda t)\frac{f(x+t)-f(x-t)}{t}\,dt = 0$$

so that we find that

$$-\lim_{y\to 0} V(x,y) = \frac{1}{\pi}\int_0^\infty \frac{f(x+t)-f(x-t)}{t}\,dt$$

and hence that

$$-\lim_{y\to 0} V(x,y) = \mathscr{H}[f(t);x].$$

In other words we have shown that for an analytic function $\Phi(z)$ of the form (3-21-14) the function $-V(x,0)$ is conjugate to the function $U(x,0)$.

For instance, by taking

$$\Phi(z) = \frac{1}{z+ia} = \frac{x}{x^2+(y+a)^2} - i\frac{y+a}{x^2+(y+a)^2}$$

with a real and positive, we see that the function $a(a^2+x^2)^{-1}$ is conjugate to the function $x(x^2+a^2)^{-1}$ in agreement with equation (3-21-11).

As an illustration of the occurrence of Hilbert transforms in mathematical physics, we consider the solution of Laplace's equation $\Delta_2 u(x,y) = 0$ in the half-plane $y \geqslant 0$, subject to the boundary condition

$$u_x(x,0) = f(x) \tag{3-21-19}$$

and the condition at infinity:

$$u(x,y) \to 0 \qquad \text{as } (x^2+y^2)^{\frac{1}{2}} \to \infty.$$

By the methods of subsec. 2-16-1(a) we can show that the solution of this boundary value problem is

$$u(x,y) = (2\pi)^{-\frac{1}{2}} \int_{-\infty}^{\infty} f(u)K(x-u,y)\,du$$

where the kernel K is defined by the equation

$$K(x,y) = \mathscr{F}^{-1}[i\xi^{-1}e^{-|\xi|y}; \xi \to x] = (2/\pi)^{\frac{1}{2}} \tan^{-1}(x/y)$$

so that

$$u(x,y) = \frac{1}{\pi}\int_{-\infty}^{\infty} f(u)\tan^{-1}\left(\frac{x-u}{y}\right) du.$$

For this solution we have the relation

$$u_y(x,0) = \frac{1}{\pi}\int_{-\infty}^{\infty} \frac{f(u)\,du}{u-x}.$$

In other words the normal derivative on the boundary $u_y(x,0)$ is the Hilbert transform of the tangential derivative $u_x(x,0)$.

This last remark can be used to find the solution of the mixed boundary value problem

$$u(x,0) = \phi(x), \qquad |x| < 1 \tag{3-21-20}$$

$$u_y(x,0) = 0, \qquad |x| > 1 \tag{3-21-21}$$

for a function harmonic in the positive half-plane $y \geqslant 0$ and tending to zero as $(x^2 + y^2)^{\frac{1}{2}} \to \infty$. From (3-21-20) we deduce that

$$u_x(x,0) = \phi'(x), \qquad |x| < 1$$

and from (3-21-21) that

$$u_y(x,0) = \chi(x)\mathrm{H}(1 - |x|)$$

where the function $\chi(x)$ is unknown and H denotes the Heaviside unit function. Now we have just shown that $u_y(x,0)$ is the Hilbert transform of $u_x(x,0)$; we use this in the inverse form that $-u_x(x,0)$ is the Hilbert transform of $u_y(x,0)$. This gives us the relation

$$\mathscr{H}[\chi(t)\mathrm{H}(1 - |t|); x] = -\phi'(x),$$

which is easily seen to be equivalent to the integral equation

$$\int_{-1}^{1} \frac{\chi(t)\,dt}{t-x} + \pi\phi'(x) = 0, \qquad |x| < 1 \tag{3-21-22}$$

which is known (Solingen, 1939) to have the solution

$$\chi(t) = C(1 - t^2)^{-\frac{1}{2}} + \pi^{-1}(1 - t^2)^{-\frac{1}{2}} \int_{-1}^{1} \frac{\phi'(x)(1 - x^2)^{\frac{1}{2}}}{t - x}\,dx \tag{3-21-23}$$

where C denotes an arbitrary constant.

3-22 THE BILATERAL LAPLACE TRANSFORM

The Laplace transform defined in sec. 3-1 is useful for transforming problems involving a function which is defined for *positive* real values of the independent variable. We shall conclude this chapter by considering briefly a transform, with the function e^{-px} again as kernel, but which can be used on functions which are defined for *all* real values of the independent variable.

If the function $f(x)$ is prescribed over the entire real line, we define its *bilateral Laplace transform* by the equation

$$\bar{f}_+(p) \equiv \mathscr{L}_+[f(x);p] = \int_{-\infty}^{\infty} f(x)e^{-px}\,dx \tag{3-22-1}$$

provided that $f(x)$ is such that the integral is convergent for some values of p. The function $\bar{f}_+(p)$ is sometimes called the *two-sided Laplace transform* of $f(x)$.

Writing the integral on the right-hand side of (3-22-1) as

$$\int_0^{\infty} f(x)e^{-px}\,dx + \int_0^{\infty} f(-x)e^{+px}\,dx$$

we see that we may consider the bilateral transform as the sum of two ordinary Laplace transforms:

$$\mathscr{L}_+[f(x);p] = \mathscr{L}[f(x);p] + \mathscr{L}[f(-x);-p]. \tag{3-22-2}$$

The first term on the right-hand side of (3-22-2) is a Laplace integral with abscissa of convergence σ_1, say; i.e., the first term exists if and only if $\operatorname{Re} p > \sigma_1$. The second term is a Laplace integral with parameter $-p$; if its abscissa of convergence is $-\sigma_2$, then this integral will exist if and only if $\operatorname{Re}(-p) > -\sigma_2$, i.e., $\operatorname{Re}(p) < \sigma_2$. If, therefore, the bilateral Laplace transform of $f(x)$ is to exist these two half-planes must overlap, i.e., we must have $\sigma_2 > \sigma_1$. When this condition is satisfied the bilateral Laplace transform converges only in the strip

$$\sigma_1 < \operatorname{Re} p < \sigma_2$$

of the complex p-plane. If $\sigma_2 = \sigma_1$, this strip contracts to the vertical line $\operatorname{Re} p = \sigma_1$, while if $\sigma_2 < \sigma_1$, the Laplace transforms on the right-hand side of (3-22-2) have no common region of convergence, and the function $f(x)$ does *not* have a bilateral Laplace transform.

For example, if $f(x) = e^{-a|x|}$, then $\mathscr{L}[f(x);p]$ converges if $\operatorname{Re} p > -a$, and $\mathscr{L}[f(-x);-p]$ converges if $\operatorname{Re}(-p) > -a$, so the bilateral Laplace transform of $e^{-a|x|}$ exists for $-a < \operatorname{Re} p < a$. By a similar argument we can show that e^{-ax} does not possess a bilateral Laplace transform, while e^{-x^2} has one for all real values of $\operatorname{Re} p$.

The inversion theorem for the bilateral Laplace transform may be established by a method similar to that used in the proof of the inversion theorem for the Laplace transform (Cf. sec. 3-10). It can be shown that if the integral

$$\bar{f}_+(p) = \int_{-\infty}^{\infty} f(x)e^{-px}\,dx \tag{3-22-3}$$

is absolutely convergent in the strip

$$\sigma_1 < \mathrm{Re}\,p < \sigma_2$$

of the p-plane, then

$$f(x) = \frac{1}{2\pi i}\int_{c-i\infty}^{c-i\infty} f_+(p)e^{px}\,dp \tag{3-22-4}$$

where $\sigma_1 < c < \sigma_2$.

An operational calculus has been based on the bilateral Laplace transform by van der Pol and Bremmer (1950). The transform pair defined by these authors

$$f(p) = p\int_{-\infty}^{\infty} e^{-pt}h(t)\,dt, \qquad \sigma_1 < \mathrm{Re}\,p < \sigma_2 \tag{3-22-5}$$

$$h(t) = \frac{1}{2\pi i}\int_{c-i\infty}^{c+i\infty} \frac{f(p)}{p}e^{pt}\,dp, \qquad \sigma_1 < c < \sigma_2 \tag{3-22-6}$$

bears an obvious relation to the bilateral Laplace transform defined above. van der Pol and Bremmer express the relationship between $h(t)$ and $f(p)$ symbolically by

$$f(p) \fallingdotseq h(t), \qquad \sigma_1 < \mathrm{Re}\,p < \sigma_2 \tag{3-22-7}$$

or alternatively

$$h(t) \fallingdotseq f(p), \qquad \sigma_1 < \mathrm{Re}\,p < \sigma_2 \tag{3-22-8}$$

and say that $f(p)$ is the *image* of $h(t)$, or alternatively, that $h(t)$ is the *original* of $f(p)$.

It should be noticed that the region of convergence *must* always be specified since two distinct originals may give rise to the same image but in a different region of the complex p-plane. For instance

$$1 \fallingdotseq \mathrm{H}(t), \qquad \mathrm{Re}\,p > 0$$

and

$$1 \fallingdotseq -\mathrm{H}(-t), \qquad \mathrm{Re}\,p < 0$$

—two different original functions have the same image 1 but in two distinct regions of the p-plane.

The operational properties of these pairs are readily established. For instance, if

$$f_1(p) \risingdotseq h_1(t), \qquad \alpha_1 < \operatorname{Re} p < \beta_1$$
$$f_2(p) \risingdotseq h_2(t), \qquad \alpha_2 < \operatorname{Re} p < \beta_2$$

then

$$f_1(p) \pm f_2(p) \risingdotseq h_1(t) \pm h_2(t), \qquad \max(\alpha_1,\alpha_2) < \operatorname{Re} p < \min(\beta_1,\beta_2) \tag{3-22-9}$$

provided, of course, that such a strip of common convergence exists.

If the relation (3-22-8) exists and $\lambda > 0$, then

$$h(\lambda t) \risingdotseq f(p/\lambda), \qquad \lambda\sigma_1 < \operatorname{Re} p < \lambda\sigma_2 \tag{3-22-10}$$

while if $\lambda < 0$,

$$h(\lambda t) \risingdotseq -f(p/\lambda), \qquad \lambda\sigma_2 < \operatorname{Re} p < \lambda\sigma_1. \tag{3-22-11}$$

Similarly if $h_1(t)$ and $h_2(t)$ are defined as in equation (3-22-9), the convolution theorem for such transforms takes the form

$$p^{-1}f_1(p)f_2(p) = \int_{-\infty}^{\infty} h_1(u)h_2(t-u)\,du \tag{3-22-11}$$

provided, again, that

$$\max(\alpha_1,\alpha_2) < \operatorname{Re} p < \min(\beta_1,\beta_2).$$

For the proofs of these results and their application to the development of an operational calculus, the reader is referred to van der Pol and Bremmer (1950). A more theoretical discussion of the bilateral Laplace transform is given in Chapter VI of Widder's treatise on the Laplace transform (Widder, 1941).

In many applications to the discussion of physical problems it is found to be easier to use the complex form of Fourier's theorem (sec. 2-11) than the bilateral Laplace transform or its van der Pol–Bremmer equivalent.

PROBLEMS

3-1 Evaluate

$$\mathscr{L}^{-1}[p^{-3}(1+p^2)^{-1};t]$$

by (*a*) expressing the function in partial fractions,
(*b*) using the convolution theorem.

3-2 Determine the first four terms in the expansion of

$$\mathscr{L}^{-1}[p^3(p^5 - p^3 - 2)^{-1};t]$$

as a power series in t.

3-3 By expanding $\tan^{-1}(p^{-1})$ in descending powers of p, and taking the Laplace transform of each term, show that

$$\mathscr{L}^{-1}\left[\tan^{-1}\left(\frac{1}{p}\right);t\right] = \frac{\sin t}{t}.$$

3-4 Evaluate

$$\mathscr{L}^{-1}[p(p^2 - a^2)^{-1}e^{-bp};t], \qquad \mathscr{L}^{-1}[(p^2 + a^2)^{-1}(1 - e^{-2\pi p/a});t].$$

3-5 Prove that

$$\mathscr{L}[f(e^{-t});p] = \mathscr{M}[f(x)\mathrm{H}(1 - x);p]$$

and deduce that if $a + b > c$

$$\mathscr{L}^{-1}\left[\frac{\Gamma(p)\Gamma(p + c)}{\Gamma(p + a)\Gamma(p + b)};t\right]$$
$$= e^{-ct}(1 - e^{-t})^{a+b-c-1}{}_2\mathrm{F}_1(a,b;a + b - c;1 - e^{-t})/\Gamma(a + b - c).$$

3-6 If the exponential integral is defined by the equation

$$E_1(t) = \int_1^\infty u^{-1}e^{-tu}\,du$$

prove that

$$\mathscr{L}[E_1(t);p] = p^{-1}\log(p + 1).$$

3-7 Show that if $\mathrm{Re}\,p > -1$

$$\mathscr{L}[\operatorname{sech} t;p] = \tfrac{1}{2}\psi(\tfrac{1}{4}p + \tfrac{3}{4}) - \tfrac{1}{2}\psi(\tfrac{1}{4}p + \tfrac{1}{4}).$$

Deduce the value of

$$\int_0^\infty \frac{\cosh(px)}{\cosh x}\,dx, \qquad 0 \leqslant \mathrm{Re}\,p < 1$$

and hence show that

$$\int_0^\infty \frac{\sinh(px)}{\cosh x}\frac{dx}{x} = \log\cot\{\tfrac{1}{4}(1 - p)\pi\}.$$

Making the substitution $p = \eta - i\xi$, with ξ and η both real, and letting $\eta \to 1$, deduce that

$$\mathscr{F}_c[x^{-1}\tanh x;\xi] = (2/\pi)^{\frac{1}{2}}\log\coth(\tfrac{1}{4}\pi\xi).$$

By a similar method show that

$$\mathscr{F}_s[x^{-1}\operatorname{sech} x;\xi] = 2(2/\pi)^{\frac{1}{2}}\tan^{-1}\{\tanh(\tfrac{1}{4}\pi\xi)\}$$

3-8 Prove that

$$\mathscr{L}^{-1}[\bar{f}(\log p);t] = \int_0^\infty \frac{t^{u-1}}{\Gamma(u)}f(u)\,du.$$

3-9 Prove that

$$(a)\ \mathscr{L}[t^{-\frac{1}{2}}\cos(at^{\frac{1}{2}});p] = (\pi/p)^{\frac{1}{2}}e^{-a^2/4p}$$

and deduce that

$$(b)\ \mathscr{L}[t^{-1}\sin(at^{\frac{1}{2}});p] = \operatorname{Erf}(\tfrac{1}{2}ap^{-\frac{1}{2}})$$

$$(c)\ \mathscr{L}[\sin(at^{\frac{1}{2}});p] = \tfrac{1}{2}a(\pi/p^3)^{\frac{1}{2}}e^{-a^2/4p}.$$

3-10 If

$$K(x,t) = \mathscr{L}^{-1}\left[\frac{e^{-x\sqrt{p}}}{\sqrt{(p)}+h};p \to t\right]$$

show that

$$\left(\frac{\partial}{\partial x} - h\right)K(x,t) = -x(4\pi t^3)^{-\frac{1}{2}}e^{-x^2/4t}$$

and hence deduce that

$$K(x,t) = -(\pi t)^{-\frac{1}{2}}e^{-x^2/4t} + he^{hx+h^2t}\operatorname{Erfc}(\tfrac{1}{2}xt^{-\frac{1}{2}} + ht^{\frac{1}{2}}).$$

3-11 Prove that

$$\int_0^\infty \frac{J_0(t) - \cos t}{t}\,dt = \log_e 2.$$

3-12 Show that if n is a positive integer

$$z^{-n}J_n(z) = (-1)^n\left(\frac{1}{z}\frac{d}{dz}\right)^n J_0(z)$$

and deduce that

$$\mathscr{L}[t^nJ_n(at);p] = (-1)^n\left(\frac{1}{a}\frac{\partial}{\partial a}\right)^n(p^2+a^2)^{-\frac{1}{2}}$$

$$= \frac{(2n)!}{2^n n!}(p^2+a^2)^{-n-\frac{1}{2}}$$

$$\mathscr{L}[t^{n+1}J_n(at);p] = \frac{(2n+1)!}{2^n n!}p(p^2+a^2)^{-n-\frac{3}{2}}.$$

3-13 Prove that

$$\sum_{n=0}^{\infty}\frac{(\frac{1}{2}t)^nJ_n(t)}{n!(2n+1)} = \frac{1-\cos t}{t}.$$

3-14 Show that if $p > 1$

$$\mathscr{L}^{-1}[(1 - 2p\cos\theta + p^2)^{-\frac{1}{2}};x] = e^{x\cos\theta}J_0(x\sin\theta)$$

and deduce that

$$\sum_{n=0}^{\infty} P_n(\cos\theta)\frac{x^n}{n!} = e^{x\cos\theta}J_0(x\sin\theta)$$

P_n denoting the Legendre polynomial of degree n

3-15 Prove that

$$\frac{d}{dt}\int_0^t \mathrm{Erf}(x^{\frac{1}{2}})I_0(t-x)\,dx = \pi^{-\frac{1}{2}}e^{-t}\int_0^t e^{2x}x^{-\frac{1}{2}}\,dx.$$

3-16 If $\mathscr{L}[f(t);p]$ can be written as the ratio of two polynomials

$$\bar{f}(p) = f_1(p)/f_2(p)$$

with

$$\bar{f}_2(p) = (p-a_1)^{m_1}(p-a_2)^{m_2}\ldots(p-a_n)^{m_n}$$

$m_1, m_2, \ldots, m_n$ being positive integers and the degree of the polynomial $f_1(p)$ being less than $m_1 + m_2 + \cdots + m_n$, show that

$$f(t) = \sum_{r=1}^{n} e^{a_r t}\sum_{s=0}^{m_r-1} A_{rs}t^s$$

where the constants A_{rs} are defined by the equation

$$A_{rs} = \frac{1}{s!(m_r-s-1)!}\left[\frac{d^{m_r-s-1}}{dp^{m_r-s-1}}(p-a_r)^{m_r}\bar{f}(p)\right]_{p=a_r}$$

3-17 Show that if $\mathrm{Re}\,p > 1$,

$$\mathscr{L}[J_\nu(t);p] = 2^{-\nu}p^{-\nu-1}{}_2F_1(\tfrac{1}{2}\nu+\tfrac{1}{2},\tfrac{1}{2}\nu+1;\nu+1;-p^{-2}).$$

Using the result

$${}_2F_1(a,a+\tfrac{1}{2};2a;z) = 2^{2a-1}(1-z)^{-\frac{1}{2}}\{1+(1-z)^{\frac{1}{2}}\}^{1-2a}$$

show that

$$\mathscr{L}[J_\nu(t);p] = (p^2+1)^{-\frac{1}{2}}\{p+\sqrt{(p^2+1)}\}^{-\nu}.$$

Deduce that

$$\mathscr{L}[tJ_\nu(t);p] = \{p+\nu\sqrt{(p^2+1)}\}(p^2+1)^{-\frac{3}{2}}\{p+\sqrt{(p^2+1)}\}^{-\nu}$$

$$\mathscr{L}[t^{-1}J_\nu(t);p] = \frac{1}{\nu}\{p^2+1)\}^{-\nu}$$

and hence that

$$\int_0^t J_\mu(t-\tau)J_\nu(\tau)\frac{d\tau}{\tau} = \frac{1}{\nu}J_{\nu+\mu}(t), \qquad (\nu > 0, \mu > 0).$$

3-18 Show that if $\mu > -1$,

$$e^{-z}I_\mu(z) = \frac{(\frac{1}{2}z)^\mu}{\Gamma(\mu+1)}\,{}_1F_1(\mu+\tfrac{1}{2};2\mu+1;-2z).$$

Deduce that

$$\mathscr{L}[x^{-\frac{1}{2}}J_\nu(2a\sqrt{x});p] = \sqrt{\frac{\pi}{p}}\,e^{-a^2/2p}I_{\frac{1}{2}\nu}(a^2/2p).$$

3-19 Show that, if $x > 0, \nu > -1$,

$$\int_0^\infty e^{-t/x}t^{\frac{1}{2}\nu}J_\nu(2\sqrt{t})\,dt = x^{\nu+1}e^{-x}$$

and deduce that, provided the integrals on both sides of the equation converge.

$$\mathscr{L}[t^{\nu-1}f(t^{-1});p] = p^{-\frac{1}{2}\nu}\int_0^\infty x^{\frac{1}{2}\nu}J_\nu(2\sqrt{px})\bar{f}(x)\,dx$$

where $\bar{f}$ denotes the Laplace transform of f.

Taking $f(t) = t^{\frac{1}{2}\nu-\frac{1}{2}}$, show that

$$\int_0^\infty J_\nu(t)\,dt = 1, \qquad (\nu > -1).$$

3-20 If m and n are positive integers such that

$$\left[\frac{d^r}{dx^r}(1-\alpha x)^n f(x)\right]_{x=0} = 0, \qquad (r = 0,1,\ldots,m-1)$$

show that

$$\mathscr{L}\left[\frac{d^m}{dx^m}(1-\alpha x)^n f(x);p\right] = p^m\left(1+\alpha\frac{d}{dp}\right)^n \bar{f}(p)$$

where $\bar{f}(p) = \mathscr{L}[f(x);p]$.

Using Rodrigues' formula

$$P_n(z) = \frac{1}{2^n n!}\frac{d^n}{dz^n}(z^2-1)^n$$

for the Legendre polynomial of degree n, show that

$$\mathscr{L}[P_n(1-x);p] = p^n\left(1+\tfrac{1}{2}\frac{d}{dp}\right)^n p^{-n-1}.$$

3-21 Using the formula

$$P_n(x) = \frac{(2x)^n\Gamma(n+\frac{1}{2})}{\Gamma(\frac{1}{2})n!}\,{}_2F_1(\tfrac{1}{2}-\tfrac{1}{2}n;\tfrac{1}{2}-n;x^{-2})$$

for the Legendre polynomial of degree n, show that the Laplace transform

$\mathscr{L}[P_n(x);p]$ is the sum of powers with negative indices in the expansion in ascending powers of p of

$$(-1)^n\left(\frac{\pi}{2p}\right)^{\frac{1}{2}} \mathrm{I}_{-n-\frac{1}{2}}(p)$$

where I_ν is the modified Bessel function of the first kind of order ν.

3-22 From the recurrence relations

$$P'_{n-1}(x) = xP'_n(x) - nP_n(x) = xP'_{n-2}(x) + (n-1)P_{n-2}(x)$$

deduce that

$$\mathscr{L}[P_{2n}(e^t);p] = \frac{(p+1)(p+3)\ldots(p+2n-1)}{p(p-2)(p-4)\ldots(p-2n)},$$

and derive a similar expression for

$$\mathscr{L}[P_{2n+1}(e^t);p]$$

3-23 A function is said to be *anti-periodic* if

$$f(t+\tau) = -f(t), \qquad (t > 0).$$

Show that

$$\mathscr{L}[f(t);p] = (1 + e^{-p\tau})^{-1}\int_0^\tau f(u)e^{-pu}\,du.$$

3-24 Calculate the Laplace transforms of the periodic functions whose graphs are shown in Figs. 3-24 and 3-25.

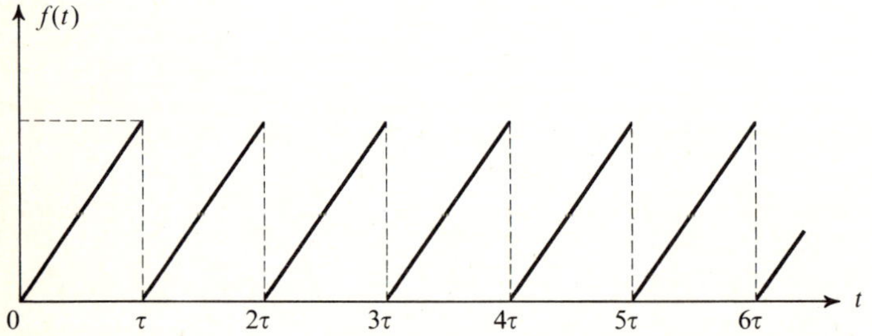

Fig. 3-24

Fig. 3-25

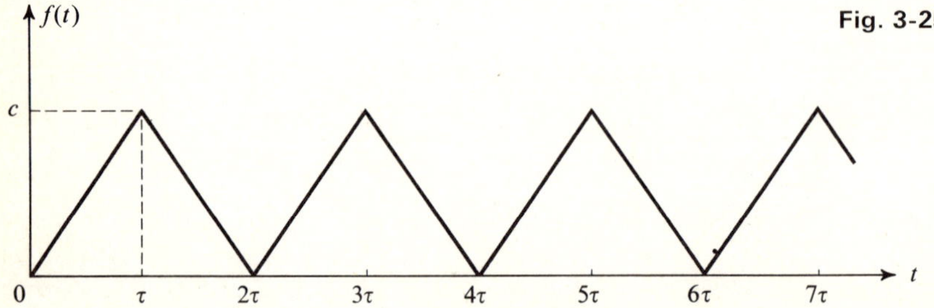

3-25 The function $\vartheta(x)$ is defined by the series

$$\vartheta(x) = 1 + 2\sum_{n=1}^{\infty} e^{-n^2\pi x}, \qquad (x > 0).$$

Show that

$$\mathscr{L}[\vartheta(x);p] = p^{-\frac{1}{2}}\coth(p^{\frac{1}{2}}).$$

Deduce the Laplace transforms of $1 + 2x\vartheta(x)$ and of

$$\int_0^x \vartheta(t)\,dt$$

and prove that

$$\int_0^x \vartheta(t)\{1 + \vartheta(x-t)\}\,dt = 1 + 2x\vartheta(x).$$

3-26 Show that the integral of the function

$$x_1^{a_1-1}x_2^{a_2-1}\ldots x_n^{a_n-1} \qquad (a_1 > 0,\ldots,a_n > 0)$$

over the region defined by the inequations

$$x_1 + x_2 + \cdots + x_n \leqslant 1, \qquad x_1 \geqslant 0, x_2 \geqslant 0,\ldots,x_n \geqslant 0$$

has the value

$$\frac{\Gamma(a_1)\Gamma(a_2)\ldots\Gamma(a_n)}{\Gamma(a_1 + a_2 + \cdots + a_n + 1)}$$

3-27 Prove that as $x \to \infty$,

$$\int_0^{\infty} \frac{e^{-xt}}{1+t^2}\,dt \sim \sum_{r=0}^{\infty} \frac{(-1)^r(2r)!}{x^{2r+1}}$$

Solve the initial value problems[1] 3-28–3-37:

3-28 $\ddot{x} + 3\dot{x} + 2x = te^{-t}, \qquad x(0) = 1,\ \dot{x}(0) = 0.$

3-29 $\dddot{x} - 2\ddot{x} - \dot{x} + 2x = e^{-t}, \qquad x(0) = \dot{x}(0) = \ddot{x}(0) = 0.$

3-30 $\ddot{x} + \omega^2 x = \sin\omega t, \qquad x(0) = \dot{x}(0) = 0.$

3-31 $\ddot{x} + 2\dot{x} + 2x = 5\sin t, \qquad x(0) = \dot{x}(0) = 0.$

3-32 $\ddot{x} + 2\dot{x} + x = 5e^t, \qquad x(0) = 1,\ \dot{x}(0) = 0.$

3-33 $\dddot{x} = x + 1, \qquad x(0) = \dot{x}(0) = \ddot{x}(0) = 0.$

3-34 $\ddot{x} + 2\dot{x} + 2x = e^{-t}\sin t, \qquad x(0) = \dot{x}(0) = 0.$

[1] Here we use the notation familiar from mechanics:

$$\dot{x} = \frac{dx}{dt}, \qquad \ddot{x} = \frac{d^2x}{dt^2}, \qquad \text{etc.}$$

3-35 $\dddot{x} + 2\ddot{x} + 5\dot{x} = \frac{5}{2}, \qquad x(0) = \dot{x}(0) = 0, \ddot{x}(0) = 1.$

3-36 $\dddot{x} + 2\ddot{x} - 4\dot{x} - 8x = 8e^{2t}, \qquad x(0) = 0, \dot{x}(0) = 1, x(0) = 2.$

3-37 $\ddot{x} + 6\dot{x} + 9x = \sin t, \qquad x(0) = 1, \dot{x}(0) = 0.$

3-38 Show that the differential equation

$$\left(\frac{d}{dx} + a\right)^n y = x^m e^{-ax}, \qquad (x > 0)$$

with n a positive integer and $m > -1$ has the particular integral

$$y(x) = \frac{\Gamma(m+1)}{\Gamma(m+n+1)} x^{m+n} e^{-ax}.$$

3-39 A mass m is attached to the lower end of a helical spring of stiffness $2\lambda^2 m$ and the upper end of the spring is made to oscillate vertically so that its displacement y from a fixed point O is given by $y = 5\cos(2\lambda t)$. If the upper end of the spring were fixed at O, the mass would rest in equilibrium at the point P. Denoting by x the displacement of the mass from P at time t, show that, if the motion of the mass is restricted by a damping force which is $2\lambda m$ times the velocity, the equation of motion is

$$\ddot{x} + 2\lambda\dot{x} + 2\lambda^2 x = 10\lambda^2 \cos(2\lambda t).$$

Show that if $x = \dot{x} = 0$ when $t = 0$, the motion of the mass is given by

$$x = e^{-\lambda t}(\cos \lambda t - 3\sin \lambda t) - \cos(2\lambda t) + 2\sin(2\lambda t).$$

3-40 Find the solution of the two-point boundary value problem

$$\frac{d^2y}{dx^2} + 9y = 8\sin x, \qquad y(0) = 0, y(\tfrac{1}{2}\pi) = -1.$$

3-41 Show that the differential equation

$$t\frac{d^2y}{dt^2} + (1-t)\frac{dy}{dt} + \lambda y = 0$$

has a solution

$$L_\lambda(t) = \mathscr{L}^{-1}[p^{-\lambda-1}(p-1)^\lambda; t].$$

Deduce that if λ is a positive integer n, then

(a) $L_n(t) = \dfrac{e^t}{n!}\dfrac{d^n}{dt^n}(t^n e^{-t})$

(b) $\sum_{n=0}^{\infty} L_n(t)x^n = (1-x)^{-1}e^{-tx/(1-x)}$

(c) $\sum_{n=0}^{\infty} (-1)^n L_{2n}(2t) = \tfrac{1}{2}e^t(\cos t + \sin t)$

(d) $\frac{d}{dx}\int_0^x L_m(t)L_n(x-t)\,dt = L_{m+n}(x)$

(e) $\mathscr{L}[t^n L_n(t);p] = n!p^{-n-1}P_n(1-2p^{-1})$.

(L_n is the Laguerre polynomial of degree n; cf. sec. 3-12.)

3-42 Show that the solution of the differential equation

$$x\frac{d^2y}{dx^2} + (1+2ax)\frac{dy}{dx} + ay = 0$$

satisfying the initial condition $y(0) = 1$ is

$$\mathscr{L}^{-1}[(p^2+2ap)^{-\frac{1}{2}};x] = e^{-ax}I_0(ax).$$

3-43 Find the solution of the differential equation

$$x\frac{d^2y}{dx^2} + 2\frac{dy}{dx} + xy = 0$$

satisfying the initial condition $y(0) = 1$.

Solve the initial value problems **3-44–3-51**:

3-44 $\dot{x} + x + y = 0,\ 2\dot{x} - \dot{y} + 4x - y = 0;\qquad x(0) = 1,\ y(0) = 0.$

3-45 $\dot{x} + x - y = 0,\ \dot{y} + 2x + 4y = 0;\qquad x(0) = 1,\ y(0) = 0.$

3-46 $\dot{x} - x - 2y = t,\ \dot{y} - 2x - y = t;\qquad x(0) = 2,\ y(0) = 4.$

3-47 $x + ky = a\sin kt,\ y - kx = a\cos kt;\qquad x(0) = 0,\ y(0) = b.$

3-48 $\dot{x} + 2\dot{y} + x - y = \cos t,\ \dot{x} - \dot{y} - x - 2y = \sin t;$
$x(0) = 1,\ y(0) = 0.$

3-49 $2\dot{x} + 3x - y = 2t,\ 3\dot{x} + \dot{y} + 4x = 3t;\qquad x(0) = y(0) = 0.$

3-50 $\dot{x} - \dot{y} - 2x + 2y = 1 - 2t,\ \ddot{x} + 2\dot{y} + x = 0;$
$x(0) = \dot{x}(0) = y(0) = 0.$

3-51 $3\dot{x} - \dot{y} - x - 3y = e^{-t},\ \dot{x} + 3\dot{y} - 3x + y = 0;$
$x(0) = y(0) = 0.$

3-52 Find the solution of the system

$$\dot{x} + y = 1,\qquad x + \dot{y} = 0$$

satisfying the initial condition $x(0) = 1$ and the condition $x(t) \to 0$ as $t \to \infty$.

3-53 Find the solution of the simultaneous equations

$$\dot{x} + \dot{y} + x + 2y = \sin t,\qquad \dot{x} + 2\dot{y} + 2y = 0$$

when $x(0) = 1$, $y(0) = 0$ and verify that as $t \to \infty$, $x(t) \sim (8/5)^{\frac{1}{2}}\sin(t-\phi)$ where $\phi = \tan^{-1}(2/3)$.

3-54 The coordinates (x,y) of a particle moving in a plane are functions of the time t and satisfy the equations

$$2\dot{x} + 8x - \dot{y} - 5y = 0, \qquad \dot{x} - 2x - \dot{y} + y = 0.$$

Initially the particle is at the point (5,7). Prove that it moves on an arc of a parabola, whose axis is parallel to $4x = 3y$, and that it steadily approaches, but never reaches, the origin.

3-55 Solve the simultaneous differential equations

$$\ddot{x} + 4\ddot{y} = x, \qquad \ddot{y} = x + y$$

under the initial conditions $x(0) = \dot{x}(0) = 0$, $y(0) = 1$, $y(0) = 0$.

Prove that $|x + 2y| < 2$ for all positive values of t.

Solve the initial value problems **3-56–3-60**:

3-56 $\ddot{x} - 3x = y, \ddot{y} + 3y = 7x; \qquad \dot{x}(0) = \dot{x}(0) = \dot{y}(0) = 0, y(0) = 1.$

3-57 $\ddot{x} + 2\dot{y} + x = \cos t, \ddot{y} + 2\dot{x} + y = 0;$

$$x(0) = \dot{x}(0) = y(0) = \dot{y}(0) = 0.$$

3-58 $\ddot{x} + 2x - \dot{y} = 1, \dot{x} + \ddot{y} + 2y = 0;$

$$x(0) = \dot{x}(0) = y(0) = \dot{y}(0) = 0.$$

3-59 $\ddot{x} - 3x - 9y = e^{2t}, \ddot{y} + x + 3y = 0;$

$$x(0) = y(0) = 1, \dot{x}(0) = \dot{y}(0) = 0.$$

3-60 $\dot{x} = 3x + y + z, \dot{y} = x + 5y + z, \dot{z} = x + y + 3z;$

$$x(0) = 3, y(0) = z(0) = 0.$$

3-61 The motion of a particle is determined by the equations

$$\dot{x} = y + z - x, \qquad \dot{y} = z + x - y, \qquad \dot{z} = x + y + z.$$

If the particle is initially at the point (1,0,0) find its coordinates at a subsequent time t and show that its trajectory has equations

$$(x + y + 2z)(x + y - z)^2 = 1, \qquad x^2 - y^2 - 2yz + 2zx = 1.$$

Show also that for large values of t the trajectory lies very close to the straight line with equations

$$\frac{x}{1} = \frac{y}{1} = \frac{z}{2}.$$

3-62 Find the solution of the system of equations

$$\tfrac{1}{2}\ddot{x} + \ddot{y} + 4y = 0, \qquad \tfrac{1}{2}\ddot{x} + \ddot{z} + 4z = 0, \qquad \ddot{x} + 4x = 9(y + z)$$

satisfying the initial conditions

$$x(0) = 0, y(0) = 1, z(0) = 0; \qquad \dot{x}(0) = \dot{y}(0) = \dot{z}(0) = 0.$$

3-63 Show that the solution of the two-point boundary value problem

$$\frac{d^2y}{dx^2} - \omega^2 y = f(x),\ (x > 0); \qquad y(0) = 0, \qquad y(x) \to 0 \text{ as } x \to \infty$$

is

$$y(x) = -\int_0^\infty G(x, t) f(t)\, dt$$

where

$$G(x, t) = \begin{cases} \omega^{-1} \sinh(\omega t) e^{-\omega x}, & 0 < t < x \\ \omega^{-1} \sinh(\omega x) e^{-\omega t}, & t > x. \end{cases}$$

3-64 If $w(x)$ is the load per unit length of an elastic beam, E is its Young's modulus and I is the moment of inertia of the cross-section of the beam, then the deflection, $y(x)$, of the beam from its unloaded position is given by the differential equation

$$EI \frac{d^4y}{dx^4} = w(x).$$

If the beam is of length a and is built-in at the end $x = 0$ so that $y(0) = y'(0) = 0$, and is freely hinged at the end $x = a$ so that $y(a) = y''(a) = 0$, show that

$$y(x) = \frac{1}{EI}\left[A\frac{x^2}{a^2} + B\frac{x^3}{a^3} + \frac{1}{6}\int_0^x (x - u)^3 w(u)\, du\right]$$

where A and B are constants given by the equations

$$A = \frac{1}{4}\int_0^a (a - u)(2a - u)uw(u)\, du,$$

$$B = -\frac{1}{12}\int_0^a (a - u)(2a^2 + 2au - u^2)w(u)\, du.$$

Derive the solution in the case in which the beam is built-in at both ends so that the end conditions are $y(0) = y'(0) = y(a) = y'(a) = 0$.

3-65 Find the solution of the equation

$$\dddot{x} + 2\ddot{x} - \dot{x} - 2x = 0$$

satisfying the initial conditions $x(0) = 1$, $x(0) = 0$ and the limiting condition $x(t) \to 0$ as $t \to \infty$.

3-66 Find the solution of the equation

$$\dddot{x} + 4\ddot{x} + \dot{x} - 6x = e^{-t}$$

satisfying the initial conditions $x(0) = x(0) = 0$ and the limiting condition $x(t) \to 0$ as $t \to \infty$.

3-67 Find the solution of the differential equation

$$t\ddot{y} + 2\dot{y} + ty = 0$$

satisfying the condition $y(t) \to 1$ as $t \to 0$.

3-68 Show that the solution $\psi(r,t)$ of the boundary value problem

$$\frac{\partial^2\psi}{\partial r^2} + \frac{2}{r}\frac{\partial\psi}{\partial r} = \frac{1}{\kappa}\frac{\partial\psi}{\partial t} \qquad (r > a, t > 0).$$

for which $\psi(a,t) = T_0$, $\psi(r,0) = 0$ for $r > a$, and $\psi(r,t) \to 0$ as $r \to \infty$, is given by the equation

$$\psi(r,t) = T_0 a r^{-1} \operatorname{Erfc}\left(\frac{r-a}{2\sqrt{(\kappa t)}}\right).$$

3-69 The temperatures u, v in two different media lying in $x < 0$, $x > 0$ satisfy the equations

$$k\frac{\partial^2 u}{\partial t^2} = \frac{\partial u}{\partial t}, \qquad \kappa\frac{\partial^2 v}{\partial t^2} = \frac{\partial v}{\partial t}$$

respectively. At the interface $x = 0$,

$$u = v, \qquad \frac{\partial u}{\partial x} = a\frac{\partial u}{\partial x}$$

(where a is a constant). Find the temperatures for all $t > 0$ given that, initially, $u = 0$, $(x < 0)$ and $v = 1$, $(x > 0)$.

3-70 Prove that the solution of the wave equation

$$c^2\frac{\partial^2 u}{\partial x^2} = \frac{\partial^2 u}{\partial t^2}$$

in the semi-infinite strip $|x| \leqslant l$, $t > 0$ satisfying the boundary conditions

$$\frac{\partial u(\pm l,t)}{\partial x} = 0, \qquad t > 0$$

and the initial conditions

$$u(x,0) = ax, \qquad \frac{\partial u(x,0)}{\partial t} = 0, \qquad |x| < l$$

may be written

$$ax - ac\mathscr{L}^{-1}[p^{-2}\sinh(px/c)\operatorname{sech}(pl/c); p \to t].$$

Hence find a series expansion for $u(x,t)$.

3-71 The function $u(x,t)$ satisfies the equation

$$\frac{\partial^4 u}{\partial x^4} + \frac{\partial^2 u}{\partial t^2} = 0, \qquad (x > 0, t > 0)$$

the boundary conditions $u(0,t) = 1$, $\partial^2 u(0,t)/\partial x^2 = 0$ and the limiting condition $u(x,t) \to 0$ as $x \to \infty$.

Show that

$$u(x,t) = \mathscr{L}^{-1}[p^{-1}\cos(x\sqrt{\tfrac{1}{2}p})\exp(-x\sqrt{\tfrac{1}{2}p});p \to t],$$

and deduce that

$$u(x,t) = 1 - \frac{1}{\sqrt{\pi}}\int_0^{x/\sqrt{(2t)}} (\cos\tfrac{1}{2}y^2 + \sin\tfrac{1}{2}y^2)\,dy.$$

3-72 The function $u(x,t)$ satisfies the equation

$$c^2\frac{\partial^2 u}{\partial x^2} = \frac{\partial^2 u}{\partial t^2}$$

in the semi-infinite strip $0 \leqslant x \leqslant l$, $t > 0$, the boundary conditions

$$u(0,t) = a\sin\omega t, \qquad u(l,t) = 0, \qquad t > 0,$$

and the initial conditions

$$u(x,0) = \frac{\partial u(x,0)}{\partial t} = 0, \qquad (0 \leqslant x \leqslant l).$$

Show that

$$u(x,t) = \frac{a\sin\omega t\sin\{\omega(l-x)/c\}}{\sin\omega l/c} + \sum_{n=1}^{\infty}\frac{2lca}{\omega^2 l^2 - n^2\pi^2c^2}\sin\frac{n\pi x}{l}\sin\frac{n\pi ct}{l}.$$

3-73 Show that the function $u(x,t)$ which satisfies the partial differential equation

$$\frac{\partial^2 u}{\partial x^2} = \frac{1}{c^2}\frac{\partial^2 u}{\partial t^2} + k$$

(c and k constants) in the semi-infinite strip $t > 0$, $0 \leqslant x \leqslant l$, the initial conditions

$$u(x,0) = \frac{\partial u(x,0)}{\partial t} = 0, \qquad (0 \leqslant x \leqslant l)$$

and the boundary conditions

$$u(0,t) = \frac{\partial u(l,t)}{\partial x} = 0, \qquad (t > 0)$$

can be expressed as

$$\mathscr{L}^{-1}\left[\frac{kc^2}{p^3}\left(\frac{\cosh\{p(l-x)/c\}}{\cosh(pl/c)} - 1\right);p \to t\right].$$

Deduce that

$$u(x,t) = \tfrac{1}{2}kx(x - 8l) +$$

$$+ \frac{16kl^2}{\pi^3} \sum_{n=0}^{\infty} \frac{(-1)^n}{(2n+1)^3} \cos\frac{(2n+1)\pi(l-x)}{2l} \cos\frac{(2n+1)\pi ct}{2l}$$

3-74 Show that the solution $u(x,t)$ of the partial differential equation

$$\frac{\partial^2 u}{\partial x^2} = \frac{\partial^2 u}{\partial t^2} + 2k\frac{\partial u}{\partial t}$$

where[1] k is a constant, defined in the half-plane $t > 0$ by the Cauchy data

$$u(x,0) = 0, \qquad \frac{\partial u(x,0)}{\partial t} = f(x)$$

is

$$u(x,t) = \int_{-\infty}^{\infty} K(x - \tau, t) f(\tau)\, d\tau$$

where

$$K(x,t) = \tfrac{1}{2}e^{-kt}\mathscr{L}^{-1}\left[\frac{1}{\sqrt{(p^2 - k^2)}} \exp\{-|x|\sqrt{(p^2 - k^2)}\}; p \to t\right].$$

3-75 The function $u(x,y)$ satisfies the equation

$$\frac{\partial^2 u}{\partial x \partial y} + u = 0$$

in the positive quadrant $x > 0$, $y > 0$, and the boundary conditions

$$u(x,0) = f(x), \ (x > 0), \qquad u(0,y) = g(y), \ (y > 0);$$

with $f(0) = g(0)$. Show that

$$u(x,y) = f(0)\mathrm{J}_0(2\sqrt{xy}) + \int_0^x f'(t)\mathrm{J}_0\{2y^{\frac{1}{2}}(x-t)^{\frac{1}{2}}\}\, dt$$

$$+ \int_0^y g'(t)\mathrm{J}_0\{2x^{\frac{1}{2}}(y-t)^{\frac{1}{2}}\}\, dt.$$

3-76 The function $u(x,t)$ satisfies the partial differential equation

$$\frac{\partial^2 u}{\partial t^2} = \frac{\partial^2 u}{\partial x^2} + \frac{1}{a^4}\frac{\partial^4 u}{\partial x^2 \partial t^2}, \qquad (x > 0,\ t > 0)$$

where a is a real constant, the initial conditions

$$u(x,0) = \frac{\partial u(x,0)}{\partial t} = 0, \qquad (x > 0),$$

and the boundary conditions

$$u(0,t) = \int_0^t f(\tau)\, d\tau, \qquad u(x,t) \to 0 \qquad \text{as } x \to \infty,\ (t > 0)$$

[1] It can be shown that $K(x,t) = \tfrac{1}{2}e^{-kt}\mathrm{I}_0\{k\sqrt{(t^2 - x^2)}\}\mathrm{H}(t - |x|)$.

where the function $f(t)$ is continuous for $t > 0$.

Prove that

$$\frac{\partial u(0,t)}{\partial x} = -a\int_0^t f(t-\tau)J_0(a\tau)\,d\tau.$$

Find the value of $\partial u(0,t)/\partial x$ if $u(0,t) = 1$ and the other conditions are unaltered.

3-77 If $u(x,t)$ is the solution of the diffusion equation

$$\frac{\partial^2 u}{\partial x^2} = \frac{1}{\kappa}\frac{\partial u}{\partial t}, \qquad x > 0,\ t > 0$$

satisfying the boundary conditions $u(0,t) = \partial u(0,t)/\partial x = 0$, $t > 0$, and the initial condition $u(x,0) = f(x)$, $x > 0$, and if $v(x,t)$ is the solution of the wave equation

$$\frac{\partial^2 v}{\partial x^2} = \frac{1}{c^2}\frac{\partial^2 v}{\partial t^2}, \qquad x > 0,\ t > 0$$

satisfying the boundary conditions $v(0,t) = \partial v(0,t)/\partial x = 0$, $t > 0$, and the initial conditions $v(x,0) = 0$, $\partial v(x,0)/\partial t = f(x)$, $x > 0$, show that

$$u(x,t) = c(4\pi\kappa t^3)^{-\frac{1}{2}}\int_0^\infty \tau v(x, \sqrt{\kappa\tau}/c)e^{-\tau^2/4t}\,d\tau.$$

3-78 If, in the last problem, the initial conditions on $v(x,t)$ are replaced by $v(x,0) = f(x)$, $\partial v(x,0)/\partial t = 0$, show that

$$u(x,t) = (\pi t)^{-\frac{1}{2}}\int_0^\infty v(x, \sqrt{\kappa\tau}/c)e^{-\tau^2/4t}\,d\tau.$$

3-79 The current $j(x,t)$ and the potential $V(x,t)$ at a point x in the semi-infinite telegraph cable $x \geqslant 0$ satisfy the equations

$$\frac{\partial V}{\partial x} = -rj, \qquad c\frac{\partial V}{\partial t} = -\frac{\partial j}{\partial x}$$

where r and c are the linear resistance and the linear capacity respectively.

The current and potential at all points of the cable are initially zero, and constant unit potential is then applied at $x = 0$ for a time T.

Show that, if $t > T$,

$$V(x,t) = \frac{x}{2}\sqrt{\frac{rc}{\pi}}\int_{t-T}^{t} e^{-rcx^2/4u}u^{-\frac{3}{2}}\,du.$$

3-80 Two functions u, v satisfy the differential equations

$$\frac{\partial u}{\partial x} + \frac{\partial v}{\partial t} + v = 0, \qquad \frac{\partial v}{\partial x} + \frac{\partial u}{\partial t} = 0$$

in $x > 0$, $t > 0$. If

$$u(x,0) = 1, \qquad v(x,0) = 0, \qquad x > 0$$
$$u(0,t) + kv(0,t) = 0, \qquad t > 0$$

where k is a positive constant, show that if $\omega^2 = p(p+1)$,

$$u(x,t) = 1 - \mathscr{L}^{-1}\left[\frac{\omega}{(kp+\omega)p} e^{-\omega x}; p \to t\right]$$

with appropriate conventions concerning $p^{\frac{1}{2}}$ and $(1+p)^{\frac{1}{2}}$.

Deduce that

$$u(0,0) = \frac{k}{k+1}, \qquad u(0,t) \sim \frac{k}{(\pi t)^{\frac{1}{2}}} \qquad \text{as } t \to \infty.$$

3-81 Solve the integral equation

$$\mathrm{J}_{n+1}(t) = \int_0^t \mathrm{J}_n(t-u) f(u)\, du$$

where n is zero or a positive integer.

3-82 Show that the integral equation

$$\int_0^x L_n(x-t) f(t)\, dt = g(x), \qquad (x > 0)$$

in which the prescribed function $g(x)$ satisfies the conditions $g(0) = g'(0) = 0$, has the solution

$$f(x) = \int_0^x L_{n+1}(t-x) e^{x-t} (\mathrm{D}_t - 1)^2 f(t)\, dt.$$

3-83 Show that the solution of the integral equation

$$\int_0^x \mathrm{J}_0\{2(x-t)^{\frac{1}{2}}\} f(t)\, dt = g(x), \qquad (x > 0)$$

in which $g(x)$ is prescribed and $g(0) = 0$ is

$$f(x) = g'(0)\mathrm{I}_0(2x^{\frac{1}{2}}) + \int_0^x \mathrm{I}_0\{2(x-t)^{\frac{1}{2}}\} g(t)\, dt.$$

3-84 The function $f(x)$ satisfies the integral equation

$$\int_0^x u^{b-1}\, {}_1F_1(a;b;u) f(x-u)\, du = g(x)$$

in which ${}_1F_1(a;b;x)$ is Kummer's confluent hypergeometric function and $g(x)$ is a prescribed function with the property that $g^{(r)}(0) = 0$, $(r = 1, 2, \ldots, n-1)$, n being a positive integer and $n > b > 0$.

Show that

$$f(x) = \frac{1}{\Gamma(b)\Gamma(n-b)} \int_0^x (x-u)^{n-b-1} {}_1F_1(-a; n-b; x-u) g^{(n)}(u)\, du.$$

3-85 Show that if the inverse transform

$$L(x) = \mathscr{L}^{-1}\left[\frac{2}{p\bar{K}(p)}; x\right]$$

exists, the integral equation

$$\int_0^x f(t) K(x^2 - t^2)\, dt = g(x), \qquad (x > 0)$$

has solution

$$f(x) = \frac{d}{dx} \int_0^x t L(x^2 - t^2) g(t)\, dt.$$

3-86 Show that the integral equation

$$\int_0^x \frac{\cos\{k(x^2 - t^2)^{\frac{1}{2}}\}}{(x^2 - t^2)^{\frac{1}{2}}} f(t)\, dt = g(x), \qquad (x > 0)$$

has the solution

$$f(x) = \frac{2}{\pi} \frac{d}{dx} \int_0^x \frac{\cosh\{k(x^2 - t^2)^{\frac{1}{2}}\}}{(x^2 - t^2)^{\frac{1}{2}}} t g(t)\, dt.$$

3-87 If $g(t)$ is prescribed and $g(0) = 0$ show that the solution of the integral equation

$$f(t) + \int_0^t \frac{f(x)\, dx}{\sqrt{(t - x)}} = g(t), \qquad (t > 0)$$

which vanishes at $t = 0$ may be written in the form

$$f(t) = \int_0^t g'(x) K(t - x)\, dx$$

where

$$K(t) = e^{\pi t} \operatorname{Erfc}(\pi^{\frac{1}{2}} t^{\frac{1}{2}}).$$

3-88 Show that the solution of the integral equation

$$f(x) - \lambda \int_0^x f(t) K(x - t)\, dt = g(x), \qquad (x > 0)$$

can be written in the form

$$f(x) = g(x) + \lambda \int_0^x g(t) L(x - t)\, dt$$

where

$$L(x) = \sum_{r=0}^{\infty} \lambda^r k_r(x), \qquad k_r(x) = \mathscr{L}^{-1}[\{\bar{K}(p)\}^{r+1}; p \to x].$$

Find $k_r(x)$ if $K(x)$ is

(*a*) $\mathrm{J}_0(2x^{\frac{1}{2}})$; (*b*) $\mathrm{J}_0(x)$.

3-89 Solve the integral equation

$$y(x) + \int_0^x e^{-t} y(x - t)\, dt = 1$$

and verify that $y(x) \to \frac{1}{2}$ as $x \to \infty$.

3-90 Find the solution of the difference equation

$$a_{n+2} - (b + c)a_{n+1} + bca_n = b^n$$

satisfying $a_0 = a_1 = 0$.

3-91 Find the solution of the difference equation

$$a_{n+2} - 2ba_{n+1} + b^2 a_n = b^n$$

satisfying $a_0 = a_1 = 0$.

3-92 Evaluate the double Laplace transforms of the functions

$$\sin(ax + by), \qquad \cos(ax + by).$$

3-93 Show that

$$\mathscr{L}_2[\mathrm{J}_0(2x^{\frac{1}{2}}y^{\frac{1}{2}}); (p, q)] = (1 + pq)^{-1}.$$

3-94 If

$$h(x, y) = \int_0^{\min(x,y)} f(t) g(x - t, y - t)\, dt$$

show that

$$\mathscr{L}_2[h(x, y); (p, q)] = \bar{f}(p + q)\bar{g}(p, q)$$

where

$$\bar{f}(p) = \mathscr{L}[f(x); p], \qquad \bar{g}(p, q) = \mathscr{L}_2[g(x, y); (p, q)].$$

3-95 Show that

$$\mathscr{L}_2\left[x\frac{\partial f}{\partial x}; (p, q)\right] = \left[\frac{\partial}{\partial a} a^{-1}\bar{f}(p/a, q)\right]_{n=1}$$

and deduce that

$$\mathscr{L}_2\left[\left(x\frac{\partial}{\partial x}\right)^m \left(y\frac{\partial}{\partial y}\right)^n f(x, y); (p, q)\right]$$

$$= (-1)^{m+n}\left(1 + p\frac{\partial}{\partial p}\right)^m \left(1 + q\frac{\partial}{\partial q}\right)^n \bar{f}(p,q).$$

3-96 Use the method of the double Laplace transform to find the solution of the equation

$$\frac{\partial u}{\partial x} + \frac{\partial u}{\partial y} = f(x,y)$$

in $x \geqslant 0$, $y \geqslant 0$, which vanishes on the coordinate axes.

3-97 Use the method of the double Laplace transform to find the solution in the positive quadrant of the partial differential equation

$$\frac{\partial^2 u}{\partial x \partial y} + u = 0$$

taking the value 1 on the coordinate axes.

3-98 Show that if the boundary value problem

$$\frac{\partial u}{\partial x} - \frac{\partial u}{\partial y} = f(x,y), \qquad (x \geqslant 0, y \geqslant 0); \qquad u(x,0) = a(x),\, u(0,y) = b(y)$$

has a solution which is analytic in the positive quadrant of the xy-plane, its boundary values must be related through the equation

$$a(x) - b(x) = \int_0^x f(t, x - t)\, dt, \qquad (x \geqslant 0).$$

3-99 Find the solution of the *Cauchy–Abel functional equation*

$$f(x + y) = f(x)f(y).$$

3-100 Show that the function f satisfying the functional equation

$$f(x + y) + f(x - y) = 2f(x)f(y), \qquad f(0) = 1$$

is an *even* function and hence determine it.

3-101 Solve the functional equations

(*a*) $f(x + y) + f(x - y) = 2f(x) + 2f(y)$;

(*b*) $f(x + y) - f(x - y) = 2f(y)$;

(*c*) $f(x + y) + f(x - y) = 2f(x)$.

3-102 Find the solution of *Jensen's functional equation*

$$f(\tfrac{1}{2}x + \tfrac{1}{2}y) = \tfrac{1}{2}f(x) + \tfrac{1}{2}f(y)$$

which is continuous.

3-103 Show that

(*a*) $\mathscr{S}[(a+x)^{-1};y] = (a-y)^{-1}\log(a/y), \qquad (a>0);$

(*b*) $\mathscr{S}[(a+x)^{-1}(b+x)^{-1};y]$

$$= \begin{cases} \dfrac{(a-b)\log(a/y)-(a-y)\log(a/b)}{(a-y)(b-y)(a-b)}, & (b \neq a), \\[2ex] \dfrac{a\log(y/a)-(a-y)}{(a-y)^2 a}, & (b=a); \end{cases}$$

(*c*) $\mathscr{S}[x^{\nu-1};y] = \pi y^{\nu-1}\operatorname{cosec}(\nu\pi), \qquad (0<\nu<1);$

(*d*) $\mathscr{S}[x^{\nu-1}(a+x)^{-1};y] = \dfrac{\pi(y^{\nu-1}-a^{\nu-1})}{(a-y)\sin(\nu\pi)} \qquad (0<\nu<1, a>0);$

(*e*) $\mathscr{S}[(a^2+x^2)^{-1};y] = (a^2+y^2)^{-1}[\tfrac{1}{2}\pi y/a - \log(y/a)], \quad (a>0);$

(*f*) $\mathscr{S}[x(a^2+x^2)^{-1};y] = (a^2+y^2)^{-1}[\tfrac{1}{2}\pi a + y\log(y/a)], \quad (a>0).$

3-104 Show that

(*a*) $\mathscr{H}[\mathrm{H}(t-a)-\mathrm{H}(t-b);x] = \dfrac{1}{\pi}\log\left|\dfrac{b-x}{a-x}\right|, \qquad (b>a>0);$

(*b*) $\mathscr{H}[t^{-1}\mathrm{H}(t-a);x] = \dfrac{1}{\pi x}\log\dfrac{a}{|a-x|}, \qquad (a>0);$

(c) $\mathscr{H}[(at+b)^{-1}\mathrm{H}(t);x]$

$$= \frac{1}{\pi(ax+b)}\log\frac{b}{a|x|}, \qquad (a,b>0, x\neq -b/a);$$

(*d*) $\mathscr{H}\left[\dfrac{\alpha t+\beta a}{t^2+a^2};x\right] = \dfrac{\alpha a-\beta x}{x^2+a^2}, \qquad (a>0).$

3-105 Suppose that the function $K(\lambda, x)$ is a solution of the self-adjoint equation

$$\frac{d}{dx}\left[p(x)\frac{dK}{dx}\right] + \{\lambda^2 u(x) - v(x)\}K = 0, \qquad (\lambda>0, x>0)$$

and that a transform[1] $\mathscr{T}$ is defined by the equation

$$\bar{f}(\lambda) \equiv \mathscr{T}[f(x);\lambda] = \int_0^\infty f(x)K(\lambda,x)\,dx.$$

[1] Transforms of this general type were introduced by Mercer (1963).

Suppose further that this transform has an inversion formula of the type

$$f(x)\,\mathscr{T} \equiv^{-1}[\bar{f}(\lambda);x] = \int_0^\infty \bar{f}(\lambda)H(\lambda,x)\,d\lambda.$$

Show that if $L(\mu,x)$ satisfies

$$\frac{d}{dx}\left[p(x)\frac{dL}{dx}\right] - \{\mu^2 u(x) + v(x)\}L = 0, \qquad (\mu > 0,\ x > 0)$$

and is such that

$$-\left[p(x)L(\mu,x)\frac{dK}{dx} - p(x)K(\lambda,x)\frac{dL}{dx}\right]_0^\infty$$

can be written in the separable form $L(\lambda)M(\mu)$, then

$$\mathscr{S}[t^{-\frac{1}{2}}L(t^{\frac{1}{2}})H(t^{\frac{1}{2}},x);t \to y] = \frac{2u(x)L(y^{\frac{1}{2}},x)}{M(y^{\frac{1}{2}})}.$$

Deduce that

$$H(\lambda,x) = \frac{i\lambda}{\pi L(\lambda)}u(x)\left[\frac{L(\lambda e^{\frac{1}{2}\pi i},x)}{M(\lambda e^{\frac{1}{2}\pi i})} - \frac{L(\lambda e^{-\frac{1}{2}\pi i},x)}{M(\lambda e^{-\frac{1}{2}\pi i})}\right].$$

Consider the special cases:

(*a*) $K(\lambda,x) = \sin(\lambda x)$;

(*b*) $K(\lambda,x) = \cos(\lambda x)$;

(*c*) $K(\lambda,x) = h\sin(\lambda x) + \lambda\cos(\lambda x)$;

(*d*) $K(\lambda,x) = J_\nu(\lambda x)$.

4

The Mellin Transform

4-1 ELEMENTARY PROPERTIES OF THE MELLIN TRANSFORM

We saw in sec. 1-7 how the Mellin transform

$$f^*(s) \equiv \mathscr{M}[f(x); s] = \int_0^\infty f(x)x^{s-1}\,dx \tag{4-1-1}$$

arises in a natural way in the solution of boundary value problems concerning an infinite wedge. In this chapter we shall consider the properties of this transform.

By making the simple change of variable $x = u/a$ in the defining integral

$$\int_0^\infty f(ax)x^{s-1}\,dx = a^{-s}\int_0^\infty f(u)u^{s-1}\,du$$

we see that

$$\mathscr{M}[f(ax); s] = a^{-s}f^*(s). \tag{4-1-2}$$

Since multiplying $f(x)$ by x^a merely results in changing s to $s + a$, we also have the relation

$$\mathscr{M}[x^a f(x);s] = f^*(s + a) \tag{4-1-3}$$

Similarly, since if $a > 0$

$$\begin{aligned}\mathscr{M}[f(x^a);s] &= \int_0^\infty f(x^a)x^{s-1}\,dx \\ &= \int_0^\infty f(u)u^{s/a-1/a}\left(\frac{1}{a}u^{1/a-1}\,du\right)\end{aligned}$$

we obtain the relation

$$\mathscr{M}[f(x^a);s] = a^{-1}f^*(s/a), \qquad (a > 0) \tag{4-1-4}$$

and by a similar method we can show that

$$\mathscr{M}[x^{-1}f(x^{-1});s] = f^*(1 - s). \tag{4-1-5}$$

If we use the well-known formula

$$\frac{d}{ds}x^{s-1} = (\log x)x^{s-1}$$

we see that

$$\mathscr{M}[\log x\, f(x);s] = \frac{d}{ds}f^*(s). \tag{4-1-6}$$

We have already computed several Mellin transforms. For instance if we replace p by s, ξ by 1 in equations (2-7-21) and (2-7-22), we obtain the formulae

$$\mathscr{M}[\cos x;s] = \Gamma(s)\cos(\tfrac{1}{2}\pi s), \qquad 0 < \operatorname{Re} s < 1 \tag{4-1-7}$$

$$\mathscr{M}[\sin x;s] = \Gamma(s)\sin(\tfrac{1}{2}\pi s), \qquad 0 < \operatorname{Re} s < 1 \tag{4-1-8}$$

Similarly we can interpret equation (2-10-7) as the formula

$$\mathscr{M}[x^{-\nu}J_\nu(x);s] = \frac{2^{s-\nu-1}\Gamma(\frac{1}{2}s)}{\Gamma(\nu - \frac{1}{2}s + 1)}, \qquad 0 < \operatorname{Re} s < 1, \qquad \nu > -\tfrac{1}{2} \tag{4-1-9}$$

From this and the rule (4-1-3) we deduce the equation

$$\mathscr{M}[J_\nu(x);s] = \frac{2^{s-1}\Gamma(\frac{1}{2}\nu + \frac{1}{2}s)}{\Gamma(\frac{1}{2}\nu - \frac{1}{2}s + 1)}, \qquad -\nu < s < \nu + 2 \tag{4-1-10}$$

We notice also that we can interpret equation (3-2-3) in the form

$$\mathscr{M}[e^{-x};s] = \Gamma(s), \qquad \operatorname{Re} s > 0 \tag{4-1-11}$$

by replacing ν by $s - 1$ and p by 1.

Another useful result can be obtained by changing the variable of integration in the integral on the left of the formula

$$\int_0^1 (1-u)^{m-1}u^{s-1}\,du = \frac{\Gamma(m)\Gamma(s)}{\Gamma(m+s)}, \qquad \mathrm{Re}\,s > 0, \qquad \mathrm{Re}\,m > 0$$

from u to x, where $u = x/(1+x)$. This gives the equation

$$\int_0^\infty \frac{x^{s-1}\,dx}{(1+x)^{m+s}} = \frac{\Gamma(m)\Gamma(s)}{\Gamma(m+s)}$$

which can be interpreted in the form

$$\mathscr{M}[(1+x)^{-a};s] = \frac{\Gamma(s)\Gamma(a-s)}{\Gamma(a)}, \qquad 0 < \mathrm{Re}\,s < \mathrm{Re}\,a \tag{4-1-12}$$

In particular if we take $a = 1$ and make use of the relation

$$\Gamma(s)\Gamma(1-s) = \frac{\pi}{\sin(\pi s)}, \qquad 0 < \mathrm{Re}\,s < 1$$

we obtain the formula

$$\mathscr{M}[(1+x)^{-1};s] = \pi\,\mathrm{cosec}\,(\pi s), \qquad 0 < \mathrm{Re}\,s < 1. \tag{4-1-13}$$

We notice, also, that if we combine equations (4-1-4) and (4-1-12) we find that

$$\mathscr{M}[(1+x^a)^{-b};s] = \frac{\Gamma(s/a)\Gamma(b-s/a)}{a\Gamma(b)}, \qquad 0 < \mathrm{Re}\,s < \mathrm{Re}\,ab \tag{4-1-14}$$

and in particular that

$$\mathscr{M}[(1+x^a)^{-1};s] = \frac{\pi}{a}\,\mathrm{cosec}\left(\frac{\pi s}{a}\right), \qquad 0 < \mathrm{Re}\,s < \mathrm{Re}\,a. \tag{4-1-15}$$

Mellin transforms of rational functions can be calculated by a method similar to that outlined in sec. 2-8 for Fourier transforms.

Suppose that $f(z)$ is a rational function which has no poles on the positive real axis and is such that, for $\sigma_1 < \mathrm{Re}\,s < \sigma_2$,

$$\lim_{z\to 0} z^s f(z) = 0 \tag{4-1-16}$$

and that

$$\lim_{z\to \infty} z^s f(z) = 0 \tag{4-1-17}$$

We assume that $f(z)$ has poles at the points $z = a_1, \ldots, a_n$. If we integrate the function

$$\phi(z) = (e^{-\pi i}z)^{s-1} f(z) \tag{4-1-18}$$

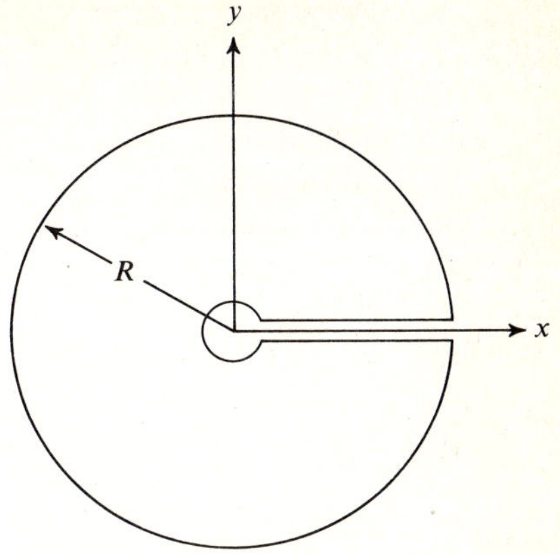

Fig. 4-1

round the contour shown in Fig. 4-1 and make use of the fact that, because of (4-1-16), the integral round the small circle tends to zero as the radius of that circle tends to zero and that, because of (4-1-17), the integral round the large circle tends to zero as $R \to \infty$, we find that, in the notation of sec. 2-8

$$\int_0^\infty (e^{-\pi i}x)^{s-1} f(x)\,dx - \int_0^\infty (e^{\pi i}x)^{s-1} f(x)\,dx = 2\pi i \sum_{r=1}^{n} \operatorname{res} \phi(a_r)$$

Since

$$(e^{-\pi i})^{s-1} - (e^{\pi i})^{s-1} = 2i \sin(s\pi)$$

we find that

$$\int_0^\infty x^{s-1} f(x)\,dx = \frac{\pi}{\sin(s\pi)} \sum_{r=1}^{n} \operatorname{res} \phi(a_r)$$

that is, that

$$\mathscr{M}[f(x); s] = \frac{\pi}{\sin(s\pi)} \sum_{r=1}^{n} \operatorname{res} \phi(a_r). \tag{4-1-19}$$

The formula is easily modified if $f(z)$ has poles $\beta_1, \ldots, \beta_m$ on the positive real axis. It is readily shown that

$$\mathscr{M}[f(x); s] = \frac{\pi}{\sin(s\pi)} \sum_{r=1}^{n} \operatorname{res} \phi(a_r) - \pi \cot(s\pi) \sum_{r=1}^{m} \beta_r^{s-1} \operatorname{res} f(\beta_r) \tag{4-1-20}$$

it being understood that the integral defining the Mellin transform is taken in the sense of its Cauchy principal value.

To illustrate the use of the formula (4-1-19) we consider the case in which $f(z) = (1 + z)^{-1}$. The conditions (4-1-16) and (4-1-17) will be satisfied if $0 < \operatorname{Re} s < 1$. Also, the function $f(z)$ has one pole only, at the point $z = e^{\pi i}$; it is easily seen that the residue of $\phi(z)$ there is 1 and so we obtain the formula (4-1-13).

Similarly, if we take $f(z) = (1 + z^2)^{-1}$, the conditions (4-1-16) and (4-1-17) will hold if $0 < \operatorname{Re} s < 2$. The only poles of $f(z)$ are at $e^{\frac{1}{2}\pi i}$, $e^{\frac{3}{2}\pi i}$ and the residues of $\phi(z)$ at these points are respectively $\frac{1}{2}e^{-\frac{1}{2}\pi s i}$, $\frac{1}{2}e^{\frac{1}{2}\pi s i}$, so that by equation (4-1-19) we have

$$\mathscr{M}[(1 + x^2)^{-1}; s] = \frac{\pi}{2 \sin(\frac{1}{2}\pi s)}, \qquad 0 < \operatorname{Re} s < 2 \tag{4-1-21}$$

in agreement with (4-1-15).

To illustrate the use of the formula (4-1-20) we take $f(z) = (1 - z)^{-1}$. Equations (4-1-16) and (4-1-17) are valid provided that $0 < \operatorname{Re} s < 1$. This function has only one pole, at $z = 1$ for which the residue of $f(z) = -1$ so that

$$\mathscr{M}[(1 - x)^{-1}; s] = \pi \cot(s\pi), \qquad 0 < \operatorname{Re} s < 1 \tag{4-1-22}$$

it being remembered that the integral on the left-hand side of this equation is a Cauchy principal value.

As with the Fourier and Laplace transforms, we introduce the inverse of the operator $\mathscr{M}$ in the sense that if

$$f^*(s) = \mathscr{M}[f(x); s]$$

we write

$$f(x) = \mathscr{M}^{-1}[f^*(s); x]$$

For instance we could rewrite equation (4-1-12) in the inverse form

$$\mathscr{M}^{-1}[\Gamma(s)\Gamma(a - s); x] = \Gamma(a)(1 + x)^{-a}. \tag{4-1-23}$$

4-2 MELLIN TRANSFORMS OF DERIVATIVES AND INTEGRALS

In the applications of Mellin transforms we often need to be able to express the Mellin transform of the derivatives of a function $f(x)$, or of some integrals involving it, in terms of the Mellin transform $f^*(s)$ of the function. We derive some of the results here.

By the definition of the Mellin transform

$$\mathscr{M}[f'(x); s] = \int_0^\infty f'(x) x^{s-1}\, dx.$$

Integrating by parts we find that

$$\mathscr{M}[f'(x);s] = [f(x)x^{s-1}]_0^\infty - (s-1)\int_0^\infty f(x)x^{s-2}\,dx.$$

Hence if there exist σ_1, σ_2 such that

$$\lim_{x\to 0} x^{s-1}f(x) = 0, \qquad \lim_{x\to\infty} x^{s-1}f(x) = 0$$

when $\sigma_1 < \operatorname{Re} s < \sigma_2$ and if $f^*(s-1)$ exists in the band, then

$$\mathscr{M}[f'(x);s] = -(s-1)f^*(s-1), \qquad \sigma_1 < \operatorname{Re} s < \sigma_2 \tag{4-2-1}$$

where $f^*(s) = \mathscr{M}[f(x);s]$. Applying this formula twice we find that

$$\begin{aligned}\mathscr{M}[f''(x);s] &= -(s-1)\mathscr{M}[f'(x);s-1]\\ &= -(s-1)\{-(s-2)f^*(s-2)\}\end{aligned}$$

i.e., that

$$\mathscr{M}[f''(x);s] = (s-1)(s-2)f^*(s-2) \tag{4-2-2}$$

We can obviously prove by induction over n that, for all values of s for which $f^*(s-n)$ exists,

$$\mathscr{M}[f^{(n)}(x);s] = (-1)^n \frac{\Gamma(s)}{\Gamma(s-n)} f^*(s-n) \tag{4-2-3}$$

provided that

$$\lim_{x\to 0} x^{s-r-1}f^{(r)}(x) = 0, \qquad r = 0,1, \ldots, n-1.$$

Since

$$\mathscr{M}[xf'(x);s] = \mathscr{M}[f'(x);s+1]$$

we deduce from (4-2-1) that

$$\mathscr{M}\left[\left(x\frac{d}{dx}\right)f(x);s\right] = -sf^*(s)$$

and by induction that

$$\mathscr{M}\left[\left(x\frac{d}{dx}\right)^n f(x);s\right] = (-s)^n f^*(s). \tag{4-2-4}$$

We note, also, that

$$\mathscr{M}[x^n f^{(n)}(x);s] = \mathscr{M}[f^{(n)}(x);s+n]$$

so that, by (4-2-3), we have the relation

$$\mathscr{M}[x^n f^{(n)}(x);s] = (-1)^n \frac{\Gamma(s+n)}{\Gamma(s)} f^*(s) \tag{4-2-5}$$

In particular we deduce by putting $n = 2$ in (4-2-4) that

$$\mathscr{M}\left[x^2 \frac{d^2f}{dx^2} + x\frac{df}{dx};s\right] = s^2 f^*(s). \tag{4-2-6}$$

For example, if we denote the two-dimensional Laplace operator by Δ_2 we find that in plane polar coordinates ρ, ϕ

$$\Delta_2 f(\rho,\phi) = \frac{\partial^2 f}{\partial\rho^2} + \frac{1}{\rho}\frac{\partial f}{\partial\rho} + \frac{1}{\rho^2}\frac{\partial^2 f}{\partial\phi^2} \tag{4-2-7}$$

so that by (4-2-6) we find that

$$\mathscr{M}[\rho^2\Delta_2 f(\rho,\phi);\rho \to s] = \left(\frac{\partial^2}{\partial\phi^2} + s^2\right) f^*(s,\phi)$$

where

$$f^*(s,\phi) = \mathscr{M}[f(\rho,\phi);\rho \to s] \tag{4-2-8}$$

Hence we deduce that

$$\mathscr{M}[\Delta_2 f(\rho,\phi);\rho \to s] = \left\{\frac{\partial^2}{\partial\phi^2} + (s-2)^2\right\} f^*(s-2,\phi) \tag{4-2-9}$$

If we define the plane biharmonic operator ${\Delta_2}^2$ by

$${\Delta_2}^2 f = \Delta_2(\Delta_2 f)$$

we find that

$$\mathscr{M}[{\Delta_2}^2 f(\rho,\phi);\rho \to s] = \left\{\frac{\partial^2}{\partial\phi^2} + (s-2)^2\right\} \mathscr{M}[\Delta_2 f(\rho,\phi);\rho \to s-2]$$

and hence that

$$\mathscr{M}[{\Delta_2}^2 f(\rho,\phi);\rho \to s] = \{{D_\phi}^2 + (s-2)^2\}\{{D_\phi}^2 + (s-4)^2\} f^*(s-4,\phi) \tag{4-2-11}$$

where

$$D_\phi = \frac{\partial}{\partial\phi} \tag{4-2-12}$$

and $f^*(s,\phi)$ is defined by (4-2-8).

Some of these results can be written conveniently in terms of integrals rather than derivatives. For instance we can write equation (4-2-1) in the form

$$\mathscr{M}[f(x);s] = -(s-1)\mathscr{M}\left[\int_0^x f(u)\,du;s-1\right].$$

Replacing s by $s + 1$ we can now write this as

$$\mathscr{M}\left[\int_0^x f(u)\,du; s\right] = -\frac{1}{s} f^*(s+1) \tag{4-2-13}$$

where $f^*(s)$ denotes the Mellin transform of $f(x)$. Repeating this process we see that

$$\begin{aligned}\mathscr{M}\left[\int_0^x dy \int_0^y f(u)\,du; s\right] &= -\frac{1}{s}\mathscr{M}\left[\int_0^y f(u)\,du; s+1\right] \\ &= -\frac{1}{s}\left\{-\frac{1}{s+1} f^*(s+2)\right\}\end{aligned}$$

showing that

$$\mathscr{M}\left[\int_0 dy \int_0 f(u)\,du; s\right] = \frac{1}{s(s+1)} f^*(s+2).$$

If we denote the n-th repeated integral of $f(x)$ by $I_n f(x)$ in the sense that

$$I_n f(x) = \int_0^x I_{n-1} f(u)\,du \tag{4-2-14}$$

we can prove by induction over n that

$$\mathscr{M}[I_n f(x); s] = (-1)^n \frac{\Gamma(s)}{\Gamma(n+s)} f^*(s+n) \tag{4-2-15}$$

Similarly, if we write equation (4-2-1) in the form

$$\mathscr{M}[f(x); s] = (s-1)\mathscr{M}\left[\int_x^\infty f(u)\,du; s-1\right]$$

and replace s by $s + 1$ we arrive at the formula

$$\mathscr{M}\left[\int_x^\infty f(u)\,du; s\right] = \frac{1}{s} f^*(s+1). \tag{4-2-16}$$

If now we introduce an n-th repeated integral $I_n{}^\infty f(x)$ of $f(x)$ defined by the recurrence formula

$$I_n{}^\infty f(x) = \int_x^\infty I_{n-1}{}^\infty f(u)\,du \tag{4-2-17}$$

we can show by induction over n that

$$\mathscr{M}[I_n{}^\infty f(x); s] = \frac{\Gamma(s)}{\Gamma(s+n)} f^*(s+n). \tag{4-2-18}$$

We shall make use of the Mellin transforms of certain integral expressions. For instance

$$\mathscr{M}\left[\int_0^\infty u^\mu f(xu)g(u)\,du; s\right] = \int_0^\infty x^{s-1}\,dx \int_0^\infty u^\mu f(xu)g(u)\,du$$
$$= \int_0^\infty u^\mu g(u)\,du \int_0^\infty x^{s-1} f(xu)\,dx$$
$$= \int_0^\infty u^{\mu-s} g(u)\,du \int_0^\infty y^{s-1} f(y)\,dy$$

showing that

$$\mathscr{M}\left[\int_0^\infty u^\mu f(xu)g(u)\,du; s\right] = f^*(s)g^*(1+\mu-s).$$

Making use of the relation (4-1-7) we obtain the equation

$$\mathscr{M}\left[x^\lambda \int_0^\infty u^\mu f(xu)g(u)\,du; s\right] = f^*(s+\lambda)g^*(1-\lambda+\mu-s). \tag{4-2-19}$$

Similarly

$$\mathscr{M}\left[\int_0^\infty u^\mu f\left(\frac{x}{u}\right) g(u)\,du; s\right] = \int_0^\infty u^\mu g(u)\,du \int_0^\infty x^{s-1} f\left(\frac{x}{u}\right) du$$
$$= \int_0^\infty u^{\mu+s} g(u)\,du \int_0^\infty y^{s-1} f(y)\,dy$$

showing that

$$\mathscr{M}\left[\int_0^\infty u^\mu f\left(\frac{x}{u}\right) g(u)\,du; s\right] = g^*(\mu+s+1)f^*(s).$$

From this we deduce that

$$\mathscr{M}\left[x^\lambda \int_0^\infty u^\mu f\left(\frac{x}{u}\right) g(u)\,du; s\right] = f^*(s+\lambda)g^*(s+\lambda+\mu+1). \tag{4-2-20}$$

Two particular cases of these results arise frequently. Taking $\lambda = \mu = 0$ in (4-2-19) we obtain the relation

$$\mathscr{M}\left[\int_0^\infty f(xu)g(u)\,du; s\right] = f^*(s)g^*(-s) \tag{4-2-21}$$

and taking $\lambda = 0$, $\mu = -1$ in (4-2-20) the relation

$$\mathscr{M}\left[\int_0^\infty f\left(\frac{x}{u}\right) g(u)\frac{du}{u}; s\right] = f^*(s)g^*(s). \tag{4-2-22}$$

This last relation is often used in its inverse form

$$\mathscr{M}^{-1}[f^*(s)g^*(s);x] = \int_0^\infty f\left(\frac{x}{u}\right) g(u)\frac{du}{u}. \tag{4-2-23}$$

As an example of the use of equation (4-2-23) we consider the Mellin transform of the *Weyl fractional integral*

$$h(x;\mu) = \mathscr{W}_\mu[f(u);x] = \frac{1}{\Gamma(\mu)}\int_x^\infty f(u)(u-x)^{\mu-1}\,du \tag{4-2-24}$$

If we write

$$F(x) = x^\mu f(x), \qquad G(x) = \frac{1}{\Gamma(\mu)}(1-x)^{\mu-1}H(1-x)$$

we can put this integral into the form

$$h(x;\mu) = \int_0^\infty F(u)G\left(\frac{x}{u}\right)\frac{du}{u}$$

from which it follows that

$$h^*(s;\mu) = \mathscr{M}[h(x;\mu);x \to s] = F^*(s)G^*(s).$$

where $F^*(s)$, $G^*(s)$ are the Mellin transforms of $F(x)$, $G(x)$ respectively. Now, by equation (4-1-7)

$$F^*(s) = f^*(s+\mu)$$

where $f^*(s)$ is the Mellin transform of $f(x)$, and

$$G^*(s) = \frac{1}{\Gamma(\mu)}\int_0^1 (1-x)^{\mu-1}x^{s-1}\,dx = \frac{\Gamma(s)}{\Gamma(\mu+s)}$$

Hence we find that

$$h^*(s;\mu) = \frac{\Gamma(s)}{\Gamma(\mu+s)}f^*(s+\mu)$$

i.e.,

$$\mathscr{M}[\mathscr{W}_\mu f(x);s] = \frac{\Gamma(s)}{\Gamma(\mu+s)}f^*(s+\mu). \tag{4-2-25}$$

This is an obvious generalization of equation (4-2-18) above.

If α has a positive real part such that

$$n-1 < \operatorname{Re}\alpha < n$$

where n is a positive integer, we may define the *fractional derivative* of order α of a function $f(x)$ by the equation

$$\mathrm{D}_\infty{}^\alpha f(x) = \frac{d^n}{dx^n}\mathscr{W}_{n-\alpha}f(x) \tag{4-2-26}$$

i.e., by the formula

$$\mathrm{D}_\infty{}^\alpha f(x) = \frac{1}{\Gamma(n-\alpha)}\frac{d^n}{dx^n}\int_x^\infty f(u)(u-x)^{n-\alpha-1}\,du,$$

$$n-1 < \operatorname{Re}\alpha < n. \tag{4-2-27}$$

The Mellin transform of such a fractional derivative can be deduced easily from equations (4-2-3) and (4-2-25). By the first of these we see that

$$\mathscr{M}[\mathrm{D}_\infty{}^\alpha f(x); s] = (-1)^n\frac{\Gamma(s)}{\Gamma(s-n)}\mathscr{M}[\mathscr{W}_{n-\alpha}f(x); s-n].$$

Using equation (4-2-25) to evaluate the Mellin transform on the right-hand side of this equation we find that

$$\mathscr{M}[\mathrm{D}_\infty{}^\alpha f(x); s] = (-1)^n\frac{\Gamma(s)}{\Gamma(s-\alpha)}f^*(s-\alpha) \tag{4-2-28}$$

which is an obvious generalization of equation (4-2-3).

4-3 THE MELLIN INVERSION THEOREM

If we let $x = e^t$, $s = c + i\xi$ in the definition of the Fourier transform

$$G(\xi) = \frac{1}{\sqrt{(2\pi)}}\int_{-\infty}^\infty g(t)e^{i\xi t}\,dt$$

we find that it becomes

$$G(ic-is) = \frac{1}{\sqrt{(2\pi)}}\int_0^\infty x^{-c}g(\log x)\cdot x^{s-1}\,dx. \tag{4-3-1}$$

If we make the same substitutions in the inversion formula

$$g(t) = \frac{1}{\sqrt{(2\pi)}}\int_{-\infty}^\infty G(\xi)e^{-i\xi t}\,d\xi$$

we obtain the relation

$$g(\log x) = \frac{1}{\sqrt{(2\pi)}i}\int_{c-i\infty}^{c+i\infty} G(ic-is)x^{c-s}\,ds. \tag{4-3-2}$$

Writing

$$f(x) = (2\pi)^{-\frac{1}{2}}x^{-c}g(\log x), \qquad f^*(s) = G(ic-is)$$

we find that equations (4-3-1) and (4-3-2) yield the pair of equations

$$f^*(s) = \int_0^\infty x^{s-1} f(x)\, dx \tag{4-3-3}$$

$$f(x) = \frac{1}{2\pi i}\int_{c-i\infty}^{c+i\infty} f^*(s) x^{-s}\, ds. \tag{4-3-4}$$

Now (4-3-3) is merely the definition of the Mellin transform of the function $f(x)$, so that (4-3-4) is, in some sense, the inversion formula for the Mellin transform. This heuristic argument is justified in:

Theorem *Suppose that $f^*(s)$ is a function of the complex variable $s = \sigma + i\tau$ which is regular in the infinite strip $S = \{s : a < \sigma < b\}$ and that for any arbitrary small positive number η, $f^*(s)$ tends to zero uniformly as $|\tau| \to \infty$ in the strip $a + \eta \leqslant \sigma \leqslant b - \eta$. Then if the integral*

$$\int_{-\infty}^{\infty} f^*(\sigma + i\tau)\, d\tau$$

is absolutely convergent for each value of σ in the open interval (a,b), and if for positive real values of x and a fixed $c \in (a,b)$ we define

$$f(x) = \frac{1}{2\pi i}\int_{c-i\infty}^{c+i\infty} x^{-s} f^*(s)\, ds$$

then, in the strip S,

$$f^*(s) = \int_0^\infty x^{s-1} f(x)\, dx$$

To prove this result we note that since $f^*(s) \to 0$, uniformly, as $|\tau| \to \infty$ in the strip $a + \eta \leqslant \sigma \leqslant b - \eta$ the path of integration in the definition of $f(x)$ may be displaced parallel to itself. In other words the function $f(x)$ is independent of the particular choice of the constant c. If we take two values of c in (a,b), say c_1 and c_2, such that $c_1 < \sigma < c_2$ (Cf. Fig. 4-2), then we may write

$$\int_0^\infty x^{s-1} f(x)\, dx = I_1 + I_2 \tag{4-3-5}$$

where the integrals I_1 and I_2 are defined by the equations

$$I_1 = \int_0^1 x^{s-1}\, dx \cdot \frac{1}{2\pi i}\int_{c_1-i\infty}^{c_1+i\infty} x^{-t} f^*(t)\, dt$$

$$I_2 = \int_1^\infty x^{s-1}\, dx \cdot \frac{1}{2\pi i}\int_{c_2-i\infty}^{c_2+i\infty} x^{-t} f^*(t)\, dt.$$

Since $c_1 < \sigma$ the integral

$$\int_0^1 x^{\sigma-c_1-1}\, dx$$

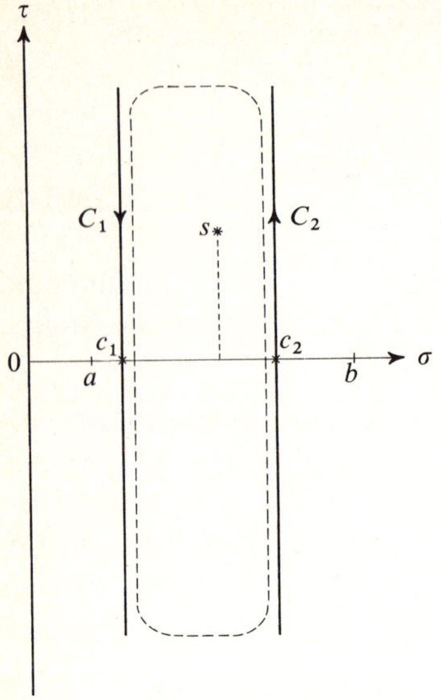

Fig. 4-2

is convergent and we find that

$$\left| \int_{c_1 - i\infty}^{c_1 + i\infty} f^*(t)\, dt \int_0^1 x^{s-t-1}\, dx \right| \leqslant \int_{-\infty}^{\infty} |F(c_1 + i\tau)|\, d\tau \int_0^1 x^{\sigma - c_1 - 1}\, dx.$$

Since the integral

$$\int_{-\infty}^{\infty} F(\sigma + i\tau)\, d\tau$$

is absolutely convergent if $\sigma \in (a,b)$ the first factor on the right of the last equation is finite and since the second one is also finite it follows that

$$\left| \int_{c_1 - i\infty}^{c_1 + i\infty} f^*(t)\, dt \int_0^1 x^{s-t-1}\, dx \right| < \infty$$

Hence we may change the order of the integrations in I_1 and obtain

$$\begin{aligned} I_1 &= \frac{1}{2\pi i} \int_{c_1 - i\infty}^{c_1 + i\infty} f^*(t)\, dt \int_0^1 x^{s-t-1}\, dx \\ &= \frac{1}{2\pi i} \int_{c_1 - i\infty}^{c_1 + i\infty} \frac{f^*(t)\, dt}{s - t} \\ &= \frac{1}{2\pi i} \int_{C_1} \frac{f^*(t)\, dt}{t - s} \end{aligned}$$

where C_1 is the line shown in Fig. 4-2.

Similarly, if $\sigma < c_2$ the integral

$$\int_1^\infty x^{\sigma - c_2 - 1}\, dx$$

is convergent and we are able to show that we can invert the order of the integrations in I_2 and obtain

$$I_2 = \frac{1}{2\pi i}\int_{C_2} \frac{f^*(t)\, dt}{t - s}$$

It follows from (4-3-5) that

$$\int_0^\infty x^{s-1} f(x)\, dx = \frac{1}{2\pi i}\left[\int_{C_1} + \int_{C_2}\right]\frac{f^*(t)\, dt}{t - s} \tag{4-3-6}$$

Since $|F(s)| \to 0$ uniformly as $|\tau| \to \infty$ it follows that each of the integrals

$$\frac{1}{2\pi i}\int_{C_1} \frac{f^*(t)\, dt}{t - s}, \qquad \frac{1}{2\pi i}\int_{C_2} \frac{f^*(t)\, dt}{t - s}$$

tends to zero as $|\tau| \to \infty$ and hence that we can replace the integral on the right-hand side of (4-3-6) by a similar integral taken round a closed contour Γ enclosing the point s. Since, by hypothesis, $f^*(s)$ is regular in the strip S, the only singularity of $(t - s)^{-1} f^*(s)$ is a simple pole at $t = s$ and it follows from Cauchy's integral theorem that the value of the integral on the right-hand side of equation (4-3-6) is $f^*(s)$.

4-4 CONVOLUTION THEOREMS FOR THE MELLIN TRANSFORM

If $f^*(s)$ and $g^*(s)$ are the Mellin transforms of the functions $f(x)$ and $g(x)$, then

$$\begin{aligned}\int_0^\infty f(x)g(x)x^{s-1}\, dx &= \int_0^\infty g(x)x^{s-1}\, dx \frac{1}{2\pi i}\int_{c-i\infty}^{c+i\infty} f^*(z)x^{-z}\, dz \\ &= \frac{1}{2\pi i}\int_{c-i\infty}^{c+i\infty} f^*(z)\, dz \int_0^\infty g(x)x^{s-z-1}\, dx \\ &= \frac{1}{2\pi i}\int_{c-i\infty}^{c+i\infty} f^*(z)g^*(s - z)\, dz\end{aligned}$$

showing that

$$\mathscr{M}[f(x)g(x); s] = \frac{1}{2\pi i}\int_{c-i\infty}^{c+i\infty} f^*(z)g^*(s - z)\, dz. \tag{4-4-1}$$

A special case of this result is obtained by taking $s = 1$:

$$\int_0^\infty f(x)g(x)\,dx = \frac{1}{2\pi i}\int_{c-i\infty}^{c+i\infty} f^*(z)g^*(1-z)\,dz \tag{4-4-2}$$

In a similar way we can show that

$$\begin{aligned}\mathscr{M}^{-1}[f^*(s)g^*(s);x] &= \frac{1}{2\pi i}\int_{c-i\infty}^{c+i\infty} f^*(s)x^{-s}\,ds\int_0^\infty g(u)u^{s-1}\,du \\ &= \int_0^\infty g(u)\frac{du}{u}\cdot\frac{1}{2\pi i}\int_{c-i\infty}^{c+i\infty} f^*(s)(x/u)^{-s}\,ds \\ &= \int_0^\infty f\left(\frac{x}{u}\right)g(u)\frac{du}{u} \end{aligned}\tag{4-4-3}$$

in agreement with equation (4-2-22).

As an example of the use of equation (4-4-1) we insert in that equation the Mellin transform pairs

$$f(x) = \frac{2(1-x^2)^{\lambda-1}H(1-x)}{\Gamma(\lambda)}, \qquad f^*(z) = \frac{\Gamma(\frac{1}{2}z)}{\Gamma(\lambda+\frac{1}{2}z)}$$

$$g(x) = \frac{2(1-a^2x^2)^{\mu-1}H(1-ax)}{\Gamma(\mu)}, \qquad g^*(z) = \frac{a^{-z}\Gamma(\frac{1}{2}z)}{\Gamma(\mu+\frac{1}{2}z)}$$

with $\lambda > 0$, $\mu > 0$, $0 < a \leqslant 1$. We then obtain the equation

$$\frac{1}{2\pi i}\int_{c-i\infty}^{c+i\infty} \frac{\Gamma(\frac{1}{2}z)\Gamma(\frac{1}{2}s-\frac{1}{2}z)a^{z-s}}{\Gamma(\lambda+\frac{1}{2}z)\Gamma(\mu+\frac{1}{2}s-\frac{1}{2}z)}\,dz = \frac{4}{\Gamma(\lambda)\Gamma(\mu)}\int_0^1 x^{s-1}(1-x^2)^{\lambda-1}(1-a^2x^2)^{\mu-1}\,dx.$$

Using the expansion

$$(1-a^2x^2)^{\mu-1} = \sum_{r=0}^\infty \frac{(1-\mu)_r}{r!}a^{2r}x^{2r}$$

and integrating term by term in this last equation we obtain the equation

$$\frac{1}{2\pi i}\int_{c-i\infty}^{c+i\infty} \frac{\Gamma(\frac{1}{2}z)\Gamma(\frac{1}{2}s-\frac{1}{2}z)a^{z-s}\,dz}{\Gamma(\lambda+\frac{1}{2}z)\Gamma(\mu+\frac{1}{2}s-\frac{1}{2}z)} = \frac{2\Gamma(\frac{1}{2}s)}{\Gamma(\lambda+\frac{1}{2}s)\Gamma(\mu)}\,{}_2F_1(1-\mu,\tfrac{1}{2}s;\lambda+\tfrac{1}{2}s;a^2).$$

With a trivial change of the variables we can rewrite this equation in the form

$$\frac{1}{2\pi i}\int_{c-i\infty}^{c+i\infty} \frac{\Gamma(\frac{1}{2}z)\Gamma(\alpha-\frac{1}{2}z)a^z\,dz}{\Gamma(\beta+\frac{1}{2}z)\Gamma(\gamma-\frac{1}{2}z)}$$

$$= \frac{2a^{2\alpha}\Gamma(\alpha)}{\Gamma(\alpha + \beta)\Gamma(\gamma - \alpha)} {}_2F_1(\alpha, 1 + \alpha - \gamma; \alpha + \beta; a^2)$$

with $0 < \alpha \leqslant 1$, $0 < \alpha < \gamma$, $\beta > 0$.

4-5 THE SOLUTION OF SOME INTEGRAL EQUATIONS

We can make use of the results of sec. 4-2 to derive the solution of some integral equations of general type.

We begin with the equation

$$\int_0^\infty f(x)K(xt)\,dx = g(t), \qquad (t > 0). \tag{4-5-1}$$

Taking the Mellin transform of both sides and using (4-2-21) we see that

$$f^*(1 - s)K^*(s) = g^*(s)$$

Replacing s by $1 - s$ we find that

$$f^*(s) = g^*(1 - s)L^*(s)$$

where

$$L^*(s) = \frac{1}{K^*(1 - s)}$$

From equation (4-2-21) we see that the solution of the integral equation (4-5-1) can be written in the form

$$f(x) = \int_0^\infty g(t)L(xt)\,dt \tag{4-5-2}$$

provided, of course, that the inverse Mellin transform

$$L(x) = \mathscr{M}^{-1}\left[\frac{1}{K^*(1 - s)}, \quad s \to x\right] \tag{4-5-3}$$

exists.

In particular, the equation (4-5-1) will have the solution

$$f(x) = \int_0^\infty g(t)K(xt)\,dt \tag{4-5-4}$$

if

$$K^*(s)K^*(1 - s) = 1. \tag{4-5-5}$$

In other words equation (4-5-5) is a necessary condition for K to be a *Fourier kernel* (see 1-1). It can easily be shown that the condition is sufficient (Cf. Problem 4-23).

From equations (4-1-7), (4-1-8), and (4-1-10) we can easily verify that

$$\sqrt{\frac{2}{\pi}}\cos x, \qquad \sqrt{\frac{2}{\pi}}\sin x, \qquad x^{\frac{1}{2}}\mathrm{J}_\nu(x)$$

are Fourier kernels.

As a further example of the use of the formulae (4-5-1) through (4-5-3) we consider the case in which

$$K(x) = x^{\frac{1}{2}}\mathrm{Y}_\nu(x)$$

where $\mathrm{Y}_\nu(x)$ is the Bessel function of the second kind of order ν. It is then easily shown [Cf. Problem 2-37(b)] that

$$K^*(s) = 2^{s-\frac{1}{2}}\frac{\Gamma(\frac{1}{4} + \frac{1}{2}s + \frac{1}{2}\nu)}{\Gamma(\frac{3}{4} - \frac{1}{2}s + \frac{1}{2}\nu)}\cot(\tfrac{3}{4}\pi - \tfrac{1}{2}s\pi + \tfrac{1}{2}\nu\pi)$$

and hence that

$$K^*(1-s) = 2^{\frac{1}{2}-s}\frac{\Gamma(\frac{3}{4} - \frac{1}{2}s + \frac{1}{2}\nu)}{\Gamma(\frac{1}{4} + \frac{1}{2}s + \frac{1}{2}\nu)}\cot(\tfrac{1}{4}\pi + \tfrac{1}{2}s\pi + \tfrac{1}{2}\nu\pi)$$

so that

$$L^*(s) = 2^{s-\frac{1}{2}}\frac{\Gamma(\frac{1}{4} + \frac{1}{2}s + \frac{1}{2}\nu)}{\Gamma(\frac{3}{4} + \frac{1}{2}s + \frac{1}{2}\nu)}\tan(\tfrac{1}{4}\pi + \tfrac{1}{2}s\pi + \tfrac{1}{2}\nu\pi).$$

This, too, is a transform we identify. Making use of (b) of Problem 2-38 we see that

$$L(x) = x^{\frac{1}{2}}\mathbf{H}_\nu(x)$$

where $\mathbf{H}_\nu(x)$ is Struve's function.

In other words we have shown that the integral equation

$$\int_0^\infty (xt)^{\frac{1}{2}}f(x)\mathrm{Y}_\nu(xt)\,dx = g(t) \tag{4-5-6}$$

has the solution

$$f(x) = \int_0^\infty (xt)^{\frac{1}{2}}g(t)\mathbf{H}_\nu(xt)\,dt \tag{4-5-7}$$

The function $g(t)$ defined by (4-5-6) is called the *Y-transform* of the function $f(x)$ and we write

$$g(t) = \mathscr{Y}_\nu[f(x);t] \tag{4-5-8}$$

Similarly the function $f(x)$ defined by (4-5-7) is called the **H**-*transform* of the function $g(t)$, and we write

$$f(x) = \mathscr{H}_\nu^*[g(t);x] \tag{4-5-9}$$

We have therefore shown that the **H**-transform is the reciprocal of the Y-transform, i.e., that

$$\mathscr{Y}_\nu^{-1} = \mathscr{H}_\nu^*$$

We can derive the solution of the integral equation

$$\int_0^\infty K_1\left(\frac{x}{y}\right) f_1(y)\,\frac{dy}{y} = g_1(x), \qquad x > 0 \tag{4-5-10}$$

in a similar way. Making use of equation (4-2-22) we see that this equation is equivalent to

$$K_1^*(s) f_1^*(s) = g_1^*(s) \tag{4-5-11}$$

where $K_1^*(s)$, $f_1^*(s)$, $g_1^*(s)$ are respectively the Mellin transforms of $K_1(x)$, $f_1(x)$, $g_1(x)$. Writing this last equation in the form

$$f_1^*(s - m\alpha + \beta) = (-1)^m L_1^*(s)\left[(-1)^m \frac{\alpha^m \Gamma(s/\alpha)}{\Gamma(s/\alpha - m)} g_1^*(s - m\alpha + \beta)\right] \tag{4-5-12}$$

where m is a positive integer and

$$L_1^*(s) = \alpha^{-m} \frac{\Gamma(s/\alpha - m)}{\Gamma(s/\alpha) K_1^*(s - m\alpha + \beta)}, \qquad (\alpha \neq 0)$$

and making use of the fact that

$$\mathscr{M}\left[\left(x^{1-\alpha}\frac{d}{dx}\right)^m x^\beta g(x); s\right] = (-1)^m \frac{\alpha^m \Gamma(s/\alpha)}{\Gamma(s/\alpha - m)} g_1^*(s - m\alpha + \beta) \tag{4-5-13}$$

we see that equation (4-5-12) is equivalent to the solution

$$f(x) = (-1)^m x^{m\alpha - \beta} \int_0^\infty L_1\left(\frac{x}{y}\right)\left(y^{1-\alpha}\frac{d}{dy}\right)^m \{y^\beta g(y)\}\,\frac{dy}{y} \tag{4-5-14}$$

provided that the inverse Mellin transform

$$L_1(x) = \alpha^{-m} \mathscr{M}^{-1}\left[\frac{\Gamma(s/\alpha - m)}{\Gamma(s/\alpha) K_1^*(s - m\alpha + \beta)}; x\right] \tag{4-5-15}$$

exists.

The integral equation

$$\int_x^1 K\left(\frac{x}{y}\right) f(y)\,\frac{dy}{y} = g(x), \qquad 0 < x < 1 \tag{4-5-16}$$

is a more interesting form of the same equation. If we make the substitutions

$$f_1(x) = f(x)\mathrm{H}(1-x), \qquad g_1(x) = g(x)\mathrm{H}(1-x),$$
$$K_1(x) = K(x)\mathrm{H}(1-x),$$

we see that equation (4-5-16) is equivalent to (4-5-10) and hence to equation (4-5-11).

To illustrate the procedure we consider the integral equation

$$\int_x^1 \frac{T_n(y/x)}{\sqrt{(y^2-x^2)}} f(y)\,dy = g(x), \qquad 0 < x < 1 \tag{4-5-17}$$

where $T_n(x)$ denotes the Tchebycheff polynomial $\cos(n\cos^{-1}x)$.

In this case

$$K_1(x) = \frac{T_n(x^{-1})}{\sqrt{(1-x^2)}}\mathrm{H}(1-x)$$

and it can be shown (Cf. Problem 4-27) that

$$K_1^*(s) = 2^{s-2}\frac{\Gamma(\tfrac{1}{2}n+\tfrac{1}{2}s)\Gamma(\tfrac{1}{2}s-\tfrac{1}{2}n)}{\Gamma(s)}, \qquad \mathrm{Re}\,s > n$$

so that

$$\frac{1}{(s-n)K_1^*(s)} = \frac{2^{1-s}\Gamma(s)}{\Gamma(\tfrac{1}{2}n+\tfrac{1}{2}s)\Gamma(1-\tfrac{1}{2}n+\tfrac{1}{2}s)}$$

and from Problem 4-26 we find that

$$L_1(x) = \mathscr{M}^{-1}\left[\frac{1}{(s-n)K_1^*(s)};x\right] = \frac{2}{\pi}\frac{T_{n-1}(x)}{\sqrt{(1-x^2)}}\mathrm{H}(1-x) \tag{4-5-18}$$

Writing

$$\frac{1}{(s-n)K_1^*(s)} = n^{-1}\frac{\Gamma(s/n-1)}{\Gamma(s/n)K_1^*(s)}$$

we see that this is the above case with $\alpha = n$, $\beta = n$, $m = 1$ so that from equation (4-5-14)

$$f(x) = -\int_0^\infty L_1\left(\frac{x}{y}\right)y^{1-n}\frac{d}{dy}\{y^n g_1(y)\}\frac{dy}{y}$$

with $L_1(x)$ defined by (4-5-18). In this way we obtain the solution

$$f(x) = -\frac{2}{\pi}\int_x^1 \frac{T_{n-1}(x/y)}{\sqrt{(y^2-x^2)}}y^{1-n}\frac{d}{dy}\{y^n g(y)\}\,dy. \tag{4-5-19}$$

4-6 THE DISTRIBUTION OF POTENTIAL IN A WEDGE

To illustrate the use of the Mellin transform in the solution of a physical problem, we consider the problem of determining a function $\psi(\rho,\phi)$ which satisfies Laplace's equation

$$\rho^2 \frac{\partial^2\psi}{\partial\rho^2} + \rho \frac{\partial\psi}{\partial\rho} + \frac{\partial^2\psi}{\partial\phi^2} = 0 \tag{4-6-1}$$

in the infinite wedge $\rho \geqslant 0$, $|\phi| \leqslant \alpha$, the boundary conditions

$$\psi(\rho,\alpha) = f(\rho), \qquad \psi(\rho,-\alpha) = g(\rho), \qquad \rho \geqslant 0 \tag{4-6-2}$$

and the limiting condition

$$\psi(\rho,\phi) \to 0, \qquad \rho \to \infty, \qquad |\phi| \leqslant \alpha. \tag{4-6-3}$$

If we apply the operator $\mathscr{M}$ to both sides of (4-6-1) and make use of (4-2-6) we obtain the differential equation

$$\left(\frac{\partial^2}{\partial\phi^2} + s^2\right) \psi^*(s,\phi) = 0 \tag{4-6-4}$$

for the determination of the Mellin transform

$$\psi^*(s,\phi) = \mathscr{M}[\psi(\rho,\phi);\rho \to s]$$

of the unknown function $\psi(\rho,\phi)$.

We notice that if we wish to determine the strip in the complex s-plane for which (4-6-4) holds we need a more precise limiting condition than (4-6-3). If we integrate by parts we see that the complete form of (4-2-6) is

$$\mathscr{M}\left[\rho^2 \frac{\partial^2\psi}{\partial\rho^2} + \rho \frac{\partial\psi}{\partial\rho}; s\right] = \left[\rho^{s+1} \frac{\partial\psi}{\partial\rho} - s\rho^s\psi\right]_{\rho=0}^{\rho=\infty} + s^2\psi^*(s,\phi)$$

so that if we have the conditions

$$\left.\begin{array}{lll} \psi(\rho,\phi) \sim \rho^{-\lambda} & \text{as} & \rho \to 0 \\ \psi(\rho,\phi) \sim \rho^{-\mu} & \text{as} & \rho \to \infty \end{array}\right\} \tag{4-6-5}$$

we see that

$$\mathscr{M}\left[\rho^2 \frac{\partial^2\psi}{\partial\rho^2} + \rho \frac{\partial\psi}{\partial\rho}; s\right] = s^2\psi^*(s,\phi)$$

provided that $\lambda < \operatorname{Re} s < \mu$, and provided that, for s lying in that strip, the Mellin transform $\psi^*(s,\phi)$ exists.

We can therefore adopt the standpoint that, at least for potential functions $\psi(\rho,\phi)$ satisfying limiting conditions of the type (4-6-5) with $\lambda < \mu$, there will be some strip in the complex s-plane in which (4-6-4) is valid.

The boundary conditions (4-6-2) may similarly be transformed to

$$\psi^*(s,\alpha) = f^*(s), \qquad \psi^*(s,-\alpha) = g^*(s) \tag{4-6-6}$$

where $f^*(s)$, $g^*(s)$ are, respectively, the Mellin transforms of $f(\rho)$ and $g(\rho)$.

The solution of the two-point boundary value problem posed by equations (4-6-4), (4-6-6) is easily seen to be

$$\psi^*(s,\alpha) = f^*(s)\,\frac{\sin s(\alpha + \phi)}{\sin 2s\alpha} + g^*(s)\,\frac{\sin s(\alpha - \phi)}{\sin 2s\alpha}. \tag{4-6-7}$$

In the symmetrical problem in which $f(\rho) \equiv g(\rho)$ this reduces to the simpler form

$$\psi^*(s,\alpha) = f^*(s)\,\frac{\cos s\phi}{\cos s\alpha}. \tag{4-6-8}$$

If, therefore, we introduce the kernel

$$K(\rho,\phi,\alpha) = \mathscr{M}^{-1}\left[\frac{\sin s\phi}{\sin 2s\alpha}; s \to \rho\right] \tag{4-6-9}$$

and make use of the convolution theorem (4-4-3) we can write the solution of the general problem in the form

$$\psi(\rho,\phi) = \int_0^\infty f(u)K(\rho/u,\alpha + \phi,\alpha)\,\frac{du}{u} + \int_0^\infty g(u)K(\rho/u,\alpha - \phi,\alpha)\,\frac{du}{u}. \tag{4-6-10}$$

Similarly, if we define a kernel L by the equation

$$L(\rho,\phi,\alpha) = \mathscr{M}^{-1}\left[\frac{\cos s\phi}{\cos s\alpha}; s \to \rho\right] \tag{4-6-11}$$

we derive the formula

$$\psi(\rho,\phi) = \int_0^\infty f(u)L(\rho/u,\phi,\alpha)\,\frac{du}{u} \tag{4-6-12}$$

for the solution of the symmetrical problem.

The evaluation of the kernels K and L for general values of α can be carried out by using the results of Problem 4-4. If we write $\alpha = \pi/(2n)$, $n > 0$, then

$$K\left(\rho,\phi,\frac{\pi}{2n}\right) = \frac{n}{\pi}\cdot\frac{\rho^n \sin(n\phi)}{1 + 2\rho^n \cos(n\phi) + \rho^{2n}}$$

Equation (4-6-10) then yields the solution

$$\psi(\rho,\phi) = \frac{n\rho^n}{\pi}\cos(n\phi)\int_0^\infty \frac{u^{n-1}f(u)\,du}{u^{2n} - 2\rho^n u^n \sin(n\phi) + \rho^{2n}}$$

$$+ \frac{n\rho^n}{\pi}\cos(n\phi)\int_0^\infty \frac{u^{n-1}g(u)\,du}{u^{2n} + 2\rho^n u^n \sin(n\phi) + \rho^{2n}},$$

$$|\phi| < \frac{\pi}{2n}. \tag{4-6-13}$$

4-7 AN APPLICATION TO THE SUMMATION OF SERIES

A useful method of summing slowly convergent series has been given by Macfarlane (1949).

If $f^*(s)$ is the Mellin transform of $f(x)$, then we may write the inversion theorem for the Mellin transform in the form of the equation

$$f(n + a) = \frac{1}{2\pi i}\int_{c-i\infty}^{c+i\infty} f^*(s)(n + a)^{-s}\,ds$$

Assuming that $0 < a \leqslant 1$ and summing both sides of this equation with respect to n from 1 to ∞ we obtain the relation

$$\sum_{n=0}^{\infty} f(n + a) = \frac{1}{2\pi i}\int_{c-i\infty}^{c+i\infty} f^*(s)\zeta(s,a)\,ds \tag{4-7-1}$$

where $\zeta(s,a)$ denotes the generalized zeta-function of Riemann, defined when $\operatorname{Re} s > 1$, by the equation

$$\zeta(s,a) = \sum_{n=0}^{\infty}(s + a)^{-n}, \qquad 0 < a \leqslant 1. \tag{4-7-2}$$

In exploiting the possibilities of equation (4-7-1) we need to know something of the properties of the function $\zeta(s,a)$. A full account of these is given in Chapter XIII of Whittaker and Watson (1927); we shall merely state them here.

First of all we notice that when $\operatorname{Re} s > 1$ the definition (4-7-2) is equivalent to the equation

$$\zeta(s,a) = -\frac{\Gamma(1 - s)}{2\pi i}\int_\infty^{(0+)} \frac{(-z)^{s-1}e^{-az}}{1 - e^{-z}}\,dz \tag{4-7-3}$$

where by $\int_\infty^{(0+)}$ we mean that the integral is taken along a path starting at "infinity" on the real axis, encircling the origin in the positive direction and

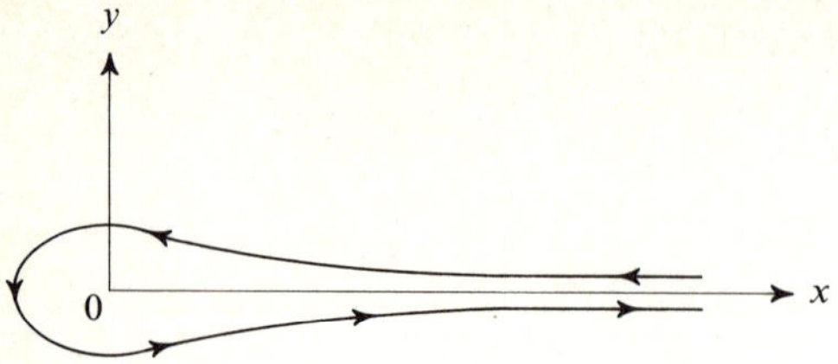

Fig. 4-3

returning to the initial point, (Cf. Fig. 4-3), and not enclosing the points $\pm 2n\pi i$ which are poles of the integrand. The integral on the right side of equation (4-7-3) is a single-valued analytic function of s for *all* values of s and so we may take that equation to define the function $\zeta(s,a)$ for *all* values of s. With that definition we see that $\zeta(s,a)$ can have singularities only at the singularities of $\Gamma(1-s)$, i.e., only at the points $s = 1, 2, \ldots$. But from the definition (4-7-2) $\zeta(s,a)$ is analytic when $\operatorname{Re} s > 1$ so that the only singularity of $\zeta(s,a)$ is at the point $s = 1$ and it is easily shown that it is a simple pole with residue 1.

It is also easily shown that, if m is a positive integer,

$$\zeta(-m,a) = -(m+1)^{-1}B_{m+1}(a), \qquad 0 < a \leqslant 1 \tag{4-7-4}$$

where $B_n(a)$ is the Bernoulli polynomial of the n-th degree defined to be the coefficient of $t^n/n!$ in the Maclaurin expansion of the function

$$\frac{e^{at}-1}{e^t-1}\,t$$

In particular

$$\zeta(0,a) = \tfrac{1}{2} - a. \tag{4-7-5}$$

Another useful result is

$$\zeta(-m,1-a) = (-1)^{m-1}\zeta(-m,a).$$

Similarly, by using the result (4-1-2) we have the representation

$$f(nx) = \mathscr{M}^{-1}[n^{-s}f^*(s);x]$$

from which we deduce the formula

$$\sum_{n=1}^{\infty} f(nx) = \mathscr{M}^{-1}[\zeta(s)f^*(s);x] \tag{4-7-6}$$

in which $\zeta(s)$ denotes the Riemann zeta-function defined for $\operatorname{Re} s > 1$ by the equation

$$\zeta(s) = \sum_{n=1}^{\infty} n^{-s} \tag{4-7-7}$$

Since $\zeta(s) = \zeta(s,1)$ many of the properties of the zeta-function may be obtained by taking $a = 1$ in those of the generalized zeta-function. For example, it is easily shown that, if m is a positive integer

$$\zeta(-2m) = 0 \tag{4-7-8}$$

$$\zeta(1 - 2m) = \frac{(-1)^m}{2m} B_m \tag{4-7-9}$$

$$\zeta(2m) = \frac{2^{2m-1}\pi^{2m}B_m}{\Gamma(2m + 1)} \tag{4-7-10}$$

where B_m is the m-th Bernoulli number, defined by the relation

$$\tfrac{1}{2}z \cot(\tfrac{1}{2}z) = 1 - B_1 \frac{z^2}{2!} - B_2 \frac{z^4}{4!} - B_3 \frac{z^6}{6!} - \cdots,$$

and, in particular,

$$B_1 = \tfrac{1}{6}, \qquad B_2 = \tfrac{1}{30}, \qquad B_3 = \tfrac{1}{42}, \qquad B_4 = \tfrac{1}{30}, \qquad B_5 = \tfrac{5}{66}.$$

Also we have

$$\zeta(0) = -\tfrac{1}{2}. \tag{4-7-11}$$

Further results can be derived from simple properties of the zeta-function. For example, since

$$\begin{aligned}\sum_{n=1}^{\infty} (-1)^{n-1} n^{-s} &= \sum_{n=1}^{\infty} (2n - 1)^{-s} - \sum_{n=1}^{\infty} (2n)^{-s} \\ &= \zeta(s) - 2 \sum_{n=1}^{\infty} (2n)^{-s} \\ &= (1 - 2^{1-s})\zeta(s)\end{aligned}$$

we have the formula

$$\sum_{n=1}^{\infty} (-1)^{n-1} f(xn) = \mathscr{M}^{-1}[(1 - 2^{1-s})\zeta(s) f^*(s); x]. \tag{4-7-12}$$

Similarly from the identities

$$\sum_{n=0}^{\infty} (2n + 1)^{-s} = (1 - 2^{-s})\zeta(s)$$

$$\sum_{n=0}^{\infty} (-1)^n (2n + 1)^{-s} = 2^{-2s}\{\zeta(s,\tfrac{1}{4}) - \zeta(s,\tfrac{3}{4})\}$$

$$\sum_{n=0}^{\infty} (-1)^n (n + 2a)^{-s} = 2^{-s}\{\zeta(s,a) - \zeta(s,1 + a)\}, \qquad 0 < a \leqslant \tfrac{1}{2}$$

we may deduce the formulae

$$\sum_{n=0}^{\infty} f\{(2n + 1)x\} = \mathscr{M}^{-1}[(1 - 2^{-s})\zeta(s) f^*(s); x] \tag{4-7-13}$$

$$\sum_{n=0}^{\infty} (-1)^n f\{(2n+1)x\} = \mathscr{M}^{-1}[\{\zeta(s,\tfrac{1}{4}) - \zeta(s,\tfrac{3}{4})\};4x] \qquad (4\text{-}7\text{-}14)$$

$$\sum_{n=0}^{\infty} (-1)^n f(n+2a) = \frac{1}{2\pi i}\int_{c-i\infty}^{c+i\infty} 2^{-s}\{\zeta(s,a) - \zeta(s,a+\tfrac{1}{2})\} f^*(s)\, ds,$$

$$0 < a \leqslant \tfrac{1}{2} \qquad (4\text{-}7\text{-}15)$$

To illustrate the use of these formulae we take

$$f(x) = x^{-1}J_1(x) \qquad f^*(s) = 2^{s-2}\frac{\Gamma(\frac{1}{2}s)}{\Gamma(2-\frac{1}{2}s)}, \qquad 0 < \operatorname{Re} s < \tfrac{5}{2}$$

—Cf. equation (4-1-10) above—in equation (4-7-14) to obtain the relation

$$\sum_{n=0}^{\infty} \frac{(-1)^n}{2n+1} J_1\{2n+1)x\}$$

$$= \frac{1}{2\pi i}\int_{c-i\infty}^{c+i\infty} \frac{\Gamma(\frac{1}{2}s)}{\Gamma(2-\frac{1}{2}s)} \{\zeta(s,\tfrac{1}{4}) - \zeta(s,\tfrac{3}{4})\} 2^{-s-2} x^{1-s}\, dx.$$

The integrand has poles only at $s = -2n$, $n = 0, 1, 2, \ldots$ and since

$$\lim_{s\to -2n} (s+2n)\Gamma(\tfrac{1}{2}s)$$

$$= \tfrac{1}{2}\lim_{s\to -2n} (\tfrac{1}{2}s+n)\cdot \frac{\Gamma(\frac{1}{2}s+n+1)}{(\frac{1}{2}s+n)(\frac{1}{2}s+n-1)\ldots\frac{1}{2}s} = \frac{(-1)^n}{2\cdot n!}$$

the residue at $s = -2n$ is

$$\tfrac{1}{4}\cdot \frac{(-1)^n}{n!(n+1)!}\{\zeta(-2n,\tfrac{1}{4}) - \zeta(-2n\tfrac{3}{4})\}(2x)^{2n+1}$$

$$= \tfrac{1}{2}\cdot \frac{(-1)^n}{n!(n+1)!}\left\{\frac{-B_{2n+1}(\frac{1}{4})}{2n+1}\right\}(2x)^{2n+1}$$

Hence we find that

$$\sum_{n=0}^{\infty} \frac{(-1)^n}{(2n+1)} J_1\{(2n+1)x\} = -\tfrac{1}{2}\sum_{n=0}^{\infty} \frac{(-1)^n B_{2n+1}(\frac{1}{4})(2x)^{2n+1}}{(2n+1)\cdot n!(n+1)!}.$$

The series on the right is very rapidly convergent for $0 < x < 1$.

4-8 A NOTE ON GENERALIZED HYPERGEOMETRIC SERIES

In calculating the Mellin transforms of some special functions it is useful to have available closed form expressions for sums of generalized hypergeometric series

$$ {}_pF_q(a_1, \ldots, a_p; b_1, \ldots, b_q; x) = \sum_{r=0}^{\infty} \frac{(a_1)_r \ldots (a_p)_r}{r!(b_1)_r \ldots (b_q)_r} x^r \qquad (4\text{-}8\text{-}1) $$

when x takes certain specific values. In the definition (4-8-1) we have used the usual notation

$$ (a)_r = a(a+1)(a+2) \ldots (a+r-1). $$

Here we shall list such theorems referring an interested reader to Slater (1966) for proofs.

The oldest result of the type we are considering is *Gauss' theorem*

$$ {}_2F_1(a,b;c;1) = \frac{\Gamma(c)\Gamma(c-a-b)}{\Gamma(c-a)\Gamma(c-b)} \qquad (4\text{-}8\text{-}2) $$

with $c > a + b$, $c > a$, $c > b$, $c > 0$.

When the argument is -1 we have *Kummer's theorem*

$$ {}_2F_1(a,b;b-a+1;-1) $$
$$ = \frac{\Gamma(1+b-a)\Gamma(1+ -b)}{\Gamma(1+b)\Gamma(1+ -b-a)} (b > -1, a < 1 + \tfrac{1}{2}b), \qquad (4\text{-}8\text{-}3) $$

When the argument is $\frac{1}{2}$ we may use *Bailey's theorem*

$$ {}_2F_1(a,1-a;b;\tfrac{1}{2}) = \frac{\Gamma(\frac{1}{2}b)\Gamma(\frac{1}{2}b+\frac{1}{2})}{\Gamma(\frac{1}{2}a+\frac{1}{2}b)\Gamma(\frac{1}{2}b-\frac{1}{2}a+\frac{1}{2})} \qquad (4\text{-}8\text{-}4) $$

with $b > a - 1$, $b > 0$.

When $p = 3$ and $q = 2$ we have the result

$$ {}_3F_2(a,b,c;d,e;1) = \frac{\Gamma(d)\Gamma(1+a-e)\Gamma(1+b-e)\Gamma(1+c-e)}{\Gamma(1-e)\Gamma(d-a)\Gamma(d-b)\Gamma(d-c)} \qquad (4\text{-}8\text{-}5) $$

provided that

(*a*) one of the parameters a, b, c is a negative integer;
(*b*) $d + e = 1 + a + b + c$.

The result (4-8-5) is called *Saalschütz's theorem* and any series satisfying the conditions (*a*), (*b*) is said to be *Saalschützian*.

The result

$$ {}_3F_2(a,b,c;1+a-b,1+a-c;1) $$
$$ = \frac{\Gamma(1+\frac{1}{2}a)\Gamma(1+\frac{1}{2}a-b-c)\Gamma(1+a-b)\Gamma(1+a-c)}{\Gamma(1+a)\Gamma(1+a-b-c)\Gamma(1+\frac{1}{2}a-b)\Gamma(1+\frac{1}{2}a-c)} \qquad (4\text{-}8\text{-}6) $$

is called *Dixon's theorem*. In particular if $c = -n$, where n is a positive integer, the result reduces to

$$
{}_3F_2(a,b,-n;1+a-b,1+a+n;1) = \frac{(1+a)_n(1+\tfrac{1}{2}a-b)_n}{(1+\tfrac{1}{2}a)_n(1+a-b)_n} \tag{4-8-7}
$$

Similarly the relation

$$
{}_3F_2(a,b,c;\tfrac{1}{2}+\tfrac{1}{2}a+\tfrac{1}{2}b,2c;1) = \frac{\Gamma(\tfrac{1}{2})\Gamma(\tfrac{1}{2}+c)\Gamma(\tfrac{1}{2}+\tfrac{1}{2}a+\tfrac{1}{2}b)\Gamma(\tfrac{1}{2}-\tfrac{1}{2}a-\tfrac{1}{2}b+c)}{\Gamma(\tfrac{1}{2}+\tfrac{1}{2}a)\Gamma(\tfrac{1}{2}+\tfrac{1}{2}b)\Gamma(\tfrac{1}{2}-\tfrac{1}{2}a+c)\Gamma(\tfrac{1}{2}-\tfrac{1}{2}b+c)},
$$
$$
(\tfrac{1}{2}+c-\tfrac{1}{2}a-\tfrac{1}{2}b>0) \tag{4-8-8}
$$

is known as *Watson's theorem* and

$$
{}_3F_2(a,1-a,c;e;1+2c-e;1) = \frac{2^{1-2c}\pi\Gamma(e)\Gamma(1+2c-e)}{\Gamma(\tfrac{1}{2}e+\tfrac{1}{2}a)\Gamma(\tfrac{1}{2}+\tfrac{1}{2}e-\tfrac{1}{2}a)\Gamma(\tfrac{1}{2}+c-\tfrac{1}{2}e+\tfrac{1}{2}a)\Gamma(1+c-\tfrac{1}{2}e-\tfrac{1}{2}a)} \tag{4-8-9}
$$

$(a > 0, e > 0, 1 + 2c > e)$, as *Whipple's theorem.*

PROBLEMS

4-1 Prove that if m is a positive integer, $\alpha \neq 0$,

(*a*) $\mathscr{M}\left[\left(x^{1-\alpha}\frac{d}{dx}\right)^m f(x);s\right] = (-1)^m\alpha^m\frac{\Gamma(s/\alpha)}{\Gamma(s/\alpha-m)}f^*(s-m\alpha)$

(*b*) $\mathscr{M}\left[\left(\frac{d}{dx}x\right)^m f(x);s\right] = (1-s)^m f^*(s).$

4-2 Show that

(*a*) $\mathscr{M}^{-1}[\Gamma(s)f^*(1-s);x] = \mathscr{L}[f(t);x],$

(*b*) $\mathscr{M}^{-1}[\sin(\tfrac{1}{2}\pi s)\Gamma(s)f^*(1-s);x] = \sqrt{(\tfrac{1}{2}\pi)}\mathscr{F}_s[f(t);x],$

(*c*) $\mathscr{M}^{-1}[\cos(\tfrac{1}{2}\pi s)\Gamma(s)f^*(1-s);x] = \sqrt{(\tfrac{1}{2}\pi)}\mathscr{F}_c[f(t);x],$

(*d*) $\mathscr{M}^{-1}\left[\frac{2^s\Gamma(\tfrac{1}{2}\nu+\tfrac{1}{2}s+\tfrac{1}{2})}{\Gamma(\tfrac{1}{2}\nu-\tfrac{1}{2}s+\tfrac{1}{2})}f^*(1-s);x\right] = x\mathscr{H}_\nu[f(\rho);x].$

4-3 Show that

$$(a)\ \mathscr{M}^{-1}\left[\frac{\Gamma(\frac{1}{2}s)}{\Gamma(\frac{1}{2}s+\frac{1}{2})}f^*(s);x\right] = \frac{2}{\sqrt{\pi}}\int_x^\infty \frac{f(t)\,dt}{\sqrt{(t^2-x^2)}},$$

$$(b)\ \mathscr{M}^{-1}\left[\frac{\Gamma(\frac{1}{2}-\frac{1}{2}s)}{\Gamma(1-\frac{1}{2}s)}f^*(s);x\right] = \frac{2}{\sqrt{\pi}}\int_0^x \frac{f(t)\,dt}{\sqrt{(x^2-t^2)}}.$$

4-4 If $-\frac{1}{2}\pi < \phi < \frac{1}{2}\pi$, show that

$$\mathscr{M}[e^{-x\cos\phi}\cos(x\sin\phi);x\to s] = \Gamma(s)\cos(s\phi), \qquad (\operatorname{Re} s > 0),$$

$$\mathscr{M}[e^{-x\cos\phi}\sin(x\sin\phi);x\to s] = \Gamma(s)\sin(s\phi), \qquad (\operatorname{Re} s > -1).$$

Deduce that if $-\frac{1}{2}\pi < \phi < \frac{1}{2}\pi$

$$\mathscr{M}^{-1}\left[\frac{\cos(s\phi)}{\sin(s\pi)};x\right] = \frac{1}{\pi}\cdot\frac{1+x\cos\phi}{1+2x\cos\phi+x^2}$$

$$\mathscr{M}^{-1}\left[\frac{\sin(s\phi)}{\sin(s\pi)};x\right] = \frac{1}{\pi}\cdot\frac{x\sin\phi}{1+2x\cos\phi+x^2}$$

and hence that if $-\dfrac{\pi}{2n} < \phi < \dfrac{\pi}{2n}, \qquad n > 0$

$$\mathscr{M}^{-1}\left[\frac{\cos(s\phi)}{\sin(s\pi/n)};x\right] = \frac{n}{\pi}\cdot\frac{1+x^n\cos(n\phi)}{1+2x^n\cos(n\phi)+x^{2n}}$$

$$\mathscr{M}^{-1}\left[\frac{\sin(s\phi)}{\sin(s\pi/n)};x\right] = \frac{n}{\pi}\cdot\frac{x^n\sin(n\phi)}{1+2x^n\cos(n\phi)+x^{2n}}.$$

4-5 Starting from the definition

$$P_n(x) = \frac{(2x)^n(\frac{1}{2})_n}{n!}\,{}_2F_1(\tfrac{1}{2}-\tfrac{1}{2}n, -\tfrac{1}{2}n; \tfrac{1}{2}-n; x^{-2})$$

of the Legendre polynomial $P_n(x)$ show that

$$\mathscr{M}[P_n(x^{-1})\mathrm{H}(1-x);s]$$

$$= \frac{2^{n-1}(\frac{1}{2})_n}{n!}\,{}_3F_2(\tfrac{1}{2}-\tfrac{1}{2}n, -\tfrac{1}{2}n, \tfrac{1}{2}s-\tfrac{1}{2}n; \tfrac{1}{2}-n, \tfrac{1}{2}s-\tfrac{1}{2}n+1; 1).$$

Summing the series on the right by means of Saalschütz's theorem show that

$$\mathscr{M}[P_n(x^{-1})\mathrm{H}(1-x);s]$$

$$= 2^{s-1}\frac{\Gamma(\frac{1}{2}s+\frac{1}{2}n+\frac{1}{2})\Gamma(\frac{1}{2}s-\frac{1}{2}n)}{\Gamma(\frac{1}{2})\Gamma(s+1)}, \qquad \operatorname{Re} s > n.$$

4-6 Show that if n is a positive integer

$$\mathscr{M}[x^n P_n(x)\mathrm{H}(1-x);s] = (n+s)^{-1}\,{}_2F_1(-n,n+1;n+s+1;\tfrac{1}{2}).$$

Evaluate the sum on the right by using Bailey's theorem and hence deduce that

$$\mathscr{M}[x^m P_n(x)\mathrm{H}(1-x);s]$$

$$= \frac{1}{2}\cdot\frac{\Gamma(\frac{1}{2}m+\frac{1}{2}s)\Gamma(\frac{1}{2}m+\frac{1}{2}s+\frac{1}{2})}{\Gamma(\frac{1}{2}m-\frac{1}{2}n+\frac{1}{2}s+\frac{1}{2})\Gamma(\frac{1}{2}m+\frac{1}{2}n+\frac{1}{2}s+1)},$$

$$(\mathrm{Re}\,s > n-m-1).$$

4-7 Defining the Gegenbauer polynomial $C_m^{\lambda}(x)$ by the formula

$$C_m^{\lambda}(x) = \sum_{r=0}^{[\frac{1}{2}m]} \frac{(-1)^r(\lambda)_{m-r}}{r!(m-2r)!}(2x)^{2r-m}, \qquad (\lambda > 0),$$

prove that

$$\mathscr{M}[C_m^{\lambda}(x^{-1})(1-x^2)^{\lambda-\frac{1}{2}}\mathrm{H}(1-x);s]$$

$$= \frac{2^{s-1}\Gamma(m+2\lambda)\Gamma(\frac{1}{2}s-\frac{1}{2}m)\Gamma(\frac{1}{2}s+\lambda+\frac{1}{2}m)}{m!\Gamma(\lambda)\Gamma(s+2\lambda)}, \qquad (\mathrm{Re}\,s > m)$$

4-8 Adopting the alternative formula

$$C_n^{\mu}(x) = \frac{(2\mu)_n}{n!}\,{}_2F_1(-n,n+2\mu;\mu+\tfrac{1}{2};\tfrac{1}{2}-\tfrac{1}{2}x), \qquad (\mu > 0)$$

for the Gegenbauer polynomial, prove that

$$\mathscr{M}[(1-x^2)^{\mu-\frac{1}{2}}C_n^{\mu}(x)\mathrm{H}(1-x);s]$$

$$= \frac{2^{-s}\Gamma(\frac{1}{2})\Gamma(\mu+\frac{1}{2})\Gamma(2\mu+n)\Gamma(s)}{n!\Gamma(2\mu)\Gamma(\frac{1}{2}n+\frac{1}{2}+\mu+\frac{1}{2}s)\Gamma(\frac{1}{2}+\frac{1}{2}s-\frac{1}{2}n)}$$

and hence that

$$\mathscr{M}[x^{n-1}(1-x^2)^{\mu-\frac{1}{2}}C_n^{\mu}(x)\mathrm{H}(1-x);s]$$

$$= \frac{2^{1-n-s}\Gamma(\frac{1}{2})\Gamma(\mu+\frac{1}{2})\Gamma(2\mu+n)\Gamma(s+n-1)}{n!\Gamma(2\mu)\Gamma(\frac{1}{2}s)\Gamma(\frac{1}{2}s+n+\mu)}, \qquad (\mathrm{Re}\,s > 0).$$

4-9 Defining the associated Legendre function $P_\nu^{-\lambda}(x)$ by the equation

$$P_\nu^{-\lambda}(x) = \frac{1}{\Gamma(\lambda+1)}\left(\frac{1-x}{1+x}\right)^{\frac{1}{2}\lambda}{}_2F_1(-\nu,\nu+1;\lambda+1;\tfrac{1}{2}-\tfrac{1}{2}x)$$

show that, if $\operatorname{Re} s > \min(0, \nu - \lambda - 1)$,

$$\mathscr{M}[(1-x^2)^{\frac{1}{2}\lambda}P_\nu^{-\lambda}(x)\mathrm{H}(1-x);s]$$
$$= \frac{2^{-\lambda-s}\Gamma(\frac{1}{2})\Gamma(s)}{\Gamma(\frac{1}{2}\lambda - \frac{1}{2}\nu + \frac{1}{2}s + \frac{1}{2})\Gamma(\frac{1}{2}\lambda + \frac{1}{2}\nu + \frac{1}{2}s + 1)}.$$

4-10 Show that, if $\operatorname{Re} s > c - a$, $\operatorname{Re} s > c - b$, $\operatorname{Re} s > 0$,

$$\mathscr{M}^{-1}\left[\frac{\Gamma(s+a-c)\Gamma(s+b-c)}{\Gamma(s)\Gamma(s+a+b-c)}; s \to x\right]$$
$$= \frac{x^{-c}}{\Gamma(c)}(1-x)^{c-1}\,{}_2F_1(a,b;c;1-x^{-1})\mathrm{H}(1-x).$$

4-11 Prove that

$$\mathscr{M}[x^\delta(1-x)^{\gamma-1}\,{}_2F_1(\alpha,\beta;\gamma;1-x)\mathrm{H}(1-x);s]$$
$$= \frac{\Gamma(\gamma)\Gamma(s+\delta)\Gamma(s-\alpha-\beta+\gamma+\delta)}{\Gamma(s-\alpha+\gamma+\delta)\Gamma(s-\beta+\gamma+\delta)}$$

and deduce that, if $\gamma > \alpha + \beta$,

$$\mathscr{M}^{-1}\left[\frac{\Gamma(s+\alpha)\Gamma(s+\beta)}{\Gamma(s)\Gamma(s+\gamma)}; x\right]$$
$$= \frac{x^{\alpha+\beta}(1-x)^{\gamma-\alpha-\beta-1}}{\Gamma(\gamma-\alpha-\beta)}\,{}_2F_1(-\alpha,-\beta;\gamma-\alpha-\beta;1-x^{-1})\mathrm{H}(1-x).$$

4-12 Prove that if n is a positive integer,

$$\mathscr{M}[(1-x)^{\gamma-1}\,{}_2F_1(-n,\alpha;\beta;x)\mathrm{H}(1-x);s]$$
$$= \frac{\Gamma(\gamma)\Gamma(s)}{\Gamma(\gamma+s)}\,{}_3F_2(-n,\alpha,s;\beta,\gamma+s;1)$$

and using Saalschütz's theorem show that

$$\mathscr{M}[(1-x)^{\gamma-1}\,{}_2F_1(-n,\beta+\gamma+n-1;\beta;x)\mathrm{H}(1-x);s]$$
$$= \frac{\Gamma(\beta)\Gamma(\gamma+n)\Gamma(s)\Gamma(\beta+n-s)}{\Gamma(\beta+n)\Gamma(\beta-s)\Gamma(\gamma+s+n)}.$$

Deduce that

$$\mathscr{M}^{-1}\left[\frac{\Gamma(\alpha+s)\Gamma(\beta+n-s)}{\Gamma(\alpha+\gamma+s+n)\Gamma(\beta-s)}; x\right] = \frac{\Gamma(\alpha+\beta+n)}{\Gamma(\alpha+\beta)}$$
$$x^\alpha(1-x)^{\gamma-1}\,{}_2F_1(-n,\alpha+\beta+\gamma+n-1;\alpha+\beta;x)\mathrm{H}(1-x)$$

4-13 Suppose that

(*a*) $f(x)$ is a real-valued function defined on $(0, \infty)$ such that its Mellin transform $f^*(s)$ exists in some strip $\sigma_1 < \operatorname{Re} s < \sigma_2$ of the complex s-plane;

(*b*) $f(x)$ has an analytic continuation, $f(z)$, in some sector $|\arg z| < \alpha$, $0 < \alpha < \pi$;

(*c*) with $\sigma_1 \leqslant \sigma_3 < \operatorname{Re} s < \sigma_4 \leqslant \sigma_2$, $|\arg z| < \alpha$

$$\lim_{|z|\to\infty} z^s f(z) = 0, \qquad \lim_{|z|\to 0} z^s f(z) = 0.$$

Prove that the integral

$$\int_{L_\theta} z^{s-1} f(z)\,dz, \qquad (\sigma_3 < \operatorname{Re} s < \sigma_4),$$

in which L_θ denotes the radial line $\arg z = \theta$, $|\theta| < \alpha$, directed outward from the origin, is independent of θ and is equal to $f^*(s)$.

Show also that

$$\mathscr{M}^{-1}[f^*(s)\cos(s\theta); s \to r] = \operatorname{Re} f(re^{i\theta}),$$
$$\mathscr{M}^{-1}[f^*(s)\sin(s\theta); s \to r] = -\operatorname{Im} f(re^{i\theta})$$

where $\sigma_3 < \operatorname{Re} s < \sigma_4$, $|\theta| < \alpha$.

4-14 Using the pair of equations immediately above calculate the inverse Mellin transforms of

$$\frac{\cos(s\theta)}{\sin(s\pi)}, \qquad \frac{\sin(s\theta)}{\sin(s\pi)}, \qquad \Gamma(s)\cos(s\theta), \qquad \Gamma(s)\sin(s\theta).$$

4-15 Prove that if l and m are positive integers, $\alpha \neq 0$, $\beta \neq 0$

(*a*) $$\mathscr{M}\left[\left(x^{1-\alpha}\frac{d}{dx}\right)^l \int_0^\infty K\left(\frac{x}{y}\right)\left(y^{1-\beta}\frac{d}{dy}\right)^m \{y^\gamma f(y)\}\frac{dy}{y}; s\right]$$
$$= (-1)^{l+m}\alpha^l\beta^m \frac{\Gamma(s/\alpha)\Gamma(s/\beta - l\alpha/\beta)}{\Gamma(s/\alpha - l)\Gamma(s/\beta - l\alpha/\beta - m)}$$
$$K^*(s - l\alpha)f^*(s - l\alpha - m\beta + \gamma);$$

(*b*) $$\mathscr{M}\left[\left(x\frac{d}{dx}\right) \int_0^\infty K\left(\frac{x}{y}\right)\left(y^{1-\beta}\frac{d}{dy}\right)^m \{y^\gamma f(y)\}\frac{dy}{y}; s\right]$$
$$= (-1)^{l+m}\beta^m s^l \frac{\Gamma(s/\beta)}{\Gamma(s/\beta - m)} K^*(s)f^*(s - m\beta + \gamma);$$

(*c*) $$\mathscr{M}\left[\left(x^{1-\alpha}\frac{d}{dx}\right)^l \int_0^\infty K\left(\frac{x}{y}\right)\left(y\frac{d}{dy}\right)^m \{y^\gamma f(y)\}\frac{dy}{y}; s\right]$$

$$= (-1)^{l+m}\alpha^l(s - l\alpha)^m \frac{\Gamma(s/\alpha)}{\Gamma(s/\alpha - l)} K^*(s - l\alpha)f^*(s - l\alpha + \gamma);$$

(d) $$\mathscr{M}\left[\left(x\frac{d}{dx}\right)^l \int_0^\infty K\left(\frac{x}{y}\right)\left(y\frac{d}{dy}\right)^m \{y^\gamma f(y)\} \frac{dy}{y};s\right]$$

$$= (-s)^{l+m}K^*(s)f^*(s + \gamma).$$

4-16 The operator $\mathrm{I}_{\eta,\alpha}$ is defined by the equation

$$\mathrm{I}_{\eta,\alpha}f(x) = \frac{2x^{-2\alpha-2\eta}}{\Gamma(\alpha)} \int_0^x (x^2 - u^2)^{\alpha-1}u^{2\eta+1}f(u)\,du.$$

Show that

$$\mathscr{M}[\mathrm{I}_{\eta,\alpha}f(x);s] = \frac{\Gamma(1 + \eta - \frac{1}{2}s)}{\Gamma(1 + \alpha + \eta - \frac{1}{2}s)} f^*(s),$$

where $f^*(s)$ is the Mellin transform of $f(x)$.

4-17 The operator $K_{x^m}{}^\alpha$ is defined for $\alpha > 0$ by the equation

$$K_{x^m}{}^\alpha f(x) = \frac{1}{\Gamma(\alpha)} \int_x^\infty (t^m - x^m)^{\alpha-1}f(t)mt^{m-1}\,dt$$

and for $\alpha < 0$ by the equation

$$K_{x^m}{}^\alpha f(x) = \left(-\frac{d}{dx^m}\right)^l K_{x^m}{}^{\alpha+l}f(x)$$

where l is a positive integer such that $0 < \alpha + l < 1$. Show that

$$\mathscr{M}[K_{x^m}{}^\alpha f(x);s] = \frac{\Gamma(s/m)}{\Gamma(s/m + \alpha)} f^*(s + m\alpha).$$

Deduce that

$$\mathscr{M}[K_{x^m}{}^\alpha x^\gamma K_{x^n}{}^\beta x^\delta f(x);s]$$

$$= \frac{\Gamma\left(\frac{s}{m}\right)\Gamma\left(\frac{s + m\alpha + \gamma}{n}\right)}{\Gamma\left(\frac{s}{m} + \alpha\right)\Gamma\left(\frac{s + m\alpha + \gamma}{n} + \beta\right)} f^*(s + m\alpha + n\beta + \gamma + \delta).$$

4-18 Show that if m and n are integers, $0 < x < 1$

$$\int_x^1 (y^2 - x^2)^{\lambda-\frac{1}{2}}(1 - y^2)^{\mu-\frac{1}{2}}C_m{}^\lambda\left(\frac{y}{x}\right) C_n{}^\mu(y)\,dy$$

$$= \frac{\Gamma(m + 2\lambda)\Gamma(n + 2\mu)}{2^{2\lambda+2\mu-1}m!n!\Gamma(\lambda)\Gamma(\mu)\Gamma(\lambda + \mu + 1)} \mathrm{J}\begin{bmatrix}\lambda, \mu \\ m, n\end{bmatrix};x$$

where

$$\mathrm{J}\begin{bmatrix}\lambda, \mu \\ m, n\end{bmatrix}; x\Bigg] = \tfrac{1}{2}\Gamma(\lambda + \mu + 1) \times$$

$$\times \mathscr{M}^{-1}\left[\frac{\Gamma(\frac{1}{2}s - \frac{1}{2}m)\Gamma(\frac{1}{2}s + \lambda + \frac{1}{2}m)}{\Gamma(\frac{1}{2}s + \frac{1}{2}n + \lambda + \mu + \frac{1}{2})\Gamma(\frac{1}{2}s - \frac{1}{2}n + \lambda + \frac{1}{2})}; x\right]$$

$$= x^{-m}(1 - x^2)^{\lambda+\mu}\, {}_2F_1(\tfrac{1}{2}n - \tfrac{1}{2}m + \mu + \tfrac{1}{2},$$

$$-\tfrac{1}{2}m - \tfrac{1}{2}n + \tfrac{1}{2}; \lambda + \mu + 1; 1 - x^2).$$

Deduce that

(*a*) if $\mu = \frac{1}{2}(m - n - 1)$, the integral has the value

$$\frac{\Gamma(m + 2\lambda)\Gamma(n + 2\mu)}{2^{2\lambda+2\mu-1}m!n!\Gamma(\lambda)\Gamma(\mu)\Gamma(\lambda + \mu + 1)} x^{-m}(1 - x^2)^{\lambda+\mu};$$

(*b*) if $m + n = 2N + 1$ where N is a positive integer,

$$\mathrm{J} = \frac{x^{2\lambda+m}}{\Gamma(N + \lambda + \mu + 1)} j(1 - x^2),$$

where

$$j(\xi) = \frac{d^N}{d\xi^N}\, \xi^{N+\lambda+\mu}(1 - \xi)^{N-m-\lambda}.$$

4-19 Prove that if m and n are positive integers, $0 < x < 1$,

$$\int_x^1 P_m(y/x)P_n(y)\, dy$$

$$= \tfrac{1}{2}x^{-m}(1 - x^2)\,{}_2F_1(1 - \tfrac{1}{2}m + \tfrac{1}{2}n, \tfrac{1}{2} - \tfrac{1}{2}m - \tfrac{1}{2}n; 2; 1 - x^2)$$

and deduce that

(*a*) $\displaystyle\int_x^1 P_m(y/x)P_{m-2}(y)\, dy = \tfrac{1}{2}x^{-m}(1 - x^2), \qquad (m > 2),$

(*b*) $\displaystyle\int_x^1 P_m(y/x)P_{m+2}(y)\, dy = \tfrac{1}{2}x^{m+1}(1 - x^2), \qquad (m > 0).$

4-20 Show that if m, n are positive integers, and $0 < x < 1$,

$$\int_x^1 \frac{T_m(y/x)T_n(y)\, dy}{\sqrt{\{(y^2 - x^2)(1 - y^2)\}}}$$

$$= \tfrac{1}{4}\pi\mathscr{M}^{-1}\left[\frac{\Gamma(\frac{1}{2}s + \frac{1}{2}m)\Gamma(\frac{1}{2}s - \frac{1}{2}m)}{\Gamma(\frac{1}{2}s + \frac{1}{2}n + \frac{1}{2})\Gamma(\frac{1}{2}s - \frac{1}{2}n + \frac{1}{2})}; x\right]$$

$$= \tfrac{1}{2}\pi\,{}_2F_1(\tfrac{1}{2} - \tfrac{1}{2}m + \tfrac{1}{2}n, \tfrac{1}{2} - \tfrac{1}{2}m - \tfrac{1}{2}n; 1; 1 - x^2)$$

and deduce that

(a) $$\int_x^1 \frac{T_m(y/x)T_{m-1}(y)\,dy}{\sqrt{\{(y^2 - x^2)(1 - y^2)\}}} = \tfrac{1}{2}\pi;$$

(b) if $m + n = 2N + 1$ where N is a positive integer the integral has the value

$$\frac{1}{2}\pi\frac{x^{2m}}{N!}\,\mathrm{J}(1 - x^2)$$

where

$$\mathrm{J}(\xi) = \frac{d^N}{d\xi^N}\,\xi^N(1 - \xi)^{N-m}.$$

4-21 If $0 < x < 1$, $0 < c < m$, show that

$$\int_x^1 y^{-m}(y - x)^{c-1}(1 - y)^{m-c-1}\,{}_2F_1(a,b;c;1 - (y/x))$$
$${}_2F_1(-a,-b;m - c;1 - y)\,dy = \frac{\Gamma(c)\Gamma(m - c)}{\Gamma(m)}\,x^{c-m}(1 - x)^{m-1}.$$

4-22 Defining the modified Bessel function of the second kind $K_\nu(x)$ by the equation

$$K_\nu(x) = \frac{\pi^{\frac{1}{2}}(\frac{1}{2}x)^\nu}{\Gamma(\nu + \frac{1}{2})}\int_1^\infty e^{-xt}(t^2 - 1)^{\nu-\frac{1}{2}}\,dt$$

show that

$$\mathscr{M}[K_\nu(x);s] = 2^{s-2}\Gamma(\tfrac{1}{2}s + \tfrac{1}{2}\nu)\Gamma(\tfrac{1}{2}s - \tfrac{1}{2}\nu), \qquad (\mathrm{Re}\,s > \nu > 0).$$

Show that

$$\int_0^1 (1 - x^2)^{\frac{1}{2}\lambda}P_\nu^{-\lambda}(x)K_{\nu+\frac{1}{2}}\left(\frac{r}{x}\right)\frac{dx}{x^{\lambda+\frac{5}{2}}} = \left(\frac{\pi}{2}\right)^{\frac{1}{2}}\cdot\frac{e^{-r}}{r^{\lambda+\frac{3}{2}}}.$$

4-23 Show that the condition $K^*(s)K^*(1 - s) = 1$ implies that

$$\int_0^\infty u^{-1}K(xu)K_1(tu)\,du = H(x) - H(x - t)$$

where

$$K_1(t) = \int_0^t K(u)\,du.$$

and hence—by Problem 1-1—is a sufficient condition for $K(x\xi)$ to be a Fourier kernel.

4-24 Prove that

$$\frac{d}{dz} z^\nu \mathbf{H}_\nu(z) = z^\nu \mathbf{H}_{\nu-1}(z)$$

and deduce that

$$(a)\ \mathscr{H}_\nu^*[t^m g(t);x] = x^{\frac{1}{2}-\nu}\left(\frac{1}{x}\frac{\partial}{\partial x}\right)^m x^{\nu+m-\frac{1}{2}}\mathscr{H}_{\nu+m}^*[g(t);x].$$

Show also that

$$(b)\ \mathscr{Y}_\nu[x^m f(x);t] = t^{\frac{1}{2}-\nu}\left(\frac{1}{t}\frac{\partial}{\partial t}\right)^m t^{\nu+m-\frac{1}{2}}\mathscr{Y}_{\nu+m}^*[f(x);t].$$

4-25 Show that if $a > 0$,

$$\mathscr{H}_\nu^*[t^{\nu+\frac{1}{2}}\mathrm{H}(a-t);x] = a^{\nu+1}x^{-\frac{1}{2}}\mathbf{H}_{\nu+1}(ax),$$

and deduce that

$$(a)\ \mathscr{H}_\nu^*[t^{-\frac{1}{2}}Y_{\nu-1}(at);x] = a^{\nu-\frac{1}{2}}x^{\nu+\frac{1}{2}}\mathrm{H}(x-a);$$

$$(b)\ \mathscr{Y}_\nu[x^{-\frac{1}{2}}\mathbf{H}_{\nu+1}(ax);t] = t^{\nu+\frac{1}{2}}a^{-\nu-1}\mathrm{H}(a-t).$$

4-26 Defining the Tchebycheff polynomial $T_n(x)$ by the equation

$$T_n(x) = {}_2F_1(-n,n;\tfrac{1}{2};\tfrac{1}{2}-\tfrac{1}{2}x)$$

show that

$$\mathscr{M}[(1-x^2)^{-\frac{1}{2}}T_n(x)\mathrm{H}(1-x);s]$$

$$= \sqrt{(\tfrac{1}{2}\pi)}\frac{\Gamma(s)}{\Gamma(s+\frac{1}{2})}\,{}_2F_1(\tfrac{1}{2}+n,\tfrac{1}{2}-n;s+\tfrac{1}{2};\tfrac{1}{2})$$

and hence that

$$\mathscr{M}[(1-x^2)^{-\frac{1}{2}}T_n(x)\mathrm{H}(1-x);s] = \frac{2^{-s}\pi\Gamma(s)}{\Gamma(\frac{1}{2}+\frac{1}{2}s+\frac{1}{2}n)\Gamma(\frac{1}{2}+\frac{1}{2}s-\frac{1}{2}n)}$$

4-27 Defining the Tchebycheff polynomial $T_n(x)$ by the equation

$$T_n(x) = \tfrac{1}{2}n\sum_{m=0}^{[\frac{1}{2}n]}(-1)^m\frac{(n-m-1)!}{m!(n-2m)!}(2x)^{n-2m}$$

show that

$$\mathscr{M}[(1-x^2)^{-\frac{1}{2}}T_n(x^{-1})\mathrm{H}(1-x);s] = 2^{n-2}\pi^{\frac{1}{2}}\frac{\Gamma(\frac{1}{2}s-\frac{1}{2}n-1)}{\Gamma(\frac{1}{2}s-\frac{1}{2}n-\frac{1}{2})}\times$$

$$\times\ {}_3F_2(\tfrac{1}{2}-\tfrac{1}{2}n,-\tfrac{1}{2}n,\tfrac{1}{2}s-\tfrac{1}{2}n-1;1-n,\tfrac{1}{2}s-\tfrac{1}{2}n-\tfrac{1}{2};1)$$

and hence that

$$\mathscr{M}[(1 - x^2)^{-\frac{1}{2}} T_n(x^{-1}) \mathrm{H}(1 - x); s] = 2^{s-2} \frac{\Gamma(\frac{1}{2}n + \frac{1}{2}s)\Gamma(\frac{1}{2}s - \frac{1}{2}n)}{\Gamma(s)}$$

4-28 Show that the integral equation

$$\int_x^1 P_n\left(\frac{y}{x}\right) f(y)\, dy = g(x)$$

with $g(1) = g'(1) = 0$ has solution

$$f(x) = \int_x^1 y^{2-n} P_{n-2}\left(\frac{x}{y}\right)\left(\frac{1}{y}\frac{d}{dy}\right)^2 \{y^n g(y)\}\, dy.$$

4-29 Show that the integral equation

$$\int_x^1 (y - x)^{c-1}\, {}_2F_1(a, b; c; 1 - x/y) f(y)\, dy = g(x),$$

in which $g(x)$ is prescribed and $g^{(r)}(1) = 0$, $(r = 0, 1, \ldots, m - 1)$, has the solution

$$f(x) = \frac{(-1)^m}{\Gamma(c)\Gamma(m - c)} \int_x^1 (y - x)^{m-c-1}\, {}_2F_1(-a, -b; m - c; 1 - y/x) g^{(m)}(y)\, dy$$

4-30 By taking

$$f(x) = x^{-\frac{2}{3}}(1 - x^{\frac{2}{3}})$$

in equation (4-7-6) show that the finite series

$$S(x) = \sum_{n=1}^{N} n^{-\frac{2}{3}}(1 - xn^{\frac{2}{3}}),$$

where $0 < x < 1$ and N is the largest integer such that $xN^{\frac{2}{3}} < 1$, has sum

$$\Gamma(\tfrac{5}{2}) x \mathscr{M}^{-1}\left[\frac{\Gamma(\frac{3}{2}s - 1)}{\Gamma(\frac{3}{2}s + \frac{1}{2})}\zeta(s); s \to x^{\frac{3}{2}}\right]$$

and deduce that as $x \to 0$

$$S(x) \sim \tfrac{3}{4}\pi x^{-\frac{1}{2}} + \zeta(\tfrac{2}{3}) + \tfrac{1}{4}x - \tfrac{1}{8}\zeta(-\tfrac{2}{3})x^2 + \cdots$$

5

Hankel Transforms

The Hankel transform arises naturally in the discussion of problems posed in cylindrical coordinates and hence, as a result of separation of variables, involving Bessel functions. The proof of the inversion theorem $\mathscr{H}_\nu^{-1} = \mathscr{H}_\nu$ (which is given in sec. 5-3) is complicated. A reader whose primary interest is in the *use* of Hankel transforms may, in the first instance, omit the reading of this section but should read the statement of the theorem to see the sense in which it justifies the pair of formulae (5-1-3) and (5-1-4) derived heuristically in the first section.

5-1 INTRODUCTION

We saw in sec. 2-13 that, if $f(r)$ is a function of $r = |\mathbf{x}|$ alone, its n-dimensional Fourier transform $F_{(n)}(\xi)$ is a function of $\rho = |\xi|$ alone. If we write

$$\mathscr{F}_{(n)}[f(r); \mathbf{x} \to \xi] = F(\rho)$$

then we see from equations (2-13-14), (2-13-15) that

$$F(\rho) = \rho^{-\nu}\mathscr{H}_\nu[r^\nu f(r);\rho], \qquad \nu = \tfrac{1}{2}n - 1. \tag{5-1-1}$$

Making use of the Fourier inversion theorem (2-12-5) we find that

$$f(r) = \mathscr{F}_{(n)}[F(\rho);\xi \to -\mathbf{x}]$$

and hence that

$$f(r) = r^{-\nu}\mathscr{H}_\nu[\rho^\nu F(\rho);r]. \tag{5-1-2}$$

If, therefore, we write

$$r^\nu f(r) = \phi(r), \qquad \rho^\nu F(\rho) = \bar{\phi}_\nu(\rho)$$

we see that equations (5-1-1) and (5-1-2) can be written in the form

$$\bar{\phi}_\nu(\rho) = \mathscr{H}_\nu[\phi(r);\rho] \tag{5-1-3}$$

$$\phi(r) = \mathscr{H}_\nu[\bar{\phi}_\nu(\rho);r]. \tag{5-1-4}$$

respectively. Hence we see that, if ν is zero or half of a positive integer (even or odd), there is a symmetrical relationship between a function and its Hankel transform of order ν in the sense that, if $\bar{\phi}_\nu(\rho)$ is the Hankel transform of order ν of an arbitrary function $\phi(r)$, then $\phi(r)$ is the Hankel transform of order ν of $\bar{\phi}_\nu(\rho)$. This result is called the *Hankel inversion theorem*.

Although in physical applications ν is usually either zero or a positive integer we shall give (in sec. 5-3) a proof of the Hankel inversion theorem for general values of ν.

5-2 ELEMENTARY PROPERTIES OF HANKEL TRANSFORMS

By making the simple change of variable $\rho = ax$ in the defining integral

$$\mathscr{H}_\nu[f(ax);\xi] = \int_0^\infty xf(ax)\mathrm{J}_\nu(x\xi)\,dx, \qquad (a > 0)$$

we find that the integral on the right becomes

$$\frac{1}{a^2}\int_0^\infty \rho f(\rho)\mathrm{J}_\nu(\rho\xi/a)\,d\rho$$

from which we deduce that

$$\mathscr{H}_\nu[f(a\rho);\xi] = a^{-2}\mathscr{H}_\nu[f(\rho);\xi/a] \tag{5-2-1}$$

Another simple relation is obtained by making use of the recurrence relation

$$\mathrm{J}_{\nu-1}(x) - \frac{2\nu}{x}\mathrm{J}_\nu(x) + \mathrm{J}_{\nu+1}(x) = 0. \tag{5-2-2}$$

Substituting $\xi\rho$ for x, multiplying both sides of the resulting equation by $\rho f(\rho)$ and integrating with respect to ρ over the positive real line we obtain the relation

$$\mathscr{H}_\nu[\rho^{-1}f(\rho);\xi] = \frac{\xi}{2\nu}\{\bar{f}_{\nu-1}(\xi) + \bar{f}_{\nu+1}(\xi)\}, \qquad (\nu \neq 0). \tag{5-2-3}$$

It is not possible to derive a simple "shift" formula—giving the transform of $f(\rho - a)$ in terms of that of $f(\rho)$—in the case of the Hankel transform. This is because the addition formula for the Bessel function even in the case in which ν is an integer takes a most complicated form. Against the simple addition formula

$$e^{x+y} = e^x \cdot e^y$$

for the exponential function (which is the basis of the shift formula for the Laplace transform) we have the Neumann–Lommel addition formula

$$\mathrm{J}_n(x + y) = \sum_{m=-\infty}^{\infty} \mathrm{J}_m(y)\mathrm{J}_{n-m}(x) \tag{5-2-4}$$

(Watson, 1944, p. 30). From the definition of the Hankel transform we have the formula

$$\mathscr{H}_n[f(\rho - a)\mathrm{H}(\rho - a);\xi] = \int_a^\infty \rho f(\rho - a)\mathrm{J}_n(\xi\rho)\,d\rho$$

$(a > 0)$ and the simple change of variable $\rho = a + u$ yields the expression

$$\int_0^\infty (u + a)f(u)\mathrm{J}_n(\xi a + \xi u)\,du$$

for the integral on the right-hand side of this last equation. Making use of the addition theorem (5-2-4) we see that in turn this reduces to

$$\sum_{m=-\infty}^{\infty} \mathrm{J}_{n-m}(\xi a) \int_0^\infty (u + a)f(u)\mathrm{J}_m(\xi u)\,du.$$

Now by equation (5-2-3) we have the formula

$$\int_0^\infty (u + a)f(u)\mathrm{J}_m(\xi u)\,du = \mathscr{H}_m[f(u);\xi] + a\mathscr{H}_m[u^{-1}f(u);\xi]$$

$$= \bar{f}_m(\xi) + \frac{a\xi}{2m}\{\bar{f}_{m-1}(\xi) + \bar{f}_{m+1}(\xi)\}$$

Now

$$\sum_{m=-\infty}^{\infty} m^{-1}\mathrm{J}_{n-m}(\xi a)\bar{f}_{m-1}(\xi) = \sum_{m=-\infty}^{\infty} (m + 1)^{-1}\mathrm{J}_{n-m-1}(\xi a)\bar{f}_m(\xi)$$

and

$$\sum_{m=-\infty}^{\infty} m^{-1}\mathrm{J}_{n-m}(\xi a)\bar{f}_{m+1}(\xi) = \sum_{m=-\infty}^{\infty} (m - 1)^{-1}\mathrm{J}_{n-m+1}(\xi a)\bar{f}_m(\xi)$$

so that finally we have the formula

$$\mathscr{H}_n[f(\rho - a)\mathrm{H}(\rho - a);\xi] = \sum_{m=-\infty}^{\infty} \alpha_m \bar{f}_m(\xi) \tag{5-2-5}$$

where

$$\alpha_m = \mathrm{J}_{n-m}(\xi a) + \tfrac{1}{2}a\xi\{(m+1)^{-1}\mathrm{J}_{n-m-1}(\xi a) + (m-1)^{-1}\mathrm{J}_{n-m+1}(\xi a)\}. \tag{5-2-6}$$

If we use the recurrence relation (5-2-2) with ν replaced by $n - m$ and x replaced by ξa we see that we can rewrite equation (5-2-6) as

$$\alpha_m = \frac{\xi a}{2(n-m)}\left[\frac{n+1}{m+1}\mathrm{J}_{n-m-1}(\xi a) + \frac{n-1}{m-1}\mathrm{J}_{n+m+1}(\xi a)\right]. \tag{5-2-7}$$

We shall not make use of this formula. It is produced here merely to convince the reader that the shift theorem for Hankel transforms is not simple!

One result we shall use depends on the fact that the function $|x^{\frac{1}{2}}\mathrm{J}_\nu(x)|$ is bounded. If we write, for x real,

$$x^{\frac{1}{2}}\mathrm{J}_\nu(x)| \leqslant A, \qquad \nu \geqslant -\tfrac{1}{2} \tag{5-2-8}$$

then

$$|\bar{f}_\nu(\xi)| = \left|\int_0^\infty x^{\frac{1}{2}}f(x)\cdot x^{\frac{1}{2}}\mathrm{J}_\nu(x\xi)\,dx\right|$$
$$\leqslant \xi^{-\frac{1}{2}}\int_0^\infty |x^{\frac{1}{2}}f(x)|\cdot|(x\xi)^{\frac{1}{2}}\mathrm{J}_\nu(x\xi)|\,dx$$
$$\leqslant A\xi^{-\frac{1}{2}}\int_0^\infty |x^{\frac{1}{2}}f(x)|\,dx$$

showing that if $\sqrt{x}f(x)$ is absolutely integrable over the positive real line $\bar{f}_\nu(\xi)$ is bounded for all positive values of ξ.

5-3 THE HANKEL INVERSION THEOREM

We shall now give a proof of the Hankel inversion theorem. The difficulty is that a valid proof for general values of ν is complicated and involves a knowledge of the properties of Bessel functions. In applications of the Hankel transform to the solution of boundary value problems in mathematical physics (for example, in potential theory and in the mathematical theory of elasticity) it is usually only the transforms of order zero and unity that are involved. For that reason a demonstration of the plausibility of the Hankel inversion theorem is given in sec. 5-7. A reader whose interest is entirely in the applications of Hankel transforms could omit this present section taking the equations (5-1-3), (5-1-4) as the statement of the Hankel inversion theorem.

For the details of the proof of the Hankel inversion theorem we need some properties of Bessel functions which are listed here for convenience. In addition to the relation (5-2-8) we need the result that the function

$$\int_0^x \sqrt{t}J_\nu(t)\,dt$$

is bounded for all positive values of x if $\nu \geqslant -\frac{1}{2}$, although the integral does not converge as $x \to \infty$. We shall write

$$\left|\int_0^x \sqrt{t}J_\nu(t)\,dt\right| \leqslant B, \qquad \nu \geqslant -\tfrac{1}{2} \tag{5-3-1}$$

From this and the relation

$$\int_\alpha^\beta \sqrt{x}J_\nu(\lambda x)\,dx = \lambda^{-\frac{3}{2}}\int_{\lambda\alpha}^{\lambda\beta} \sqrt{t}J_\nu(t)\,dt$$

we deduce that if $\beta > \alpha \geqslant 0$, $\lambda > 0$

$$\left|\int_\alpha^\beta \sqrt{x}J_\nu(\lambda x)\,dx\right| \leqslant 2B\lambda^{-\frac{3}{2}} \tag{5-3-2}$$

A relation of a similar kind may be deduced from (5-2-8). Since if $\beta \geqslant \alpha \geqslant 0$, $\lambda > 0$, we may write

$$\int_\alpha^\beta \omega(x)\sqrt{x}J_\nu(\lambda x)\,dx = \lambda^{-\frac{1}{2}}\int_\alpha^\beta \omega(x)\sqrt{(\lambda x)}J_\nu(\lambda x)\,dx$$

we conclude that

$$\left|\int_\alpha^\beta \omega(x)\sqrt{x}J_\nu(\lambda x)\,dx\right| \leqslant A\lambda^{-\frac{1}{2}}\int_\alpha^\beta |\omega(x)|\,dx. \tag{5-3-3}$$

We also require the dominant term in the asymptotic expansion of the Bessel function of the first kind of order ν:

$$J_\nu(z) \sim \left(\frac{2}{\pi z}\right)^{\frac{1}{2}} \cos\left(z - \tfrac{1}{2}\nu\pi - \tfrac{1}{4}\pi\right), \qquad |z| \to \infty, \tag{5-3-4}$$

$\nu \geqslant -\frac{1}{2}$, $|\arg z| < \pi$ (Watson, 1944, p. 199).

Two integrals involving Bessel functions arise in the proof. If we carry out the differentiation on the left and make use of well-known recurrence relations for $J_\nu(z)$ we find that

$$\frac{\partial}{\partial u}\left[x\cdot u^{\nu+1}J_{\nu+1}(xu)\cdot u^{-\nu}J_\nu(yu) - yu^{-\nu}J_\nu(xu)\cdot u^{\nu+1}J_{\nu+1}(yu)\right]$$
$$= (x^2 - y^2)uJ_\nu(xu)J_\nu(yu).$$

Recalling that

$$\lim_{u\to 0} uJ_\nu(xu)J_\nu(yu) = 0$$

if $\nu > -\frac{1}{2}$ we find, on integrating both sides of this last equation with respect to u from 0 to λ, that

$$\int_0^\lambda u\mathrm{J}_\nu(xu)\mathrm{J}_\nu(yu)\,du = h_\nu(x,y;\lambda), \qquad \nu > -\tfrac{1}{2} \tag{5-3-5}$$

where

$$h_\nu(x,y;\lambda) = \frac{\lambda x\mathrm{J}_{\nu+1}(\lambda x)\mathrm{J}_\nu(\lambda y) - \lambda y\mathrm{J}_\nu(\lambda x)\mathrm{J}_{\nu+1}(\lambda y)}{x^2 - y^2}. \tag{5-3-6}$$

We *assume* that the value of Weber's second exponential integral

$$\mathrm{J}_\nu(\rho,\tau,p) = \int_0^\infty ue^{-p^2u^2}\mathrm{J}_\nu(\rho u)\mathrm{J}_\nu(ru)\,du \tag{5-3-7}$$

is known to be

$$\frac{1}{2p^2}\,\mathrm{I}_\nu\left(\frac{\tau\rho}{2p^2}\right)\exp\left(-\frac{r^2+\rho^2}{2p^2}\right)$$

(Watson, 1944, p. 395) where I_ν denotes the modified Bessel function of the first kind of order ν. If we also make use of the result that for small values of p,

$$\mathrm{I}_\nu\left(\frac{r\rho}{2p^2}\right) = \frac{p}{\sqrt{(\pi r\rho)}}\{1 + 0(p^2)\}\exp\left(\frac{r\rho}{2p^2}\right)$$

we see that

$$\mathrm{J}_\nu(\rho,r,p) = \frac{1}{2p\sqrt{(\pi r\rho)}}\{1 + 0(p^2)\}\exp\left[-\frac{(r-\rho)^2}{4p^2}\right]$$

and hence that

$$\int_r^{r+\delta}\rho^{\frac{1}{2}}\mathrm{J}_\nu(\rho,r,p)\,d\rho = \frac{1}{\sqrt{(\pi r)}}\{1 + 0(p^2)\}\int_0^{\delta/2p} e^{-z^2}\,dz$$

($\delta > 0$). From this we deduce that if $\delta > 0$

$$\lim_{p\to 0}\int_r^{r+\delta}(r\rho)^{\frac{1}{2}}\mathrm{J}_\nu(\rho,r,p)\,d\rho = \tfrac{1}{2}. \tag{5-3-8}$$

Similarly, if $\delta > 0$,

$$\lim_{p\to 0}\int_{r-\delta}^{r}(r\rho)^{\frac{1}{2}}\mathrm{J}_\nu(\rho,r,p)\,d\rho = \tfrac{1}{2}.$$

The proof of the Hankel inversion we shall now give is basically that of Watson (1944, pp. 456–465) which is itself substantially Hankel's proof. As in the proof of the Fourier inversion theorem we break the proof up by establishing a number of lemmas. These are more complicated than the corresponding ones in the Fourier case because the properties of Bessel functions are more complicated than those of the circular functions.

We begin with a lemma which is the analogue of the Riemann–Lebesgue lemma.

Lemma 1 (the Riemann–Lebesgue lemma) *If $\sqrt{x}f(x)$ is piecewise continuous and absolutely integrable in* (a,b), $(b \geqslant a \geqslant 0)$, *then as* $\lambda \to \infty$

$$\int_a^b xf(x)J_\nu(\lambda x)\,dx = o(\lambda^{-\frac{1}{2}}), \qquad \nu > -\tfrac{1}{2}. \tag{5-3-10}$$

To prove this lemma we note that if b is finite we may divide the interval (a,b) into n subintervals $(x_{s-1},x_s)(s = 1,2, \ldots, n)$ with $x_0 = a$, $x_{n+1} = b$, each of length δ, n being chosen so large that

$$\sum_{s=1}^{n} (M_s - m_s)\delta < \varepsilon \tag{5-3-11}$$

where $m_s \leqslant \sqrt{x}f(x) \leqslant M_s$ for $x \in (x_{s-1},x_s)$. Under these circumstances we may write

$$\sqrt{x}f(x) = \sqrt{x_{s-1}}f(x_{s-1}) + \omega_s(x) \tag{5-3-12}$$

with

$$|\omega_s(x)| \leqslant M_s - m_s \tag{5-3-13}$$

and

$$\int_a^b xf(x)J_\nu(\lambda x)\,dx = I_1 + I_2 \tag{5-3-14}$$

where I_1 and I_2 are defined by the equations

$$I_1 = \sum_{s=1}^{n} \sqrt{x_{s-1}}f(x_{s-1}) \int_a^b \sqrt{x}J_\nu(\lambda x)\,dx \tag{5-3-15}$$

$$I_2 = \sum_{s=1}^{n} \int_{x_{s-1}}^{x_s} \sqrt{x}\omega_s(x)J_\nu(\lambda x)\,dx \tag{5-3-16}$$

respectively.

For the first of these integrals we have from equation (5-3-2) that

$$|I_1| \leqslant 2B\lambda^{-\frac{3}{2}} \sum_{s=1}^{n} |\sqrt{x_{s-1}}f(x_{s-1})| \leqslant 2nBM\lambda^{-\frac{3}{2}}$$

where M is the upper bound of $\sqrt{x}f(x)$.

Also from (5-3-3) we find that

$$\begin{aligned} |I_2| &\leqslant \lambda^{-\frac{1}{2}}A \sum_{s=1}^{n} \int_{x_{s-1}}^{x_s} |\omega_s(x)|\,dx \\ &\leqslant \lambda^{-\frac{1}{2}}A \sum_{s=1}^{n} (M_s - m_s)\delta \\ &< \lambda^{-\frac{1}{2}}A\varepsilon \end{aligned}$$

by equation (5-3-11).

Hence we have the estimate

$$\left|\int_a^b xf(x)\mathrm{J}_\nu(\lambda x)\,dx\right| \leqslant 2nBM\lambda^{-\frac{3}{2}} + \lambda^{-\frac{1}{2}}A\varepsilon. \tag{5-3-17}$$

Having prescribed ε we can find n; for this fixed value of n we can now choose λ sufficiently large that $2nBM\lambda^{-1} \leqslant A\varepsilon$ and hence that

$$\left|\int_a^b xf(x)\mathrm{J}_\nu(\lambda x)\,dx\right| \leqslant 2\lambda^{-\frac{1}{2}}A\varepsilon \tag{5-3-18}$$

and the lemma follows.

The case in which b is infinite can be deduced from this case by noticing that we can write

$$\left|\int_a^\infty xf(x)\mathrm{J}_\nu(\lambda x)\,dx\right| \leqslant \left|\int_a^b xf(x)\mathrm{J}_\nu(\lambda x)\,dx\right| + \lambda^{-\frac{1}{2}}A\int_b^\infty \sqrt{x}|f(x)|\,dx$$

for arbitrary b and then choosing b large enough that

$$\int_b^\infty \sqrt{x}|f(x)|\,dx < \varepsilon.$$

The next lemma shows that we can invert the order of the integrations in the repeated Hankel integral:

Lemma 2 *If $\sqrt{x}f(x)$ is piecewise continuous and absolutely integrable on $(0,\infty)$, and $\bar{f}_\nu(u) = \mathscr{H}_\nu[f(x);u]$, then*

$$\int_0^\infty u\bar{f}_\nu(u)\mathrm{J}_\nu(ru)\,du = \lim_{\lambda\to\infty}\int_0^\infty \rho f(\rho)h_\nu(\rho,r;\lambda)\,d\rho, \qquad \nu > -\tfrac{1}{2}$$

provided that the limit on the right side of this equation exists.

To simplify the writing out of the proof of this lemma we write

$$F(u,\rho;r) = \rho u f(\rho)\mathrm{J}_\nu(\rho u)\mathrm{J}_\nu(ru)$$

so that from equation (5-2-8)

$$|F(u,\rho;r)| \leqslant A^2\rho^{\frac{1}{2}}r^{-\frac{1}{2}}|f(\rho)|. \tag{5-3-19}$$

Then

$$\left|\int_0^\infty d\rho\int_0^\lambda F(u,\rho;r)\,du - \int_0^\lambda du\int_0^\infty F(u,\rho;r)\,d\rho\right| \leqslant F_1 + F_2$$

where F_1 and F_2 are defined respectively by the equations

$$F_1 = \int_c^\infty d\rho\int_0^\lambda |F(u,\rho;r)|\,du, \qquad F_2 = \int_0^\lambda du\int_c^\infty |F(u,\rho;r)|\,d\rho$$

where $c > 0$. This result is true for arbitrary positive values of c, but, given any arbitrary positive number ε, we now choose c to be such that

$$\int_c^\infty \sqrt{x}|f(x)|\,dx < \frac{\varepsilon\sqrt{r}}{2A^2\lambda} \tag{5-3-20}$$

where again A is the constant introduced in equation (5-2-8). Then from (5-3-19) and (5-3-20) we have

$$F_1 \leqslant \int_c^\infty d\rho\lambda A^2\rho^{\frac{1}{2}}r^{-\frac{1}{2}}|f(\rho)|\,d\rho = \frac{A^2\lambda}{\sqrt{r}}\int_c^\infty \sqrt{\rho}|f(\rho)|\,d\rho < \tfrac{1}{2}\varepsilon$$

Similarly we find that

$$F_2 \leqslant \int_0^\lambda du \frac{A^2}{\sqrt{r}}\int_c^\infty \sqrt{\rho}|f(\rho)|\,d\rho < \int_0^\lambda \frac{\varepsilon}{2\lambda}\,du = \tfrac{1}{2}\varepsilon$$

and hence, since ε is arbitrary, that

$$\int_0^\infty d\rho \int_0^\lambda F(u,\rho;r)\,du = \int_0^\lambda du \int_0^\infty F(u,\rho;r)\,d\rho.$$

If the integral on the left has a limit as $\lambda \to \infty$ it follows from the definition of an infinite integral that

$$\int_0^\infty du \int_0^\infty F(u,\rho;r)\,d\rho = \lim_{\lambda\to\infty}\int_0^\infty d\rho \int_0^\lambda F(u,\rho;r)\,du$$

and if we insert the form for $F(u,\rho;r)$ we get the desired lemma.

The next lemma shows that the only part of the ρ-range of integration which contributes to the value of Hankel's repeated integral is in the immediate vicinity of the point $\rho = r$:

Lemma 3 *If $\sqrt{x}f(x)$ is piecewise continuous and absolutely integrable on the positive real line and r does not belong to the interval (a,b), then*

$$\int_0^\infty u\mathrm{J}_\nu(ur)\,du \int_a^b \rho f(\rho)\mathrm{J}_\nu(\rho u)\,d\rho = 0, \qquad \nu > -\tfrac{1}{2}.$$

To prove this result we use Lemma 2 to obtain

$$\int_0^\infty u\mathrm{J}_\nu(ur)\,du \int_a^b \rho f(\rho)\mathrm{J}_\nu(\rho u)\,d\rho = \lim_{\lambda\to\infty}\int_a^b \rho f(\rho)h_\nu(\rho,r;\lambda)\,d\rho.$$

Substituting the expression (5-3-6) for $h_\nu(\rho,r;\lambda)$ we see that the right-hand side of this equation is equal to $f_1 - f_2$ where

$$f_1 = \lim_{\lambda\to\infty}\mathrm{J}_\nu(\lambda r)\int_a^b \frac{\rho^2 f(\rho)}{\rho^2 - r^2}\mathrm{J}_{\nu+1}(\lambda\rho)\,d\rho$$

and

$$f_2 = \lim_{\lambda\to\infty} r\mathrm{J}_{\nu+1}(\lambda r)\int_a^b \frac{\rho f(\rho)}{\rho^2 - r^2}\mathrm{J}_\nu(\lambda\rho)\,d\rho.$$

By our hypotheses the function $\rho^{\frac{3}{2}}f(\rho)/(\rho^2 - r^2)$ is absolutely integrable over (a,b) so that by the Riemann–Lebesgue lemma $f_1 = 0$. Similarly it is easily shown that $f_2 = 0$.

It should be noted that this lemma is not restricted to finite b.

To establish the contribution from the immediate vicinity of $\rho = r$ we need:

Lemma 4 *The repeated integral*

$$\int_a^b \sqrt{\rho}d\rho \int_0^\lambda uJ_\nu(u\rho)J_\nu(ur)\,du, \qquad \nu > -\tfrac{1}{2}$$

in which a and b have any finite positive values, remains bounded as $\lambda \to \infty$.

If we replace the Bessel functions by the dominant terms (5-3-4) of their asymptotic expansions we find that

$$I_1 = \frac{2}{\pi\sqrt{r}}\int_a^b d\rho \int_0^\lambda \cos(u\rho - \tfrac{1}{2}\nu\pi - \tfrac{1}{4}\pi)\cos(ur - \tfrac{1}{2}\nu\pi - \tfrac{1}{4}\pi)\,du$$

$$= \frac{1}{\pi\sqrt{r}}\int_a^b \left[\frac{\sin\{\lambda(\rho - r)\}}{\rho - r} - \frac{\cos\{\lambda(\rho + r) - \nu\pi\} - \cos\nu\pi}{\rho + r}\right] d\rho$$

$$= \frac{1}{\pi\sqrt{r}}\left[\int_{\lambda(a-r)}^{\lambda(b-r)} \frac{\sin x}{x}\,dx - \int_{\lambda(a+r)}^{\lambda(b+r)} \frac{\cos(x - \nu\pi)}{x}\,dx \right.$$
$$\left. + \cos(\nu\pi)\log\frac{b + r}{a + r}\right]$$

showing that this integral is bounded and tends to the limit

$$\frac{1}{\pi\sqrt{r}}\lim_{\lambda\to\infty}\left[\int_{\lambda(a-r)}^{\lambda(b-r)} \frac{\sin x}{x}\,dx + \cos(\nu\pi)\log\frac{b + r}{a + r}\right]$$

provided, of course, that this limit exists.

Similarly by considering the terms $O(u^{-2})$ in the asymptotic expansion of the function $a_\nu(\rho,r,u)$

$$a_\nu(r,\rho,u) = \tfrac{1}{2}\pi u\sqrt{(r\rho)}J_\nu(ur)J_\nu(u\rho) - \cos(u\rho - \tfrac{1}{2}\nu\pi - \tfrac{1}{4}\pi)\times$$
$$\times\cos(ur - \tfrac{1}{2}\nu\pi - \tfrac{1}{4}\pi)$$

we can show that the integral

$$I_2 = \frac{2}{\pi\sqrt{r}}\int_a^b d\rho \int_\lambda^\infty a_\nu(\rho,r,u)\,du$$

also remains bounded as $\lambda \to \infty$.

Now the integral

$$I_3 = \frac{2}{\pi\sqrt{r}} \int_a^b d\rho \int_0^\infty a_\nu(\rho,r,u)\, du$$

is the integral with respect to ρ of an integral which is uniformly convergent for positive values of ρ and r, and so it is a continuous (and hence bounded) function of r when r is positive.

Observing that

$$\int_a^b \sqrt{\rho}\, d\rho \int_0^\lambda u\mathrm{J}_\nu(u\rho)\mathrm{J}_\nu(ur)\, du = I_1 + I_2 + I_3$$

we conclude that the integral is bounded as $\lambda \to \infty$ and converges to a limit if

$$\lim_{\lambda\to\infty} \int_{\lambda(a-r)}^{\lambda(b-r)} \frac{\sin x}{x}\, dx$$

exists. Hence the lemma holds not only when a and b are finite positive numbers but also when they are functions of λ, either or both of which tend to r as $\lambda \to \infty$.

With the establishment of this result we can calculate the contribution made to the repeated integral by the immediate vicinity of the point $\rho = r$. This is contained in:

Lemma 5 *If $x(\rho) = \rho^{\frac{1}{2}}f(\rho)$ is piecewise continuous and absolutely integrable over the positive real line and δ is chosen such that $|x(\rho) - x(r+)| < \varepsilon$ whenever $r < \rho < r + \delta$, then as $\lambda \to \infty$*

$$\int_0^\lambda u\mathrm{J}_\nu(ur)\, du \int_r^{r+\delta} \rho f(\rho)\mathrm{J}_\nu(u\rho)\, d\rho \to \tfrac{1}{2}f(r+), \qquad \nu > -\tfrac{1}{2}. \tag{5-3-21}$$

Similarly if $|x(\rho) - x(r-)| < \varepsilon$ for $r - \delta < \rho < r$ we have that

$$\int_0^\lambda u\mathrm{J}_\nu(ur)\, du \int_{r-\delta}^{r} \rho f(\rho)\mathrm{J}_\nu(\rho u)\, d\rho = \tfrac{1}{2}f(r-), \qquad \nu > -\tfrac{1}{2} \tag{5-3-22}$$

as $\lambda \to \infty$.

$$\begin{aligned}
&\int_0^\lambda u\mathrm{J}_\nu(ur)\, du \int_r^{r+\delta} \rho f(\rho)\mathrm{J}_\nu(u\rho)\, d\rho \\
&\qquad = \int_r^{r+\delta} \rho^{\frac{1}{2}}x(\rho)\, d\rho \int_0^\lambda u\mathrm{J}_\nu(ur)\mathrm{J}_\nu(u\rho)\, du \\
&\qquad = f(r+) \int_r^{r+\delta} (\rho r)^{\frac{1}{2}}\, d\rho \int_0^\lambda u\mathrm{J}_\nu(ur)\mathrm{J}_\nu(u\rho)\, du \\
&\qquad\qquad + \{x(r-\delta) - x(r+)\} \int_{r+\xi}^{r+\delta} \rho^{\frac{1}{2}}\, d\rho \int_0^\lambda u\mathrm{J}_\nu(ur)\mathrm{J}_\nu(u\rho)\, du.
\end{aligned} \tag{5-3-23}$$

Now by Lemma 4 the first term on the right tends to a limit as $\lambda \to \infty$ which may be interpreted as

$$\lim_{p \to 0} f(r+) \int_r^{r+\delta} (\rho r)^{\frac{1}{2}} \mathrm{J}_\nu(\rho,r,p)\, d\rho = \tfrac{1}{2} f(r+)$$

by equation (5-3-8). By Lemma 4 the repeated integral in the second term on the right-hand side of equation (5-3-23) remains bounded as $\lambda \to \infty$. If we denote the upper bound of the modulus of this repeated integral by κ we therefore deduce that the absolute value of the second term is always less than $\kappa\varepsilon$ and hence establish the equation (5-3-21).

Theorem 1 (the Hankel Inversion theorem) *If $\sqrt{x} f(x)$ is piecewise continuous and absolutely integrable on the positive real line, then if $\nu > -\frac{1}{2}$, $\bar{f}_\nu(u) = \mathscr{H}_\nu[f(x);u]$ exists and*

$$\int_0^\infty u\bar{f}_\nu(u)\mathrm{J}_\nu(ur)\, du = \tfrac{1}{2}[f(r+) + f(r-)].$$

The proof follows immediately from Lemmas 3 and 5, for if we write

$$I(\alpha, \beta) = \lim_{\lambda \to \infty} \int_0^\lambda u\mathrm{J}_\nu(ur)\, du \int_\alpha^\beta \rho f(\rho)\mathrm{J}_\nu(\rho u)\, d\rho$$

it follows from Lemma 3 that, since r is contained in neither of the intervals $(0,r - \delta)$, $(r + \delta,\infty)$,

$$I(0,r - \delta) = 0, \qquad I(r + \delta,\infty) = 0$$

Also, by Lemma 5

$$I(r - \delta,r) = \tfrac{1}{2} f(r-), \qquad I(r,r + \delta) = \tfrac{1}{2} f(r+).$$

The theorem follows at once from the observation that

$$\begin{aligned} \int_0^\infty u\bar{f}_\nu(u)\mathrm{J}_\nu(ur)\, du &= \lim_{\lambda \to \infty} \int_0^\lambda u\mathrm{J}_\nu(ur)\, du \int_0^\infty \rho f(\rho)\mathrm{J}_\nu(\rho u)\, d\rho \\ &= I(0,r - \delta) + I(r - \delta,r) + I(r,r + \delta) + I(r + \delta,\infty). \end{aligned}$$

Corollary *If $f(\rho)$ is continuous at the point $\rho = r$,*

$$\int_0^\infty u\bar{f}_\nu(u)\mathrm{J}_\nu(ur)\, du = f(r)$$

i.e.,

$$\mathscr{H}_\nu[\bar{f}_\nu(u);r] = f(r). \tag{5-3-24}$$

5-4 HANKEL TRANSFORMS OF DERIVATIVES OF FUNCTIONS

In the application of Hankel transforms to physical problems and in the calculation of the Hankel transforms of some elementary functions, it is useful to have available some formulae connecting the Hankel transforms of derivatives of functions.

Using the definition of a Hankel transform and the formula for integrating by parts we find that

$$\mathscr{H}_\nu\left[\rho^{\nu-1}\frac{\partial}{\partial\rho}\{\rho^{1-\nu}f(\rho)\};\xi\right]$$

$$= \int_0^\infty \rho^\nu J_\nu(\xi\rho)\frac{\partial}{\partial\rho}\{\rho^{1-\nu}f(\rho)\}\,d\rho$$

$$= [\rho f(\rho)J_\nu(\xi\rho)]_0^\infty - \int_0^\infty \rho^{1-\nu}f(\rho)\frac{\partial}{\partial\rho}\{\rho^\nu J_\nu(\xi\rho)\}\,d\rho.$$

The first term vanishes provided that $f(\rho)$ is such that

$$\lim_{\rho\to 0}\rho^{\nu+1}f(\rho) = 0, \qquad \lim_{\rho\to\infty}\rho^{\frac{1}{2}}f(\rho) = 0.$$

It follows from the argument of sec. 5-2 that the second of these conditions holds for any $f(\rho)$ whose Hankel transform exists. Hence the term in the square bracket vanishes if

$$f(\rho) = o(\rho^{-\nu-1}), \qquad \rho\to 0 \tag{5-4-1}$$

If in the second term we make use of the recurrence relation

$$\frac{\partial}{\partial\rho}\{\rho^\nu J_\nu(\xi\rho)\} = \xi\rho^\nu J_{\nu-1}(\xi\rho)$$

we see that the second term reduces to

$$-\xi\int_0^\infty \rho f(\rho)J_{\nu-1}(\xi\rho)\,d\rho$$

so that, provided $f(\rho)$ satisfies the condition (5-4-1) and is such that its Hankel transform of orders ν and $\nu - 1$ exist, we have the relation

$$\mathscr{H}_\nu\left[\rho^{\nu-1}\frac{\partial}{\partial\rho}\{\rho^{1-\nu}f(\rho)\};\xi\right] = -\xi\mathscr{H}_{\nu-1}[f(\rho);\xi] \tag{5-4-2}$$

In particular if we put $\nu = 1$ we obtain the relation

$$\mathscr{H}_1[f'(\rho);\xi] = -\xi\mathscr{H}_0[f(\rho);\xi] \tag{5-4-3}$$

where $f(\rho) = o(\rho^{-2})$ as $\rho\to 0$.

In a similar way we have that

$$\mathscr{H}_\nu\left[\rho^{-\nu-1}\frac{\partial}{\partial\rho}\{\rho^{\nu+1}f(\rho)\};\xi\right]$$

$$= \int_0^\infty \rho^{-\nu}\mathrm{J}_\nu(\xi\rho)\frac{\partial}{\partial\rho}\{\rho^{\nu+1}f(\rho)\}\,d\rho$$

$$= [\rho f(\rho)\mathrm{J}_\nu(\rho\xi)]_0^\infty - \int_0^\infty \rho^{\nu+1}f(\rho)\frac{\partial}{\partial\rho}\{\rho^{-\nu}\mathrm{J}_\nu(\rho\xi)\}\,d\rho.$$

The terms in the square bracket vanish if $f(\rho)$ satisfies the same conditions as before, and if we make use of the recurrence relation

$$\frac{\partial}{\partial\rho}\{\rho^{-\nu}\mathrm{J}_\nu(\xi\rho)\} = -\xi\rho^{-\nu}\mathrm{J}_{\nu+1}(\xi\rho)$$

in the second term we find that it reduces to

$$\xi\int_0^\infty \rho f(\rho)\mathrm{J}_{\nu+1}(\xi\rho)\,d\rho.$$

In this way we find that, if $f(\rho)$ satisfies (5-4-1),

$$\mathscr{H}_\nu\left[\rho^{-\nu-1}\frac{\partial}{\partial\rho}\{\rho^{\nu+1}f(\rho)\};\xi\right] = \xi\mathscr{H}_{\nu+1}[f(\rho);\xi]. \tag{5-4-4}$$

Corresponding to $\nu = 0$ we have the special case

$$\mathscr{H}_0\left[\frac{1}{\rho}\frac{\partial}{\partial\rho}\{\rho f(\rho)\};\xi\right] = \xi\mathscr{H}_1[f(\rho);\xi]. \tag{5-4-5}$$

provided that $f(\rho) = \mathrm{o}(\rho^{-1})$ as $\rho \to 0$.

We shall now consider some results involving the differential operator

$$\mathscr{B}_\nu = \frac{\partial^2}{\partial\rho^2} + \frac{1}{\rho}\frac{\partial}{\partial\rho} - \frac{\nu^2}{\rho^2}. \tag{5-4-6}$$

From the definition of a Hankel transform of order zero we have

$$\mathscr{H}_0[\mathscr{B}_0 f;\xi] = \mathscr{H}_0\left[\frac{1}{\rho}\frac{\partial}{\partial\rho}\left(\rho\frac{\partial}{\partial\rho}\right);\xi\right]$$

Using (5-4-5) we see that provided

$$\frac{\partial f}{\partial\rho} = \mathrm{o}(\rho^{-1}), \qquad \rho \to 0,$$

$$\mathscr{H}_0[\mathscr{B}_0 f;\xi] = \xi\mathscr{H}_1\left[\frac{\partial f}{\partial\rho};\xi\right]$$

Now if

$$f = o(\rho^{-2}), \qquad \rho \to 0,$$

we see from (5-4-3) that

$$\mathscr{H}_0[\mathscr{B}_0 f;\xi] = -\xi^2 \mathscr{H}_0[f;\xi]. \tag{5-4-7}$$

This method is easily generalized to the case in which $\nu \neq 0$. We begin by noting that

$$\mathscr{B}_\nu f = \rho^{-\nu-1}\frac{\partial}{\partial\rho}\left\{\rho^{2\nu+1}\frac{\partial}{\partial\rho}(\rho^{-\nu}f)\right\}$$

$$= \rho^{-\nu-1}\frac{\partial}{\partial\rho}\left[\rho^{\nu+1}\left\{\rho^{\nu}\frac{\partial}{\partial\rho}(\rho^{-\nu}f)\right\}\right]$$

so that from (5-4-4)

$$\mathscr{H}_\nu[\mathscr{B}_\nu f;\xi] = \xi\mathscr{H}_{\nu+1}\left\{\rho^{\nu}\frac{\partial}{\partial\rho}(\rho^{-\nu}f)\right\}.$$

We can now use (5-4-2) with ν replaced by $\nu + 1$ to evaluate the term on the right-hand side and obtain the result

$$\mathscr{H}_\nu[\mathscr{B}_\nu f;\xi] = -\xi^2\mathscr{H}_\nu[f;\xi]. \tag{5-4-8}$$

If we use cylindrical coordinates (ρ,ϕ,z), the Laplacian operator in three dimensions becomes

$$\Delta_3 = \frac{\partial^2}{\partial\rho^2} + \frac{1}{\rho}\frac{\partial}{\partial\rho} + \frac{1}{\rho^2}\frac{\partial^2}{\partial\phi^2} + \frac{\partial^2}{\partial z^2} \tag{5-4-9}$$

and in problems in which there is axial symmetry we may, by choosing the axis of symmetry to be the z-axis, take the Laplacian operator to be

$$\Delta_a = \frac{\partial^2}{\partial\rho^2} + \frac{1}{\rho}\frac{\partial}{\partial\rho} + \frac{\partial^2}{\partial z^2}.$$

It is immediately obvious from these definitions that

$$\Delta_3 u(\rho,z)e^{i\nu\phi} = (\mathscr{B}_\nu + \mathrm{D}^2)u(\rho,z)e^{i\nu\phi}$$

$$\Delta_a u(\rho,z) = (\mathscr{B}_0 + \mathrm{D}^2)u(\rho,z)$$

where D denotes the operator $\partial/\partial z$, and then from equations (5-4-7) and (5-4-8) that

$$\mathscr{H}_\nu[\Delta_3 u(\rho,z)e^{i\nu\phi};\rho\to\xi] = (\mathrm{D}^2 - \xi^2)\bar{u}_\nu(\xi,z)e^{i\nu\phi} \tag{5-4-11}$$

$$\mathscr{H}_0[\Delta_a u(\rho,z);\rho\to\xi] = (\mathrm{D}^2 - \xi^2)\bar{u}_0(\xi,z) \tag{5-4-12}$$

where

$$\bar{u}_\nu(\xi,z) = \mathscr{H}_\nu[u(\rho,z);\rho \to \xi] \tag{5-4-13}$$

is the Hankel transform of order ν of $u(\rho,z)$ with respect to the variable ρ.

In the discussion of problems in hydrodynamics and in the theory of elasticity we need the results

$$\mathscr{H}_\nu[\Delta_3{}^2 u(\rho,z)e^{i\nu\phi};\rho \to \xi] = (\mathrm{D}^2 - \xi^2)^2 \bar{u}_\nu(\xi,z)e^{i\nu\phi} \tag{5-4-14}$$

$$\mathscr{H}_0[\Delta_a{}^2 u(\rho,z);\rho \to \xi] = (\mathrm{D}^2 - \xi^2)^2 \bar{u}_0(\xi,z) \tag{5-4-15}$$

obtained by repeated applications of equations (5-4-11) and (5-4-12) respectively. It will be recalled that $\Delta_3{}^2$ is the biharmonic operator $\Delta_3\Delta_3$.

5-5 THE HANKEL TRANSFORMS OF SOME ELEMENTARY FUNCTIONS

In this section we shall calculate the Hankel transforms of some elementary functions.

Expanding the Bessel function in the definition of the Hankel transform and assuming that $\mu > \nu \geqslant 0$ we have

$$\begin{aligned}\mathscr{H}_\nu[x^\nu(a^2 - x^2)^{\mu-\nu-1}\mathrm{H}(a-x);\xi]&\\ &= \int_0^a x^{\nu+1}(a^2 - x^2)^{\mu-\nu-1}\mathrm{J}_\nu(x\xi)\,dx\\ &= \sum_{r=0}^{\infty} \frac{(-1)^r(\frac{1}{2}\xi)^{\nu+2r}}{r!\Gamma(r+\nu+1)} \int_0^a x^{2\nu+2r+1}(a^2 - x^2)^{\mu-\nu-1}\,dx.\end{aligned}$$

The integral on the right-hand side of this last equation is easily seen to have the value

$$\frac{1}{2}a^{2\mu+2r}\frac{\Gamma(r+\nu+1)\Gamma(\mu-\nu)}{\Gamma(r+\mu+1)}$$

so that

$$\begin{aligned}\mathscr{H}_\nu[x^\nu(a^2 - x^2)^{\mu-\nu-1}\mathrm{H}(a-x);\xi]&\\ &= 2^{\mu-\nu-1}\Gamma(\mu-\nu)a^\mu\xi^{\nu-\mu}\sum_{r=0}^{\infty}\frac{(-1)^r(\frac{1}{2}\xi a)^{\mu+2r}}{r!\Gamma(r+\mu+1)}\end{aligned}$$

from which we deduce that

$$\mathscr{H}_\nu[x^\nu(a^2 - x^2)^{\mu-\nu-1}\mathrm{H}(a-x);\xi] = 2^{\mu-\nu-1}\Gamma(\mu-\nu)a^\mu\xi^{\nu-\mu}\mathrm{J}_\mu(\xi a). \tag{5-5-1}$$

Now using the Hankel inversion theorem we find that if $\mu > \nu \geqslant 0$

$$2^{\mu-\nu-1}\Gamma(\mu-\nu)a^\mu\mathscr{H}_\nu[\xi^{\nu-\mu}\mathrm{J}_\mu(\xi a);x] = x^\nu(a^2 - x^2)^{\mu-\nu-1}\mathrm{H}(a-x)$$

a result which can be written in the alternative form

$$\mathscr{H}_\nu[x^{\nu-\mu}J_\mu(ax);\xi] = \frac{\xi^\nu(a^2-\xi^2)^{\mu-\nu-1}}{2^{\mu-\nu-1}\Gamma(\mu-\nu)a^\mu}H(a-\xi) \tag{5-5-2}$$

with $a > 0$, $\mu > \nu \geqslant 0$, i.e.,

$$\int_0^\infty x^{1-\mu+\nu}J_\mu(ax)J_\nu(\xi x)\,dx = \frac{\xi^\nu(a^2-\xi^2)^{\mu-\nu-1}}{2^{\mu-\nu-1}\Gamma(\mu-\nu)a^\mu}H(a-\xi) \tag{5-5-3}$$

with the same restrictions on a, μ, and ν.

If we put $\mu = \nu + 1$ in (5-5-1) and (5-5-2) we obtain the special cases

$$\mathscr{H}_\nu[x^\nu H(a-x);\xi] = \frac{a^{\nu+1}}{\xi}J_{\nu+1}(\xi a), \qquad a > 0 \tag{5-5-4}$$

$$\mathscr{H}_\nu[x^{-1}J_{\nu+1}(ax);\xi] = \frac{\xi^\nu}{a^{\nu+1}}H(a-\xi), \qquad a > 0. \tag{5-5-5}$$

Similarly if we put $\mu = \nu + \frac{1}{2}$ in the same pair of equations we find that

$$\mathscr{H}_\nu\left[\frac{x^\nu H(a-x)}{\sqrt{(a^2-x^2)}};\xi\right] = \sqrt{\frac{\pi}{2\xi}}\,a^{\nu+\frac{1}{2}}J_{\nu+\frac{1}{2}}(\xi a) \tag{5-5-6}$$

$$\mathscr{H}_\nu[x^{-\frac{1}{2}}J_{\nu+\frac{1}{2}}(ax);\xi] = \sqrt{\frac{2}{\pi}}\frac{\xi^\nu H(a-\xi)}{a^{\nu+\frac{1}{2}}\sqrt{(a^2-\xi^2)}} \tag{5-5-7}$$

$a > 0$, $\nu \geqslant 0$.

If we take $\nu = 0$ in (5-5-5) and (5-5-7) we find that

$$\mathscr{H}_0[x^{-1}J_1(ax);\xi] = \frac{H(a-\xi)}{a}, \qquad (a > 0) \tag{5-5-8}$$

$$\mathscr{H}_0[x^{-1}\sin(ax);\xi] = \frac{H(a-\xi)}{\sqrt{(a^2-\xi^2)}}, \qquad (a > 0). \tag{5-5-9}$$

Similarly if we take $\nu = \frac{1}{2}$ in (5-5-7) we obtain the result

$$\int_0^\infty x^{\frac{1}{2}}J_1(ax)J_{\frac{1}{2}}(x\xi)\,dx = \sqrt{\frac{2}{\pi}}\cdot\frac{\xi^{\frac{1}{2}}H(a-\xi)}{a\sqrt{(a^2-\xi^2)}}.$$

Interchanging a and ξ we see that this result can be written in the form

$$\mathscr{H}_1[x^{-1}\sin(ax);\xi] = \frac{aH(\xi-a)}{\xi\sqrt{(\xi^2-a^2)}} \tag{5-5-10}$$

If we write out the left-hand side of equation (5-5-8) we find that

$$\int_0^\infty J_1(ax)J_0(\xi x)\,dx = \frac{H(a-\xi)}{a}. \tag{5-5-11}$$

Using the results

$$\int_0^a J_1(ax)\,da = \frac{1 - J_0(ax)}{x}$$

$$\int_0^a \frac{H(a-\xi)}{a}\,da = H(a-\xi)\log\left(\frac{a}{\xi}\right)$$

we find on integrating both sides of (5-5-11) with respect to a from 0 to a that

$$\int_0^\infty \frac{1 - J_0(ax)}{x} J_0(\xi x)\,dx = H(a-\xi)\log\left(\frac{a}{\xi}\right)$$

i.e., that

$$\mathscr{H}_0[x^{-2}\{1 - J_0(ax)\};\xi] = H(a-\xi)\log\left(\frac{a}{\xi}\right). \tag{5-5-12}$$

We shall now consider the Hankel transforms of some functions containing a simple exponential factor.

By the definition of the Hankel transform we have that

$$\mathscr{H}_\nu[e^{-px}f(x);\xi] = \int_0^\infty \{xf(x)J_\nu(\xi x)\}e^{-px}\,dx$$

from which it follows that

$$\mathscr{H}_\nu[e^{-px}f(x);\xi] = \mathscr{L}[xf(x)J_\nu(\xi x);p]. \tag{5-5-13}$$

For example from equations (3-7-3), (3-7-5), and Problem 3-18 we deduce immediately that if $\nu > -1$

$$\mathscr{H}_\nu[x^{\nu-1}e^{-px};\xi] = \frac{2^\nu\xi^\nu\Gamma(\nu+\frac{1}{2})}{\sqrt{\pi}(p^2+\xi^2)^{\nu+\frac{1}{2}}} \tag{5-5-14}$$

$$\mathscr{H}_\nu[x^{\nu}e^{-px};\xi] = \frac{2^{\nu+1}\xi^\nu p\Gamma(\nu+\frac{3}{2})}{\sqrt{\pi}(p^2+\xi^2)^{\nu+\frac{3}{2}}} \tag{5-5-15}$$

$$\mathscr{H}_\nu[x^{-1}e^{-px};\xi] = \frac{1}{\sqrt{(p^2+\xi^2)}}\left\{\frac{\xi}{p+\sqrt{(p^2+\xi^2)}}\right\}^\nu \tag{5-5-16}$$

$$\mathscr{H}_\nu[e^{-px};\xi] = \frac{p+\nu\sqrt{(p^2+\xi^2)}}{(p^2+\xi^2)^{\frac{3}{2}}}\left\{\frac{\xi}{p+\sqrt{(p^2+\xi^2)}}\right\}^\nu. \tag{5-5-17}$$

Similarly we have the relation

$$\mathscr{H}_\nu[e^{-\frac{1}{4}px^2}f(x);\xi] = \int_0^\infty xf(x)e^{-\frac{1}{4}px^2}J_\nu(x\xi)\,dx.$$

If we change the variable of integration in the integral on the right from x

to u where $u = \frac{1}{4}x^2$ we find that it assumes the form

$$2\int_0^\infty f(2u^{\frac{1}{2}})J_\nu(2\xi u^{\frac{1}{2}})e^{-pu}\,du$$

from which we deduce that

$$\mathscr{H}_\nu[e^{-\frac{1}{4}px^2}f(x);\xi] = 2\mathscr{L}[f(2\sqrt{x})J_\nu(2\xi\sqrt{x});p] \tag{5-5-18}$$

If we take $f(x) = x^\nu$ then

$$\mathscr{H}_\nu[x^\nu e^{-\frac{1}{4}px^2};\xi] = 2^{\nu+1}\mathscr{L}[x^{\frac{1}{2}\nu}J_\nu(2\xi\sqrt{x});p]$$

and using the result (3-7-8) we find that

$$\mathscr{H}_\nu[x^\nu e^{-\frac{1}{4}px^2};\xi] = \frac{2^{\nu+1}\xi^\nu}{p^{\nu+1}}\,e^{-\xi^2/p}$$

which can be written in the form

$$\mathscr{H}_\nu[x^\nu e^{-x^2/a^2};\xi] = (\tfrac{1}{2}a^2)^{\nu+1}\xi^\nu e^{-\frac{1}{4}\xi^2a^2}. \tag{5-5-19}$$

In particular, if we take $a^2 = 2$ we obtain the result

$$\mathscr{H}_\nu[x^\nu e^{-\frac{1}{2}x^2};\xi] = \xi^\nu e^{-\frac{1}{2}\xi^2} \tag{5-5-20}$$

showing that $x^\nu e^{-\frac{1}{2}x^2}$ is self-reciprocal under the Hankel transform of order ν.

Similarly if we take $f(x) = x^{-1}$ in equation (5-5-18) we find that

$$\mathscr{H}_\nu[x^{-1}e^{-\frac{1}{4}px^2};\xi] = \mathscr{L}[x^{-\frac{1}{2}}J_\nu(2\xi\sqrt{x});p]$$

and using the result of Problem 3-19 we find that

$$\mathscr{H}_\nu[x^{-1}e^{-\frac{1}{4}p^2x^2};\xi] = \sqrt{\frac{\pi}{p}}\,e^{-\xi^2/2p}I_{\frac{1}{2}\nu}\left(\frac{\xi^2}{2p}\right) \tag{5-5-21}$$

Analogous to equation (5-5-13) we have a relation involving the Mellin transform. By the definition of the Hankel transform we have

$$\mathscr{H}_\nu[x^{s-2}f(x);\xi] = \int_0^\infty x^{s-1}f(x)J_\nu(\xi x)\,dx.$$

Interpreting the right-hand side as a Mellin transform we obtain the relation

$$\mathscr{H}_\nu[x^{s-2}f(x);\xi] = \mathscr{M}[f(x)J_\nu(\xi x);s]. \tag{5-5-22}$$

As a special case of this equation we have

$$\mathscr{H}_\nu[x^{s-1};\xi] = \mathscr{M}[J_\nu(\xi x);s+1]$$

and evaluating the Mellin transform by means of equation (4-1-10) we see that

$$\mathscr{H}_\nu[x^{s-1};\xi] = \frac{2^s\Gamma(\frac{1}{2}s + \frac{1}{2}\nu + \frac{1}{2})}{\xi^{s+1}\Gamma(\frac{1}{2}\nu - \frac{1}{2}s + \frac{1}{2})} \tag{5-5-23}$$

with $-1 - \nu < s < 1 + \nu$.

5-6 THE PARSEVAL RELATION FOR HANKEL TRANSFORMS

Suppose that the functions $f(x)$ and $g(x)$ have, respectively, the Hankel transforms of order ν, $\bar{f}_\nu(\xi)$ and $\bar{g}_\nu(\xi)$. Then, proceeding formally, we have

$$\int_0^\infty \xi\bar{f}_\nu(\xi)\bar{g}_\nu(\xi)\,d\xi = \int_0^\infty \xi\bar{f}_\nu(\xi)\,d\xi \int_0^\infty xg(x)J_\nu(\xi x)\,dx$$

$$= \int_0^\infty xg(x)\,dx \int_0^\infty \xi\bar{f}_\nu(\xi)J_\nu(\xi x)\,d\xi$$

so that we obtain the Parseval relation

$$\int_0^\infty \xi\bar{f}_\nu(\xi)\bar{g}_\nu(\xi)\,d\xi = \int_0^\infty xf(x)g(x)\,dx. \tag{5-6-1}$$

A systematic discussion of this formula has been given by Macauley-Owen (1939) who proved the following theorem:

Theorem 2 *If the functions $f(x)$ and $g(x)$ satisfy the conditions of Theorem 1 and if $\bar{f}_\nu(\xi)$ and $\bar{g}_\nu(\xi)$ denote their Hankel transforms of order $\nu \geqslant -\frac{1}{2}$, then*

$$\int_0^\infty xf(x)g(x)\,dx = \int_0^\infty \xi\bar{f}_\nu(\xi)\bar{g}_\nu(\xi)\,d\xi.$$

The usefulness of this result in the evaluation of integrals is shown by the evaluation of $\mathscr{H}_\nu[x^{-2}J_\nu(ax);\xi]$, $\quad \nu > -\frac{1}{2}$.

If we take

$$f(x) = x^\nu H(a - x), \qquad g(x) = x^\nu H(b - x),\ \nu > -\tfrac{1}{2}$$

with $a > 0$, $b > 0$, we have

$$\bar{f}_\nu(\xi) = \frac{a^{\nu+1}}{\xi}J_{\nu+1}(\xi a), \qquad \bar{g}_\nu(\xi) = \frac{b^{\nu+1}}{\xi}J_{\nu+1}(\xi b)$$

so that by (5-6-1)

$$(ab)^{\nu+1}\int_0^\infty \xi^{-1}J_{\nu+1}(\xi a)J_{\nu+1}(\xi b)\,d\xi = \int_0^{\min(a,b)} x^{2\nu+1}\,dx.$$

Suppose that we take $0 < a < b$. Then this last equation gives the result

$$\int_0^\infty \xi^{-1} J_{\nu+1}(\xi a) J_{\nu+1}(\xi b)\, d\xi = \frac{1}{2(\nu+1)}\left(\frac{a}{b}\right)^{\nu+1}, \qquad 0 < a < b,\ \nu > -\tfrac{1}{2}, \tag{5-6-2}$$

which can be written in the alternative form

$$\mathscr{H}_\nu[x^{-2}J_\nu(ax);\xi] = \begin{cases} \dfrac{1}{2\nu}\left(\dfrac{\xi}{a}\right)^\nu, & 0 < \xi < a, \\ \dfrac{1}{2\nu}\left(\dfrac{a}{\xi}\right)^\nu, & \xi > a \end{cases} \tag{5-6-3}$$

where $\nu > \frac{1}{2}$.

5-7 RELATIONS BETWEEN FOURIER AND HANKEL TRANSFORMS

The well-known integrals

$$\int_0^x \frac{\cos(\xi t)\, dt}{\sqrt{(x^2 - t^2)}} = \tfrac{1}{2}\pi J_0(\xi x), \qquad (x > 0, \xi > 0) \tag{5-7-1}$$

$$\int_x^\infty \frac{\sin(\xi t)\, dt}{\sqrt{(t^2 - x^2)}} = \tfrac{1}{2}\pi J_0(\xi x), \qquad (x > 0, \xi > 0) \tag{5-7-2}$$

suggest that there are close relations between Fourier and Hankel transforms. To investigate these relations it is convenient to define *Abel transforms* $\mathscr{A}_1$ and $\mathscr{A}_2$ through the equations

$$\hat{f}_1(x) \equiv \mathscr{A}_1[f(t);x] = \sqrt{\frac{2}{\pi}}\int_0^x \frac{f(t)\, dt}{\sqrt{(x^2 - t^2)}}, \qquad x > 0 \tag{5-7-3}$$

$$\hat{f}_2(x) \equiv \mathscr{A}_2[f(t);x] = \sqrt{\frac{2}{\pi}}\int_x^\infty \frac{f(t)\, dt}{\sqrt{(t^2 - x^2)}}, \qquad x > 0. \tag{5-7-4}$$

If we regard (5-7-3) as an integral equation for $f(t)$, $\hat{f}_1(x)$ being prescribed, then its solution is given by equation (3-16-22) in the form

$$f(t) = \sqrt{\frac{2}{\pi}}\frac{d}{dt}\int_0^t \frac{x\hat{f}_1(x)\, dx}{\sqrt{(t^2 - x^2)}}, \qquad t > 0$$

which may be rewritten as

$$f(t) \equiv \mathscr{A}^{-1}[\hat{f}_1(x);t] = D_t \mathscr{A}_1[x\hat{f}_1(x);t] \tag{5-7-5}$$

where, as usual, D_t denotes the differential operator d/dt. Similarly by using equation (3-16-24) we can write (5-7-4) in the inverse form

$$f(t) \equiv \mathscr{A}_2^{-1}[\hat{f}_2(x);t] = -D_t[x\hat{f}_2(x);t] \tag{5-7-6}$$

Many standard integrals can be interpreted as statements involving these transforms. For instance, we can interpret (5-7-1) and (5-7-2) in the forms

$$\mathscr{A}_1[\cos(\xi t);x] = \sqrt{(\tfrac{1}{2}\pi)}J_0(\xi x) \tag{5-7-7}$$

$$\mathscr{A}_2[\sin(\xi t);x] = \sqrt{(\tfrac{1}{2}\pi)}J_0(\xi x) \tag{5-7-8}$$

respectively. If we differentiate both sides of (5-7-7) with respect to ξ we obtain the formula

$$\mathscr{A}_1[t\sin(\xi t);x] = \sqrt{(\tfrac{1}{2}\pi)}xJ_1(\xi x) \tag{5-7-9}$$

and using this result in (5-7-5) we find that

$$\mathscr{A}_1^{-1}[\sin(\xi x);t] = \sqrt{(\tfrac{1}{2}\pi)}\xi tJ_0(\xi t) \tag{5-7-10}$$

Similarly from (5-7-6) and (5-7-8) we find that

$$\begin{aligned}\mathscr{A}_2^{-1}\left[\frac{\sin(\xi x)}{x};t\right] &= -D_t\mathscr{A}_2[\sin(\xi x);t]\\ &= -\sqrt{(\tfrac{1}{2}\pi)}D_tJ_0(\xi t)\end{aligned}$$

from which we immediately deduce the relation

$$\mathscr{A}_2^{-1}\left[\frac{\sin(\xi x)}{x};t\right] = \sqrt{(\tfrac{1}{2}\pi)}\xi J_1(\xi t) \tag{5-7-11}$$

The proof that $\mathscr{H}_\nu^{-1} = \mathscr{H}_\nu$ for $\nu = 0,1$ follows from two general identities which are simply established.

From the definition (5-7-3) we have, for any arbitrary integrable kernel $K(\xi,x)$,

$$\int_0^\infty K(\xi,x)\hat{f}_1(x)\,dx = \sqrt{\frac{2}{\pi}}\int_0^\infty K(\xi,x)\,dx\int_0^x \frac{f(t)\,dt}{\sqrt{(x^2-t^2)}}$$

Reversing the order in which we perform the integrations we see that the double integral on the right may be written in the form

$$\sqrt{\frac{2}{\pi}}\int_0^\infty f(t)\,dt\int_t^\infty \frac{K(\xi,x)\,dx}{\sqrt{(x^2-t^2)}}$$

which shows that

$$\int_0^\infty K(\xi,x)\hat{f}_1(x)\,dx = \int_0^\infty f(x)\hat{K}_2(\xi,x)\,dx \tag{5-7-12}$$

where

$$\hat{K}_2(\xi,x) = \mathscr{A}_2[K(\xi,t); t \to x]$$

In a similar way we can show that

$$\int_0^\infty K(\xi,x)\hat{f}_2(x)\,dx = \int_0^\infty f(x)\hat{K}_1(\xi,x)\,dx \tag{5-7-13}$$

where

$$\hat{K}_1(\xi,x) = \mathscr{A}_1[K(\xi,t); t \to x] \tag{5-7-14}$$

If, in equation (5-7-12) we put $K(\xi,x) = (2/\pi)^{\frac{1}{2}} \sin(\xi x)$ and make use of equation (5-7-8) we obtain the relation

$$\mathscr{F}_s[\hat{f}_1(x); \xi] = \int_0^\infty f(x) \mathrm{J}_0(\xi x)\,dx$$

which may be written in the alternative form

$$\mathscr{F}_s[\mathscr{A}_1\{tf(t); x\}; \xi] = \bar{f}_0(\xi) \equiv \mathscr{H}_0[f(x); \xi] \tag{5-7-15}$$

Applying the operator $\mathscr{F}_s^{-1} \equiv \mathscr{F}_s$ to both sides of this equation we find that

$$\mathscr{A}_1[tf(t); x] = \sqrt{\frac{2}{\pi}} \int_0^\infty \bar{f}_0(\xi) \sin(\xi x)\,d\xi$$

and hence that

$$f(t) = \sqrt{\frac{2}{\pi}}\, t^{-1} \int_0^\infty \bar{f}_0(\xi) \mathscr{A}_1^{-1}[\sin(\xi x); t]\,d\xi$$

Using equation (5-7-10) we see that this last equation is equivalent to the relation

$$f(t) = \int_0^\infty \xi \bar{f}_0(\xi) \mathrm{J}_0(\xi t)\,d\xi.$$

In other words $\mathscr{H}_0^{-1} = \mathscr{H}_0$.

Similarly from equations (5-7-13) and (5-7-9) we see that

$$\begin{aligned}\mathscr{F}_s[x\hat{f}_2(x); \xi] &= \sqrt{\frac{2}{\pi}} \int_0^\infty f(x) \mathscr{A}_1[t \sin(\xi t); x]\,dx \\ &= \int_0^\infty x f(x) \mathrm{J}_1(\xi x)\,dx\end{aligned}$$

showing that

$$\mathscr{F}_s[x\hat{f}_2(x); \xi] = \bar{f}_1(\xi) \equiv \mathscr{H}_1[f(x); \xi]. \tag{5-7-16}$$

From this we deduce that

$$\hat{f}_2(x) = \sqrt{\frac{2}{\pi}}\, x^{-1} \int_0^\infty \bar{f}_1(\xi) \sin(\xi x)\, d\xi$$

and hence that

$$f(t) = \sqrt{\frac{2}{\pi}} \int_0^\infty \bar{f}_1(\xi) \mathscr{A}_2^{-1}\left[\frac{\sin(\xi x)}{x}; t\right] d\xi$$

$$= \int_0^\infty \xi \bar{f}_1(\xi) J_1(\xi t)\, d\xi$$

by equation (5-7-11). Hence $\mathscr{H}_1^{-1} = \mathscr{H}_1$.

We can get another set of relations if, instead of starting from the integrals (5-7-1) and (5-7-2) we take

$$\mathscr{F}_c[J_0(\xi\rho); \xi \to x] = \sqrt{\frac{2}{\pi}} \cdot \frac{H(\rho - x)}{\sqrt{(\rho^2 - x^2)}} \tag{5-7-17}$$

$$\mathscr{F}_s[J_0(\xi\rho); \xi \to x] = \sqrt{\frac{2}{\pi}} \cdot \frac{H(x - \rho)}{\sqrt{(x^2 - \rho^2)}} \tag{5-7-18}$$

as our basic relations.

For instance we can write

$$\mathscr{F}_c[\bar{f}_0(\xi); x] = \int_0^\infty \rho f(\rho) \mathscr{F}_c[J_0(\xi\rho); \xi \to x]\, d\rho$$

$$= \sqrt{\frac{2}{\pi}} \int_x^\infty \frac{\rho f(\rho)\, d\rho}{\sqrt{(\rho^2 - x^2)}}$$

and

$$\mathscr{F}_s[\bar{f}_0(\xi); x] = \int_0^\infty \rho f(\rho) \mathscr{F}_s[J_0(\xi\rho); \xi \to x]\, d\rho$$

$$= \sqrt{\frac{2}{\pi}} \int_0^x \frac{\rho f(\rho)\, d\rho}{\sqrt{(x^2 - \rho^2)}}$$

relations which may be written in the form

$$\mathscr{F}_c[\bar{f}_0(\xi); x] = \mathscr{A}_2[\rho f(\rho); x] \tag{5-7-19}$$

$$\mathscr{F}_s[\bar{f}_0(\xi); x] = \mathscr{A}_1[\rho f(\rho); x] \tag{5-7-20}$$

On the other hand, if $F_c(\xi) = \mathscr{F}_c[f(t);\xi]$, we can write

$$\mathscr{H}_0[\xi^{-1}F_c(\xi);\rho] = \sqrt{\frac{2}{\pi}}\int_0^\infty J_0(\xi\rho)\,d\xi \int_0^\infty f(t)\cos(\xi t)\,dt$$
$$= \int_0^\infty f(t)\mathscr{F}_c[J_0(\xi\rho);\xi \to t]\,dt$$
$$= \sqrt{\frac{2}{\pi}}\int_0^\rho \frac{f(t)\,dt}{\sqrt{(\rho^2 - t^2)}}$$

and if $F_s(\xi) = \mathscr{F}_s[f(t);\xi]$, we can write

$$\mathscr{H}_0[\xi^{-1}F_s(\xi);\rho] = \sqrt{\frac{2}{\pi}}\int_0^\infty J_0(\xi\rho)\,d\xi \int_0^\infty f(t)\sin(\xi t)\,dt$$
$$= \int_0^\infty f(t)\mathscr{F}_s[J_0(\xi\rho);\xi \to t]\,dt$$
$$= \sqrt{\frac{2}{\pi}}\int_\rho^\infty \frac{f(t)\,dt}{\sqrt{(t^2 - \rho^2)}}$$

and hence we have the relations

$$\mathscr{H}_0[\xi^{-1}F_c(\xi);\rho] = \mathscr{A}_1[f(t);\rho] \tag{5-7-21}$$
$$\mathscr{H}_0[\xi^{-1}F_s(\xi);\rho] = \mathscr{A}_2[f(t);\rho]. \tag{5-7-22}$$

Since

$$\mathscr{F}_s[f'(t);\xi] = -\xi F_c(\xi)$$

we find from equation (5-7-22) that

$$\mathscr{H}_0[F_c(\xi);\rho] = -\mathscr{A}_2[f'(t);\rho] \tag{5-7-23}$$

Also if $f(0) = 0$, then

$$\mathscr{F}_c[f'(t);\xi] = \xi F_s(\xi)$$

so that equation (5-7-22) yields the formula

$$\mathscr{H}_0[F_s(\xi);\rho] = \mathscr{A}_1[f'(t);\rho]. \tag{5-7-24}$$

Special cases of these formulae are of special interest. If we take $f(t) = g(t)H(a - t)$, $(a > 0)$, in equations (5-7-21) through (5-7-24) we obtain the formulae

$$\mathscr{H}_0[\xi^{-1}\mathscr{F}_c\{g(t)H(a - t);\xi\};\rho] = \sqrt{\frac{2}{\pi}}\int_0^{\min(\rho,a)} \frac{g(t)\,dt}{\sqrt{(\rho^2 - t^2)}} \tag{5-7-25}$$

$$\mathscr{H}_0[\xi^{-1}\mathscr{F}_s\{g(t)\mathrm{H}(a-t);\xi\};\rho] = \sqrt{\frac{2}{\pi}}\,\mathrm{H}(a-\rho)\int_\rho^a \frac{g(t)\,dt}{\sqrt{(t^2-\rho^2)}} \tag{5-7-26}$$

$$\mathscr{H}_0[\mathscr{F}_c\{g(t)\mathrm{H}(a-t);\xi\};\rho] = -\sqrt{\frac{2}{\pi}}\,\mathrm{H}(a-\rho)\int_\rho^a \frac{g'(t)\,dt}{\sqrt{(t^2-\rho^2)}} \tag{5-7-27}$$

$$\mathscr{H}_0[\mathscr{F}_s\{g(t)\mathrm{H}(a-t);\xi\};\rho] = \sqrt{\frac{2}{\pi}}\int_0^{\min(\rho,a)} \frac{g'(t)\,dt}{\sqrt{(\rho^2-t^2)}},\ g(0)=0. \tag{5-7-28}$$

In a similar way we can deduce the relations

$$\mathscr{F}_c[\mathscr{H}_0\{t^{-1}g(t)\mathrm{H}(a-t);\xi\};x] = \sqrt{\frac{2}{\pi}}\,\mathrm{H}(a-x)\int_x^a \frac{g(t)\,dt}{\sqrt{(t^2-x^2)}} \tag{5-7-29}$$

$$\mathscr{F}_s[\mathscr{H}_0\{t^{-1}g(t)\mathrm{H}(a-t);\xi\};x] = \sqrt{\frac{2}{\pi}}\int_0^{\min(x,a)} \frac{g(t)\,dt}{\sqrt{(x^2-t^2)}} \tag{5-7-30}$$

from equations (5-7-19) and (5-7-20) respectively.

5-8 FORMULAE OF BELTRAMI TYPE

In certain physical applications it is desirable to express the function $\mathscr{H}_\nu[\xi^\alpha\bar{\phi}_\mu(\xi);\rho]$ in terms of the original function $\phi(\rho)$. The general case is complicated (see Sneddon, 1965) but it is possible to derive simple expressions for $\mathscr{H}_0[\xi^{\pm 1}\bar{\phi}_0(\xi);\rho]$ and these are the ones which occur most frequently in applications. Since the formula for $\mathscr{H}_0[\xi\bar{\phi}_0(\xi);\rho]$ is equivalent to a result in potential theory originally due to Beltrami (see sec. 5-10 below) we shall call relations of this kind "formulae of Beltrami type."

If

$$F_s(\xi) = \bar{\phi}_0(\xi) \tag{5-8-1}$$

then by equation (5-7-20)

$$f(t) = \mathscr{F}_s\mathscr{H}_0\phi = \mathscr{A}_1[x\phi(x);t]. \tag{5-8-2}$$

Substituting from (5-8-1) and (5-8-2) into equations (5-7-22) we obtain the relation

$$\mathscr{H}_0[\xi^{-1}\bar{\phi}_0(\xi);\rho] = \mathscr{A}_2[\mathscr{A}_1\{x\phi(x);t\};\rho] \tag{5-8-3}$$

which when written out assumes the form

$$\mathscr{H}_0[\xi^{-1}\bar{\phi}_0(\xi);\rho] = \frac{2}{\pi}\int_\rho^\infty \frac{dt}{\sqrt{(t^2-\rho^2)}}\int_0^t \frac{x\phi(x)\,dx}{\sqrt{(t^2-x^2)}}. \tag{5-8-4}$$

If we apply the Hankel inversion theorem to the results

$$\int_0^\infty \cos(\rho\xi)\mathrm{J}_0(x\xi)\,d\xi = (x^2-\rho^2)^{-\frac{1}{2}}\mathrm{H}(x-\rho)$$

$$\int_0^\infty \sin(\rho\xi)\mathrm{J}_0(x\xi)\,d\xi = (\rho^2-x^2)^{-\frac{1}{2}}\mathrm{H}(\rho-x)$$

we obtain the integrals

$$\int_\rho^\infty \frac{x\mathrm{J}_0(\xi x)\,dx}{\sqrt{(x^2-\rho^2)}} = \xi^{-1}\cos(\rho\xi), \qquad \int_0^\rho \frac{x\mathrm{J}_0(x\xi)\,dx}{\sqrt{(\rho^2-x^2)}} = \xi^{-1}\sin(\xi\rho)$$

which may be written as

$$\left.\begin{aligned}\mathscr{A}_2[x\mathrm{J}_0(\xi x);\rho] &= (2/\pi)^{\frac{1}{2}}\xi^{-1}\cos(\rho\xi)\\ \mathscr{A}_1[x\mathrm{J}_0(\xi x);\rho] &= (2/\pi)^{\frac{1}{2}}\xi^{-1}\sin(\rho\xi).\end{aligned}\right\} \tag{5-8-5}$$

Hence

$$\begin{aligned}\mathscr{A}_1\mathscr{A}_2\rho\mathscr{H}_0[\xi\bar{f}_0(\xi);\rho] &= \mathscr{A}_1\mathscr{A}_2\rho\int_0^\infty \xi^2\bar{f}_0(\xi)\mathrm{J}_0(\xi\rho)\,d\xi\\ &= \mathscr{A}_1\mathscr{F}_c[\xi\bar{f}_0(\xi)]\\ &= \sqrt{\frac{2}{\pi}}\int_0^\infty \xi\bar{f}_0(\xi)\mathscr{A}_1[\cos(\xi t);x]\,d\xi\\ &= \int_0^\infty \xi\bar{f}_0(\xi)\mathrm{J}_0(\xi x)\,d\xi\end{aligned}$$

by equation (5-7-7). Hence, using the Hankel inversion theorem we derive the relation

$$\mathscr{A}_1\mathscr{A}_2\rho\mathscr{H}_0[\xi\bar{f}_0(\xi);\rho] = f(x). \tag{5-8-6}$$

Applying the operator $\mathscr{A}_2^{-1}\mathscr{A}_1^{-1}$ to both sides of this equation we find that

$$\mathscr{H}_0[\xi\bar{f}_0(\xi);\rho] = \rho^{-1}\mathscr{A}_2^{-1}\mathscr{A}_1^{-1}f(x)$$

which when written out takes the form

$$\mathscr{H}_0[\xi\bar{f}_0(\xi);\rho] = -\frac{2}{\pi\rho}\frac{d}{d\rho}\int_\rho^\infty \frac{t\,dt}{\sqrt{(t^2-\rho^2)}}\cdot\frac{d}{dt}\int_0^t \frac{xf(x)\,dx}{\sqrt{(t^2-x^2)}}. \tag{5-8-7}$$

5-9 THE MODIFIED OPERATOR OF HANKEL TRANSFORMS

In certain theoretical investigations it is more convenient to use a modified operator of Hankel transforms $S_{\eta,\alpha}$ instead of the operator $\mathscr{H}_\nu$. We define the modified operator by the formula

$$S_{\eta,\alpha}[f(t);x] = 2^\alpha x^{-\alpha}\mathscr{H}_{2\eta+\alpha}[t^{-\alpha}f(t);x], \tag{5-9-1}$$

so that

$$S_{\eta,\alpha}[f(t);x] = 2^\alpha x^{-\alpha}\int_0^\infty t^{1-\alpha}f(t)J_{2\eta+\alpha}(xt)\,dt. \tag{5-9-2}$$

If we write

$$\bar{f}_{\eta,\alpha}(x) = S_{\eta,\alpha}[f(t);x]$$

then

$$\mathscr{H}_{2\eta+\alpha}[t^{-\alpha}f(t);x] = 2^{-\alpha}x^\alpha\bar{f}_{\eta,\alpha}(x)$$

and using the Hankel inversion theorem we find that

$$f(t) = 2^{-\alpha}t^\alpha\mathscr{H}_{2\eta+\alpha}[x^\alpha\bar{f}_{\eta,\alpha}(x);t]$$

an equation which we may write as

$$f(t) = S_{\eta+\alpha,-\alpha}[\bar{f}_{\eta,\alpha}(x);t].$$

Since we may also write

$$f(t) = S_{\eta,\alpha}{}^{-1}[\bar{f}_{\eta,\alpha}(x);t],$$

we see that this last equation may be written in the operator form

$$S_{\eta,\alpha}{}^{-1} = S_{\eta+\alpha,-\alpha}. \tag{5-9-3}$$

5-10 THE USE OF HANKEL TRANSFORMS IN THE SOLUTION OF PARTIAL DIFFERENTIAL EQUATIONS

In this section we shall describe briefly how Hankel transforms may be used to determine the solution of certain partial differential equations which occur in mathematical physics.

We begin with a simple problem in potential theory.

5-10-1 AXISYMMETRIC DIRICHLET PROBLEM FOR A HALF-SPACE

Suppose that we wish to determine the solution of Laplace's equation

$$\Delta_a u(\rho,z) = 0, \tag{5-10-1}$$

in the half-space $z \geqslant 0$ satisfying the boundary condition

$$u(\rho,0) = f(\rho) \tag{5-10-2}$$

and the condition that $u(\rho,z) \to 0$ as $r = \sqrt{(\rho^2 + z^2)} \to \infty$, $z \geqslant 0$.

If we take the Hankel transform of order zero of both sides of (5-10-1) we find that this equation is equivalent to the equation

$$(\mathrm{D}^2 - \xi^2)\bar{u}(\xi,z) = 0 \tag{5-10-3}$$

where $\mathrm{D} = \partial/\partial z$ and

$$\bar{u}(\xi,z) = \mathscr{H}_0[u(\rho,z);\rho \to \xi]. \tag{5-10-4}$$

Also, because of the boundary condition (5-10-2) and the condition at infinity we have the additional conditions

$$\bar{u}(\xi,0) = \bar{f}_0(\xi), \qquad \bar{u}(\xi,z) \to 0 \qquad \text{as } z \to \infty \tag{5-10-5}$$

on the function $\bar{u}(\xi,z)$; in the first of these equations $\bar{f}_0(\xi)$ denotes $\mathscr{H}_0[f(\rho);\xi]$.

The solution of the equation (5-10-3) satisfying the condition (5-10-5) is obviously

$$\bar{u}(\xi,z) = \bar{f}_0(\xi)e^{-\xi z}.$$

Inverting this result by means of the Hankel inversion theorem we obtain the solution

$$u(\rho,z) = \mathscr{H}_0[\bar{f}_0(\xi)e^{-\xi z};\xi \to \rho]. \tag{5-10-6}$$

From this form of the solution we easily deduce that

$$\frac{\partial u(\rho,0)}{\partial z} = -\mathscr{H}_0[\xi\bar{f}_0(\xi);\rho].$$

Making use of equation (5-8-7) we find that

$$\frac{\partial u(\rho,0)}{\partial z} = \frac{2}{\pi\rho}\frac{d}{d\rho}\int_\rho^\infty \frac{t\,dt}{\sqrt{(t^2-\rho^2)}}\cdot\frac{d}{dt}\int_0^t \frac{xf(x)\,dx}{\sqrt{(t^2-x^2)}}. \tag{5-10-7}$$

This relation was originally due to Beltrami. It is for this reason that we called the relations derived in sec. 5-8 "of Beltrami type."

5-10-2 AXISYMMETRIC DIRICHLET PROBLEM FOR A THICK PLATE

We now consider the problem of determining the function $u(\rho,z)$ which is harmonic in the thick plate $|z| \leqslant b$ and satisfies the boundary conditions

$$u(\rho,b) = f(\rho), \qquad u(\rho,-b) = g(\rho). \tag{5-10-8}$$

The appropriate solution of equation (5-10-3) is

$$\bar{u}(\xi,z) = A(\xi)\sinh\{\xi(b+z)\} + B(\xi)\sinh\{\xi(b-z)\}.$$

Putting $z = b$ and using the first of the conditions (5-10-8) we find that

$$A(\xi)\sinh(2\xi b) = \bar{f}_0(\xi) \equiv \mathscr{H}_0[f(\rho);\xi].$$

Similarly, by putting $z = -b$ and using the second of the conditions (5-10-8) we obtain the relation

$$B(\xi)\sinh(2\xi b) = \bar{g}_0(\xi) \equiv \mathscr{H}_0[g(\rho);\xi]$$

so that the appropriate expression for $\bar{u}(\xi,z)$ is

$$\bar{u}(\xi,z) = \operatorname{cosech}(2\xi b)[\bar{f}_0(\xi)\sinh\{\xi(b+z)\} + \bar{g}_0(\xi)\sinh\{\xi(b-z)\}].$$

Applying the Hankel inversion theorem we then obtain the solution

$$u(\rho,z) = \mathscr{H}_0\left[\bar{f}_0(\xi)\frac{\sinh\{\xi(b+z)\}}{\sinh(2\xi b)} + \bar{g}_0(\xi)\frac{\sinh\{\xi(b-z)\}}{\sinh(2\xi b)};\xi \to \rho\right]. \tag{5-10-9}$$

From this solution we easily derive the formulae

$$u(\rho,0) = \mathscr{H}_0[\tfrac{1}{2}\{\bar{f}_0(\xi) + \bar{g}_0(\xi)\}\operatorname{sech}(\xi b);\xi \to \rho] \tag{5-10-10}$$

$$\frac{\partial u(\rho,0)}{\partial z} = \mathscr{H}_0[\tfrac{1}{2}\xi\{\bar{f}_0(\xi) - \bar{g}_0(\xi)\}\operatorname{cosech}(\xi b);\xi \to \rho]. \tag{5-10-11}$$

These expressions take a much simpler form in the symmetrical case in which $f(\rho) \equiv g(\rho)$. In this case

$$\bar{u}(\xi,z) = \bar{f}_0(\xi)\frac{\cosh(\xi z)}{\cosh(\xi b)}$$

so that the relevant solution is

$$u(\rho,z) = \mathscr{H}_0\left[\bar{f}_0(\xi)\frac{\cosh(\xi z)}{\cosh(\xi b)};\xi \to \rho\right] \tag{3-10-12}$$

which is such that

$$u(\rho,0) = \mathscr{H}_0[\bar{f}_0(\xi)\operatorname{sech}(\xi b);\xi \to \rho] \tag{5-10-13}$$

$$\frac{\partial u(\rho,0)}{\partial z} = 0. \tag{5-10-14}$$

5-10-3 THE BIHARMONIC EQUATION

A similar method may be used for the solution of the biharmonic equation

$$\Delta_a\Delta_a u(\rho,z) = 0. \tag{5-10-15}$$

It is easily seen that in this case equation (5-10-3) is replaced by the equation

$$(\mathrm{D}^2 - \xi^2)^2\bar{u}(\xi,z) = 0 \tag{5-10-16}$$

where, as before, $\bar{u}(\xi,z) = \mathscr{H}_0[u(\rho,z);\rho \to \xi]$.

For example, if we consider a function which is biharmonic in the half-space $z \geqslant 0$, and satisfies the conditions

$$u(\rho,0) = f(\rho), \qquad \frac{\partial u(\rho,0)}{\partial z} = 0 \tag{5-10-17}$$

$u(\rho,z) \to 0$ as $r \to \infty$. The corresponding boundary conditions on $\bar{u}(\xi,z)$ are

$$\bar{u}(\xi,0) = \bar{f}_0(\xi), \qquad \frac{\partial \bar{u}(\xi,0)}{\partial z} = 0$$

where $\bar{f}_0(\xi) = \mathscr{H}_0[f(\rho);\xi]$, and $\bar{u}(\xi,z) \to 0$ as $z \to \infty$. The solution of the differential equation (5-10-16) corresponding to these boundary values is easily seen to be

$$\bar{u}(\xi,z) = \bar{f}_0(\xi)(1 + \xi z)e^{-\xi z} \tag{5-10-18}$$

so that the required solution is

$$u(\rho,z) = \mathscr{H}_0[\bar{f}_0(\xi)(1 + \xi z)e^{-\xi z};\xi \to \rho]. \tag{5-10-19}$$

For example, if $f(\rho) = C(a^2 + \rho^2)^{-\frac{1}{2}}$, where C is a constant, $\bar{f}_0(\xi) = C\xi^{-1}e^{-a\xi}$, so that

$$u(\rho,z) = C\mathscr{H}_0[\xi^{-1}(1 + \xi z)e^{-\xi(z+a)};\xi \to \rho].$$

This Hankel transform is easily calculated and we find that

$$u(\rho,z) = C\frac{\rho^2 + (z + a)(2z + a)}{\{\rho^2 + (z + a)^2\}^{\frac{3}{2}}} \tag{5-10-20}$$

Problems involving biharmonic functions usually have much more complicated boundary conditions than (5-10-17) but the same method is applicable in these cases. (Cf., for example, Problems 5-16 through 5-18 below.)

5-10-4 THE SYMMETRICAL VIBRATIONS OF A LARGE MEMBRANE

The first illustration of the use of Hankel transforms in the solution of vibration problems is provided by the calculation of the transverse displacement $z(\rho,t)$ of a large membrane which is deformed symmetrically. The displacement $z(\rho,t)$ is governed by the nonhomogeneous wave equation

$$\frac{\partial^2 z}{\partial \rho^2} + \frac{1}{\rho}\frac{\partial z}{\partial \rho} = \frac{1}{c^2}\frac{\partial^2 z}{\partial t^2} - \frac{p(\rho,t)}{T} \tag{5-10-21}$$

where T is the tension in the membrane, $p(\rho,t)$ is the symmetrical pressure applied (normally) to the membrane and $c^2 = T/\sigma$, where σ is the density per unit area of the membrane.

If we introduce the Hankel transforms

$$\bar{z}(\xi,t) = \mathscr{H}_0[z(\rho,t);\rho \to \xi], \qquad \bar{p}(\xi,t) = \mathscr{H}_0[p(\rho,t);\rho \to \xi]$$

it follows from equation (5-4-8) that

$$\frac{\partial^2 \bar{z}}{\partial t^2} + c^2\xi^2\bar{z} = \frac{\bar{p}(\xi,t)}{\sigma}. \tag{5-10-22}$$

If the initial conditions are

$$z(\rho,0) = f(\rho), \qquad \frac{\partial z(\rho,0)}{\partial t} = g(\rho) \tag{5-10-23}$$

it follows that the initial conditions on $\bar{z}(\xi,t)$ are

$$\bar{z}(\xi,0) = \bar{f}(\xi), \qquad \frac{\partial \bar{z}(\xi,0)}{\partial t} = \bar{g}(\xi)$$

where $\bar{f}(\xi)$ and $\bar{g}(\xi)$ are the Hankel transforms of order zero of $f(\rho)$ and $g(\rho)$ respectively. The solution of the differential equation (5-10-22) satisfying these initial conditions is

$$\bar{z}(\xi,t) = \bar{f}(\xi)\cos(ct\xi) + (c\xi)^{-1}\bar{g}(\xi)\sin(ct\xi) + (c\sigma\xi)^{-1}\int_0^t \bar{p}(\xi,\tau)\sin\{c\xi(t-\tau)\}\,d\tau. \tag{5-10-24}$$

For example, if there is no applied pressure and the initial conditions are

$$z(\rho,0) = \varepsilon(1 + \rho^2/a^2)^{-\frac{1}{2}}, \qquad \frac{\partial z(\rho,0)}{\partial t} = 0$$

we may take $\bar{p}(\xi,t) \equiv 0$, $\bar{g}(\xi) \equiv 0$ and

$$\bar{f}(\xi) = \varepsilon a \xi^{-1} e^{-\xi a}$$

in equation (5-10-24) to obtain the solution

$$\bar{z}(\xi,t) = \varepsilon a \xi^{-1}\cos(ct\xi)e^{-\xi a}$$

so that by Hankel's inversion theorem, we deduce that

$$\begin{aligned} z(\rho,t) &= \varepsilon a \mathscr{H}_0[\xi^{-1}\cos(ct\xi)e^{-\xi a};\xi \to \rho] \\ &= \varepsilon a \operatorname{Re}\int_0^\infty e^{-\xi(a+ict)}J_0(\xi\rho)\,d\xi \\ &= \varepsilon a \operatorname{Re}\{\rho^2 + (a + ict)^2\}^{-\frac{1}{2}} \\ &= \varepsilon a R^{-1}\cos\varphi \end{aligned}$$

where R and φ are defined by the equations

$$R^4 = (a^2 + \rho^2 - c^2t^2)^2 + 4a^2c^2t^2,$$
$$\tan(2\varphi) = 2act/(a^2 + \rho^2 - c^2t^2).$$

5-10-5 THE SYMMETRICAL VIBRATIONS OF A THIN ELASTIC PLATE

As a second example of the application of the theory of Hankel transforms to the solution of vibration problems we consider the symmetrical vibrations of a thin elastic plate.

The symmetrical free vibrations of an elastic plate are described by the equation

$$b^2\left(\frac{\partial^2}{\partial\rho^2}+\frac{1}{\rho}\frac{\partial}{\partial\rho}\right)^2 w(\rho,t)+\frac{\partial^2 w}{\partial t^2}=0 \tag{5-10-25}$$

where b is defined in terms of the flexural rigidity D, the density σ and the thickness $2h$ of the plate by the equation $b^2 = D/(2\sigma h)$. We begin by considering the vibrations of a plate of infinite radius resulting from the initial conditions

$$w(\rho,0)=f(\rho), \qquad \frac{\partial w(\rho,0)}{\partial t}=0. \tag{5-10-26}$$

From equation (5-4-8) it follows immediately that the Hankel transform of order zero

$$\bar{w}(\xi,t)=\mathscr{H}_0[w(\rho,t);\rho\to\xi]$$

of the transverse displacement $w(\rho,t)$ of the central surface of the plate satisfies the equation

$$\frac{\partial^2\bar{w}}{\partial t^2}+b^2\xi^4\bar{w}=0. \tag{5-10-27}$$

Similarly the initial conditions (5-10-26) transform to

$$\bar{w}(\xi,0)=\bar{f}(\xi), \qquad \frac{\partial\bar{w}(\xi,0)}{\partial t}=0,$$

where $\bar{f}(\xi)=\mathscr{H}_0[f(\rho);\xi]$. The solution of (5-10-27) subject to these initial conditions is

$$\bar{w}(\xi,t)=\bar{f}(\xi)\cos(bt\xi^2).$$

Applying the Hankel inversion theorem to this result we see that

$$w(\rho,t)=\int_0^\infty \xi\bar{f}(\xi)J_0(\xi\rho)\cos(bt\xi^2)\,d\xi. \tag{5-10-28}$$

Inserting the definition

$$\int_0^\infty uf(u)J_0(\xi u)\,du$$

for the Hankel transform $\bar{f}(\xi)$ and interchanging the order of the integrations we find that

$$w(\rho,t)=\int_0^\infty uf(u)K(\rho,t;u)\,du \tag{5-10-29}$$

where

$$K(\rho,t;u) = \int_0^\infty \xi J_0(\xi\rho)J_0(\xi u)\cos(bt\xi^2)\,d\xi.$$

Using the result stated in Problem 5-4 we see that

$$K(\rho,t;u) = \frac{1}{2bt} J_0\left(\frac{u\rho}{2bt}\right)\sin\left(\frac{u^2+\rho^2}{4bt}\right) \tag{5-10-30}$$

so that equation (5-10-29) reduces to

$$w(\rho,t) = \frac{1}{2bt}\int_0^\infty uf(u)J_0\left(\frac{u\rho}{2bt}\right)\sin\left(\frac{u^2+\rho^2}{4bt}\right)du. \tag{5-10-31}$$

In any given case it might be more convenient to use equation (5-10-28) than equation (5-10-31). For example, if the initial displacement of the central surface is given by

$$f(\rho) = \varepsilon e^{-\rho^2/a^2}$$

where ε is a (small) constant,

$$\bar{f}(\xi) = \tfrac{1}{2}\varepsilon a^2 e^{-\frac{1}{4}\xi^2a^2}$$

and so from equation (5-10-28) we deduce that

$$w(\rho,t) = \tfrac{1}{2}\varepsilon a^2 \int_0^\infty e^{-\frac{1}{4}\xi^2a^2}\cos(bt\xi^2)J_0(\xi\rho)\,d\xi. \tag{5-10-32}$$

Using the first result stated in Problem 5-2 we see that, for these particular initial conditions

$$w(\rho,t) = \frac{\varepsilon}{1+\tau^2}\exp\left[-\frac{r^2}{1+\tau^2}\right]\left(\cos\frac{r^2\tau}{1+\tau^2} + \tau\sin\frac{r^2\tau}{1+\tau^2}\right)$$

where $r = \rho/a$ and $\tau = 4bt/a^2$.

We can discuss forced vibrations in a similar way. In the present account we shall consider the case in which the intensity of the transverse force applied to the plate is of magnitude $16\sigma hbf(\rho)\psi'(t)$, where $f(\rho)$ and $\psi(t)$ are prescribed functions of ρ and t respectively. The equation of motion then becomes

$$b^2\left(\frac{\partial^2}{\partial\rho^2}+\frac{1}{\rho}\frac{\partial}{\partial\rho}\right)^2 w + \frac{\partial^2 w}{\partial t^2} = 8bf(\rho)\psi'(t) \tag{5-10-33}$$

which may be transformed immediately to

$$\frac{\partial^2\bar{w}}{\partial t^2} + b^2\xi^4\bar{w} = 8b\bar{f}(\xi)\psi'(t)$$

where $\bar{f}(\xi)$ denotes the Hankel transform of order zero of $f(\rho)$. We may take the solution of this equation in the form

$$\bar{w}(\xi,t) = 8b\bar{f}(\xi)\int_{-\infty}^{t} \psi(\tau)\cos\{b(t-\tau)\xi^2\}\,d\tau \tag{5-10-34}$$

so that, by the Hankel inversion theorem, we derive the solution

$$w(\rho,t) = 8b\int_{-\infty}^{t} \psi(\tau)L(t-\tau,\rho)\,d\tau$$

where the kernel $L(t,\rho)$ is defined by the equation

$$\begin{aligned} L(t,\rho) &= \int_0^\infty \xi\bar{f}(\xi)\mathrm{J}_0(\xi\rho)\cos(bt\xi^2)\,d\xi \\ &= \int_0^\infty uf(u)K(\rho,t;u)\,du. \end{aligned}$$

Hence we derive the solution of the prescribed problem in the form

$$w(\rho,t) = 8b\int_{-\infty}^{t} \psi(\tau)\,d\tau \int_0^\infty uf(u)K(\rho,t-\tau;u)\,du \tag{5-10-35}$$

where the kernel $K(\rho,t;u)$ is defined by equation (5-10-30).

In dealing with special problems it is often simpler to make use of the solution in the form (5-10-34) instead of the more elegant result (5-10-35). For example, if the total force $16\sigma hb\psi'(t)$ is distributed uniformly over a circle of radius a, then we may take

$$f(\rho) = \frac{\mathrm{H}(a-\rho)}{\pi a^2}$$

so that

$$\bar{f}(\xi) = \frac{\mathrm{J}_1(a\xi)}{\pi a\xi}$$

and (5-10-34) reduces to

$$w(\rho,t) = \frac{8b}{\pi a}\int_{-\infty}^{t} \psi(\tau)\,d\tau \int_0^\infty \mathrm{J}_0(\xi\rho)\mathrm{J}_1(\xi a)\cos\{b(t-\tau)\xi^2\}\,d\xi. \tag{5-10-36}$$

From this it follows that the transverse displacement of the center of the plate is

$$w(0,t) = \frac{8b}{\pi a}\int_{-\infty}^{t} \psi(\tau)\,d\tau \int_0^\infty \mathrm{J}_1(\xi a)\cos\{b(t-\tau)\xi^2\}\,d\xi.$$

Making use of the third result stated in Problem 5-1 we see that

$$w(0,t) = \frac{8b}{\pi a^2}\int_{-\infty}^{t} \psi(\tau)\left[1 - \cos\frac{a^2}{4b(t-\tau)}\right]d\tau.$$

Changing the variable of integration from τ to v where

$$v = \frac{a^2}{4b(t-\tau)}$$

we obtain the formula

$$w(0,t) = \frac{2}{\pi}\int_0^\infty \psi\left(t - \frac{a^2}{4bv}\right)\frac{1-\cos v}{v^2}\,dv.$$

We obtain the solution for the case in which the force is concentrated at the center by letting $a \to 0$ in (5-10-36). In this way we obtain the solution

$$w(\rho,t) = \frac{4b}{\pi}\int_{-\infty}^{t} \psi(\tau)\,d\tau \int_0^\infty \xi J_0(\xi\rho)\cos\{b(t-\tau)\xi^2\}\,d\xi$$

which may be written in the form

$$w(\rho,t) = \frac{4b}{\pi}\int_{-\infty}^{t} \psi(\tau)K(\rho,t-\tau;0)\,d\tau$$

so that making use of the form (5-10-30) we find that

$$w(\rho,t) = \frac{2}{\pi}\int_{-\infty}^{t} \psi(\tau)\sin\left(\frac{\rho^2}{4b(t-\tau)}\right)\frac{d\tau}{t-\tau}.$$

Changing the variable of integration from τ to ξ where

$$\xi = \frac{\rho^2}{4b(t-\tau)}$$

we see that this equation can be written in the form

$$w(\rho,t) = \frac{2}{\pi}\int_0^\infty \psi\left(t - \frac{\rho^2}{4b\xi}\right)\frac{\sin\xi}{\xi}\,d\xi \qquad (5\text{-}10\text{-}37)$$

which is Boussinesq's solution (see p. 470 of Boussinesq, 1885) for the case of a concentrated force.

5-10-6 MOTION OF A VISCOUS FLUID UNDER A SURFACE LOAD

In the discussion of the plastic recoil of the earth after the disappearance of the Pleistocene ice sheets a boundary value problem in the theory of viscous fluids arises (Cf. Haskell, 1935) which can be solved by means of the theory of Hankel transforms. The curvature of the earth is neglected, and as a model we treat the motion of a semi-infinite, incompressible, viscous fluid under the action of a radially symmetrical pressure applied to the free surface. Since, in the case of the earth, we are dealing with extremely small accelerations and very high viscosity, we may neglect the inertial terms in the equations of motion in comparison with those arising from viscous forces.

With this assumption the equations of axisymmetric motion of a viscous fluid in a gravitational field may be written in the form

$$\left(\mathscr{B}_1 + \frac{\partial^2}{\partial z^2}\right) v_\rho = \frac{1}{\mu}\frac{\partial P}{\partial \rho} \tag{5-10-38}$$

$$\left(\mathscr{B}_0 + \frac{\partial^2}{\partial z^2}\right) v_z = \frac{1}{\mu}\frac{\partial P}{\partial z} \tag{5-10-39}$$

where, in cylindrical coordinates, the fluid velocity has components $[v_\rho(\rho,z), 0, v_z(\rho,z)]$, and P is related to the pressure $p(\rho,z)$ through the equation $P = p(\rho,z) - g\sigma z$, σ denoting the density of the fluid. Similarly the equation of continuity assumes the form

$$\frac{1}{\rho}\frac{\partial}{\partial \rho}(\rho v_\rho) + \frac{\partial v_z}{\partial z} = 0. \tag{5-10-40}$$

The components of stress associated with the z-direction are

$$\sigma_{zz} = -p + 2\mu\frac{\partial v_z}{\partial z} \tag{5-10-41}$$

and

$$\sigma_{\rho z} = \mu\left(\frac{\partial v_\rho}{\partial z} + \frac{\partial v_z}{\partial \rho}\right). \tag{5-10-42}$$

The boundary conditions are that, on the free surface, the shearing stress $\sigma_{\rho z}$ is zero and that the normal component of stress σ_{zz} is equal to the applied pressure. In addition it is assumed that at infinity the components of the velocity vector and of the stress tensor are zero. If we suppose that the equation of the free surface is $z = \zeta(\rho,t)$ and that the equation of the undisturbed free surface is $z = 0$, then we take ζ to be small in comparison with the other distances which enter into the problem, such as, for instance, the radius of the circle to which the load is applied. As in the case of the theory of surface waves on water, we may take, at least in first approximation, the free surface to be $z = 0$. Thus we replace the value of $\partial v_z/\partial z$ at $z = \zeta$ by its value at $z = 0$, and similarly with the other quantities except $g\sigma z = g\sigma\zeta(\rho,t)$. If we denote the applied pressure by $f(\rho,t)$, we have $\sigma_{zz} = -f(\rho,t)$ when $z = \zeta(\rho,t)$, or, by means of equation (5-10-41) with the "surface wave approximation,"

$$P + g\sigma\zeta(\rho,t) - 2\mu\frac{\partial v_z}{\partial z} = f(\rho,t) \qquad \text{on } z = 0. \tag{5-10-43}$$

The relation between ζ and v_z is that, at the free surface, the rate of change of ζ is equal to v_z; hence we have

$$\frac{\partial \zeta}{\partial t} = v_z(\rho,0). \tag{5-10-44}$$

Finally, the condition that the shearing stress on the free surface is zero, becomes, in this approximation,

$$\sigma_{\rho z}(\rho,0) = 0. \tag{5-10-45}$$

If we take the Hankel transform of order unity (with respect to ρ) of equation (5-10-38) we find that it is equivalent to

$$\mu(\mathrm{D}^2 - \xi^2)\bar{v}_\rho = -\xi \bar{P} \tag{5-10-46}$$

where $\mathrm{D} = \partial/\partial z$ and

$$\bar{v}_\rho = \mathscr{H}_1[v_\rho(\rho,z);\rho \to \xi], \qquad \bar{P} = \mathscr{H}_0[P(\rho,z);\rho \to \xi].$$

Similarly, we can show that equations (5-10-39) and (5-10-40) are equivalent to

$$\mu(\mathrm{D}^2 - \xi^2)\bar{v}_z = \mathrm{D}\bar{P} \tag{5-10-47}$$

$$\xi\bar{v}_\rho + \mathrm{D}\bar{v}_z = 0. \tag{5-10-48}$$

with $\bar{v}_z = \mathscr{H}_0[v_z(\rho,z);\rho \to \xi]$. Eliminating $\bar{P}$ and $\bar{v}_\rho$ from the equations (5-10-46) through (5-10-48) we find that $\bar{v}_z$ satisfies the equation

$$(\mathrm{D}^2 - \xi^2)^2\bar{v}_z = 0.$$

We take the solution of this equation in the form

$$\bar{v}_z = \{A(\xi) + B(\xi)\xi z\}e^{-\xi z}. \tag{5-10-49}$$

From (5-10-48) we deduce that

$$\bar{v}_\rho = \{A(\xi) - B(\xi) + B(\xi)\xi z\}e^{-\xi z}.$$

Equations (5-10-42) and (5-10-45) together imply the condition

$$\mathrm{D}\bar{v}_\rho(\xi,0) - \xi\bar{v}_z(\xi,0) = 0$$

and it is easily seen that this is equivalent to the condition $B(\xi) = A(\xi)$. From (5-10-46) we deduce that

$$\bar{P} = 2\mu\xi A(\xi)e^{-\xi z}.$$

In this way we obtain the solutions

$$\left.\begin{aligned} v_\rho &= z\mathscr{H}_1[\xi A(\xi)e^{-\xi z};\xi \to \rho] \\ v_z &= \mathscr{H}_0[A(\xi)(1 + \xi z)e^{-\xi z};\xi \to \rho] \\ P &= 2\mu\mathscr{H}_0[\xi A(\xi)e^{-\xi z};\xi \to \rho] \end{aligned}\right\} \tag{5-10-50}$$

of the equations of motion. It only remains to determine the function $A(\xi)$ in terms of the prescribed surface pressure $f(\rho,t)$. From equation (5-10-43) we deduce the relation

$$2\mu\mathscr{H}_0[\xi\dot{A}(\xi);\rho] + g\sigma\dot{\zeta}(\rho,t) = \dot{f}(\rho,t)$$

the dot denoting differentiation with respect to t, while equation (5-10-44) gives

$$\dot{\zeta}(\rho,t) = \mathscr{H}_0[A(\xi);\rho].$$

Eliminating $\dot{\zeta}(\rho,t)$ from these two equations and applying the Hankel inversion theorem we then obtain the differential equation

$$2\mu\xi\dot{A}(\xi) + g\sigma A(\xi) = S(\xi,t)$$

for the determination of $A(\xi)$ in terms of

$$S(\xi,t) = \mathscr{H}_0[\dot{f}(\rho,t);\rho \to \xi].$$

We therefore obtain the relation

$$A(\xi) = K(\xi)e^{-g\sigma t/2\mu\xi} + \frac{1}{2\mu\xi}\int_0^t S(\xi,\tau)e^{g\sigma(\tau-t)/2\mu\xi}\,d\tau \qquad (5\text{-}10\text{-}51)$$

where the function $K(\xi)$ is determined by the initial conditions. If we substitute the expression for $S(\xi,t)$ into the integral on the right of this equation, change the order of the integrations and use the rule for integrating by parts in the inner integral, we obtain the alternative expression

$$\begin{aligned}A(\xi) = K(\xi)e^{-g\sigma t/2\mu\xi} &+ \frac{\xi}{2\mu}\int_0^\infty uf(u,t)\mathrm{J}_0(\xi u)\,du \\ &- \frac{\xi}{2\mu}e^{-g\sigma t/2\mu\xi}\int_0^\infty uf(u,0)\mathrm{J}_0(\xi u)\,du \\ &- \frac{g\sigma}{4\mu^2}e^{-g\sigma t/2\mu\xi}\int_0^t e^{g\sigma\tau/2\mu\xi}\,d\tau\int_0^\infty uf(u,\tau)\mathrm{J}_0(\xi u)\,du. \qquad (5\text{-}10\text{-}52)\end{aligned}$$

To consider a specific example we suppose that

$$f(\rho,t) = \begin{cases}0 & t \leqslant 0 \\ F\mathrm{H}(a-\rho), & t > 0\end{cases}$$

with the fluid initially at rest. For this loading $K(\xi) = 0$ and

$$A(\xi) = \frac{aF}{2\mu\xi^2}e^{-\sigma g t/2\mu\xi}\mathrm{J}_1(\xi a). \qquad (5\text{-}10\text{-}53)$$

Substituting this form for $A(\xi)$ into equations (5-10-50) we obtain the solution

$$v_\rho = \frac{Faz}{2\mu}\mathscr{H}_1[\xi^{-1}e^{-g\sigma t/2\mu\xi-\xi z}\mathrm{J}_1(\xi a);\xi \to \rho]$$

$$v_z = \frac{Fa}{2\mu}\mathscr{H}_0[\xi^{-2}e^{-g\sigma t/2\mu\xi-\xi z}(1+\xi z)\mathrm{J}_1(\xi a);\xi \to \rho]$$

$$p = g\sigma z + Fa\mathscr{H}_0[\xi^{-1}e^{-g\sigma t/2\mu\xi-\xi z}\mathrm{J}_1(\xi a);\xi \to \rho].$$

It does not appear to be possible to obtain simple closed forms for the integrals occurring in these expressions, but they may readily be put into a dimensionless form suitable for numerical integration.

As a special case we also consider the problem when there is no surface load, i.e., $f(\rho,t) \equiv 0$. In this case

$$A(\xi) = K(\xi)e^{-\sigma gt/2\mu\xi} \tag{5-10-54}$$

where the function $K(\xi)$ has yet to be determined. This function can be determined either from the initial velocity of the surface or the initial configuration of the surface. Substituting from (5-10-54) into the second of the equations (5-10-50) we find that at $t = 0$,

$$v_z(\rho,0) = \mathscr{H}_0[K(\xi);\rho]$$

so that if at $t = 0$, $v_z(\rho,0) = g(\rho)$,

$$K(\xi) = \mathscr{H}_0[g(\rho);\xi]. \tag{5-10-55}$$

On the other hand when the initial displacement is prescribed we make use of the integrated form of (5-10-44) in the form

$$\zeta(\rho,t) = \zeta(\rho,0) + \int_0^t v_z(\rho,0,\tau)\,d\tau$$

to obtain the relation

$$\zeta(\rho,t) = \zeta(\rho,0) + \frac{2\mu}{g\sigma}\mathscr{H}_0[\xi(1 - e^{-g\sigma t/2\mu\xi})K(\xi);\xi \to \rho].$$

The condition $\zeta(\rho,t) \to 0$ as $t \to \infty$ is therefore equivalent to the relation

$$\zeta(\rho,0) = -\frac{2\mu}{g\sigma}\mathscr{H}_0[\xi K(\xi);\rho]. \tag{5-10-56}$$

Hence $K(\xi)$ may be determined by the formula

$$K(\xi) = -\frac{g\sigma}{2\mu}\xi^{-1}\mathscr{H}_0[\zeta(\rho,0);\xi]. \tag{5-10-57}$$

In geophysical applications the function $g(\rho)$ may be represented with sufficient accuracy by the exponential function $g(\rho) = -\beta e^{-b^2\rho^2}$, where b and β are constants. From (5-10-55) we deduce that $K(\xi) = -\beta\mathscr{H}_0[e^{-b^2\rho^2};\xi]$.

Putting $\mu = \nu = 0$ in the first result in Problem 5-1 we find that $K(\xi) = -\beta/(2b^2)e^{-\xi^2/4b^2}$. Inserting this form for $K(\xi)$ into (5-10-56) we find that the surface displacement is given by the equation

$$\zeta(\rho,0) = \frac{\beta\mu}{g\sigma b^2}\mathscr{H}_0[\xi e^{-\xi^2/4b^2};\rho].$$

Putting $\nu = 0$, $\mu = 1$ in the first result in Problem 5-1 we therefore find that

$$\zeta(\rho,0) = \frac{2\sqrt{\pi}b\beta\mu}{g\sigma}\,{}_1F_1(\tfrac{3}{2};1;-b^2\rho^2). \tag{5-10-58}$$

Numerical values of this function can be found in Haskell (1936). Haskell also uses the formula $\zeta(0,0) = 2\sqrt{\pi}b\beta\mu/(g\sigma)$ to estimate the kinematic viscosity of the earth.

5-11 DUAL INTEGRAL EQUATIONS AND MIXED BOUNDARY VALUE PROBLEMS

One of the classic problems of mathematical physics is that of determining the electrostatic potential of a circular disk which is charged to a prescribed potential. If we use cylindrical coordinates (ρ,ϕ,z), we may take the disk to be specified by $0 \leqslant \rho \leqslant a$, $z = 0$. In the axisymmetric case the problem is to determine a function $u(\rho,z)$ which is harmonic in the whole space, which tends to zero as $r = \sqrt{(\rho^2 + z^2)} \to \infty$ and which satisfies the condition

$$u(\rho,0) = f(\rho), \qquad 0 \leqslant \rho \leqslant a. \tag{5-11-1}$$

As the solution of this problem we may take

$$u(\rho,z) = \begin{cases} u_+(\rho,z), & z \geqslant 0 \\ u_-(\rho,z), & z \leqslant 0 \end{cases}$$

where $u_+(\rho,z)$ and $u_-(\rho,z)$ are harmonic in the half-spaces $z \geqslant 0$ and $z \leqslant 0$ respectively. Since the surface charge density on the plane $z = 0$ is given by the formula

$$\sigma(\rho) = -\frac{1}{4\pi}\left[\frac{\partial u_+(\rho,0)}{\partial z} - \frac{\partial u_-(\rho,0)}{\partial z}\right] \tag{5-11-2}$$

and since there is no electric charge on the plane $z = 0$ outside of the disk $0 \leqslant \rho \leqslant a$, we have the condition

$$\frac{\partial u_+(\rho,0)}{\partial z} = \frac{\partial u_-(\rho,0)}{\partial z}, \qquad \rho > a. \tag{5-11-3}$$

On the other hand, the potential function $u(\rho,z)$ must be continuous across the plane $z = 0$ so that we have the condition

$$u_+(\rho,0) = u_-(\rho,0). \tag{5-11-4}$$

From equation (5-10-6) we see that any such pair of harmonic functions $u_+(\rho,z)$, $u_-(\rho,z)$ satisfying the continuity condition (5-11-4) must take the form

$$u_+(\rho,z) = \mathscr{H}_0[\xi^{-1}A(\xi)e^{-\xi z};\xi \to \rho] \tag{5-11-5}$$

$$u_-(\rho,z) = \mathscr{H}_0[\xi^{-1}A(\xi)e^{\xi z};\xi \to \rho]. \tag{5-11-6}$$

Now for these expressions

$$\frac{\partial u_+(\rho,0)}{\partial z} = -\mathscr{H}_0[A(\xi);\rho]$$

$$\frac{\partial u_-(\rho,0)}{\partial z} = \mathscr{H}_0[A(\xi);\rho]$$

so that the boundary conditions (5-11-1), (5-11-4) are equivalent to the relations

$$\mathscr{H}_0[\xi^{-1}A(\xi);\rho] = f(\rho), \qquad 0 \leqslant \rho \leqslant a \tag{5-11-7}$$

$$\mathscr{H}_0[A(\xi);\rho] = 0, \qquad \rho > a \tag{5-11-8}$$

for the unknown function $A(\xi)$.

We also observe that the above expressions for the normal derivatives on the boundary together with equation (5-11-2) leads to the expression

$$\sigma(\rho) = (2\pi)^{-1}\mathscr{H}_0[A(\xi);\rho], \qquad 0 \leqslant \rho \leqslant a, \tag{5-11-9}$$

for the surface charge density on the disk.

Relations of the kind (5-11-7), (5-11-8) i.e.

$$\int_0^\infty A(\xi)\mathrm{J}_0(\xi\rho)d\xi = f(\rho), \qquad 0 \leqslant \rho \leqslant a$$

$$\int_0^\infty \xi A(\xi)\mathrm{J}_0(\xi\rho)d\xi = 0, \qquad \rho > a$$

involving an unknown function under the integral signs with data of differing kinds prescribed on the subintervals $[0,a]$, (a,∞) of the positive real line are called *dual integral equations.*

Dual integral equations can arise in another way. Suppose that we wish to find a function $u(\rho,z)$ which is harmonic in the half-space $z \geqslant 0$ and which satisfies the boundary conditions

$$u(\rho,0) = f(\rho), \qquad 0 \leqslant \rho \leqslant a$$

$$\frac{\partial u(\rho,0)}{\partial z} = 0, \qquad \rho > a$$

then it is obvious that we may take

$$u(\rho,z) = \mathscr{H}_0[\xi^{-1}A(\xi)e^{-\xi z};\xi \to \rho]$$

where, again, $A(\xi)$ satisfies the dual integral equations (5-11-7) and (5-11-8). This boundary value problem for the determination of the function $u(\xi,z)$ is called a *mixed boundary value problem*, because of the fact that the boundary conditions are of mixed type, the function u being prescribed over part of the boundary and its normal derivative over the remaining part.

Mixed boundary value problems arise naturally in several branches of physics (see, e.g., Muskhelishvili, 1953; Noble, 1958; Sneddon, 1966) and one way of solving them consists in reducing the problem to solving a pair of dual integral equations. For that reason a good deal of attention has been given in recent years to the study of dual integral equations. We shall not give an account of that work here (the interested reader should consult Chapter IV of Sneddon, 1966) but content ourselves with showing how some of the simpler problems may be solved by using results derived in sec. 5-7.

For example, from equation (5-7-27) we see that, if $\rho > a$

$$\mathscr{H}_0[\mathscr{F}_c\{g(t)\mathrm{H}(a-t);\xi\};\rho] = 0$$

so that a suitable representation for a function $A(\xi)$ satisfying the condition (5-11-8) is

$$A(\xi) = \sqrt{\frac{2}{\pi}} \int_0^a g(t)\cos(\xi t)\, dt. \tag{5-11-10}$$

For this form of $A(\xi)$ it follows from equation (5-7-25) that

$$\mathscr{H}_0[\xi^{-1}A(\xi);\rho] = \sqrt{\frac{2}{\pi}} \int_0^{\min(\rho,a)} \frac{g(t)\, dt}{\sqrt{(\rho^2 - t^2)}}$$

so that the condition (5-11-7) is satisfied if $g(t)$ is chosen to be the solution of the integral equation

$$\sqrt{\frac{2}{\pi}} \int_0^{\rho} \frac{g(t)\, dt}{\sqrt{(\rho^2 - t^2)}} = f(\rho), \qquad 0 \leqslant \rho \leqslant a$$

that is, if we take

$$g(t) = \sqrt{\frac{2}{\pi}} \frac{d}{dt} \int_0^t \frac{\rho f(\rho)\, d\rho}{\sqrt{(t^2 - \rho^2)}}. \tag{5-11-11}$$

Similar methods exist for the solution of the corresponding problems in two dimensions. Again we shall illustrate the method by discussing a specific problem.

We consider the determination of a function $u(x, y)$ which is harmonic in the half-plane $y \geqslant 0$ and satisfies the mixed boundary conditions

$$\frac{\partial u(x,0)}{\partial y} = -v(x), \qquad |x| < a$$

$$u(x,0) = 0, \qquad |x| > a,$$

where $v(x)$ is an *even* function of x and $u(x, y) \to 0$ as $\rho = \sqrt{(x^2 + y^2)} \to \infty$. This boundary value problem arises in the discussion of the problem of determining the stress field in the neighborhood of a Griffith crack opened up by a symmetrically distributed internal pressure. (See Sneddon and Lowengrub, 1969, p. 72.)

A function which is harmonic in $y \geqslant 0$ and tends to zero as $\rho \to \infty$ may be represented by an expression of the type

$$u(x, y) = \mathscr{F}_c[A(\xi)e^{-\xi y}; \xi \to x]. \tag{5-11-12}$$

For this function

$$\frac{\partial u(x, y)}{\partial y} = -\mathscr{F}_c[\xi A(\xi)e^{-\xi y}; \xi \to x]$$

and this, in turn, can be written in the form

$$\frac{\partial u(x, y)}{\partial y} = -\mathrm{D}_x \mathscr{F}_s[A(\xi)e^{-\xi y}; \xi \to x]$$

where D_x denotes the differential operator $\partial/\partial x$. The boundary values are therefore given by the equations

$$u(x,0) = \mathscr{F}_c[A(\xi); x]$$

$$\frac{\partial u(x,0)}{\partial y} = -\mathrm{D}_x \mathscr{F}_s[A(\xi); x].$$

The unknown function $A(\xi)$ is therefore determined by the dual integral equations

$$\mathscr{F}_s[A(\xi); x] = V(x), \qquad 0 \leqslant x \leqslant a \tag{5-11-13}$$

$$\mathscr{F}_c[A(\xi); x] = 0, \qquad x > a \tag{5-11-14}$$

where

$$V(x) = \int_0^x v(t)\, dt.$$

From equation (5-7-29) we see that if we write

$$A(\xi) = \int_0^a g(t) \mathrm{J}_0(\xi t)\, dt \tag{5-11-15}$$

then equation (5-11-14) is automatically satisfied whatever the form of the function $g(t)$, and from equation (5-7-30) we see that if we are to satisfy equation (5-11-13) we must take $g(t)$ such that

$$\sqrt{\frac{2}{\pi}} \int_0^x \frac{g(t)\,dt}{\sqrt{(x^2 - t^2)}} = V(x).$$

Since $V(0) = 0$, $V'(x) = v(x)$, we may take the solution of this integral equation in the form

$$g(t) = \sqrt{\frac{2}{\pi}}\, t \int_0^t \frac{v(x)\,dx}{\sqrt{(t^2 - x^2)}}. \qquad (5\text{-}11\text{-}16)$$

The solution of the stated mixed boundary value problem is therefore given by equations (5-11-12), (5-11-15), and (5-11-16).

For example, if $v(x) \equiv 1$, then

$$g(t) = \sqrt{(\tfrac{1}{2}\pi)}\,t$$

and

$$A(\xi) = \sqrt{(\tfrac{1}{2}\pi)} \int_0^a t\mathrm{J}_0(\xi t)\,dt = \sqrt{(\tfrac{1}{2}\pi)}\,a\xi^{-1}\mathrm{J}_1(a\xi)$$

so that

$$u(x,y) = a \int_0^\infty \xi^{-1}\mathrm{J}_1(a\xi)\cos(\xi x)e^{-\xi y}\,d\xi. \qquad (5\text{-}11\text{-}17)$$

The evaluation of the integral on the right-hand side of this equation is well known, and numerical tables are available (George, 1962) by means of which the numerical value of the function u can be derived at any point of the half-plane.

5-12 GENERALIZED AXISYMMETRIC POTENTIALS

If we seek solutions of Laplace's equation in n dimensions

$$\Delta_n u \equiv \frac{\partial^2 u}{\partial x_1^2} + \cdots + \frac{\partial^2 u}{\partial x_n^2} = 0, \qquad (n \geqslant 3)$$

which are functions of $\rho = (x_1^2 + x_2^2 + \cdots + x_{n-1}^2)$ and $z = x_n$ alone, then since

$$\frac{\partial u}{\partial x_i} = \frac{x_i}{\rho}\frac{\partial u}{\partial \rho}, \qquad \frac{\partial^2 u}{\partial x_i^2} = \frac{x_i^2}{\rho^2}\frac{\partial^2 u}{\partial \rho^2} + \frac{1}{\rho}\left(1 - \frac{x_i^2}{\rho^2}\right)\frac{\partial u}{\partial \rho},$$
$$(i = 1,2,\ldots,n-1)$$

it follows that $u(\rho,z)$ must satisfy the equation

$$\frac{\partial^2 u}{\partial \rho^2} + \frac{n-2}{\rho}\frac{\partial u}{\partial \rho} + \frac{\partial^2 u}{\partial z^2} = 0. \tag{5-12-1}$$

If we let

$$u(\rho,z) = \rho^{\nu} v(\rho,z) \tag{5-12-2}$$

where

$$\nu = -\tfrac{1}{2}(n-3), \tag{5-12-3}$$

we find that (5-12-1) is equivalent to

$$\left(\mathscr{B}_\nu + \frac{\partial^2}{\partial z^2}\right) v(\rho,z) = 0 \tag{5-12-4}$$

where $\mathscr{B}_\nu$ is the Bessel operator defined by equation (5-4-6). From equation (5-4-8) we see that this equation is equivalent to the ordinary differential equation

$$\left(\frac{\partial^2}{\partial z^2} - \xi^2\right) \bar{v}_\nu(\xi,z) = 0 \tag{5-12-5}$$

for the Hankel transform

$$\bar{v}_\nu(\xi,z) = \mathscr{H}_\nu[v(\rho,z);\rho \to \xi] = \mathscr{H}_\nu[\rho^{-\nu} u(\rho,z);\rho \to \xi]. \tag{5-12-6}$$

Inverting this result by means of the Hankel inversion theorem we find that

$$u(\rho,z) = \rho^{\nu}\mathscr{H}_\nu[\bar{v}_\nu(\xi,z);\xi \to \rho]. \tag{5-12-7}$$

For instance, if we are seeking a form of $u(\rho,z)$ suitable for the solution of problems in the region $z \geqslant 0$ and for which $u(\rho,z) \to 0$ as $z \to \infty$, we may take a solution of (5-12-5) in the form

$$\bar{v}_\nu(\xi,z) = A(\xi)e^{-\xi z}$$

which is equivalent to the solution

$$u(\rho,z) = \rho^{\nu}\mathscr{H}_\nu[A(\xi)e^{-\xi z};\xi \to \rho], \qquad z \geqslant 0 \tag{5-12-8}$$

of equation (5-12-1) with $\nu = -\tfrac{1}{2}(n-3)$.

PROBLEMS

5-1 Prove that

$$\mathscr{H}_\nu[x^{\mu}e^{-px^2};\xi]$$

$$= \frac{\xi^{\nu}\Gamma(1+\tfrac{1}{2}\mu+\tfrac{1}{2}\nu)}{2^{\nu+1}p^{1+\frac{1}{2}\mu+\frac{1}{2}\nu}\Gamma(1+\nu)}\, {}_1F_1(1+\tfrac{1}{2}\mu+\tfrac{1}{2}\nu;\nu+1;-\xi^2/4p).$$

Deduce that

$$\mathscr{H}_1[x^{-1}e^{-px^2};\xi] = \xi^{-1}(1 - e^{-\xi^2/4p})$$

and hence that

$$\mathscr{H}_1[x^{-1}\cos(tx^2);\xi] = \xi^{-1}\{1 - \cos(\xi^2/4t)\}$$
$$\mathscr{H}_1[x^{-1}\sin(tx^2);\xi] = \xi^{-1}\sin(\xi^2/4t).$$

5-2 Show that

$$\mathscr{H}_0[e^{-\frac{1}{4}\xi^2 a^2}\cos(bt\xi^2);\xi \to \rho]$$
$$= \frac{2}{a^2(1+\tau^2)}\exp\left[-\frac{r^2}{1+\tau^2}\right]\left(\cos\frac{r^2\tau}{1+\tau^2} + \tau\sin\frac{r^2\tau}{1+\tau^2}\right)$$
$$\mathscr{H}_0[e^{-\frac{1}{4}\xi^2 a^2}\sin(bt\xi^2);\xi \to \rho]$$
$$= \frac{2}{a^2(1+\tau^2)}\exp\left[-\frac{r^2}{1+\tau^2}\right]\left(\tau\cos\frac{r^2\tau}{1+\tau^2} - \sin\frac{r^2\tau}{1+\tau^2}\right)$$

where $r = \rho/a$, $\tau = 4bt/a^2$.

5-3 Prove *Sonine's first integral*

$$\mathrm{J}_{\mu+\nu+1}(z)$$
$$= \frac{z^{\nu+1}}{2^\nu\Gamma(\nu+1)}\int_0^{\frac{1}{2}\pi} \mathrm{J}_\mu(z\sin\theta)\sin^{\mu+1}\theta\cos^{2\nu+1}\theta\,d\theta \qquad (\mu > -\tfrac{1}{2})$$

and deduce that if $\mu > -\frac{1}{2}$, $\nu > -\frac{1}{2}$, $x > 0$, $u > 0$,

$$\int_0^\infty y^{-\nu}\mathrm{J}_\nu(uy)\mathrm{J}_{\mu+\nu+1}(xy)\,dy = \frac{u^\mu x^{-\mu-\nu-1}}{2^\nu\Gamma(\nu+1)}(x^2 - u^2)^\nu \mathrm{H}(x-u).$$

5-4 Using the formula

$$\mathrm{J}_0(\xi u)\mathrm{J}_0(\xi\rho) = \frac{1}{\pi}\int_0^\pi \mathrm{J}_0(\xi R)\,d\theta$$

where $R^2 = u^2 + \rho^2 - 2u\rho\cos\theta$, show that

$$\mathscr{H}_0[e^{-p\xi^2}\mathrm{J}_0(\xi u);\xi \to \rho] = \frac{1}{2p}\exp\left(-\frac{u^2+\rho^2}{4p}\right)\mathrm{I}_0\left(\frac{u\rho}{2p}\right).$$

Deduce that

$$\mathscr{H}_0[\cos(bt\xi^2)\mathrm{J}_0(\xi u);\xi \to \rho] = \frac{1}{2bt}\mathrm{J}_0\left(\frac{u\rho}{2bt}\right)\sin\left(\frac{u^2+\rho^2}{4bt}\right)$$

$$\mathscr{H}_0[\sin(bt\xi^2)\mathrm{J}_0(\xi u);\xi \to \rho] = \frac{1}{2bt}\mathrm{J}_0\left(\frac{u\rho}{2bt}\right)\cos\left(\frac{u^2+\rho^2}{4bt}\right).$$

5-5 Prove that if $m > -1$, $n > -1$,

$$\int_0^\gamma u^{\frac{1}{2}m} J_m(\alpha\sqrt{u})(\gamma - u)^{\frac{1}{2}n} J_n(\beta\sqrt{\gamma - u})\, du$$

$$= \frac{2\alpha^m \beta^n \gamma^{\frac{1}{2}(m+n+1)}}{(\alpha^2 + \beta^2)^{\frac{1}{2}(m+n+1)}} J_{m+n+1}\{[\gamma(a^2 + \beta^2)]^{\frac{1}{2}}\}.$$

Deduce *Sonine's second integral*

$$\int_0^\infty t^{\mu+1}(t^2 + z^2)^{-\frac{1}{2}\nu} J_\nu\{a\sqrt{(t^2 + z^2)}\} J_\mu(xt)\, dt$$

$$= a^{-\nu} x^\mu z^{-\nu+\mu+1}(a^2 - x^2)^{\frac{1}{2}(\nu-\mu-1)} J_{\nu-\mu-1}\{z\sqrt{(a^2 - x^2)}\} H(a - x),$$

$$(\nu > \mu > -1).$$

Show that this result can be written in the alternative form

$$\int_z^\infty (\tau^2 - z^2)^{\frac{1}{2}\nu - \frac{1}{2}r - \frac{1}{2}} \tau^{-\nu+1} J_{\nu-r-1}\{x\sqrt{(\tau^2 - z^2)}\} J_\nu(a\tau)\, d\tau$$

$$= a^{-\nu} x^{\nu-r-1} z^{-r}(a^2 - x^2)^{\frac{1}{2}r} J_r\{z\sqrt{(a^2 - x^2)}\} H(a - x),$$

$$(\nu > r > -1)$$

and by letting $x \to 0$ prove that

$$\int_z^\infty (\tau^2 - z^2)^{\nu-r-1} \tau^{1-\nu} J_\nu(a\tau)\, d\tau = 2^{\nu-r-1}\Gamma(\nu - r) a^{r-\nu} z^{-r} J_r(az),$$

$$(a > 0, \nu > r > \tfrac{1}{2}\nu - \tfrac{3}{4}).$$

5-6 Defining the Erdélyi–Köber operators of fractional integration $I_{\eta,\alpha}$ and $K_{\eta,\alpha}$ by the equations

$$I_{\eta,\alpha} f(x) = \frac{2x^{-2\alpha-2\eta}}{\Gamma(\alpha)} \int_0^x (x^2 - u^2)^{\alpha-1} u^{2\eta+1} f(u)\, du$$

$$K_{\eta,\alpha} f(x) = \frac{2x^{2\eta}}{\Gamma(\alpha)} \int_x^\infty (u^2 - x^2)^{\alpha-1} u^{-2\alpha-2\eta+1} f(u)\, du$$

with $\alpha > 0$, $\eta > -1$, prove that

(*a*) $I_{\eta+\alpha,\beta} S_{\eta,\alpha} = S_{\eta,\alpha+\beta}$

(*b*) $S_{\eta+\alpha,\beta} S_{\eta,\alpha} = I_{\eta,\alpha+\beta}$

(*c*) $S_{\eta+\alpha,\beta} I_{\eta,\alpha} = S_{\eta,\alpha+\beta}$

(*d*) $K_{\eta,\alpha} S_{\eta+\alpha,\beta} = S_{\eta,\alpha+\beta}$

(*e*) $S_{\eta,\alpha} S_{\eta+\alpha,\beta} = K_{\eta,\alpha+\beta}$

(*f*) $S_{\eta,\alpha} K_{\eta+\alpha,\beta} = S_{\eta,\alpha+\beta}$

where $S_{\eta,\alpha}$ is the modified operator of Hankel transforms defined by equation (5-9-1).

5-7 Deduce from equation (5-5-14) that, if $a > 0$, $b > 0$, $\nu > -1$,

$$\int_0^\infty \frac{x^{2\nu+1}\,dx}{(a^2 + x^2)^{\nu+\frac{1}{2}}(b^2 + x^2)^{\nu+\frac{1}{2}}} = \frac{\pi^{\frac{1}{2}}\Gamma(\nu)}{2\Gamma(\nu + \frac{1}{2})(a + b)^{2\nu}}.$$

5-8 Show that

$$\mathscr{A}_1[\sin(\xi t); t \to x] = (\tfrac{1}{2}\pi)^{\frac{1}{2}}\mathbf{H}_0(\xi x)$$

where $\mathbf{H}_0$ denotes Struve's function of order zero.

5-9 Defining the Bessel function of the second kind of order zero by its integral representation

$$\mathrm{Y}_0(x) = -\frac{2}{\pi}\int_1^\infty \frac{\cos(xt)\,dt}{\sqrt{(t^2 - 1)}}$$

show that

$$\mathscr{A}_2[\cos(\xi t); x] = -(\tfrac{1}{2}\pi)^{\frac{1}{2}}\mathrm{Y}_0(\xi x)$$

and deduce that

$$\mathscr{A}_2^{-1}[\sin(\xi x); t] = (\tfrac{1}{2}\pi)^{\frac{1}{2}}\xi t\,\mathrm{Y}_0(\xi t).$$

5-10 If the operators $\mathscr{H}_0^*$ and $\mathscr{Y}_0$ are defined respectively by equations (4-5-7) and (4-5-8), show that

$$(\mathscr{H}_0^*)^{-1} = \mathscr{Y}_0.$$

5-11 The small transverse oscillations of a heavy chain of uniform line density σ, suspended vertically from one end are determined by the equation

$$\sigma\frac{\partial^2 y}{\partial t^2} = g\sigma\frac{\partial}{\partial x}\left(x\frac{\partial y}{\partial x}\right) + p(x,t)$$

where the origin of the x-coordinate is taken at the position of equilibrium of the *lower* end and the x-axis is taken along the equilibrium position of the chain, pointing vertically *upward*. The function $p(x,t)$ denotes the intensity of the external transverse force.

(*a*) If, in the absence of external force, the chain is initially released from rest in the position $y = f(x)$, show that

$$y(x,t) = \mathscr{H}_0[F(\xi)\cos(\xi t); \xi \to 2(x/g)^{\frac{1}{2}}]$$

where

$$F(\xi) = \mathscr{H}_0[f(\tfrac{1}{4}g\rho^2); \rho \to \xi].$$

In particular, if $f(x) = \varepsilon(1 + x/a)^{-\frac{1}{2}}$, show that

$$y = \varepsilon R^{-\frac{1}{2}}\cos(\tfrac{1}{2}\varphi)$$

where R and φ are defined by the equations

$$R^2 = \left(1 + \frac{x}{a} - \frac{gt^2}{4a}\right)^2 + \frac{gt^2}{a}, \qquad \tan\varphi = \frac{(gt^2/a)^{\frac{1}{2}}}{1 + x/a - (gt^2/4a)}.$$

(*b*) If, initially, $y(x,0) = 0$, $\dfrac{\partial y(x,0)}{\partial t} = 0$, show that

$$y(x,t) = \mathscr{H}_0\left[\xi^{-1}\int_0^t P(\xi,\tau)\sin\{\xi(t-\tau)\}\,d\tau;\xi \to 2(x/g)^{\frac{1}{2}}\right]$$

where

$$\sigma P(\xi,t) = \mathscr{H}_0[p(\tfrac{1}{4}g\rho^2,t);\rho \to \xi].$$

5-12 Show that if we wish to give the center of a thin elastic plate the prescribed motion $w(0,t) = \psi(t)$, that this can be achieved by giving the plate an initial displacement

$$w(\rho,0) = \frac{2}{\pi}\int_0^\infty \frac{\psi(t)}{t}\sin\left(\frac{\rho^2}{4bt}\right)dt.$$

Deduce that if we wish the center of the plate to have the displacement

$$w(0,t) = \begin{cases} V(T-t), & 0 \leqslant t \leqslant T \\ 0, & t > T \end{cases}$$

then we must choose

$$w(\rho,0) = \frac{2VT}{\pi}F\left(\frac{\rho^2}{4bT}\right)$$

where the function $F(x)$ is defined by the equation

$$F(x) = \tfrac{1}{2}\pi - \mathrm{Si}(x) - \sin x + x\mathrm{Ci}(x)$$

with the usual definitions

$$\mathrm{Si}(x) = \int_0^x \frac{\sin u}{u}\,du, \qquad \mathrm{Ci}(x) = -\int_x^\infty \frac{\cos u}{u}\,du.$$

5-13 The vibrations produced in a thin elastic plate by a Gaussian distribution of pressure may be represented by the equation

$$b^2\left(\frac{\partial^2}{\partial\rho^2} + \frac{1}{\rho}\frac{\partial}{\partial\rho}\right)^2 w + \frac{\partial^2 w}{\partial t^2} = \frac{2A}{s^2}e^{-\rho^2/s^2}\psi'(t)$$

where A and s are constants and the function $\psi(t)$ is prescribed.

Show that

$$w(\rho,t) = A\int_{-\infty}^t \psi(\tau)Q(\rho,t-\tau)\,d\tau$$

where

$$Q(\rho,t) = \frac{2p^2 e^{-pq/(p^2+t^2)}}{s^2(p^2+t^2)}\left[\cos\left(\frac{qt}{p^2+t^2}\right) + \frac{t}{p}\sin\left(\frac{qt}{p^2+t^2}\right)\right]$$

with $p = s^2/4b$, $q = \rho^2/4b$.

Deduce that

$$w(0,t) = \frac{\pi A}{b}\int_{-\infty}^{t} \psi'(\tau)R(t-\tau)\,d\tau$$

where

$$R(x) = 1 - \frac{2}{\pi}\tan^{-1}\left(\frac{s}{4bx}\right).$$

5-14 In the axisymmetric flow of a perfect fluid the velocity components $u_\rho(\rho,z)$, $u_z(\rho,z)$ satisfy the first-order equations

$$\frac{1}{\rho}\frac{\partial}{\partial\rho}(\rho u_\rho) + \frac{\partial u_z}{\partial z} = 0, \qquad \frac{\partial u_\rho}{\partial z} - \frac{\partial u_z}{\partial \rho} = 0.$$

If the fluid is introduced to the half-space $z \geqslant 0$ through the circular aperture $\rho \leqslant a$, $z = 0$ in such a way that

$$u_z(\rho,0) = f(\rho)\mathrm{H}(a-\rho)$$

and if u_ρ and u_z both tend to zero as $\sqrt{(\rho^2+z^2)} \to \infty$, show that

$$u_\rho(\rho,z) = \mathscr{H}_1[\bar{f}_0(\xi)e^{-\xi z};\xi\to\rho], \qquad u_z(\rho,z) = \mathscr{H}_0[\bar{f}_0(\xi)e^{-\xi z};\xi\to\rho]$$

where

$$\bar{f}_0(\xi) = \int_0^a \rho f(\rho)\mathrm{J}_0(\xi\rho)\,d\rho.$$

Evaluate $u_\rho(\rho,z)$ and $u_z(\rho,z)$ in the case in which $f(\rho) = c(a^2-\rho^2)^{-\frac{1}{2}}$.

5-15 The flow of a jet of perfect fluid through a circular aperture, of radius a, in a plane rigid screen is determined by a function $\psi(\rho,z)$ which is harmonic in the half-space $z \geqslant 0$ and satisfies the mixed boundary conditions

$$\psi(\rho,0) = C, \qquad 0 \leqslant \rho \leqslant a,$$

$$\frac{\partial\psi(\rho,0)}{\partial z} = 0, \qquad \rho > a,$$

where C is a constant. Show that

$$\psi(\rho,z) = \frac{2C}{\pi}\mathscr{H}_0[\xi^{-2}\sin(\xi a)e^{-\xi z};\xi\to\rho].$$

5-16 Show that the stress-strain relations

$$\sigma_{\rho\rho} = \lambda\Delta + 2\mu\frac{\partial u_\rho}{\partial\rho}, \qquad \sigma_{\phi\phi} = \lambda\Delta + 2\mu\frac{u_\rho}{\rho}, \qquad \sigma_{zz} = \lambda\Delta + 2\mu\frac{\partial u_z}{\partial z}$$

$$\sigma_{\rho z} = \mu\left(\frac{\partial u_z}{\partial\rho} + \frac{\partial u_\rho}{\partial z}\right)$$

in which λ and μ are Lamé's elastic constants and

$$\Delta = \frac{1}{\rho}\frac{\partial}{\partial\rho}(\rho u_\rho) + \frac{\partial u_z}{\partial z}$$

may be written in the alternative form

$$\sigma_{\rho\rho} = \{(\lambda + \mu)\mathscr{H}_0 - \mu\mathscr{H}_2\}[\xi\bar{u}_\rho;\xi \to \rho] + \lambda\mathrm{D}\mathscr{H}_0[\bar{u}_z;\xi \to \rho]$$

$$\sigma_{\phi\phi} = \{(\lambda + \mu)\mathscr{H}_0 + \mu\mathscr{H}_2\}[\xi\bar{u}_\rho;\xi \to \rho] + \lambda\mathrm{D}\mathscr{H}_0[\bar{u}_z;\xi \to \rho]$$

$$\sigma_{zz} = \mathscr{H}_0[\lambda\xi\bar{u}_\rho + (\lambda + 2\mu)\mathrm{D}\bar{u}_z;\xi \to \rho]$$

$$\sigma_{\rho z} = \mathscr{H}_1[\mathrm{D}\bar{u}_\rho - \xi\bar{u}_z;\xi \to \rho]$$

where $\mathrm{D} = \partial/\partial z$ and

$$\bar{u}_\rho(\rho,z) = \mathscr{H}_1[u_\rho(\rho,z);\rho \to \xi], \qquad \bar{u}_z(\rho,z) = \mathscr{H}_0[u_z(\rho,z);\rho \to \xi].$$

It can be shown that the equations of equilibrium are satisfied if we take

$$u_\rho = -\frac{\lambda + \mu}{\mu}\frac{\partial^2\psi}{\partial\rho\partial z}, \qquad u_z = \frac{\lambda + 2\mu}{\mu}\Delta_a\psi - \frac{\lambda + \mu}{\mu}\frac{\partial^2\psi}{\partial z^2}$$

where $\psi(\rho,z)$ is a solution of the biharmonic equation $\Delta_a\Delta_a\psi = 0$. Show that

$$u_\rho(\rho,z) = \frac{\lambda + \mu}{\mu}\mathscr{H}_1[\xi\mathrm{D}\bar{\psi}(\xi,z);\xi \to \rho]$$

$$u_z(\rho,z) = \mathscr{H}_0\left[\left(\mathrm{D}^2 - \frac{\lambda + 2\mu}{\mu}\xi^2\right)\bar{\psi}(\xi,z);\xi \to \rho\right]$$

where

$$(\mathrm{D}^2 - \xi^2)^2\bar{\psi}(\xi,z) = 0$$

and deduce the corresponding expressions for $\sigma_{\rho\rho}$, $\sigma_{\phi\phi}$, σ_{zz}, $\sigma_{\rho z}$.

5-17 Show that the solution of the equations of equilibrium in the half-space $z \geqslant 0$ satisfying the conditions

$$\sigma_{zz} = -p(\rho), \qquad \sigma_{\rho z} = 0$$

on the boundary plane $z = 0$ is given by the expressions of the last problem with

$$\bar{\psi}(\xi, z) = -\frac{\lambda}{2(\lambda + \mu)^2}\xi^{-3}\left(1 + \frac{\lambda + \mu}{\mu}\xi z\right)e^{-\xi z}\mathscr{H}_0[p(\rho);\xi].$$

5-18 The contact problem for a rigid punch indenting the half-space $z \geqslant 0$ can be reduced to solving the equations of equilibrium in $z \geqslant 0$, subject to the mixed boundary conditions (*a*) $\sigma_{\rho z}(\rho, 0) = 0$, $\rho \geqslant 0$; (*b*) $u_z(\rho, 0) = f(\rho)$, $0 \leqslant \rho \leqslant a$; (*c*) $\sigma_{zz}(\rho, 0) = 0$, $\rho > a$. Show that the solution is given by

$$u_\rho(\rho, z) = -\frac{1}{\lambda + 2\mu}\mathscr{H}_1[\{\mu - (\lambda + \mu)\xi z\}\xi^{-1}\psi(\xi)e^{-\xi z};\xi \to \rho]$$

$$u_z(\rho, z) = \frac{1}{\lambda + 2\mu}\mathscr{H}_0[\{\lambda + 2\mu + (\lambda + \mu)\xi z\}\xi^{-1}\psi(\xi)e^{-\xi z};\xi \to \rho]$$

where $\psi(\xi)$ is the solution of the dual integral equations

$$\mathscr{H}_0[\xi^{-1}\psi(\xi);\rho] = f(\rho), \qquad 0 \leqslant \rho \leqslant a$$

$$\mathscr{H}_0[\psi(\xi);\rho] = 0, \qquad \rho \geqslant a.$$

Write down the solution of these equations in the case in which $f(\rho) = \varepsilon$, a constant, and verify that

$$\psi(\rho, 0) = \frac{2\varepsilon}{\pi}\sin^{-1}(a/\rho), \qquad \rho > a.$$

5-19 Show that the dual integral equations

$$\mathscr{H}_0[A(\xi);\rho] = f(\rho), \qquad 0 \leqslant \rho \leqslant a$$

$$\mathscr{H}_0[\xi^{-1}A(\xi);\rho] = 0, \qquad \rho > a$$

have the solution

$$A(\xi) = \int_0^a g(t)\sin(\xi t)\,dt,$$

where

$$g(t) = \frac{2}{\pi}\int_0^t \frac{\rho f(\rho)\,d\rho}{\sqrt{(t^2 - \rho^2)}}.$$

5-20 A function $u(\rho, z)$ is harmonic in the half-space $z \geqslant 0$ and satisfies the mixed boundary conditions

$$\frac{\partial u(\rho, 0)}{\partial z} = -f(\rho), \qquad 0 \leqslant \rho \leqslant a$$

$$u(\rho, 0) = 0, \qquad \rho > a$$

and the limiting condition $u(\rho,z) \to 0$ as $r \to \infty$. Show that

$$u(\rho,z) = \mathscr{H}_0[\xi^{-1}A(\xi)e^{-\xi z};\xi \to \rho]$$

where the function $A(\xi)$ satisfies the dual integral equations of Problem 5-19. Deduce that $u(\rho,z)$ is the imaginary part of

$$\int_0^a \frac{g(t)\,dt}{\sqrt{\{\rho^2 + (z - it)^2\}}}$$

where

$$g(t) = \frac{2}{\pi}\int_0^t \frac{\rho f(\rho)\,d\rho}{\sqrt{(t^2 - \rho^2)}}.$$

5-21 Two functions $f_2(x,y), f_3(\rho,z)$ are related through the equations

$$\mathscr{F}_c[f_2(x,y);x \to \xi] = g(\xi,y)\mathscr{F}_c[p_2(x);\xi]$$

$$\mathscr{H}_0[f_3(\rho,z);\rho \to \xi] = g(\xi,z)\mathscr{H}_0[p_3(\rho);\xi]$$

where

$$p_2(x) = \sqrt{\frac{2}{\pi}}\int_x^\infty \frac{\rho p_3(\rho)\,d\rho}{\sqrt{(\rho^2 - x^2)}}.$$

Show that

$$f_3(\rho,z) = -\frac{2}{\pi}\frac{1}{\rho}\frac{\partial}{\partial\rho}\int_\rho^\infty \frac{x f_2(x,z)\,dx}{\sqrt{(x^2 - \rho^2)}}.$$

If $f_2(x,y)$ and $f_3(\rho,z)$ are related through the equations

$$\mathscr{F}_s[f_2(x,y);x \to \xi] = h(\xi,y)\mathscr{F}_c[p_2(x);\xi]$$

$$\mathscr{H}_1[f_3(\rho,z);\rho \to \xi] = h(\xi,z)\mathscr{H}_0[p_3(\rho);\xi]$$

where $p_2(x)$ and $p_3(\rho)$ are related in the same way as before, show that

$$f_3(\rho,z) = -\sqrt{\frac{2}{\pi}}\,\rho\frac{\partial}{\partial\rho}\int_\rho^\infty \frac{f_2(x,z)\,dx}{x\sqrt{(x^2 - \rho^2)}}.$$

5-22 The function $f_3(\rho,z)$ satisfies the axisymmetric Laplace equation in the half-space $z \geqslant 0$ and the conditions $f_3(\rho,0) = p_3(\rho)$, $f_3(\rho,z) \to 0$ as $r \to \infty$. Show that

$$f_3(\rho,z) = -\sqrt{\frac{2}{\pi}}\frac{1}{\rho}\frac{\partial}{\partial\rho}\int_\rho^\infty \frac{x f_2(x,z)\,dx}{\sqrt{(x^2 - \rho^2)}}$$

where $f_2(x,y)$ is the function which is harmonic in the upper half-plane

$y \geqslant 0$ and satisfies the conditions

$$f_2(x,0) = \sqrt{\frac{2}{\pi}} \int_x^\infty \frac{\rho p_3(\rho)\, d\rho}{\sqrt{(\rho^2 - x^2)}}$$

$f_2(x,y) \to 0$ as $\sqrt{(x^2 + y^2)} \to \infty$.

5-23 The function $f_3(\rho,z)$ satisfies the axisymmetric Laplace equation in the thick plate $0 \leqslant z \leqslant a$ and the boundary conditions

$$f_3(\rho,0) = p_3{}^{(1)}(\rho), \qquad f_3(\rho,a) = p_3{}^{(2)}(\rho).$$

Show that

$$f_3(\rho,z) = -\sqrt{\frac{2}{\pi}}\frac{1}{\rho}\frac{\partial}{\partial\rho}\int_\rho^\infty \frac{x f_2(x,z)\, dx}{\sqrt{(x^2 - \rho^2)}}$$

where $f_2(x,y)$ is the function which is harmonic in the infinite strip $0 \leqslant y \leqslant a$ and which satisfies the boundary conditions

$$f_2(x,0) = \sqrt{\frac{2}{\pi}} \int_x^\infty \frac{\rho p_3{}^{(1)}(\rho)\, d\rho}{\sqrt{(\rho^2 - x^2)}}, \qquad f_2(x,a) = \sqrt{\frac{2}{\pi}} \int_x^\infty \frac{\rho p_3{}^{(2)}(\rho)\, d\rho}{\sqrt{(\rho^2 - x^2)}}.$$

5-24 A function $u(\rho,z)$ is a generalized axisymmetric potential in the half-space $z \geqslant 0$ satisfying $u(\rho,z) \to 0$ as $\sqrt{(\rho^2 + z^2)} \to \infty$. Show that

$$\lim_{\rho \to 0} \rho^{-2\nu} u(\rho,z) = \frac{2z}{B(\nu + 1, \frac{1}{2})} \int_0^\infty \frac{\rho u(\rho,0)\, d\rho}{(\rho^2 + z^2)^{\nu + \frac{3}{2}}}.$$

6
The Kontorovich–Lebedev Transform

There has been considerable interest in recent years in the use of Kontorovich–Lebedev transforms which arise naturally through the use of the method of separation of variables to derive solutions of boundary value problems formulated in terms of a system of cylindrical coordinates. In this chapter we give a brief account of the properties of this transform.

An understanding of the Kontorovich–Lebedev transform rests on a knowledge of the elementary properties of $K_\nu(x)$, the modified Bessel function of the second kind and of $K_{i\tau}(x)$, the Macdonald function; these are outlined in sec. 6-1. The Kontorovich–Lebedev transform is introduced in sec. 6-2 which also contains a *sketch* of Lebedev's proof of the inversion theorem. The proof of the inversion is usually found difficult by students of engineering or physics, and, for that reason, an "operational" proof of the inversion formula is given in sec. 6-3 in the hope that it will at least demonstrate to such readers the plausibility of the formula. The remaining sections of the chapter refer briefly to applications.

6-1 MODIFIED BESSEL FUNCTIONS AND THE MACDONALD FUNCTION

If we define the function $I_\nu(z)$ by the integral formula

$$I_\nu(z) = \frac{1}{\pi}\int_0^\pi e^{z\cos t}\cos(\nu t)\,dt \tag{6-1-1}$$

and the function $K_\nu(z)$ by the formula

$$K_\nu(z) = \int_0^\infty e^{-z\cosh t}\cosh(\nu t)\,dt \tag{6-1-2}$$

then by differentiating under the integral sign twice in each case and making use of the rule for integrating by parts, we can show that each of these functions satisfies the modified Bessel equation

$$\frac{d^2w}{dz^2} + \frac{1}{z}\frac{dw}{dz} - \left(1 + \frac{\nu^2}{z^2}\right)w = 0. \tag{6-1-3}$$

These functions are known as the *modified Bessel functions* of order ν of the first and second kinds respectively. (See Watson, 1944.)

From (6-1-2) we see that, in particular,

$$K_0(z) = \int_0^\infty e^{-z\cosh u}\,du. \tag{6-1-4}$$

The Fourier cosine transform of $K_0(at)$ is easily deduced. From the equations

$$\begin{aligned}\mathscr{F}_c[K_0(at);x] &= \frac{2}{\pi}\int_0^\infty \cos(xt)\,dt\int_0^\infty e^{-at\cosh u}\,du\\ &= \frac{2}{\pi}\int_0^\infty du\int_0^\infty e^{-at\cosh u}\cos(xt)\,dt\\ &= \frac{2}{\pi}\int_0^\infty \frac{a\cosh u\,du}{x^2 + a^2\cosh^2 u}\\ &= \frac{2}{\pi}\int_0^\infty \frac{dv}{x^2 + a^2 + v^2}\end{aligned}$$

we deduce that

$$\mathscr{F}_c[K_0(at);x] = \left[\frac{\pi}{2(a^2 + x^2)}\right]^{\frac{1}{2}} \tag{6-1-5}$$

If, in equation (6-1-2) we replace ν by $i\tau$ where τ is real, we obtain the function

$$K_{i\tau}(x) = \int_0^\infty e^{-x\cosh t}\cos(\tau t)\,dt \tag{6-1-6}$$

which is a solution of the differential equation

$$\frac{d^2y}{dx^2} + \frac{1}{x}\frac{dy}{dx} - \left(1 - \frac{\tau^2}{x^2}\right)y = 0. \tag{6-1-7}$$

The function $K_{i\tau}(x)$ is known as the *Macdonald function.*

It follows immediately from the definition (6-1-6) that in the notation of the Fourier cosine transform

$$K_{i\tau}(x) = (\tfrac{1}{2}\pi)^{\frac{1}{2}}\mathscr{F}_c[e^{-x\cosh t}, t \to \tau]. \tag{6-1-8}$$

Making use of the Fourier cosine inversion theorem we find that

$$\mathscr{F}_c[K_{i\tau}(x); \tau \to t] = (\tfrac{1}{2}\pi)^{\frac{1}{2}} e^{-x\cosh t} \tag{6-1-9}$$

which when written in integral form produces the equation

$$\int_0^\infty K_{i\tau}(x)\cos(t\tau)\,d\tau = \tfrac{1}{2}\pi e^{-x\cosh t}. \tag{6-1-10}$$

Differentiating both sides of this equation with respect to t we find that

$$\int_0^\infty \tau K_{i\tau}(x)\sin(\tau t)\,d\tau = \tfrac{1}{2}\pi x \sinh t e^{-x\cosh t} \tag{6-1-11}$$

and differentiating again we find that

$$\int_0^\infty \tau^2 K_{i\tau}(x)\cos(\tau t)\,d\tau = \tfrac{1}{2}\pi x(\cosh t - x\sinh t\cosh t)e^{-x\cosh t}. \tag{6-1-12}$$

In general we have

$$\int_0^\infty \tau^{2n} K_{i\tau}(x)\cos(\tau t)\,d\tau = \tfrac{1}{2}\pi(-1)^n \mathrm{D}_t^{2n} e^{-x\cosh t} \tag{6-1-13}$$

$(\mathrm{D}_t = d/dt)$.

In particular, if we put $t = 0$ in equations (6-1-10), (6-1-11), and (6-1-12) we obtain the equations

$$\int_0^\infty K_{i\tau}(x)\,dt = \tfrac{1}{2}\pi e^{-x} \tag{6-1-14}$$

$$\int_0^\infty \tau^2 K_{i\tau}(x)\,dt = \tfrac{1}{2}\pi x e^{-x} \tag{6-1-15}$$

$$\int_0^\infty \tau^{2n} K_{i\tau}(x)\,d\tau = \tfrac{1}{2}\pi(-1)^n[\mathrm{D}^{2n} e^{-x\cosh t}]_{t=0} \tag{6-1-16}$$

If we multiply both sides of equation (6-1-10) by $f(t)$ and integrate with respect to τ from 0 to ∞ we obtain the equation

$$\int_0^\infty K_{i\tau}(x)f(\tau)\,d\tau = \sqrt{(\tfrac{1}{2}\pi)}\int_0^\infty e^{-x\cosh t}F_c(t)\,dt \tag{6-1-17}$$

where $F_c(t)$ is the Fourier cosine transform of $f(\tau)$.

For example, if we take

$$f(\tau) = e^{-a\tau}, \qquad F_c(t) = \sqrt{\frac{2}{\pi}} \frac{a}{a^2 + t^2}, \qquad (a > 0)$$

we obtain the equation

$$\int_0^\infty K_{i\tau}(x) e^{-a\tau}\, d\tau = a \int_0^\infty (a^2 + t^2)^{-1} e^{-x\cosh t}\, dt, \qquad (a > 0). \tag{6-1-18}$$

If we interchange this pair of Fourier cosine transforms we obtain the equation

$$\int_0^\infty K_{i\tau}(x) \frac{d\tau}{a^2 + \tau^2} = \frac{\pi}{2a} \int_0^\infty e^{-at - x\cosh t}\, dt, \qquad (a > 0). \tag{6-1-19}$$

Similarly if we take

$$f(\tau) = K_0(a\tau), \qquad F_c(t) = \left\{\frac{\pi}{2(a^2 + t^2)}\right\}^{\frac{1}{2}}, \qquad (a > 0) \tag{6-1-20}$$

we obtain the equation

$$\int_0^\infty K_{i\tau}(x) K_0(a\tau)\, d\tau = \tfrac{1}{2}\pi \int_0^\infty (a^2 + t^2)^{-\frac{1}{2}} e^{-x\cosh t}\, dt \tag{6-1-21}$$

and interchanging this pair of Fourier cosine transforms we find that

$$\int_0^\infty K_{i\tau}(x) \frac{d\tau}{\sqrt{(a^2 + \tau^2)}} = \int_0^\infty K_0(at) e^{-x\cosh t}\, dt. \tag{6-1-22}$$

In a similar way we can deduce from equation (6-1-11) that

$$\int_0^\infty \tau K_{i\tau}(x) f(\tau)\, d\tau = \sqrt{(\tfrac{1}{2}\pi)} \int_0^\infty F_s(t) \sinh t e^{-x\cosh t}\, dt \tag{6-1-23}$$

where $F_s(t)$ is the Fourier sine transform of $f(\tau)$. In terms of the operators $\mathscr{L}$, $\mathscr{F}_s$ of the Laplace transform end of the Fourier sine transform respectively, we can write this last equation in the form

$$\int_0^\infty \tau K_{i\tau}(x) f(\tau)\, d\tau$$
$$= \sqrt{(\tfrac{1}{2}\pi)} \mathscr{L}[H(u - 1) \mathscr{F}_s\{f(\tau); \tau \to \sqrt{(u^2 - 1)}\}; u \to x]. \tag{6-1-24}$$

If we take $t = a + ib$ in equation (6-1-10) and equate real and imaginary parts we find that

$$\int_0^\infty K_{i\tau}(x) \cos(a\tau) \cosh(b\tau)\, d\tau$$
$$= \tfrac{1}{2}\pi \exp(-x \cosh a \cos b) \cos(x \sinh a \sin b) \tag{6-1-25}$$

and that

$$\int_0^\infty K_{i\tau}(x) \sin(a\tau) \sinh(b\tau) \quad d\tau$$
$$= \tfrac{1}{2}\pi \exp(-x \cosh a \cos b) \sin(x \sinh a \sin b) \qquad (6\text{-}1\text{-}26)$$

and by replacing a by $2a$ in these equations we find in turn that

$$\int_0^\infty K_{i\tau}(x) \cosh(at) \cosh(b\tau)\, d\tau$$
$$= \tfrac{1}{2}\pi \exp(-x \cos a \cos b) \cosh(x \sin a \sin b) \qquad (6\text{-}1\text{-}27)$$

and that

$$\int_0^\infty K_{i\tau}(x) \sinh(at) \sinh(b\tau)\, d\tau$$
$$= \tfrac{1}{2}\pi \exp(-x \cos a \cos b) \sinh(x \sin a \sin b). \qquad (6\text{-}1\text{-}28)$$

An interesting result is obtained by substituting

$$f(u) = e^{-\cosh u}, \qquad F_c(\tau) = (2/u)^{\frac{1}{2}} K_{i\tau}(x)$$

$$g(u) = e^{-y\cosh u}, \qquad G_c(\tau) = (2/\pi)^{\frac{1}{2}} K_{i\tau}(y)$$

in the convolution theorem

$$\int_0^\infty F_c(\tau) G_c(\tau) \cos(\xi\tau)\, d\tau = \tfrac{1}{2} \int_0^\infty g(u)\{f(|\xi - u|) + f(\xi + u)\}\, du$$

for Fourier cosine transforms. We find that

$$\frac{2}{\pi} \int_0^\infty K_{i\tau}(x) K_{i\tau}(y) \cos(\xi\tau)\, d\tau$$
$$= \tfrac{1}{2} \int_0^\infty \{e^{-y\cosh u - x\cosh(\xi - u)} + e^{-y\cosh u - x\cosh(\xi + u)}\}\, du$$
$$= \tfrac{1}{2} \int_{-\infty}^\infty e^{-y\cosh u - x\cosh(\xi - u)}\, du$$
$$= \tfrac{1}{2} \int_{-\infty}^\infty e^{-R\cosh(u - \varepsilon)}\, du$$

where

$$R \cosh \varepsilon = y + x \cosh \xi, \qquad R \sinh \varepsilon = x \sinh \xi.$$

Hence

$$\frac{2}{\pi} \int_0^\infty K_{i\tau}(x) K_{i\tau}(y) \cos(\xi\tau)\, d\tau = \tfrac{1}{2} \int_{-\infty}^\infty e^{-R\cosh v}\, dv$$

and so we have the relation

$$\frac{2}{\pi}\int_0^\infty K_{i\tau}(x)K_{i\tau}(y)\cos(\xi\tau)\,d\tau = K_0(R). \tag{6-1-29a}$$

where

$$R^2 = x^2 + 2xy\cosh\xi + y^2. \tag{6-1-29b}$$

If we let $\xi = 0$ we obtain the relation

$$\int_0^\infty K_{i\tau}(x)K_{i\tau}(y)\,d\tau = \tfrac{1}{2}\pi K_0(x+y), \qquad (x > 0, y > 0) \tag{6-1-30}$$

and if we let $\xi = i\alpha$ we find that

$$\int_0^\infty K_{i\tau}(x)K_{i\tau}(y)\cosh(\alpha\tau)\,d\tau = \tfrac{1}{2}\pi K_0\{(x^2 + 2xy\cos\alpha + y^2)^{\frac{1}{2}}\}. \tag{6-1-31}$$

Differentiating both sides of this last equation with respect to α we obtain the equation

$$\int_0^\infty \tau K_{i\tau}(x)K_{i\tau}(y)\sinh(\alpha\tau)\,d\tau = \tfrac{1}{2}\pi xy\sin(x^2 + 2xy\cos\alpha + y^2)^{-\frac{1}{2}} \times \times K_1\{(x^2 + 2xy\cos\alpha + y^2)^{\frac{1}{2}}\}. \tag{6-1-32}$$

On the other hand if we invert equation (6-1-29a) by means of the Fourier cosine inversion theorem we find that

$$K_{i\tau}(x)K_{i\tau}(y) = \int_0^\infty K_0(R)\cos(\xi\tau)\,d\xi \tag{6-1-33}$$

where R is given by equation (6-1-29b). The result (6-1-33) is due to Lebedev (1949).

We can also calculate the Laplace transform of $K_{i\tau}(x)$ directly from equation (6-1-6). We have

$$\mathscr{L}[K_{i\tau}(x);p] = \int_0^\infty \cos(\tau t)\,dt\int_0^\infty e^{-(p+\cosh t)x}\,dx.$$

The x-integration is elementary and we have

$$\mathscr{L}[K_{i\tau}(x);p] = \int_0^\infty \frac{\cos(\tau t)\,dt}{\cosh t + \cosh\alpha}$$

where we have written $p = \cosh\alpha$. The value of this integral is known (Cf. Problem 2-21),

$$\frac{\pi\sin(\alpha\tau)}{\sinh\alpha\cdot\sinh(\pi\tau)}$$

from which it follows that

$$\mathscr{L}[K_{i\tau}(x);\cosh\alpha] = \frac{\pi\sin(\alpha\tau)}{\sinh\alpha\cdot\sinh(\pi\tau)} \tag{6-1-34}$$

This result can be written in the alternative form

$$\mathscr{L}^{-1}\left[\frac{\sin(\tau\cosh^{-1}p)}{\sqrt{(p^2-1)}};x\right] = \pi^{-1}\sinh(\pi\tau)K_{i\tau}(x). \tag{6-1-35}$$

6-2 THE KONTOROVICH–LEBEDEV TRANSFORM

We define the Kontorovich–Lebedev transform $\tilde{f}(\tau)$ of a function $f(x)$, defined on the positive real line, by the formula

$$\mathscr{K}[f(x);\tau] \equiv \tilde{f}(\tau) = \int_0^\infty x^{-1}f(x)K_{i\tau}(x)\,dx \tag{6-2-1}$$

and we write

$$\tilde{f}(\tau) = \mathscr{K}[f(x);\tau] \tag{6-2-2}$$

If f is a function of several variables $x_1, x_2, \ldots, x_n$ we denote the integral

$$\int_0^\infty f(x_1,x_2,\ldots,x_n)K_{i\tau}(x_1)\,dx_1$$

by the symbol

$$\tilde{f}(\tau,x_2,\ldots,x_n) = \mathscr{K}[f(x_1,x_2,\ldots,x_n);x_1\to\tau]. \tag{6-2-3}$$

From equations (6-1-6) and (6-1-10) we deduce immediately that

$$|K_{i\tau}(x)| \leqslant K_0(x) \tag{6-2-4}$$

so that

$$|\tilde{f}(\tau)| \leqslant \int_0^\infty x^{-1}|f(x)|K_0(x)\,dx. \tag{6-2-5}$$

It follows immediately that, provided $f(x)$ is such that the integral on the right-hand side of the inequality (6-2-5) converges, the integral (6-2-1) converges uniformly with respect to τ and therefore defines a continuous function of τ.

In these circumstances, the integral

$$F(T,x) = \frac{2}{\pi^2}\int_0^T \tau\sinh(\pi\tau)K_{i\tau}(x)\tilde{f}(\tau)\,d\tau \tag{6-2-6}$$

has a precise meaning. If we substitute from equation (6-2-1) into equation (6-2-6) and change the order of the integrations we find that

$$F(T,x) = \frac{2}{\pi^2}\int_0^\infty u^{-1}f(u)\,du\int_0^T K_{i\tau}(x)K_{i\tau}(u)\tau\sinh(\pi\tau)\,d\tau.$$

If we make the substitution

$$u = xe^{\theta}$$

we find that

$$F(T,x) = \int_{-\infty}^{\infty} f(xe^{\theta})G(x,\theta,T)\,d\theta \tag{6-2-7}$$

where

$$G(x,\theta,T) = \frac{2}{\pi^2}\int_0^T K_{i\tau}(x)K_{i\tau}(xe^{\theta})\tau\sinh(\pi\tau)\,d\tau. \tag{6-2-8}$$

We now make use of the identity

$$\frac{2}{\pi}K_{i\tau}(x)K_{i\tau}(xe^{\theta})\sinh(\pi\tau) = \int_{|\theta|}^{\infty} \mathrm{J}_0(xe^{\frac{1}{2}\theta}\psi^{\frac{1}{2}})\sin\tau\alpha\,d\alpha$$

with $\psi = 2(\cosh\alpha - \cosh\theta)$ to obtain

$$G(x,\theta,T) = \frac{1}{\pi}\int_0^T \tau\,d\tau \int_{|\theta|}^{\infty} \mathrm{J}_0(xe^{\frac{1}{2}\theta}\psi^{\frac{1}{2}})\sin(\tau\alpha)\,d\alpha.$$

Now

$$\int_0^T \tau\sin(\tau\alpha)\,d\tau = \frac{\sin(\alpha T)}{\alpha^2} - \frac{T\cos(\alpha T)}{\alpha} = -\frac{\partial}{\partial\alpha}\frac{\sin(\alpha T)}{\alpha}$$

so that we may write

$$G(x,\theta,T) = -\frac{1}{\pi}\int_{|\theta|}^{\infty} \mathrm{J}_0(xe^{\frac{1}{2}\theta}\psi^{\frac{1}{2}})\frac{\partial}{\partial\alpha}\left(\frac{\sin\alpha T}{\alpha}\right)d\alpha$$

and hence

$$F(T,x) = -\frac{1}{\pi}\int_{-\infty}^{\infty} f(xe^{\theta})\,d\theta \int_{|\theta|}^{\infty} \mathrm{J}_0(xe^{\frac{1}{2}\theta}\psi^{\frac{1}{2}})\frac{\partial}{\partial\alpha}\left(\frac{\sin\alpha T}{\alpha}\right)d\alpha.$$

It can be shown that this double integral is absolutely convergent so that we may change the order of integration to obtain the equation

$$F(T,x) = -\frac{1}{\pi}\int_0^{\infty} \frac{\partial}{\partial\alpha}\left(\frac{\sin\alpha T}{\alpha}\right)d\alpha \int_{-\alpha}^{\alpha} f(xe^{\theta})\mathrm{J}_0(xe^{\frac{1}{2}\theta}\psi^{\frac{1}{2}})\,d\theta.$$

If we now use the rule for integration by parts to carry out the α-integration we find that we can write this equation in the alternative form

$$F(T,x) = \frac{1}{\pi}\int_0^{\infty} w(\alpha)\frac{\sin\alpha T}{\alpha}\,dx \tag{6-2-9}$$

with

$$w(\alpha) = f(xe^{\alpha})e^{\alpha} + f(xe^{-\alpha})e^{-\alpha} + \int_{-\alpha}^{\alpha} f(xe^{\theta})e^{\theta} \frac{\partial}{\partial x} J_0(xe^{-\frac{1}{2}\theta}\psi^{\frac{1}{2}})\, d\theta. \tag{6-2-10}$$

It now follows from the use of Dirichlet's integral that

$$\lim_{T\to\infty} F(T,x) = \tfrac{1}{2}[f(x+) + f(x-)]$$

and hence we have the inversion theorem (Lebedev, 1946) in the form:

Theorem *If $f(x)$ is a function defined on the positive real line such that $x^{-1}f(x)$ is continuously differentiable and $xf(x)$ and $x\dfrac{d}{dx}\{x^{-1}f(x)\}$ are absolutely integrable over the positive real line, then*

$$f(x) = \frac{2}{\pi^2}\int_0^{\infty} K_{i\tau}(x)\tau \sinh(\pi\tau)\tilde{f}(\tau)\, d\tau \tag{6-2-11}$$

where

$$\tilde{f}(\tau) = \mathscr{K}[f(x);\tau]. \tag{6-2-12}$$

6-3 AN OPERATIONAL PROOF OF THE INVERSION FORMULA

Even if, as we have done, we omit the details of the analysis, the proof of the inversion formula for the Kontorovich–Lebedev transform is complicated. The following argument, though lacking the rigor of a formal proof, may be an acceptable demonstration of the validity for formulae (6-2-13) and (6-2-14) to readers whose interest is in applications.

We may write equation (6-1-6) in the alternative form

$$\mathscr{F}_c[K_{i\tau}(x);\tau \to t] = \sqrt{(\tfrac{1}{2}\pi)}\, e^{-x\cosh t}$$

from which it follows that

$$\mathscr{F}_c[\tilde{f}(\tau);t] = \sqrt{(\tfrac{1}{2}\pi)}\int_0^{\infty} e^{-x\cosh t}\, x^{-1}f(x)\, dx$$

a result which may be written in the form

$$\mathscr{L}[x^{-1}f(x);p] = \frac{2}{\pi}\int_0^{\infty} \tilde{f}(\tau)\cos(\tau\cosh^{-1}p)\, d\tau \tag{6-3-1}$$

Using the well-known result

$$\mathscr{L}[f(x);p] = -\frac{\partial}{\partial p}\mathscr{L}[x^{-1}f(x);p]$$

we find that

$$\mathscr{L}[f(x);p] = \frac{2}{\pi}\int_0^\infty \tau\tilde{f}(\tau)\frac{\sin(\tau\cosh^{-1}p)}{\sqrt{(p^2-1)}}\,d\tau$$

Applying the operator $\mathscr{L}^{-1}$ to both sides of this equation and making use of equation (6-1-35) we obtain the inversion formula

$$f(x) = \frac{2}{\pi^2}\int_0^\infty \tau\sinh(\pi\tau)K_{i\tau}(x)\tilde{f}(\tau)\,d\tau. \tag{6-3-2}$$

6-4 PARSEVAL RELATION FOR KONTOROVICH–LEBEDEV TRANSFORMS

Suppose that

$$\tilde{f}(\tau) = \mathscr{K}[f(x);\tau], \qquad \tilde{g}(\tau) = \mathscr{K}[g(x);\tau].$$

Then

$$\begin{aligned}\frac{2}{\pi^2}\int_0^\infty &\tau\sinh(\pi\tau)\tilde{f}(\tau)\tilde{g}(\tau)\,d\tau \\ &= \frac{2}{\pi^2}\int_0^\infty \tau\sinh(\pi\tau)\tilde{f}(\tau)\,d\tau\int_0^\infty x^{-1}g(x)K_{i\tau}(x)\,dx \\ &= \int_0^\infty x^{-1}g(x)\,dx\,\frac{2}{\pi^2}\int_0^\infty \tau\sinh(\tau\pi)\tilde{f}(\tau)K_{i\tau}(x)\,d\tau \\ &= \int_0^\infty x^{-1}f(x)g(x)\,dx\end{aligned}$$

so that we have the Parseval relation

$$\int_0^\infty x^{-1}f(x)g(x)\,dx = \frac{2}{\pi^2}\int_0^\infty \tau\sinh(\pi\tau)\tilde{f}(\tau)\tilde{g}(\tau)\,d\tau. \tag{6-4-1}$$

In particular, if we take

$$g(x) = x\sin\alpha\, e^{-x\cos\alpha}, \qquad \tilde{g}(t) = \frac{\sinh(\alpha\tau)}{\sinh(\pi\tau)}$$

we find that

$$\frac{2}{\pi}\int_0^\infty \sinh(\alpha\tau)\tilde{f}(\tau)\,d\tau = \sin\alpha\int_0^\infty e^{-x\cos\alpha}f(x)\,dx \tag{6-4-2}$$

and letting $\alpha \to 0$ we find that

$$\frac{2}{\pi}\int_0^\infty \tau\tilde{f}(\tau)\,d\tau = \int_0^\infty e^{-x}f(x)\,dx. \tag{6-4-3}$$

If we put

$$f(x) = e^{-x\cosh t}, \quad \tilde{f}(\tau) = \frac{\pi \cos(\tau t)}{\tau \sinh(\pi\tau)}$$

we find that

$$2\int_0^\infty \frac{\cos(\tau t)\, d\tau}{\sinh(\pi\tau)} = \int_0^\infty e^{-x(1+\cosh t)}\, dx$$

showing that

$$\mathscr{F}_c[\operatorname{cosech}(\pi\tau); \tau \to t] = \frac{1}{2\sqrt{(2\pi)}} \operatorname{sech}^2(\tfrac{1}{2}t). \tag{6-4-4}$$

6-5 FUNCTIONS WHICH ARE HARMONIC IN AN INFINITE WEDGE

If we seek solutions of Laplace's equation

$$\frac{\partial^2 u}{\partial\rho^2} + \frac{1}{\rho}\frac{\partial u}{\partial\rho} + \frac{1}{\rho^2}\frac{\partial^2 u}{\partial\phi^2} + \frac{\partial^2 u}{\partial z^2} = 0 \tag{6-5-1}$$

in the form

$$u(\rho,\phi,z) = e^{\pm\phi\tau \pm i\sigma z} R(\rho),$$

then $R(\rho)$ satisfies the equation

$$\frac{d^2 R}{d\rho^2} + \frac{1}{\rho}\frac{dR}{d\rho} + \left(\frac{\tau^2}{\rho^2} - \sigma^2\right) R = 0.$$

One solution of this equation is $K_{i\tau}(\rho|\sigma|)$ so that we have harmonic functions of the form

$$u(\rho,\phi,z) = e^{\pm\phi\tau \pm i\sigma z} K_{i\tau}(\rho|\sigma|). \tag{6-5-2}$$

By the principle of linear superposition we can construct more sophisticated solutions such as

$$u(\rho,\phi,z) = \sqrt{\frac{2}{\pi^5}} \int_{-\infty}^\infty e^{-i\sigma z}\, d\sigma \int_0^\infty \tau U(\sigma,\tau) \begin{matrix}\cosh\\ \sinh\end{matrix} (\phi\tau) K_{i\tau}(\rho|\sigma|) \sinh(\pi\tau)\, d\tau \tag{6-5-3}$$

or

$$u(\rho,\phi,z) = \sqrt{\frac{8}{\pi^5}} \int_0^\infty \cos(\sigma z)\, d\sigma \int_0^\infty \tau U(\sigma,\tau) \begin{matrix}\cosh\\ \sinh\end{matrix} (\phi\tau) K_{i\tau}(\rho\sigma) \sinh(\pi\tau)\, d\tau \tag{6-5-4}$$

or

$$u(\rho,\phi,z) = \sqrt{\frac{8}{\pi^5}} \int_0^\infty \sin(\sigma z)\, d\sigma \int_0^\infty \tau U(\sigma,\tau) \frac{\cosh}{\sinh}(\phi\tau) K_{i\tau}(\rho\sigma)\, d\tau \sinh(\pi\tau)\, d\tau \tag{6-5-5}$$

We can rewrite these equations in the forms

$$u(\rho,\phi,z) = \mathscr{F}^{-1}\left[\mathscr{K}^{-1}\left\{U(\sigma,\tau)\frac{\cosh}{\sinh}(\phi\tau);\tau \to \rho|\sigma|\right\};\sigma \to z\right] \tag{6-5-6}$$

$$u(\rho,\phi,z) = \mathscr{F}_c\left[\mathscr{K}^{-1}\left\{U(\sigma,\tau)\frac{\cosh}{\sinh}(\phi\tau);\tau \to \rho\sigma\right\};\sigma \to z\right] \tag{6-5-7}$$

$$u(\rho,\phi,z) = \mathscr{F}_s\left[\mathscr{K}^{-1}\left\{U(\sigma,\tau)\frac{\cosh}{\sinh}(\phi\tau);\tau \to \rho\sigma\right\};\sigma \to z\right] \tag{6-5-8}$$

Corresponding to the solutions (6-5-7) and (6-5-8) we have the formulae

$$u_z(\rho,\phi,z) = -\mathscr{F}_s\left[\mathscr{K}^{-1}\left\{\sigma U(\sigma,\tau)\frac{\cosh}{\sinh}(\phi\tau);\tau \to \rho\right\};\sigma \to z\right] \tag{6-5-9}$$

$$u_z(\rho,\phi,z) = \mathscr{F}_c\left[\mathscr{K}^{-1}\left\{\sigma U(\sigma,\tau)\frac{\cosh}{\sinh}(\phi\tau);\tau \to \rho\right\};\sigma \to z\right] \tag{6-5-10}$$

so that the solution (6-5-8) has the property that

$$u(\rho,\phi,0) = 0$$

and the solution (6-5-6) has the property that

$$u_z(\rho,\phi,0) = 0.$$

As an example of the use of these formulae suppose that we wish to find the function $u(\rho,\phi,z)$ which is harmonic in the infinite wedge $0 \leqslant \rho < \infty$, $|\phi| \leqslant \alpha$, $-\infty < z < \infty$ and which satisfies the boundary conditions

$$u(\rho,-\alpha,z) = f_1(\rho,z), \qquad u(\rho,\alpha,z) = f_2(\rho,z) \tag{6-5-11}$$

From the basic solution (6-5-6) we construct the solution

$$u(\rho,\phi,z) = \mathscr{F}^{-1}\left[\mathscr{K}\left\{F_1(\sigma\tau)\frac{\sinh(\alpha\tau - \phi\tau)}{\sinh(2\alpha\tau)} + F_2(\sigma,\tau)\frac{\sinh(\alpha\tau + \phi\tau)}{\sinh(2\alpha\tau)};\tau \to \rho|\sigma|\right\};\sigma \to z\right]. \tag{6-5-12}$$

Where the functions $F_1(\sigma,\tau)$, $F_2(\sigma,\tau)$ are chosen to be such that

$$f_1(\rho,z) = \mathscr{F}^{-1}[\mathscr{K}\{F_1(\sigma,\tau);\tau \to \rho|\sigma|\};\sigma \to z]$$
$$f_2(\rho,z) = \mathscr{F}^{-1}[\mathscr{K}\{F_2(\sigma,\tau);\tau \to \rho|\sigma|\};\sigma \to z].$$

From the first of these equations we have

$$\mathscr{F}[f_1(\rho,z);z\to\sigma]=\mathscr{K}^{-1}\{F_1(\sigma,\tau);\tau\to\rho|\sigma|\}$$

from which we deduce that

$$F_1(\sigma,\tau)=\mathscr{K}\left[\mathscr{F}\left\{f_1\left(\frac{\rho}{|\sigma|},z\right);z\to\sigma\right\};\rho\to\tau\right] \tag{6-5-13}$$

Writing this out in full we see that

$$F_1(\sigma,\tau)=\frac{1}{\sqrt{(2\pi)}}\int_0^\infty \rho^{-1}K_{i\tau}(\rho|\sigma|)\,d\rho\int_{-\infty}^\infty f_1(\rho,z)e^{i\sigma z}\,dz \tag{6-5-14}$$

In an exactly similar way we could show that

$$F_2(\sigma,\tau)=\frac{1}{\sqrt{(2\pi)}}\int_0^\infty \rho^{-1}K_{i\tau}(\rho|\sigma|)\,d\rho\int_{-\infty}^\infty f_2(\rho,z)e^{i\sigma z}\,dz. \tag{6-5-15}$$

The solution of the boundary value problem (6-5-11) is therefore given by equation (6-5-12) with the functions $F_1(\sigma,\tau)$, $F_2(\sigma,\tau)$ given by the equations (6-5-14) and (6-5-15) respectively.

In particular, if

$$u(\rho,\pm\alpha,z)=f(\rho,z) \tag{6-5-16}$$

we find that

$$u(\rho,\phi,z)=\mathscr{F}^{-1}\left[\mathscr{K}^{-1}\left\{F(\sigma,\tau)\frac{\cosh(\phi\tau)}{\cosh(\alpha\tau)};\tau\to\rho|\sigma|\right\},\sigma\to z\right] \tag{6-5-17}$$

where

$$F(\sigma,\tau)=\frac{1}{\sqrt{(2\pi)}}\int_0^\infty \rho^{-1}K_{i\tau}(\rho|\sigma|)\,dp\int_{-\infty}^\infty f(\rho,z)e^{i\sigma z}\,dz \tag{6-5-18}$$

Similarly if

$$u(\rho,\pm\alpha,z)=\pm f(\rho,z) \tag{6-5-19}$$

we find that

$$u(\rho,\phi,z)=\mathscr{F}^{-1}\left[\mathscr{K}^{-1}\left\{F(\sigma,\tau)\frac{\sinh(\phi\tau)}{\sinh(\alpha\tau)};\tau\to\rho|\sigma|\right\};\sigma\to z\right] \tag{6-5-20}$$

where $F(\sigma,\tau)$ is again given by equation (6-5-18).

If $u(\rho,\phi,z)$ is harmonic in the same infinite wedge and satisfies the boundary conditions

$$\left(\frac{\partial u}{\partial\phi}\right)_{\phi=-\alpha}=f_1(\rho,z),\quad \left(\frac{\partial u}{\partial\phi}\right)_{\phi=\alpha}=f_2(\rho,z) \tag{6-5-21}$$

then it is given by the expression

$$u(\rho,\phi,z) = \mathscr{F}^{-1}\left[\mathscr{K}^{-1}\left\{\tau^{-1}F_2(\sigma,\tau)\frac{\cosh(\alpha\tau + \phi\tau)}{\sinh(2\alpha\tau)}\right.\right.$$

$$\left.\left. - \tau^{-1}F_1(\sigma,\tau)\frac{\cosh(\alpha\tau - \phi\tau)}{\sinh(2\alpha\tau)};\tau \to \sigma|\sigma|\right\};\sigma \to z\right] \tag{6-5-22}$$

where $F_1(\sigma,\tau)$ and $F_2(\sigma,\tau)$ are defined by equations (6-5-14) and (6-5-15) respectively.

The solution of the symmetric case

$$\left(\frac{\partial u}{\partial \phi}\right)_{\phi=\pm\alpha} = f(\rho,z) \tag{6-5-23}$$

is given by

$$u(\rho,\phi,z) = \mathscr{F}^{-1}\left[\mathscr{K}^{-1}\left\{\tau^{-1}F(\sigma,\tau)\frac{\sinh(\phi\tau)}{\cosh(\alpha\tau)};\tau \to \rho|\sigma|\right\};\sigma \to z\right] \tag{6-5-24}$$

where $F(\sigma,\tau)$ is defined by equation (6-5-18) and that of the antisymmetric case

$$\left(\frac{\partial u}{\partial \phi}\right)_{\phi=\pm\alpha} = \pm f(\rho,z) \tag{6-5-25}$$

is given by

$$u(\rho,\phi,z) = \mathscr{F}^{-1}\left[\mathscr{K}^{-1}\left\{\tau^{-1}F(\sigma,\tau)\frac{\cosh(\phi\tau)}{\sinh(\alpha\tau)};\tau \to \rho|\sigma|\right\};\sigma \to z\right] \tag{6-5-26}$$

$F(\sigma,\tau)$ being defined by the same equation.

In the case in which the function $u(\rho,\phi,z)$ is harmonic in the infinite wedge $|\phi| \leqslant \alpha$ and satisfies the boundary conditions

$$\left(\frac{\partial u}{\partial \phi}\right)_{\phi=-\alpha} = f_1(\rho,z), \qquad u(\rho,\alpha,z) = f_2(\rho,z) \tag{6-5-27}$$

we see that the solution is

$$u(\rho,\phi,z) = \mathscr{F}^{-1}\left[\mathscr{K}^{-1}\left\{F_2(\sigma,\tau)\frac{\cosh(\alpha\tau + \phi\tau)}{\cosh 2\alpha}\right.\right.$$

$$\left.\left. - \tau^{-1}F_1(\sigma,\tau)\frac{\sinh(\alpha\tau - \phi\tau)}{\cosh 2\alpha};\tau \to \rho|\sigma|\right\};\sigma \to \mathbf{z}\right] \tag{6-5-28}$$

where the functions $F_1(\sigma,\tau)$ and $F_2(\sigma,\tau)$ are defined respectively by equations (6-5-14) and (6-5-15) respectively.

6-6 FUNCTIONS WHICH ARE HARMONIC IN A WEDGE OF FINITE THICKNESS

To illustrate the use of the Kontorovich–Lebedev transform in a wedge of finite thickness we consider the problem of determining a function $u(\rho,\phi,z)$ which is harmonic in the wedge

$$\rho \geqslant 0, \qquad 0 \leqslant \phi \leqslant \alpha, \qquad 0 \leqslant z \leqslant a$$

and satisfies the condition

$$u(p,\alpha,z) = f(\rho,z). \tag{6-6-1}$$

while on the other boundaries it vanishes identically.

From equation (6-5-2) we construct solutions of the type

$$u(\rho,\phi,z) = \frac{2}{\pi^2}\sum_{n=1}^{\infty} \sin\frac{n\pi z}{a}\int_0^{\infty} \tau F_n(\tau)\frac{\sinh(\tau\phi)}{\sinh(\tau\alpha)}K_{i\tau}\left(\frac{n\pi\rho}{a}\right)\sinh\pi\tau\, d\tau \tag{6-6-2}$$

which obviously vanishes on the boundaries $z = 0$, $z = a$ and $\phi = 0$. It will provide a solution of the stated boundary value problem provided that we can find functions $F_n(\tau)$ such that

$$f(\rho,z) = \sum_{n=1}^{\infty} \sin\left(\frac{n\pi z}{a}\right)\mathscr{K}^{-1}\left[F_n(\tau);\frac{n\pi\rho}{a}\right]$$

i.e., such that

$$\mathscr{K}^{-1}\left[F_n(\tau);\frac{n\pi\rho}{a}\right] = \frac{2}{a}\int_0^a f(\rho,y)\sin\left(\frac{n\pi y}{a}\right)dy.$$

Replacing ρ by $ar/n\pi$ we see that this condition is equivalent to

$$\mathscr{K}^{-1}[F_n(\tau);r] = \frac{2}{a}\int_0^a f\left(\frac{ar}{n\pi},y\right)\sin\frac{n\pi y}{a}\,dy$$

and hence to

$$F_n(\tau) = \frac{2}{a}\int_0^a \sin\left(\frac{n\pi y}{a}\right)\mathscr{K}\left[f\left(\frac{ar}{n\pi},y\right);r\to\tau\right]dy.$$

Now

$$\mathscr{K}\left[f\left(\frac{ar}{n\pi},y\right);r\to\tau\right] = \int_0^{\infty} r^{-1}f\left(\frac{ar}{n\pi},y\right)K_{i\tau}(r)\,dr$$

and a trivial change of the variable of integration shows that this integral can be written in the form

$$\int_0^{\infty} x^{-1}f(x,y)K_{i\tau}\left(\frac{n\pi x}{a}\right)dx.$$

Hence we have the formula

$$F_n(\tau) = \frac{2}{a}\int_0^a \sin\left(\frac{n\pi y}{a}\right) dy \int_0^\infty x^{-1} f(x,y) K_{i\tau}\left(\frac{n\pi x}{a}\right) dx \tag{6-6-3}$$

by means of which we may calculate the coefficients of the series (6-6-2).

PROBLEMS

6-1 Show that

$$\mathscr{K}[f(x);\tau] = \sqrt{\frac{\pi}{2}}\,\mathscr{F}_c[\mathscr{L}\{x^{-1}f(x);\cosh u\};u \to \tau].$$

6-2 Prove that

$$(a)\ \mathscr{K}[xe^{-x\cos t};x \to \tau] = \frac{\pi \sinh(t\tau)}{\sinh(\pi\tau)\sin t}$$

and deduce that

$$(b)\ \mathscr{K}[x;\tau] = \tfrac{1}{2}\pi \operatorname{sech}(\tfrac{1}{2}\pi\tau)$$

$$(c)\ \mathscr{K}[xe^{-x};\tau] = \pi\tau \operatorname{cosech}(\pi\tau).$$

6-3 Prove that

$$(a)\ \mathscr{K}[e^{-x\cos t};x \to \tau] = \frac{\pi \cosh(t\tau)}{\tau \sinh(\pi\tau)}$$

and deduce that

$$(b)\ \mathscr{K}[1;\tau] = \tfrac{1}{2}\pi\tau^{-1} \operatorname{cosech}(\tfrac{1}{2}\pi\tau)$$

$$(c)\ \mathscr{K}[e^{-x};\tau] = \pi\tau^{-1} \operatorname{cosech}(\pi\tau)$$

$$(d)\ \mathscr{K}[e^{x};\tau] = \pi\tau^{-1} \coth(\pi\tau).$$

6-4 Prove that

$$(a)\ \mathscr{K}[e^{-x\cosh t};x \to \tau] = \frac{\pi \cos(t\tau)}{\tau \sinh(\pi\tau)}$$

$$(b)\ \mathscr{K}[xe^{-x\cosh t};x \to \tau] = \frac{\pi \sin(t\tau)}{\sinh(\pi\tau)\sinh t}.$$

7
The Mehler–Fock Transform

7-1 INTRODUCTION

In recent years there has been considerable work on the use of Mehler–Fock transforms in the solution of boundary value problems in the mathematical theory of elasticity, particularly those concerned with the analysis of stress in the vicinity of an "external" crack (of the type shown in Fig. 7-2 below). The object of this chapter is to provide—in a form suitable for those whose interest lies in applications—a short account of the basic properties of the Mehler–Fock transform. Emphasis is put on the transform which corresponds to axisymmetric potential problems not because these problems are in themselves intrinsically either more significant or more interesting but simply because not many transforms of particular functions have been calculated.[1]

[1] A useful table of transforms is given in Oberhettinger and Higgins (1961).

A brief account is given in sec. 7-2 of the basic properties of the associated Legendre function of the first kind since it is convenient to have listed there precisely those properties of that function needed in subsequent sections.

7-2 LEGENDRE'S ASSOCIATED EQUATION

In discussing Mehler–Fock transforms we need to know one or two of the basic properties of a certain solution of Legendre's associated equation. We list these properties here for convenience.

To obtain solutions of Legendre's associated equation

$$(1 - x^2)\frac{d^2y}{dx^2} - 2x\frac{dy}{dx} + \left\{\nu(\nu + 1) - \frac{m^2}{1 - x^2}\right\} y = 0 \tag{7-2-1}$$

we make the substitution

$$y = (x^2 - 1)^{+\frac{1}{2}m} w(x)$$

to obtain the equation

$$(1 - x^2)\frac{d^2w}{dx^2} - 2(1 - m)x\frac{dw}{dx} + (\nu + m)(\nu - m + 1)w = 0. \tag{7-2-2}$$

If we now change the independent variable from x to ξ, where

$$\xi = \frac{1 - x}{2},$$

we find that

$$\xi(1 - \xi)\frac{d^2w}{d\xi^2} + (m + 1)(1 - 2\xi)\frac{dw}{d\xi} + (\nu - m)(\nu + m + 1)w = 0$$

which is identical with the hypergeometric equation

$$\xi(1 - \xi)\frac{d^2w}{d\xi^2} + [c - (a + b + 1)\xi]\frac{dw}{d\xi} - abw = 0$$

with

$$a = m - \nu, \qquad b = m + \nu + 1, \qquad c = m + 1$$

so that equation (7-2-2) has a solution of the form

$${}_2F_1(m - \nu, m + \nu + 1; m + 1; \tfrac{1}{2} - \tfrac{1}{2}x),$$

and hence that equation (7-2-1) has the solution

$$P_\nu^m(x) = \frac{\Gamma(\nu + m + 1)}{2^m\Gamma(m + 1)\Gamma(\nu - m + 1)} \times$$
$$\times (x^2 - 1)^{-\frac{1}{2}m} {}_2F_1(m - \upsilon, m + \upsilon + 1; m + 1; \tfrac{1}{2} - \tfrac{1}{2}x). \tag{7-2-3}$$

If we make use of Kummer's relation

$$ {}_2F_1(a,b;c;\xi) = (1-\xi)^{c-a-b}\,{}_2F_1(c-a,c-b;c;\xi) $$

we find that

$$ {}_2F_1\left(m-\nu,m+\nu+1;m+1;\frac{1-x}{2}\right) \times $$

$$ = 2^m(x-1)^{-m}\,{}_2F_1\left(-\nu,\nu+1;m+1;\frac{1-x}{2}\right) $$

so that we obtain the alternative expression

$$ P_\nu^m(x) = \frac{\Gamma(\nu+m+1)}{\Gamma(m+1)\Gamma(\nu-m+1)} \left(\frac{x-1}{x+1}\right)^{\frac{1}{2}m} {}_2F_1\left(-\nu,\nu+1;m+1;\frac{1-x}{2}\right). \tag{7-2-4} $$

The function $P_\nu^m(x)$ defined by equation (7-2-3) or its equivalent (7-2-4) is called an *associated Legendre function* of the first kind. When $m = 0$ it reduces to the *Legendre function*

$$ P_\nu(x) = {}_2F_1\left(-\nu,\nu+1;1;\frac{1-x}{2}\right). \tag{7-2-5} $$

In most applications m is a positive integer though ν is not. In this case we can easily show that

$$ \frac{d^m}{dx^m}P_\nu(x) = \frac{\Gamma(\nu+m+1)}{2^m m!\,\Gamma(\nu-m+1)} \times $$

$$ \times\,{}_2F_1\left(-\nu+m,\nu+m+1;m+1;\frac{1-x}{2}\right). $$

Hence, when m is a positive integer, we can define the associated Legendre function $P_\nu^m(x)$ by the equation

$$ P_\nu^m(x) = (x^2-1)^{\frac{1}{2}m}\frac{d^m}{dx^m}P_\nu(x). \tag{7-2-6} $$

Since Legendre's associated equation is unaltered if we replace m by $-m$, we should expect the function defined by the equation

$$ P_\nu^{-m}(x) = (x^2-1)^{-\frac{1}{2}m}\int_1^x d\xi_1\int_1^{\xi_1} d\xi_2 \ldots \int_1^{\xi_{m-1}} P_\nu(\xi_m)\,d\xi_m \tag{7-2-7} $$

also to be a solution. It is easily shown that this is in fact so and using the relation

$$\int_1^x d\xi_1 \int_1^{\xi_1} d\xi_2 \ldots \int_1^{\xi_{m-1}} \left(\frac{1-\xi_m}{2}\right)^r d\xi_m$$

$$= \frac{(x-1)^m}{m!} \cdot \frac{r!}{(m+1)_r}\left(\frac{1-x}{2}\right)^r$$

we see that

$$P_\nu^{-m}(x) = \left(\frac{x-1}{x+1}\right)^{\frac{1}{2}m} \frac{1}{m!}\, {}_2F_1\left(-\nu,\nu+1;m+1;\frac{1-x}{2}\right). \tag{7-2-8}$$

When m is not an integer, we can use (7-2-8) to *define* the function $P_\nu^{-m}(x)$.

Comparing equations (7-2-4) and (7-2-8) we find that

$$P_\nu^{-m}(x) = \frac{\Gamma(\nu - m + 1)}{\Gamma(\nu + m + 1)} P_\nu^m(x). \tag{7-2-9}$$

Putting $x = \cos\theta$ in equations (7-2-3) and (7-2-8), we find that if m is a positive integer

$$P_\nu^m(\cos\theta) = \frac{e^{\frac{1}{2}m\pi i}\Gamma(\nu+m+1)}{2^m m!\,\Gamma(\nu-m+1)} \times$$

$$\times (\sin\theta)^m\, {}_2F_1(m-\nu, m+\nu+1; m+1; \sin^2\tfrac{1}{2}\theta) \tag{7-2-10}$$

$$P_\nu^{-m}(\cos\theta) = \frac{e^{\frac{1}{2}m\pi i}}{m!}(\tan\tfrac{1}{2}\theta)^m\, {}_2F_1(-\nu,\nu+1;m+1:\sin^2\tfrac{1}{2}\theta). \tag{7-2-11}$$

It is useful to have an integral expression for the associated Legendre function. Starting from the identity

$$\cos(\nu+\tfrac{1}{2})\phi = \cos(\tfrac{1}{2}\phi)\, {}_2F_1(-\nu,\nu+1;\tfrac{1}{2};\sin^2\tfrac{1}{2}\phi)$$

and integrating term by term we can easily show that if m is a positive integer

$$\int_0^\theta \cos\{(\nu+\tfrac{1}{2})\phi\}(\cos\phi - \cos\theta)^{m-\frac{1}{2}}\, d\phi$$

$$= \frac{2^{m-\frac{1}{2}}\Gamma(m+\frac{1}{2})\Gamma(\frac{1}{2})}{\Gamma(m+1)}(\sin\tfrac{1}{2}\theta)^{2m}\, {}_2F_1(-\nu,\nu+1;m+1;\sin^2\tfrac{1}{2}\theta).$$

Making use of equation (7-2-11) we see that

$$P_\nu^{-m}(\cos\theta) = \sqrt{\frac{2}{\pi}}\frac{e^{\frac{1}{2}m\pi i}(\sin\theta)^{-m}}{\Gamma(m+\frac{1}{2})} \times$$

$$\times \int_0^\theta \cos\{(\nu+\tfrac{1}{2})\phi\}(\cos\phi - \cos\theta)^{m-\frac{1}{2}}\, d\phi. \tag{7-2-12}$$

In particular by taking $m = 0$ we find that

$$P_\nu(\cos\theta) = \frac{\sqrt{2}}{\pi}\int_0^\theta \frac{\cos\{(\nu+\frac{1}{2})\phi\}}{\sqrt{(\cos\phi - \cos\theta)}}\, d\phi. \tag{7-2-13}$$

This result is due to Mehler (1881); it is known in the literature as the *Mehler–Dirichlet formula.* We shall return to these formulae in sec. 7-4 below once we have seen the precise manner in which the associated Legendre functions enter into the solution of Laplace's equation expressed in toroidal coordinates.

7-3 LAPLACE'S EQUATION IN TOROIDAL COORDINATES

If the cylindrical coordinates (ρ,ϕ,z) are expressed in terms of new coordinates (α,β,ϕ) through the relation

$$\rho + iz = f(\alpha + i\beta)$$

where the function f is a holomorphic function of the complex variable $\alpha + i\beta$, then it is easily shown that Laplace's equation assumes the form

$$\frac{\partial}{\partial\alpha}\left(\rho\frac{\partial\psi}{\partial\alpha}\right) + \frac{\partial}{\partial\beta}\left(\rho\frac{\partial\psi}{\partial\beta}\right) + \frac{|f'|^2}{\rho}\frac{\partial^2\psi}{\partial\phi^2} = 0. \tag{7-3-1}$$

In the case in which

$$\rho + iz = a\tanh\tfrac{1}{2}(\alpha + i\beta) \tag{7-3-2}$$

the coordinates (α,β,ϕ) are called *toroidal coordinates.* Separating real and imaginary parts in (7-3-2), we find that ρ and z are given by the equations

$$\rho = a\frac{\sinh\alpha}{\cosh\alpha + \cos\beta}, \qquad z = a\frac{\sin\beta}{\cosh\alpha + \cos\beta}. \tag{7-3-3}$$

Also, from (7-3-2), we have the relation

$$|f'|^2 = a^2(\cosh\alpha + \cos\beta)^{-2}. \tag{7-3-4}$$

Substituting from (7-3-3) and (7-3-4) into (7-3-1), we find that in terms of toroidal coordinates Laplace's equation assumes the form

$$\frac{\partial}{\partial\alpha}\left\{\frac{\sinh\alpha}{\cosh\alpha + \cos\beta}\frac{\partial\psi}{\partial\alpha}\right\} + \frac{\partial}{\partial\beta}\left\{\frac{\sinh\alpha}{\cosh\alpha + \cos\beta}\frac{\partial\psi}{\partial\beta}\right\} + \frac{1}{\sinh\alpha(\cosh\alpha + \cos\beta)}\frac{\partial^2\psi}{\partial\phi^2} = 0. \tag{7-3-5}$$

We notice from (7-3-3) that if $0 < \beta < \pi$, then $z > 0$, and that if $-\pi < \beta < 0$, then $z < 0$.

The surface $\beta = \beta_1$ has equation

$$\rho^2 + (z + a\cot\beta_1)^2 = a^2\operatorname{cosec}^2\beta_1 \tag{7-3-6}$$

so that if $0 < \beta_1 < \pi$, the surface $\beta = \beta_1$ is that part of the sphere with centre $(0,0,-a\cot\beta_1)$ and radius $a\operatorname{cosec}\beta_1$ which lies *above* the xy-plane. On the

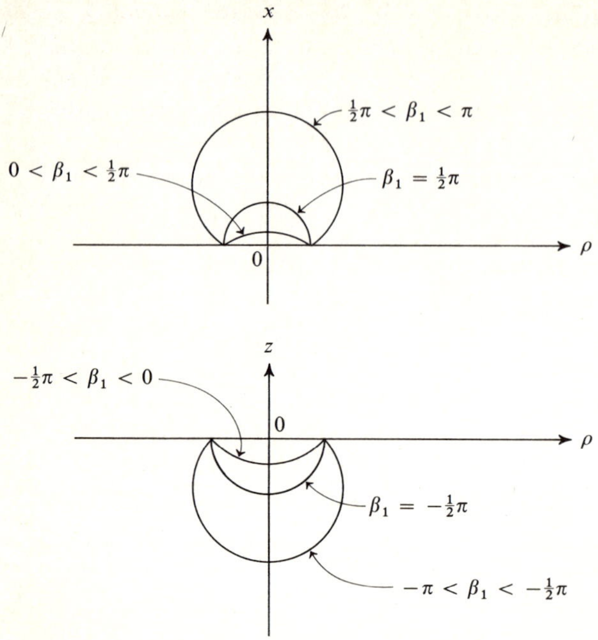

Fig. 7-1

other hand, if $-\pi < \beta_1 < 0$, the surface is that part of the sphere with center $(0,0,-a \cot \beta_1)$ and radius $a \operatorname{cosec} \beta_1$ which lies *below* the xy-plane. The sections by a plane through the z-axis of some typical surfaces of this kind are shown in Fig. 7-1.

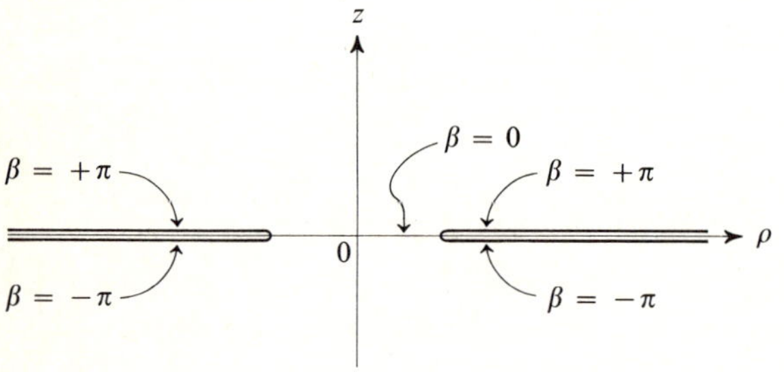

Fig. 7-2

It should be noticed that, whatever the value of β_1 the circle C with equation $\rho = a$, $z = 0$ lies on the surface $\beta = \beta_1$.

If $\beta = 0$, $z = 0$ and $0 \leqslant \rho \leqslant a$, so that $\beta = 0$ is the interior of the circle C. Also, if $\beta = \pi$, $z = 0+$, $\rho \geqslant a$ so that the surface $\beta = \pi$ is the upper part (i.e., the part lying above the xy-plane) of the external groove, or "external crack," formed by cutting the space along the z-plane and then joining the two resultant half-spaces at all interior points of the circle C. Similarly, the surface $\beta = -\pi$ is the lower part of the same groove. These surfaces are shown diagrammatically in Fig. 7-2. The rim C of the external crack may be

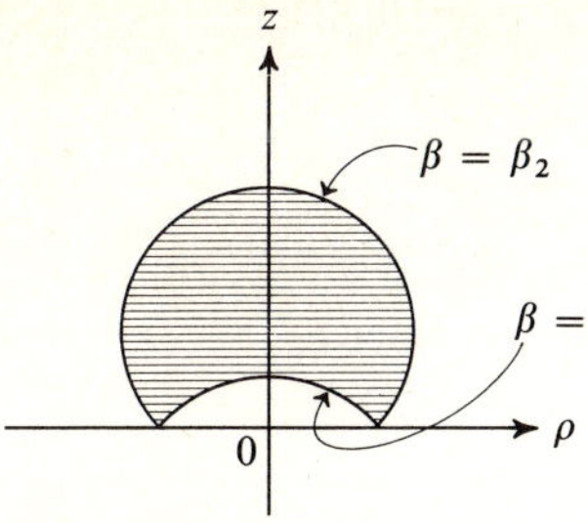

Case (a): $0 < \beta_1 < \frac{1}{2}\pi < \beta_2 < \pi$

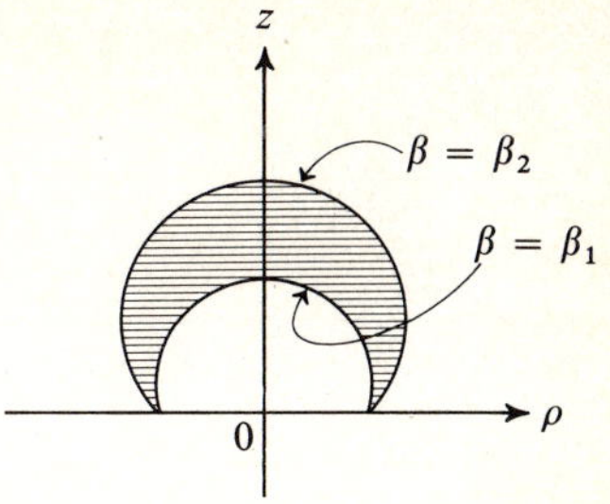

Case (b): $\frac{1}{2}\pi < \beta_1 < \beta_2 < \pi$

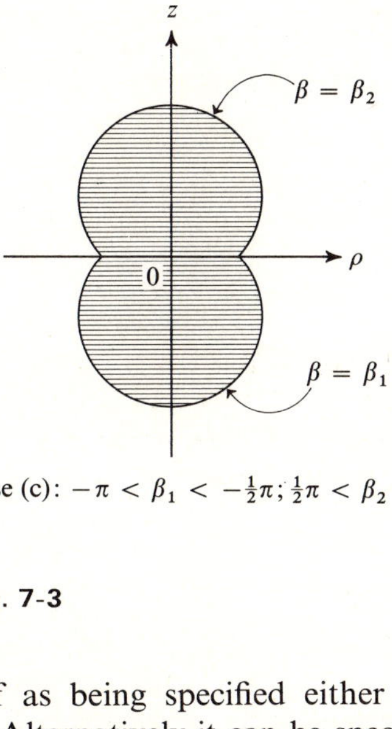

Case (c): $-\pi < \beta_1 < -\frac{1}{2}\pi; \frac{1}{2}\pi < \beta_2 < \pi$

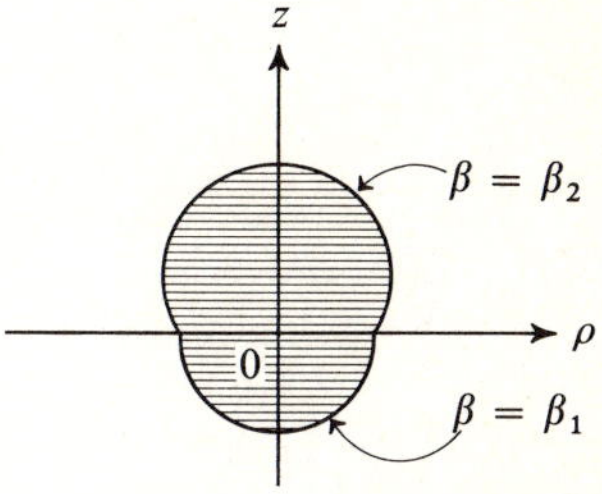

Case (d): $-\frac{1}{2}\pi < \beta_1 < 0, \frac{1}{2}\pi < \beta_2 < \pi$

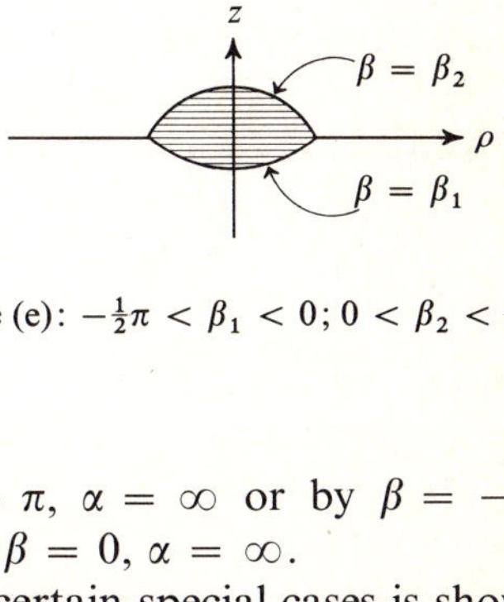

Case (e): $-\frac{1}{2}\pi < \beta_1 < 0; 0 < \beta_2 < \frac{1}{2}\pi$

Fig. 7-3

thought of as being specified either by $\beta = \pi$, $\alpha = \infty$ or by $\beta = -\pi$, $\alpha = -\infty$. Alternatively it can be specified by $\beta = 0$, $\alpha = \infty$.

The form of the region $\beta_1 < \beta < \beta_2$ in certain special cases is shown in Fig. 7-3.

In a similar way we can show that the surface $\alpha = \alpha_1$ has constraint equation

$$(\rho - a \coth \alpha_1)^2 + z^2 = a^2 \operatorname{cosech}^2 \alpha_1,$$

so that it is the surface of the torus formed by rotating Fig. 7-4 about the z-axis.

If $\alpha = 0$, we have $\rho = 0$, $z = \frac{1}{2} \sin \beta \sec^2 \frac{1}{2}\beta$ so that $\alpha = 0$, $0 < \beta < \pi$ is the positive z-axis, and $\alpha = 0$, $-\pi < \beta < 0$ is the negative z-axis. The surface $\alpha = \alpha_1$ degenerates into the circle $\rho = a$, $z = 0$ as $\alpha_1 \to \infty$, so that the interior points of the torus whose surface is $\alpha = \alpha_1$ have $\alpha > \alpha_1$.

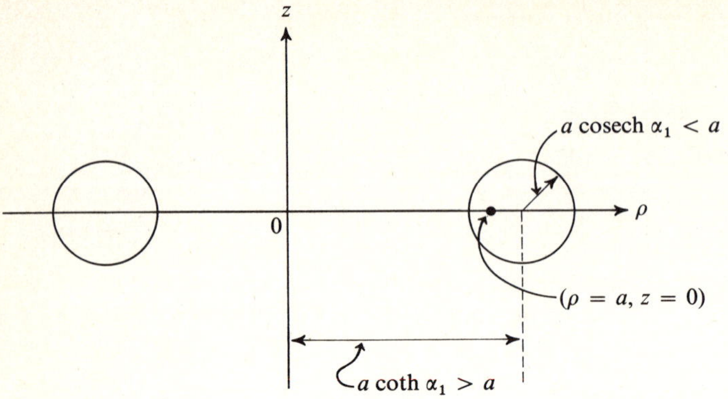

Fig. 7-4

In the discussion of boundary value problems we need expressions for the derivatives

$$\frac{\partial\psi}{\partial\rho}, \qquad \frac{\partial\psi}{\partial z}$$

in terms of the derivatives

$$\frac{\partial\psi}{\partial\alpha}, \qquad \frac{\partial\psi}{\partial\beta}.$$

To calculate $\partial\psi/\partial\rho$ through the formula

$$\frac{\partial\psi}{\partial\rho} = \frac{\partial\psi}{\partial\alpha}\frac{\partial\alpha}{\partial\rho} + \frac{\partial\psi}{\partial\beta}\frac{\partial\beta}{\partial\rho} \tag{7-3-7}$$

we must first calculate $\partial\alpha/\partial\rho$ and $\partial\beta/\partial\rho$.

If we differentiate the equations (7-3-3) with respect to ρ we obtain the pair of equations

$$(\cosh\alpha + \cos\beta) + \rho\left(\sinh\alpha\frac{\partial\alpha}{\partial\rho} - \sin\beta\frac{\partial\beta}{\partial\rho}\right) = a\cosh\alpha\frac{\partial\alpha}{\partial\rho},$$

$$z\left(\sinh\alpha\frac{\partial\alpha}{\partial\rho} - \sin\beta\frac{\partial\beta}{\partial\rho}\right) = a\cos\beta\frac{\partial\beta}{\partial\rho},$$

which may be solved to give

$$\frac{\partial\alpha}{\partial\rho} = \frac{1}{a}(1 + \cosh\alpha\cos\beta), \qquad \frac{\partial\beta}{\partial\rho} = \frac{1}{a}\sinh\alpha\sin\beta.$$

If we substitute these expressions into (7-3-7), we find that

$$a\frac{\partial\psi}{\partial\rho} = (1 + \cosh\alpha\cos\beta)\frac{\partial\psi}{\partial\alpha} + \sinh\alpha\sin\beta\frac{\partial\psi}{\partial\beta}. \tag{7-3-8}$$

In a similar way we can show that

$$a\frac{\partial\psi}{\partial z} = -\sinh\alpha\sin\beta\frac{\partial\psi}{\partial\alpha} + (1 + \cosh\alpha\cos\beta)\frac{\partial\psi}{\partial\beta}. \tag{7-3-9}$$

If we apply the formula (7-3-9) twice, we obtain the equation

$$\begin{aligned} a^2\frac{\partial^2\psi}{\partial z^2} &= \sinh^2\alpha\sin^2\beta\frac{\partial^2\psi}{\partial\alpha^2} - 2\sinh\alpha\sin\beta(1 + \cosh\alpha\cos\beta)\frac{\partial^2\psi}{\partial\alpha\partial\beta} \\ &\quad + (1 + \cosh\alpha\cos\beta)^2\frac{\partial^2\psi}{\partial\beta^2} \\ &\quad - \{(1 + \cosh\alpha\cos\beta)\cos\beta - \cosh\alpha\sin^2\beta\}\sinh\alpha\frac{\partial\psi}{\partial\alpha} \\ &\quad - \{(1 + \cosh\alpha\cos\beta)\cosh\alpha + \sinh^2\alpha\cos\beta\}\sin\beta\frac{\partial\psi}{\partial\beta}. \end{aligned} \tag{7-3-10}$$

The forms of these expressions when $\beta = \pm\pi$ are particularly useful:

$$a\left(\frac{\partial\psi}{\partial\rho}\right)_{\beta=\pm\pi} = (1 - \cosh\alpha)\left(\frac{\partial\psi}{\partial\alpha}\right)_{\beta=\pm\pi} \tag{7-3-11}$$

$$a\left(\frac{\partial\psi}{\partial z}\right)_{\beta=\pm\pi} = (1 - \cosh\alpha)\left(\frac{\partial\psi}{\partial\beta}\right)_{\beta=\pm\pi} \tag{7-3-12}$$

$$\begin{aligned} a^2\left(\frac{\partial^2\psi}{\partial z^2}\right)_{\beta=\pm\pi} &= (1 - \cosh\alpha)^2\left(\frac{\partial^2\psi}{\partial\beta^2}\right)_{\beta=\pm\pi} \\ &\quad + (\cosh\alpha - 1)\sinh\alpha\left(\frac{\partial\psi}{\partial\alpha}\right)_{\beta=\pm\pi}. \end{aligned} \tag{7-3-13}$$

It should also be noticed that if $\psi(\alpha,\beta)$ is of the form

$$\psi(\alpha,\beta) = \sqrt{(\cosh\alpha + \cos\beta)}\psi_1(\alpha,\beta),$$

then

$$\frac{\partial\psi}{\partial\beta} = (\cosh\alpha + \cos\beta)\frac{\partial\psi_1}{\partial\beta} - \frac{\sin\beta}{2\sqrt{(\cosh\alpha + \cos\beta)}}\psi_1$$

so that the condition

$$\left(\frac{\partial\psi}{\partial\beta}\right)_{\beta=\pm\pi} = 0$$

implies the condition

$$\left(\frac{\partial\psi_1}{\partial\beta}\right)_{\beta=\pm\pi} = 0.$$

Hence the condition

$$\left(\frac{\partial \psi}{\partial z}\right)_{\beta = \pm\pi} = 0$$

implies the condition

$$\left(\frac{\partial \psi_1}{\partial \beta}\right)_{\beta = \pm\pi} = 0.$$

We also need the result obtained by putting $\beta = 0$ in equation (7-3-9), namely

$$a\left(\frac{\partial \psi}{\partial z}\right)_{\beta = 0} = (1 + \cosh\alpha)\left(\frac{\partial \psi}{\partial \beta}\right)_{\beta = 0}. \tag{7-3-14}$$

We now consider the solution of Laplace's equation in toroidal coordinates.

If we put

$$\psi = \sqrt{(\cosh\alpha + \cos\beta)}\,v(\alpha,\beta,\phi) \tag{7-3-15}$$

in Laplace's equation (7-3-5), we find that the function v satisfies the equation

$$\sinh\alpha\,\frac{\partial^2 v}{\partial \alpha^2} + \cosh\alpha\,\frac{\partial y}{\partial \alpha} + \sinh\alpha\,\frac{\partial^2 v}{\partial \beta^2} + \tfrac{1}{4}v\sinh\alpha + \frac{1}{\sinh\alpha}\frac{\partial^2 v}{\partial \phi^2} = 0 \tag{7-3-16}$$

which has solutions of the form

$$v(\alpha,\beta,\phi) = A(\alpha)e^{\pm\tau\beta \pm im\phi} \tag{7-3-17}$$

for functions $A(\alpha)$ satisfying the ordinary differential equation

$$\sinh\alpha\,\frac{d^2A}{d\alpha^2} + \cosh\alpha\,\frac{dA}{d\alpha} + (\tfrac{1}{4} + \tau^2)\sinh\alpha\,A - \frac{m^2}{\sinh\alpha}A = 0.$$

If we now change the independent variable from α to μ where $\mu = \cosh\alpha$, we find that this equation becomes

$$(1 - \mu^2)\frac{d^2A}{d\mu^2} - 2\mu\frac{dA}{d\mu} - \left[\frac{m^2}{1 - \mu^2} + (\tfrac{1}{4} + \tau^2)\right]A = 0. \tag{7-3-18}$$

Now if we write

$$v = -\tfrac{1}{2} + i\tau \tag{7-3-19}$$

we see that

$$v(v + 1) = -(\tfrac{1}{4} + \tau^2)$$

and that we may write (7-3-18) as

$$(1 - \mu^2)\frac{d^2A}{d\mu^2} - 2\mu\frac{dA}{d\mu} + \left[\nu(\nu + 1) - \frac{m^2}{1 - \mu^2}\right]A = 0. \qquad (7\text{-}3\text{-}20)$$

This equation has a solution of the form

$$A = cP_\nu^{-m}(\mu)$$

where c is an arbitrary constant. Hence equation (7-3-16) has solutions of the form

$$v(\alpha,\beta,\phi) = P_{-\frac{1}{2}+i\tau}^{-m}(\cosh\alpha)e^{\pm\tau\beta\pm im\phi}. \qquad (7\text{-}3\text{-}21)$$

It will be noticed that since, in physical applications, we are interested only in solutions with the property that

$$\psi(\alpha,\beta,\phi + 2n\pi) = \psi(\alpha,\beta,\phi), \qquad n = \pm 1, \pm 2, \pm 3, \ldots \qquad (7\text{-}3\text{-}22)$$

we must take m to be an integer in solutions of the type (7-3-21). In this way we derive solutions of Laplace's equation of the type

$$\psi(\alpha,\beta,\phi) = \sqrt{(\cosh\alpha + \cos\beta)}P_{-\frac{1}{2}+i\tau}^{-m}(\cosh\alpha)e^{\pm\tau\beta\pm im\phi} \qquad (7\text{-}3\text{-}23)$$

where to ensure the property (7-3-22) we must take m to be an integer. Putting $m = 0$, we deduce the axisymmetric solutions

$$\psi(\alpha,\beta) = \sqrt{(\cosh\alpha + \cos\beta)}P_{-\frac{1}{2}+i\tau}(\cosh\alpha)e^{\pm\tau\beta}. \qquad (7\text{-}3\text{-}24)$$

From (7-3-23) we can, by a simple superposition process, construct solutions of Laplace's equation of the form

$$\psi(\alpha,\beta,\phi) = \sqrt{(\cosh\alpha + \cos\beta)} \times \sum_{m=-\infty}^{\infty} e^{im\phi}\Phi_m^{-1}[A_m(\tau)\cosh(\beta\tau) + B_m(\tau)\sinh(\beta\tau); \tau \to \alpha] \qquad (7\text{-}3\text{-}25)$$

where the operator Φ_m^{-1} is defined by the equation

$$\Phi_m^{-1}[\chi_m(\tau); \tau \to \alpha] = (-1)^m \int_0^\infty \tau\tanh(\pi\tau)P_{-\frac{1}{2}+i\tau}^{-m}(\cosh\alpha)\chi_m(\tau)\,d\tau. \qquad (7\text{-}3\text{-}26)$$

Similarly, in the axisymmetric case, we have solutions of the form

$$\psi(\alpha,\beta) = \sqrt{(\cosh\alpha + \cos\beta)} \times \Phi_0^{-1}[A(\tau)\cosh(\beta\tau) + B(\tau)\sinh(\beta\tau); \tau \to \alpha] \qquad (7\text{-}3\text{-}27)$$

where

$$\Phi_0^{-1}[\chi(\tau); \tau \to \alpha] = \int_0^\infty \tau\tanh(\pi\tau)P_{-\frac{1}{2}+i\tau}(\cosh\alpha)\chi(\tau)\,d\tau. \qquad (7\text{-}3\text{-}28)$$

For instance, suppose that we wish to find a solution of Laplace's equation $\Delta_3\psi = 0$ in the half-space $z > 0$ when ψ is prescribed on the disk

$0 < \rho < a$, $z = 0$ and $\partial\psi/\partial z = 0$ in the region of the boundary outside of that disk. This is equivalent to finding a harmonic function $\psi(\alpha,\beta,\phi)$ in the region $0 < \beta < \pi$ satisfying the boundary conditions

$$\psi(\alpha,0,\phi) = f(\alpha,\phi) \tag{7-3-29}$$

$$\left(\frac{\partial\psi}{\partial z}\right)_{\beta=\pi} = 0 \tag{7-3-30}$$

with the function $f(\alpha,\phi)$ prescribed for $0 < \phi < 2\pi$, $\alpha > 0$. By equation (7-3-27) we see that the condition (7-3-30) may be replaced by the condition

$$\left(\frac{\partial\psi}{\partial \beta}\right)_{\beta=\pi} = 0. \tag{7-3-31}$$

If we take a solution of the form

$$\psi(\alpha,\beta,\phi) = \sqrt{(\cosh\alpha + \cos\beta)} \times \sum_{m=-\infty}^{\infty} \Phi_m^{-1}\left[A_m(\tau)\frac{\cosh(\pi\tau - \beta\tau)}{\cosh(\pi\tau)};\tau \to \alpha\right]e^{im\phi} \tag{7-3-32}$$

we see that the condition (7-3-31) is automatically satisfied. The condition (7-3-29) will be satisfied if we can find a sequence of functions $\{A_m(\tau)\}$ with the property that

$$f(\alpha,\phi) = \sqrt{2}\sinh(\tfrac{1}{2}\alpha)\sum_{m=-\infty}^{\infty}\Phi_m^{-1}[A_m(\tau);\tau \to \alpha]e^{im\phi}$$

that is, such that

$$\Phi_m^{-1}[A_m(\tau);\tau \to \alpha] = 2^{-\frac{3}{2}}\pi^{-1}\operatorname{cosech}(\tfrac{1}{2}\alpha)\int_0^{2\pi} f(\alpha,\phi)e^{-im\phi}\,d\phi.$$

Operating on both sides of this equation by Φ_m we find formally that

$$A_m(\tau) = 2^{-\frac{3}{2}}\pi^{-1}\int_0^{2\pi} e^{-im\phi}\Phi_m[\operatorname{cosech}(\tfrac{1}{2}\alpha)f(\alpha,\phi);\alpha \to \tau]\,d\phi. \tag{7-3-33}$$

The problem therefore reduces to that of finding the operator Φ_m which is the inverse of the operator Φ_m^{-1} defined by equation (7-3-26).

The solution is even simpler in the axisymmetric case when the boundary condition (7-3-29) is replaced by the symmetric one

$$\psi(\alpha,0) = f(\alpha). \tag{7-3-34}$$

The required harmonic function is then given by the equation

$$\psi(\alpha,\beta) = \sqrt{(\cosh\alpha + \cos\beta)}\Phi_0^{-1}\left[A_0(\tau)\frac{\cosh(\pi\tau - \beta\tau)}{\cosh(\pi\tau)};\tau \to \alpha\right] \tag{7-3-35}$$

where the function $A_0(\tau)$ is defined by the equation

$$A_0(\tau) = 2^{-\frac{1}{2}}\Phi_0[\operatorname{cosech}(\tfrac{1}{2}\alpha)f(\alpha); \alpha \to \tau]. \tag{7-3-36}$$

7-4 PROPERTIES OF THE FUNCTION $P_{-\frac{1}{2}+i\tau}$ (cosh α)

We saw in the last section how the function $P_{-\frac{1}{2}+i\tau}{}^m(\cosh\alpha)$, in which m is zero or an integer and τ is real, enters into the solution of Laplace's equation when it is expressed in toroidal coordinates. We shall now derive some properties of this function in the special case in which $m = 0$. If we put $\nu = -\frac{1}{2} + i\tau$, $\theta = i\alpha$ in the Mehler–Dirichlet formula (7-2-12) we find that if m is a positive integer

$$P_{-\frac{1}{2}+i\tau}{}^{-m}(\cosh\alpha) = \sqrt{\frac{2}{\pi}}\frac{(\sinh\alpha)^{-m}}{\Gamma(m+\frac{1}{2})}\int_0^\alpha \cos(\tau t)(\cosh\alpha - \cosh t)^{m-\frac{1}{2}}\,dt \tag{7-4-1}$$

from which, using equation (7-2-9), we deduce that

$$P_{-\frac{1}{2}+i\tau}{}^{m}(\cosh\alpha) = \sqrt{\frac{2}{\pi}}\frac{\Gamma(\frac{1}{2}+i\tau+m)(\sinh\alpha)^{-m}}{\Gamma(\frac{1}{2}+i\tau-m)\Gamma(m+\frac{1}{2})}\int_0^\alpha \cos(\tau t)(\cosh\alpha - \cosh t)^{m-\frac{1}{2}}\,dt \tag{7-4-2}$$

and, in particular, that

$$P_{-\frac{1}{2}+i\tau}(\cosh\alpha) = \frac{\sqrt{2}}{\pi}\int_0^\alpha \frac{\cos(\tau t)\,dt}{\sqrt{(\cosh\alpha - \cosh t)}}. \tag{7-4-3}$$

In a similar way (Cf. Lagrange, 1939, p. 47) we can show that

$$P_{-\frac{1}{2}+i\tau}(\cosh\alpha) = \frac{\sqrt{2}}{\pi}\coth(\pi\tau)\int_\alpha^\infty \frac{\sin(\tau t)\,dt}{\sqrt{(\cosh t - \cosh\alpha)}}. \tag{7-4-4}$$

Writing the right side of (7-4-3) as

$$\mathscr{F}_c\left[\frac{1}{\sqrt{\pi}}\cdot\frac{H(\alpha - t)}{\sqrt{(\cosh\alpha - \cosh t)}}; t \to \tau\right]$$

we see on applying the Fourier cosine inversion theorem that

$$\mathscr{F}_c[P_{-\frac{1}{2}+i\tau}(\cosh\alpha); \tau \to t] = \frac{H(\alpha - t)}{\sqrt{[\pi(\cosh\alpha - \cosh t)]}}$$

which when written out becomes

$$\int_0^\infty P_{-\frac{1}{2}+i\tau}(\cosh\alpha)\cos(\tau t)\,d\tau = \frac{H(\alpha - t)}{\sqrt{[2(\cosh\alpha - \cosh t)]}}. \tag{7-4-5}$$

On the other hand if we regard (7-4-3) as an integral equation for $\cos(t\tau)$ in terms of $P_{-\frac{1}{2}+i\tau}(\cosh\alpha)$, we find from equations (3-16-13) and (3-16-15) that

$$\cos(\tau t) = \frac{1}{\sqrt{2}}\frac{d}{dt}\int_0^t P_{-\frac{1}{2}+i\tau}(\cosh\alpha)\frac{\sinh\alpha\, d\alpha}{\sqrt{(\cosh t - \cosh\alpha)}}. \tag{7-4-6}$$

Similarly we deduce immediately from (7-4-4) that

$$\mathscr{F}_s[\tanh(\pi\tau)P_{-\frac{1}{2}+i\tau}(\cosh\alpha); \tau \to t) = \frac{H(t-\alpha)}{\sqrt{[\pi(\cosh t - \cosh\alpha)]}} \tag{7-4-7}$$

which is equivalent to

$$\int_0^\infty \tanh(\pi\tau)P_{-\frac{1}{2}+i\tau}(\cosh\alpha)\sin(\tau t)\, d\tau = \frac{H(t-\alpha)}{\sqrt{[2(\cosh t - \cosh\alpha)]}}. \tag{7-4-8}$$

Similarly, by applying the formulas (3-16-14) and (3-16-16), we obtain the equation

$$\sin(\tau t) = -\frac{1}{\sqrt{2}}\tanh(\pi\tau)\ \times$$

$$\times\ \frac{d}{dt}\int_t^\infty (\cosh\alpha - \cosh t)^{-\frac{1}{2}}P_{-\frac{1}{2}+i\tau}(\cosh\alpha)\sinh\alpha\, d\alpha \tag{7-4-9}$$

from equation (7-4-4). If we use the results

$$\cos(\tau t) = \frac{d}{dt}\left[\frac{\sin(\tau t)}{\tau}\right], \qquad \sin(\tau t) = -\frac{d}{dt}\left[\frac{\cos(\tau t)}{\tau}\right]$$

we see that equations (7-4-6) and (7-4-7) are equivalent to the pair of equations

$$\frac{\sin(\tau t)}{\tau} = \frac{1}{\sqrt{2}}\int_0^t (\cosh t - \cosh\alpha)^{-\frac{1}{2}}P_{-\frac{1}{2}+i\tau}(\cosh\alpha)\sinh\alpha\, d\alpha \tag{7-4-10}$$

$$\frac{\cos(\tau t)}{\tau} = \frac{1}{\sqrt{2}}\tanh(\pi\tau)\ \times$$

$$\times\int_t^\infty (\cosh\alpha - \cosh t)^{-\frac{1}{2}}P_{-\frac{1}{2}+i\tau}(\cosh\alpha)\sinh\alpha\, d\alpha. \tag{7-4-11}$$

From equation (7-4-3) we see that

$$\int_0^\infty P_{-\frac{1}{2}+i\tau}(\cosh\alpha)\sinh\alpha\, g(\cosh\alpha)\, d\alpha$$

$$= \frac{\sqrt{2}}{\pi}\int_0^\infty \cos(\tau t)\, dt\int_t^\infty \frac{\sinh\alpha\, g(\cosh\alpha)\, d\alpha}{\sqrt{(\cosh\alpha - \cosh t)}}.$$

Now

$$\frac{1}{\sqrt{\pi}}\int_t^\infty \frac{\sinh\alpha\, g(\cosh\alpha)\, d\alpha}{\sqrt{(\cosh\alpha - \cosh t)}} = \frac{1}{\sqrt{\pi}}\int_{\cosh t}^\infty \frac{g(x)\, dx}{\sqrt{(x - \cosh t)}}.$$

In terms of the Weyl fractional integral (4-2-24) we can write this last integral as

$$\mathscr{W}_{\frac{1}{2}}[g(x);\cosh t]$$

so that we obtain the relation

$$\int_0^\infty P_{-\frac{1}{2}+i\tau}(\cosh\alpha)g(\cosh\alpha)\sinh\alpha\, d\alpha = \mathscr{F}_c\{\mathscr{W}_{\frac{1}{2}}[g(x);\cosh t];t\to\tau\}. \tag{7-4-12}$$

This can obviously also be written in the form

$$\int_1^\infty P_{-\frac{1}{2}+i\tau}(x)g(x)\, dx = \mathscr{F}_c[\mathscr{W}_{\frac{1}{2}}\{g(x);\cosh t\};t\to\tau]. \tag{7-4-13}$$

For example, if we take

$$g(x) = \frac{1}{x+k}$$

where k is a positive constant, then

$$\mathscr{W}_{\frac{1}{2}}[g(x);y] = \frac{1}{\sqrt{\pi}}\int_y^\infty \frac{dx}{(x+k)\sqrt{(x-y)}} = \frac{2}{\sqrt{\pi}}\int_0^\infty \frac{du}{u^2+(k+y)}$$

so that

$$\mathscr{W}_{\frac{1}{2}}[g(x);y] = \left[\frac{\pi}{k+y}\right]^{\frac{1}{2}}$$

Now it can be shown that

$$\mathscr{F}_c[(\cosh t + k)^{-\frac{1}{2}};t\to\tau] = \sqrt{\pi}\,\mathrm{sech}\,(\pi\tau)P_{-\frac{1}{2}+i\tau}(k), \qquad |k|<1 \tag{7-4-14}$$

(Cf. Erdelyi *et al* (1954), Vol. 1, p. 30) so that we obtain the integral

$$\int_0^\infty P_{-\frac{1}{2}+i\tau}(x)\frac{dx}{x+k} = \pi\,\mathrm{sech}\,(\pi\tau)P_{-\frac{1}{2}+i\tau}(k), \qquad |k|<1 \tag{7-4-15}$$

a result which can be written in the alternative form

$$\int_0^\infty P_{-\frac{1}{2}+i\tau}(\cosh\alpha)\frac{\sinh\alpha\, d\alpha}{\cosh\alpha+\cos\beta} = \pi\,\mathrm{sech}\,(\pi\tau)P_{-\frac{1}{2}+i\tau}(\cos\beta) \tag{7-4-16}$$

with β real, $-\pi<\beta<\pi$.

We notice also that if we apply the Fourier cosine inversion theorem to equation (7-4-14), then we obtain the integral

$$\int_0^\infty \frac{\cos(\tau t)}{\cosh(\pi\tau)} P_{-\frac{1}{2}+i\tau}(\cos\beta)\, d\tau = \frac{1}{\sqrt{[2\pi(\cosh t + \cos\beta)]}} \tag{7-4-17}$$

In a similar way equation (7-4-4) leads to the relation

$$\int_0^\infty P_{-\frac{1}{2}+i\tau}(\cosh\alpha) g(\cosh\alpha) \sinh\alpha\, d\alpha$$
$$= \coth(\pi\tau)\mathscr{F}_s[\mathscr{R}_{\frac{1}{2}}\{g(x);\cosh t\};t \to \tau] \tag{7-4-18}$$

where $\mathscr{R}_\mu$ is the operator of the Riemann–Liouville fractional integral defined by the equation

$$\mathscr{R}_\mu[g(x);y] = \frac{1}{\Gamma(\mu)} \int_0^y g(x)(y-x)^{\mu-1}\, dx, \qquad \operatorname{Re}\mu > 0. \tag{7-4-19}$$

On the other hand if we substitute for the Legendre function from equation (7-4-3), we find that

$$\int_0^\infty g(\tau) P_{-\frac{1}{2}+i\tau}(\cosh\alpha)\, d\tau$$
$$= \frac{\sqrt{2}}{\pi} \int_0^\infty g(\tau)\, d\tau \int_0^\alpha \frac{\cos(\tau t)\, dt}{\sqrt{(\cosh\alpha - \cosh t)}}$$
$$= \frac{1}{\sqrt{\pi}} \int_0^\alpha \frac{dt}{\sqrt{(\cosh\alpha - \cosh t)}} \cdot \sqrt{\frac{2}{\pi}} \int_0^\infty g(\tau)\cos(t\tau)\, d\tau$$

so that if we denote the Fourier cosine transform of $g(\tau)$ by $G_c(t)$, we obtain the relation

$$\int_0^\infty g(\tau) P_{-\frac{1}{2}+i\tau}(\cosh\alpha)\, d\tau = \frac{1}{\sqrt{\pi}} \int_0^\alpha \frac{G_c(t)\, dt}{\sqrt{(\cosh\alpha - \cosh t)}}. \tag{7-4-20}$$

If we take

$$g(\tau) = \int_0^a g_1(s)\cos(s\tau)\, ds, \qquad (a > 0)$$

then

$$G_c(t) = \sqrt{\frac{\pi}{2}}\, g_1(t) \mathrm{H}(a - t)$$

and equation (7-4-20) implies that

$$\int_0^\infty P_{-\frac{1}{2}+i\tau}(\cosh\alpha)\,d\tau \int_0^a g_1(s)\cos(\tau s)\,ds$$

$$= \frac{1}{\sqrt{2}} \int_0^{\min(a,\alpha)} \frac{g_1(s)\,ds}{\sqrt{(\cosh\alpha - \cosh s)}}. \qquad (7\text{-}4\text{-}21)$$

In a similar way we can show that

$$\int_0^\infty P_{-\frac{1}{2}+i\tau}(\cosh\alpha)\,d\tau \int_a^\infty g_1(s)\cos(s\tau)\,ds$$

$$= \frac{H(\alpha - a)}{\sqrt{2}} \int_a^\alpha \frac{g_1(s)\,ds}{\sqrt{(\cosh\alpha - \cosh s)}}. \qquad (7\text{-}4\text{-}22)$$

If we use the interpretation (7-4-4) for the Legendre function, we find that

$$\int_0^\infty \tanh(\pi\tau) g(\tau) P_{-\frac{1}{2}+i\tau}(\cosh\alpha)\,d\tau = \frac{1}{\sqrt{\pi}} \int_\alpha^\infty \frac{G_s(t)\,dt}{\sqrt{(\cosh t - \cosh\alpha)}} \qquad (7\text{-}4\text{-}23)$$

where $G_s(t)$ denotes the Fourier sine transform of $g(\tau)$.
For example, if we take

$$g(\tau) = \frac{\cosh(\beta\tau)}{\sinh(\pi\tau)}, \qquad (-\pi < \beta < \pi)$$

and make use of the second formula in Problem 2-21 we have

$$G_s(t) = (2\pi)^{-\frac{1}{2}} \frac{\sinh t}{\cosh t + \cos\beta},$$

so we find that equation (7-4-23) with the integration

$$\int_\alpha^\infty \frac{\sinh t\,dt}{(\cosh t + \cos\beta)\sqrt{(\cosh t - \cosh\alpha)}} = 2\int_0^\infty \frac{du}{u^2 + \cosh\alpha + \cos\beta}$$

$$= \frac{\pi}{\sqrt{(\cosh\alpha + \cos\beta)}}$$

yields the formula

$$\int_0^\infty \frac{\cosh(\beta\tau)}{\cosh(\pi\tau)} P_{-\frac{1}{2}+i\tau}(\cosh\alpha)\,d\tau$$

$$= \frac{1}{\sqrt{2(\cosh\alpha + \cos\beta)}}, \qquad -\pi < \beta < \pi. \qquad (7\text{-}4\text{-}24)$$

Some special cases of the use of equation (7-4-13) arise frequently so we shall derive them here.

Since

$$\mathscr{W}_{\frac{1}{2}}[e^{-ax};y] = a^{-\frac{1}{2}}e^{-ay} \qquad (\operatorname{Re} a > 0),$$

we find from equation (7-4-13) that

$$\int_0^\infty P_{-\frac{1}{2}+i\tau}(x)e^{-ax}\,dx = a^{-\frac{1}{2}}\mathscr{F}_c[e^{-a\cosh t};\tau].$$

Now the modified Bessel function of the second kind of order ν may be defined by the equation

$$K_\nu(z) = \int_0^\infty e^{-z\cosh t}\cosh(\nu t)\,dt \qquad |\arg z| < \tfrac{1}{2}\pi$$

so that

$$\mathscr{F}_c[e^{-a\cosh t};\tau] = \sqrt{\frac{2}{\pi}}\,K_{i\tau}(a), \qquad |\arg a| < \tfrac{1}{2}\pi$$

showing that

$$\int_1^\infty P_{-\frac{1}{2}+i\tau}(x)e^{-ax}\,dx = \sqrt{\frac{2}{\pi a}}\,K_{i\tau}(a), \qquad |\arg a| < \tfrac{1}{2}\pi. \tag{7-4-25}$$

On the other hand, since if, $\operatorname{Re} v > \frac{1}{2}$,

$$\mathscr{W}_{\frac{1}{2}}[x^{-\nu};y] = \frac{\Gamma(\nu - \frac{1}{2})}{\Gamma(\nu)}\,y^{\frac{1}{2}-\nu}$$

we find that equation (7-4-13) gives the result

$$\int_1^\infty P_{-\frac{1}{2}+i\tau}(x)x^{-\nu}\,dx = \frac{\Gamma(\nu - \frac{1}{2})}{\Gamma(\nu)}\mathscr{F}_c[(\operatorname{sech} t)^{\nu-\frac{1}{2}};\tau].$$

Using formula (5) on p. 30 of Vol. 1 of Erdelyi *et al.* (1954), we find that if $\nu > \frac{1}{2}$,

$$\int_1^\infty P_{-\frac{1}{2}+i\tau}(x)x^{-\nu}\,dx = \frac{2^{\nu-1}}{\sqrt{\pi}\,\Gamma(\nu)}\Gamma(\tfrac{1}{2}\nu - \tfrac{1}{4} + \tfrac{1}{2}i\tau)\Gamma(\tfrac{1}{2}\nu - \tfrac{1}{4} - \tfrac{1}{2}i\tau). \tag{7-4-26}$$

In particular, if n is a positive integer, we may use formulas (3) and (4) on the same page to show that

$$\int_1^\infty P_{-\frac{1}{2}+i\tau}(x)x^{-2n-\frac{1}{2}}\,dx$$

$$= \frac{2^{2n-3}}{\Gamma(2n + \frac{1}{2})}\sqrt{(2\pi)}\,\tau\operatorname{cosech}(\tfrac{1}{2}\pi\tau)\prod_{r=1}^{n=1}(\tfrac{1}{4}\tau^2 + r^2), \qquad (n \geqslant 2) \tag{7-4-27}$$

$$\int_1^\infty P_{-\frac{1}{2}+i\tau}(x)x^{-2n-\frac{3}{2}}\,dx$$

$$= \frac{2^{2n-1}}{\Gamma(2n+\frac{3}{2})}\sqrt{(2\pi)}\,\mathrm{sech}\,(\tfrac{1}{2}\pi\tau)\prod_{r=1}^{n}[\tfrac{1}{4}\tau^2 + (r-\tfrac{1}{2})^2]. \tag{7-4-28}$$

In addition we have the special cases

$$\int_0^\infty P_{-\frac{1}{2}+i\tau}(x)x^{-\frac{3}{2}}\,dx = \sqrt{2}\,\mathrm{sech}\,(\tfrac{1}{2}\pi\tau) \tag{7-4-29}$$

$$\int_1^\infty P_{-\frac{1}{2}+i\tau}(x)x^{-\frac{5}{2}}\,dx = \frac{2\sqrt{2}}{3}\tau\,\mathrm{cosech}\,(\tfrac{1}{2}\pi\tau) \tag{7-4-30}$$

resulting from formulas (1) and (2) on the same page of the book by Erdelyi *et al.*

In certain cases it is more convenient to use the representation

$$P_{-\frac{1}{2}+i\tau}{}^m(\cosh\alpha) = \sqrt{\frac{2}{\pi}}\frac{(\sinh\alpha)^m}{\Gamma(\frac{1}{2}-m)}\cosh(\pi\tau)\int_0^\infty \frac{\cos(\tau t)\,dt}{(\cosh\alpha + \cosh t)^{m+\frac{1}{2}}} \tag{7-4-31}$$

instead of equation (7-4-2)—cf. Lagrange (1939), p. 47. The special case

$$P_{-\frac{1}{2}+i\tau}(\cosh\alpha) = \frac{\sqrt{2}}{\pi}\cosh(\pi\tau)\int_0^\infty \frac{\cos(\tau t)\,dt}{\sqrt{(\cosh\alpha + \cosh t)}} \tag{7-4-32}$$

is also useful.

From equation (7-4-31) we can derive yet another representation of the function $P_{-\frac{1}{2}+i\tau}{}^m(\cosh\alpha)$. We consider the integral

$$\begin{aligned}\int_0^\infty t^{m-\frac{1}{2}}e^{-t\cosh\alpha}K_{i\tau}(t)\,dt &= \int_0^\infty t^{m-\frac{1}{2}}e^{-t\cosh\alpha}\,dt\int_0^\infty e^{-t\cosh u}\cos(\tau u)\,du\\ &= \int_0^\infty \cos(\tau u)\,du\int_0^\infty t^{m-\frac{1}{2}}e^{-t(\cosh\alpha+\cosh u)}\,dt\\ &= \Gamma(m+\tfrac{1}{2})\int_0^\infty \frac{\cos(\tau u)\,du}{(\cosh u + \cosh\alpha)^{m+\frac{1}{2}}}.\end{aligned}$$

As a result of equation (7-4-31), we therefore derive the integral representation

$$P_{-\frac{1}{2}+i\tau}{}^m(\cosh\alpha)$$

$$= (-1)^m 2^{\frac{1}{2}}\pi^{-\frac{3}{2}}\cosh(\pi\tau)(\sinh\alpha)^m\int_0^\infty t^{m-\frac{1}{2}}e^{-t\cosh\alpha}K_{i\tau}(t)\,dt \tag{7-4-33}$$

where, it will be recalled, m is a positive integer.

As an example of the use of the integral representation (7-4-32) we have the relation

$$\int_0^\infty g(\tau)\,\operatorname{sech}(\pi\tau)P_{-\frac{1}{2}+i\tau}(\cosh\alpha)\,d\tau$$

$$= \frac{\sqrt{2}}{\pi}\int_0^\infty g(\tau)\,d\tau \int_0^\infty \frac{\cos(\tau t)\,dt}{\sqrt{(\cosh\alpha + \cosh t)}}$$

$$= \frac{1}{\sqrt{\pi}}\int_0^\infty \frac{G_c(t)\,dt}{\sqrt{(\cosh\alpha + \cosh t)}} \tag{7-4-34}$$

where $G_c(t)$ denotes the Fourier cosine transform of the function $g(\tau)$.

7-5 FOCK'S THEOREMS

We shall begin by finding the form of the operator Φ_0. Suppose that

$$\Phi_0^{-1}[f_0^*(\tau);\tau \to \alpha] = f(\alpha): \tag{7-5-1}$$

then, by equation (7-3-28) we see that this is equivalent to the relation

$$\int_0^\infty \tau\tanh(\pi\tau)P_{-\frac{1}{2}+i\tau}(\cosh\alpha)f_0^*(\tau)\,d\tau = f(\alpha). \tag{7-5-2}$$

Determining the operator Φ_0 is therefore equivalent to finding the solution

$$f_0^*(\tau) = \Phi_0[f(\alpha);\alpha \to \tau] \tag{7-5-3}$$

of this integral equation. If we insert the formula (7-4-3) for the function $P_{-\frac{1}{2}+i\tau}(\cosh\alpha)$, we find that

$$f(\alpha) = \frac{\sqrt{2}}{\pi}\int_0^\infty \tau f_0^*(\tau)\,d\tau \int_\alpha^\infty \frac{\sin(\tau t)\,dt}{\sqrt{(\cosh t - \cosh\alpha)}}$$

$$= \frac{1}{\sqrt{\pi}}\int_\alpha^\infty \frac{dt}{\sqrt{(\cosh t - \cosh\alpha)}}\cdot\sqrt{\frac{2}{\pi}}\int_0^\infty \tau f_0^*(\tau)\sin(\tau t)\,d\tau$$

a relation which we may write in the alternative form

$$f(\alpha) = \frac{1}{\sqrt{\pi}}\int_\alpha^\infty \mathscr{F}_s[\tau f^*(\tau);t]\frac{dt}{\sqrt{(\cosh t - \cosh\alpha)}}.$$

Regarding this as an integral equation for $\mathscr{F}_s[\tau f^*(\tau);t]$, we see from equations (3-16-14) and (3-16-16) that it has the solution

$$\mathscr{F}_s[\tau f^*(\tau);t] = -\frac{1}{\sqrt{\pi}}\frac{d}{dt}\int_t^\infty \frac{f(\alpha)\,d\alpha}{\sqrt{(\cosh\alpha - \cosh t)}}.$$

Inverting this result by means of the Fourier sine inversion theorem, we find that

$$\tau f_0^*(\tau) = -\frac{1}{\sqrt{\pi}}\mathscr{F}_s\left[\frac{d}{dt}\int_t^\infty \frac{f(\alpha)\sinh\alpha\, d\alpha}{\sqrt{(\cosh\alpha - \cosh t)}}; t \to \tau\right]$$

$$= \frac{\tau}{\sqrt{\pi}}\mathscr{F}_c\left[\int_t^\infty \frac{f(\alpha)\sinh\alpha\, d\alpha}{\sqrt{(\cosh\alpha - \cosh t)}}; t \to \tau\right].$$

Hence we find that

$$f_0^*(\tau) = \frac{\sqrt{2}}{\pi}\int_0^\infty \cos(\tau t)\, d\tau \int_t^\infty \frac{f(\alpha)\sinh\alpha\, d\alpha}{\sqrt{(\cosh\alpha - \cosh t)}}$$

$$= \int_0^\infty f(\alpha)\sinh\alpha\, d\alpha \cdot \frac{\sqrt{2}}{\pi}\int_0^\alpha \frac{\cos(\tau t)\, d\tau}{\sqrt{(\cosh\alpha - \cosh t)}}$$

Using equation (7-4-2) we have finally that

$$f_0^*(\tau) = \Phi_0[f(\alpha);\tau] = \int_0^\infty f(\alpha)P_{-\frac{1}{2}+i\tau}(\cosh\alpha)\sinh\alpha\, d\alpha. \tag{7-5-4}$$

We can put the pair of formulas (7-5-2) and (7-5-4) in a slightly different form. If we replace $f(\alpha)$ by $g(\cosh\alpha)$ and if

$$g(x) = \int_0^\infty P_{-\frac{1}{2}+i\tau}(x)g_0^*(\tau)\, d\tau, \qquad x \in [1,\infty) \tag{7-5-5}$$

then

$$g_0^*(\tau) = \tau\tanh(\pi\tau)\int_1^\infty P_{-\frac{1}{2}+i\tau}(x)g(x)\, dx. \tag{7-5-6}$$

We have proceeded in a purely formal way. However, it is a simple matter to prove the following two theorems:

Theorem 1 *If a function $g(x)$ defined on $[1,\infty)$ is such that the function*

$$h(t) = 2\sinh(\tfrac{1}{2}t)g(\cosh t)$$

has a first derivative which is absolutely integrable on $[0,\infty)$, and a second derivative absolutely integrable in any finite interval, and if $h(0) = h(+\infty) = 0$, then $g(x)$ can be written as the integral (7-5-5) where $g_0^*(\tau)$ is defined by equation (7-5-6).

$g_0^*(0) = 0$, then $g_0^*(\tau)$ can be written in the form (7-5-6), where $g(x)$ is

Theorem 2 *If a function $g_0{}^*(\tau)$, absolutely integrable in $[0,\infty)$, has a derivative which is absolutely integrable in any finite interval and if $g_0^*(0) = 0$, then $g_0^*(\tau)$ can be written in the form (7-5-6), where $g(x)$ is defined by equation (7-5-5).*

See Fock (1943). Another form of the theorem, making weaker restrictions on the function $g(x)$, is also due to Fock (1943):

Theorem 3 *If a function $g(x)$ defined on $[1,\infty)$ is such that the integral*

$$\int_1^\infty \frac{|g(x)|\,dx}{\sqrt{(x+1)}}$$

exists, then at every point x in whose neighborhood $g(x)$ has a bounded variation

$$\int_0^\infty P_{-\frac{1}{2}+i\tau}(x)g_0^*(\tau)\,d\tau = \tfrac{1}{2}[g(x+) + g(x-)]$$

where

$$g_0^*(\tau) = \tau \tanh(\pi\tau) \int_1^\infty P_{-\frac{1}{2}+i\tau}(x)g(x)\,dx.$$

For other forms of the theorem the reader should consult Lebedev (1950).

7-6 MEHLER–FOCK TRANSFORMS OF ZERO ORDER

We can write Fock's theorems as the inversion theorem for an integral transform. If we define the *Mehler–Fock transform of order m* of a function $f(\alpha)$ by the equation

$$f_m^*(\tau) = \Phi_m[f(\alpha);\tau] = \int_0^\infty f(\alpha)P_{-\frac{1}{2}+i\tau}{}^m(\cosh\alpha)\sinh\alpha\,d\alpha$$

then the transform of order zero is defined by the equation

$$f_0^*(\tau) = \Phi_0[f(\alpha);\tau] = \int_0^\infty f(\alpha)P_{-\frac{1}{2}+i\tau}(\cosh\alpha)\sinh\alpha\,d\alpha \tag{7-6-1}$$

and Fock's theorems state that the inverse transform is given by the formula

$$f(\alpha) = \Phi_0^{-1}[f_0^*(\tau);\alpha] = \int_0^\infty \tau f_0^*(\tau)\tanh(\pi\tau)P_{-\frac{1}{2}+i\tau}(\cosh\alpha)\,d\tau. \tag{7-6-2}$$

The results obtained in sec. 7-4 can now be restated as rules for these transforms. Thus equations (7-4-10), (7-4-11), (7-4-16), (7-4-21), (7-4-22), and (7-4-24) can be written respectively in the forms

$$\Phi_0[H(t-\alpha)(\cosh t - \cosh\alpha)^{-\frac{1}{2}};\alpha\to\tau] = \sqrt{2}\tau^{-1}\sin(t\tau) \tag{7-6-3}$$

$$\Phi_0[H(\alpha-t)(\cosh\alpha - \cosh t)^{-\frac{1}{2}};\alpha\to\tau] = \sqrt{2}\tau^{-1}\coth(\pi\tau)\cos(\tau t) \tag{7-6-4}$$

$$\Phi_0[(\cosh\alpha + \cos\beta)^{-1};\alpha\to\tau] = \pi\,\mathrm{sech}(\pi\tau)P_{-\frac{1}{2}+i\tau}(\cos\beta), \qquad (-\pi < \beta < \pi) \tag{7-6-5}$$

$$\Phi_0^{-1}\left[\tau^{-1}\coth(\pi\tau)\int_0^a g_1(s)\cos(s\tau)\,ds;\tau\to\alpha\right]$$
$$=\frac{1}{\sqrt{2}}\int_0^{\min(a,\alpha)}\frac{g_1(s)\,ds}{\sqrt{(\cosh\alpha-\cosh s)}} \tag{7-6-6}$$

$$\Phi_0^{-1}\left[\tau^{-1}\coth(\pi\tau)\int_a^\infty g_1(s)\cos(s\tau)\,ds;\tau\to\alpha\right]$$
$$=\frac{H(\alpha-a)}{\sqrt{2}}\int_a^\alpha\frac{g_1(s)\,ds}{\sqrt{(\cosh\alpha-\cosh s)}} \tag{7-6-7}$$

$$\Phi_0^{-1}\left[\frac{\cosh(\beta\tau)}{\tau\tanh(\pi\tau)};\alpha\right]=\frac{1}{\sqrt{[2(\cosh\alpha+\cos\beta)]}}. \tag{7-6-8}$$

Inverting this last result by Fock's theorem, we find that

$$\Phi_0[(\cosh\alpha+\cos\beta)^{-\frac{1}{2}};\tau]=\frac{\sqrt{2}}{\tau}\cdot\frac{\cosh(\beta\tau)}{\sinh(\pi\tau)},\qquad(-\pi<\beta<\pi). \tag{7-6-9}$$

Putting $\beta=0$, we find that

$$\Phi_0[\operatorname{sech}(\tfrac{1}{2}\alpha);\tau]=2\tau^{-1}\operatorname{cosech}(\pi\tau) \tag{7-6-10}$$

and putting $\beta=\frac{1}{2}\pi$ that

$$\Phi_0[\operatorname{sech}^{\frac{1}{2}}(\alpha);\tau]=2^{-\frac{1}{2}}\tau^{-1}\operatorname{cosech}(\tfrac{1}{2}\pi\tau). \tag{7-6-11}$$

Similarly, the results (7-4-12) and (7-4-18) can be written in the form

$$\Phi_0[f(\cosh\alpha);\tau]=\mathscr{F}_c[\mathscr{W}_{\frac{1}{2}}\{f(x);\cosh t\};t\to\tau] \tag{7-6-12}$$
$$=\coth(\pi\tau)\mathscr{F}_s[\mathscr{R}_{\frac{1}{2}}\{f(x);\cosh t\};t\to\tau] \tag{7-6-13}$$

and the pair (7-4-20) and (7-4-23) can be written in the form

$$\Phi_0^{-1}[\tau^{-1}\coth(\pi\tau)g(\tau);\alpha]=\frac{1}{\sqrt{\pi}}\int_0^\alpha\frac{G_c(t)\,dt}{\sqrt{(\cosh\alpha-\cosh t)}} \tag{7-6-14}$$

$$\Phi_0^{-1}[\tau^{-1}g(\tau);\alpha]=\frac{1}{\sqrt{\pi}}\int_\alpha^\infty\frac{G_s(t)\,dt}{\sqrt{(\cosh t-\cosh\alpha)}} \tag{7-6-15}$$

where $G_c(t)$ and $G_s(t)$ are respectively the Fourier cosine and the Fourier sine transforms of the function $g(\tau)$.

It is sometimes convenient to use an alternative notation. We shall write

$$\tilde{f}_0(\tau)=\phi_0[f(x);\tau]=\int_1^\infty f(x)P_{-\frac{1}{2}+i\tau}(x)\,dx \tag{7-6-16}$$

and the inverse operator ϕ_0^{-1} will then be defined by the equation

$$f(x) = \phi_0^{-1}[\tilde{f}_0(\tau); x] = \int_0^\infty \tau \tanh(\pi\tau)\tilde{f}_0(\tau)P_{-\frac{1}{2}+i\tau}(x)\, d\tau. \tag{7-6-17}$$

It will be observed that

$$\phi_0[f(x); \tau] = \Phi_0[f(\cosh\alpha); \tau] \tag{7-6-18}$$

and that

$$\phi_0^{-1}[\tilde{f}_0(\tau); x] = \Phi_0^{-1}[\tilde{f}_0(\tau); \cosh^{-1} x]. \tag{7-6-19}$$

For example we can write equations (7-4-13), (7-4-18) as rules for calculating $\phi_0 g$:

$$\phi_0[g(x); \tau] = \mathscr{F}_c[\mathscr{W}_{\frac{1}{2}}\{g(x); \cosh t\}; t \to \tau] \tag{7-6-20}$$

$$= \coth(\pi\tau)\mathscr{F}_s[\mathscr{R}_{\frac{1}{2}}\{g(x); \cosh t\}; t \to \tau] \tag{7-6-21}$$

and the special results (7-4-25), (7-4-29), (7-4-30), (7-4-27), (7-4-28) in the forms

$$\phi_0[e^{-ax}; \tau] = \sqrt{\frac{2}{\pi a}}\, K_{i\tau}(a), \qquad |\arg a| < \tfrac{1}{2}\pi \tag{7-6-22}$$

$$\phi_0[x^{-\frac{3}{2}}; \tau] = \sqrt{2}\, \operatorname{sech}(\tfrac{1}{2}\pi\tau) \tag{7-6-23}$$

$$\phi_0[x^{-\frac{5}{2}}; \tau] = \frac{2\sqrt{2}}{3}\tau \operatorname{cosech}(\tfrac{1}{2}\pi\tau) \tag{7-6-24}$$

$$\phi_0[x^{-2n-\frac{1}{2}}; \tau] = \frac{2^{2n-3}}{\Gamma(2n+\frac{1}{2})}\sqrt{(2\pi)}\,\tau \operatorname{cosech}(\tfrac{1}{2}\pi\tau)\prod_{r=1}^{n}(\tfrac{1}{4}\tau^2 + r^2), \qquad (n \geqslant 2) \tag{7-6-25}$$

$$\phi_0[x^{-2n-\frac{3}{2}}; \tau] = \frac{2^{2n-1}}{\Gamma(2n+\frac{3}{2})}\sqrt{(2\pi)}\operatorname{sech}(\tfrac{1}{2}\pi\tau)\prod_{r=1}^{n}[\tfrac{1}{4}\tau^2 + (r-\tfrac{1}{2})^2] \tag{7-6-26}$$

respectively. We can also write equation (7-4-15) as

$$\phi_0[(x+t)^{-1}; x \to \tau] = \pi \operatorname{sech}(\pi\tau)P_{-\frac{1}{2}+i\tau}(t), \qquad |t| < 1. \tag{7-6-27}$$

Applying the inversion theorem to this formula we find that

$$\pi\int_0^\infty \tau \frac{\sinh(\pi\tau)}{\cosh^2(\pi\tau)} P_{-\frac{1}{2}+i\tau}(x)P_{-\frac{1}{2}+i\tau}(t)\, d\tau = \frac{1}{x+t}, \qquad |t| < 1. \tag{7-6-28}$$

Corresponding to equation (7-4-34) we have the relation

$$\phi_0^{-1}[\tau^{-1}\operatorname{cosech}(\pi\tau)g(\tau);\tau\to\alpha] = \frac{1}{\sqrt{\pi}}\int_0^\infty \frac{G_c(t)\,dt}{\sqrt{(\cosh\alpha+\cosh t)}} \tag{7-6-29}$$

where, as before, $G_c(t)$ is the Fourier cosine transform of $g(\tau)$. For example, if in this result we take

$$g(\tau) = \frac{\cosh(\pi\tau-\beta\tau)}{\cosh(\pi\tau)}$$

so that

$$G_c(t) = \sqrt{\frac{2}{\pi}}\,\frac{\sin\frac{1}{2}\beta\cosh\frac{1}{2}t}{\cosh t-\cos\beta}$$

(see, for example, formula (12) on p. 31 of Vol. 1 of Erdelyi *et al.* (1954)), we find that we obtain the equation

$$\begin{aligned}\Phi_0^{-1}&\left[2\tau^{-1}\frac{\cosh(\pi\tau-\beta\tau)}{\sinh(2\pi\tau)};\tau\to\alpha\right]\\ &= \frac{\sqrt{2}}{\pi}\sin(\tfrac{1}{2}\beta)\int_0^\infty\frac{\cosh(\frac{1}{2}t)\,dt}{(\cosh t-\cos\beta)\sqrt{(\cosh\alpha+\cosh t)}}\\ &= \frac{1}{\pi}\operatorname{cosec}(\tfrac{1}{2}\beta)\int_0^\infty\frac{du}{(1+k^2u^2)\sqrt{(u^2+1)}}\end{aligned}$$

where $k = \cosh(\frac{1}{2}\alpha)\operatorname{cosec}(\frac{1}{2}\beta)$. Putting $u = (v^2-1)^{-\frac{1}{2}}$, we find that

$$\begin{aligned}\int_0^\infty\frac{du}{(1+k^2u^2)\sqrt{(u^2+1)}} &= \int_1^\infty\frac{dv}{v^2+(k^2-1)}\\ &= (k^2-1)^{-\frac{1}{2}}\tan^{-1}\sqrt{(k^2-1)}\end{aligned}$$

and hence that

$$\begin{aligned}\Phi_0^{-1}&\left[2\tau^{-1}\frac{\cosh(\pi\tau-\beta\tau)}{\sinh(2\pi\tau)};\tau\to\alpha\right]\\ &= \frac{\sqrt{2}}{\pi}(\cosh\alpha+\cos\beta)^{-\frac{1}{2}}\tan^{-1}\left[\frac{\cosh\alpha+\cos\beta}{1-\cos\beta}\right]^{\frac{1}{2}}.\end{aligned} \tag{7-6-30}$$

7-7 A RELATION OF PARSEVAL TYPE

It is a simple matter to establish formally a relation for Mehler–Fock transforms (of order zero) which is the analogue of Parseval's relation for Fourier transforms. (Cf., Lebedev, 1949; Lowndes, 1961.)

If

$$\tilde{f}_0(\tau) = \phi_0[f(x);\tau], \qquad \tilde{g}_0(\tau) = \phi_0[g(x);\tau]$$

then

$$\int_0^\infty \tau \tanh(\pi\tau)\tilde{f}_0(\tau)\tilde{g}_0(\tau)\,d\tau$$

$$= \int_0^\infty \tau \tanh(\pi\tau)\tilde{g}_0(\tau)\,d\tau \int_1^\infty f(x)P_{-\frac{1}{2}+i\tau}(x)\,dx$$

$$= \int_1^\infty f(x)\,dx \int_0^\infty \tau \tanh(\pi\tau)g_0(\tau)P_{-\frac{1}{2}+i\tau}(x)\,d\tau.$$

By the inversion theorem the inner integral is equal to $g(x)$ so that we obtain the Parseval relation

$$\int_0^\infty \tau \tanh(\pi\tau)\tilde{f}_0(\tau)\tilde{g}_0(\tau)\,d\tau = \int_1^\infty f(x)g(x)\,dx. \tag{7-7-1}$$

For example, if we take

$$f(x) = \frac{1}{x+s}, \qquad g(x) = \frac{1}{x+t}$$

then, from equation (7-6-27) we have

$$\tilde{f}_0(\tau) = \pi \operatorname{sech}(\pi\tau)P_{-\frac{1}{2}+i\tau}(s), \ |s| < 1,$$

$$\tilde{g}_0(\tau) = \pi \operatorname{sech}(\pi\tau)P_{-\frac{1}{2}+i\tau}(t), \ |t| < 1.$$

Now if $s \neq t$,

$$\int_1^\infty \frac{dx}{(x+s)(x+t)} = \frac{1}{t-s}\log\frac{1+t}{1+s}$$

while

$$\int_1^\infty \frac{dx}{(x+s)^2} = \frac{1}{s+1}$$

so that from (7-7-1) we have the result

$$\int_0^\infty \frac{\tau \sinh(\pi\tau)}{\cosh^3(\pi\tau)} P_{-\frac{1}{2}+i\tau}(s)P_{-\frac{1}{2}+i\tau}(t)\,d\tau = \begin{cases} \dfrac{1}{\pi^2(t-s)}\log\dfrac{1+t}{1+s}, & (t \neq s) \\[2ex] \dfrac{1}{\pi^2(1+s)} & (t = s) \end{cases}$$

where $|s| < 1$, $|t| < 1$.

If we take

$$f(x) = e^{-ax}, \qquad g(x) = e^{-bx},$$

where a and b are positive real constants, then by equation (7-6-22)

$$\tilde{f}_0(\tau) = \sqrt{\frac{2}{\pi a}}\,K_{i\tau}(a), \qquad \tilde{g}_0(\tau) = \sqrt{\frac{2}{\pi b}}\,K_{i\tau}(b).$$

Also

$$\int_1^\infty e^{-(a+b)x}\,dx = \frac{e^{-a-b}}{a+b}$$

so that from (7-7-1) we have the result

$$\int_0^\infty \tau \tanh(\pi\tau) K_{i\tau}(a) K_{i\tau}(b)\,d\tau = \frac{\pi\sqrt{(ab)}}{2(a+b)}\,e^{-(a+b)}. \tag{7-7-2}$$

If we take

$$f(x) = x^{-\frac{3}{2}}, \qquad g(x) = x^{-\frac{5}{2}}$$

so that, by equations (7-6-23) and (7-6-24), we must take

$$\tilde{f}_0(\tau) = \sqrt{2}\operatorname{sech}(\tfrac{1}{2}\pi\tau), \qquad \tilde{g}_0(\tau) = \frac{2\sqrt{2}}{3}\,\tau \operatorname{cosech}(\tfrac{1}{2}\pi\tau)$$

then, since

$$\int_1^\infty x^{-4}\,dx = \tfrac{1}{3},$$

we find that

$$\int_0^\infty \tau^2 \operatorname{sech}(\pi\tau)\,d\tau = \tfrac{1}{2}.$$

As a final example, we take

$$f(x) = e^{-ax}, \qquad g(x) = \frac{1}{x+t}$$

and hence

$$\tilde{f}_0(\tau) = \sqrt{\frac{2}{\pi a}}\,K_{i\tau}(a), \qquad \tilde{g}_0(\tau) = \pi \operatorname{sech}(\pi\tau) P_{-\frac{1}{2}+i\tau}(t)$$

to obtain the equation

$$\int_0^\infty \tau \tanh(\pi\tau)\operatorname{sech}(\pi\tau) P_{-\frac{1}{2}+i\tau}(t) K_{i\tau}(a)\,d\tau = \sqrt{\frac{a}{2\pi}}\int_1^\infty \frac{e^{-ax}\,dx}{x+t}.$$

In terms of the exponential integral

$$ei(x) = \int_x^\infty e^{-u}u^{-1}\,du$$

we can, therefore, write

$$\int_0^\infty \tau \tanh(\pi\tau)\operatorname{sech}(\pi\tau) P_{-\frac{1}{2}+i\tau}(t) K_{i\tau}(a)\,d\tau = \sqrt{\frac{a}{2\pi}}\,e^{at}\,ei(a+at).$$

We can write this result in the form

$$\phi_0^{-1}\left[\sqrt{\frac{2\pi}{a}}\operatorname{sech}(\pi\tau)K_{i\tau}(a);\tau\to x\right] = e^{ax}ei(a+ax)$$

or, alternatively, in the form

$$\phi_0[e^{ax}ei(a+ax);x\to\tau] = \sqrt{\frac{2\pi}{a}}\operatorname{sech}(\pi\tau)K_{i\tau}(a). \tag{7-7-3}$$

7-8 THE SOLUTION OF BOUNDARY VALUE PROBLEMS USING MEHLER–FOCK TRANSFORMS OF ZERO ORDER

In this section we shall describe briefly how Mehler–Fock transforms may be used to derive the solution of boundary value problems in potential theory which are of significance in the mathematical theory of elasticity. This method is exploited fully in Uflyand (1963) to which the reader is referred.

We consider first the problem of determining a function ψ which is harmonic in the half-space $z \geqslant 0$ and which satisfies the conditions

$$\psi = F(\rho), \qquad 0 \leqslant \rho \leqslant a \tag{7-8-1}$$

$$\frac{\partial\psi}{\partial z} = 0, \qquad \rho > a \tag{7-8-2}$$

on the boundary plane $z = 0$. This is the boundary value problem which arises in the contact problem for a half-space in the theory of elasticity.

We saw in sec. 7-3 that if we write

$$\frac{1}{\sqrt{2}}\operatorname{sech}(\tfrac{1}{2}\alpha)F(a\tanh^2\tfrac{1}{2}\alpha) \equiv f(\alpha) \tag{7-8-3}$$

then this boundary value problem has the solution

$$\psi(\alpha,\beta) = \sqrt{(\cosh\alpha + \cos\beta)}\,\Phi_0^{-1}\left[A_0(\tau)\frac{\cosh(\pi\tau-\beta\tau)}{\cosh(\pi\tau)};\tau\to\alpha\right] \tag{7-8-4}$$

where $\Phi_0^{-1}[A_0(\tau);\alpha] = f(\alpha)$, so that

$$\psi(\alpha,\beta) = \sqrt{(\cosh\alpha + \cos\beta)}\,\Phi_0^{-1}\left[f_0^*(\tau)\frac{\cosh(\pi\tau-\beta\tau)}{\cosh(\pi\tau)};\tau\to\alpha\right] \tag{7-8-5}$$

where $f_0^*(\tau)$ is the Mehler–Fock transform of order zero of the function $f(\alpha)$ defined by equation (7-8-3).

For example, suppose that $F(\alpha) = \varepsilon$, a constant, then

$$f(\alpha) = \frac{\varepsilon}{\sqrt{2}} \operatorname{sech}(\tfrac{1}{2}\alpha) \tag{7-8-6}$$

and

$$f_0^*(\tau) = \frac{\varepsilon}{\sqrt{2}} \Phi_0[\operatorname{sech}(\tfrac{1}{2}\alpha);\tau].$$

By equation (7-6-10) we find that, in this particular case,

$$A_0(\tau) = \sqrt{2}\varepsilon\tau^{-1} \operatorname{cosech}(\pi\tau). \tag{7-8-7}$$

Substituting from (7-8-7) into (7-8-4) we obtain the solution

$$\psi(\alpha,\beta) = \varepsilon\sqrt{[2(\cosh\alpha + \cos\beta)]}\Phi_0^{-1}\left[\tau^{-1}\frac{2\cosh(\pi\tau - \beta\tau)}{\sinh(2\pi\tau)};\tau \to \alpha\right].$$

Hence, by equation (7-6-29), we have finally

$$\psi(\alpha,\beta) = \frac{2\varepsilon}{\pi}\tan^{-1}\left[\frac{\cosh\alpha + \cos\beta}{1 - \cos\beta}\right]^{\frac{1}{2}}. \tag{7-8-8}$$

A quantity which is of physical interest in this problem is the number

$$P = -2\pi\int_0^a \rho\left(\frac{\partial\psi}{\partial z}\right)_{z=0} d\rho. \tag{7-8-9}$$

Now

$$\frac{\partial\psi}{\partial\beta} = -\frac{\sqrt{(2)}\varepsilon}{\pi}\cos(\tfrac{1}{2}\beta)(\cosh\alpha + \cos\beta)^{-\frac{1}{2}}$$

so that, from equation (7-3-14), we derive the relation

$$\left(\frac{\partial\psi}{\partial z}\right)_{z=0} = -\frac{2\varepsilon}{\pi a}\cosh(\tfrac{1}{2}\alpha).$$

Also on $z = 0$, $0 \leqslant \rho \leqslant a$, we have $\rho = a\tanh(\tfrac{1}{2}\alpha)$, $0 < \alpha < \infty$, so that

$$P = 2\varepsilon a\int_0^\infty \tanh(\tfrac{1}{2}\alpha)\operatorname{sech}(\tfrac{1}{2}\alpha)\,d\alpha.$$

The integration is elementary and we get

$$P = 4\varepsilon a. \tag{7-8-10}$$

Next we consider the problem of determining a function ψ which is harmonic in the half-space $z \geqslant 0$ and which satisfies the conditions

$$\psi = 0, \qquad 0 \leqslant \rho \leqslant a, \tag{7-8-11}$$

$$\frac{\partial\psi}{\partial z} = -G(\rho), \qquad \rho > a, \tag{7-8-12}$$

on the boundary $z = 0$. This is the boundary value problem which arises in the analysis (within the framework of the classical theory of elasticity) of the axisymmetric deformation of an infinite body weakened by an external circular crack—cf. Sneddon and Lowengrub (1969), sec. 3-9.

If we introduce a new function g defined by the equation

$$g(\alpha) = \frac{a}{2\sqrt{2}} \operatorname{cosech}^3 (\tfrac{1}{2}\alpha) G(a \coth \tfrac{1}{2}\alpha) \tag{7-8-13}$$

then it follows from equation (7-3-12) that we can replace the pair of equations (7-8-11) and (7-8-12) by the pair

$$\psi(\alpha,0) = 0, \tag{7-8-14}$$

$$\left(\frac{\partial \psi}{\partial \beta}\right)_{\beta=\pi} = \sqrt{2} \sinh (\tfrac{1}{2}\alpha) g(\alpha). \tag{7-8-15}$$

Taking a representation of $\psi(\alpha,\beta)$ of the form

$$\psi(\alpha,\beta) = \sqrt{(\cosh \alpha + \cos \beta)}\,\Phi_0^{-1}\left[\tau^{-1} A_0(\tau) \frac{\sinh(\beta\tau)}{\cosh(\pi\tau)}; \alpha\right] \tag{7-8-16}$$

we find that $\psi(\alpha,0) = 0$ and that

$$\left(\frac{\partial \psi}{\partial \beta}\right)_{\beta=\pi} = \sqrt{2} \sinh (\tfrac{1}{2}\alpha) \Phi_0^{-1}[A_0(\tau); \alpha].$$

Hence the boundary condition (7-8-15) is satisfied if we take $A_0(\tau) = g_0^*(\tau)$, i.e., the required solution is

$$\psi(\alpha,\beta) = \sqrt{(\cosh \alpha + \cos \beta)}\,\Phi_0^{-1}\left[\tau^{-1} g_0^*(\tau) \frac{\sinh(\beta\tau)}{\cosh(\pi\tau)}; \alpha\right], \tag{7-8-17}$$

where $g_0^*(\tau)$ is the Mehler–Fock transform of the function $g(\alpha)$ defined by equation (7-8-13).

We note that for this solution

$$\left(\frac{\partial \psi}{\partial \beta}\right)_{\beta=0} = \sqrt{2} \cosh (\tfrac{1}{2}\alpha) \Phi_0^{-1}[g_0^*(\tau) \operatorname{sech}(\pi\tau); \alpha] \tag{7-8-18}$$

and hence, from equation (7-3-14), that, when $0 \leqslant \rho \leqslant a$,

$$\left(\frac{\partial \psi}{\partial z}\right)_{z=0} = \frac{2\sqrt{2}}{a} \cosh^3 (\tfrac{1}{2}\alpha) \Phi_0^{-1}[g_0^{-1}[g_0^*(\tau) \operatorname{sech}(\pi\tau); \alpha] \tag{7-8-19}$$

where $\rho = a \tanh (\tfrac{1}{2}\alpha)$.

For example, if

$$G(\rho) = ka^3(b^2 + \rho^2)^{-\frac{3}{2}} \tag{7-8-20}$$

where k and b are constants, then

$$g(\alpha) = \frac{ka}{2\sqrt{2}} \operatorname{cosech}^3(\tfrac{1}{2}\alpha)(\lambda^2 + \coth^2 \tfrac{1}{2}\alpha)^{-\frac{3}{2}}, \qquad \lambda = b/a \tag{7-8-21}$$

and

$$g_0^*(\tau) = \sqrt{2}\, ka \sin(\tfrac{1}{2}\xi) \tan(\tfrac{1}{2}\xi) \frac{\sinh(\pi\tau - \xi\tau)}{\sinh(\pi\tau)} \tag{7-8-22}$$

with $\sin \frac{1}{2}\xi = (1 + \lambda^2)^{-\frac{1}{2}} = a(a^2 + b^2)^{-\frac{1}{2}}$.

The solution is then given by equations (7-8-17) and (7-8-22). As a further example, see Problem 7-7 below.

The problem of determining the axisymmetric stresses in an elastic body weakened by a penny-shaped crack can be reduced to the problem of determining the solution ψ in the half-space $z \geqslant 0$ of Laplace's equation satisfying the conditions

$$\left(\frac{\partial \psi}{\partial z}\right)_{z=0} = -p(\rho), \qquad 0 \leqslant \rho \leqslant a$$

$$\psi_{z=0} = 0, \qquad \rho > a$$

(see Sneddon and Lowengrub, 1969, sec. 3-2). This is equivalent to the problem of determining a harmonic function $\psi(\alpha,\beta)$ satisfying the boundary conditions

$$\frac{\partial \psi(\alpha,0)}{\partial \beta} = -2a \operatorname{sech}^2(\tfrac{1}{2}\alpha) p(a \tanh \tfrac{1}{2}\alpha) \tag{7-8-23}$$

$$\psi(\alpha,\pi) = 0. \tag{7-8-24}$$

To satisfy Laplace's equation and the boundary condition (7-8-24) we take

$$\psi(\alpha,\beta) = \sqrt{(\cosh \alpha + \cos \beta)}\, \Phi_0^{-1}\left[\tau^{-1} f^*(\tau) \frac{\sinh(\pi\tau - \beta\tau)}{\cosh(\pi\tau)}; \alpha\right] \tag{7-8-25}$$

and to satisfy equation (7-8-23) we take

$$f_0^*(\tau) = \Phi_0[f(\alpha); \tau]$$

where the function $f(\alpha)$ is defined by the equation

$$f(\alpha) = \sqrt{(2)}\, a \operatorname{sech}^3(\tfrac{1}{2}\alpha) p(a \tanh \tfrac{1}{2}\alpha). \tag{7-8-26}$$

For instance, if $p(\rho) = p_0$, a constant, then

$$f(\alpha) = \sqrt{(2)} p_0 a \operatorname{sech}^3(\tfrac{1}{2}\alpha)$$

and we find that

$$f_0^*(\tau) = 8\sqrt{(2)}p_0 a\tau \operatorname{cosech}(\pi\tau)$$

—see Problem 7-2 below. Hence we have the solution

$$\psi(\alpha,\beta) = 16\sqrt{(\cosh^2 \tfrac{1}{2}\alpha - \sin^2 \tfrac{1}{2}\beta)}p_0 a \Phi_0^{-1}\left[\frac{2\sinh(\pi\tau - \beta\tau)}{\sinh(2\pi\tau)};\alpha\right].$$

In particular

$$\begin{aligned}\psi(\alpha,0) &= 16\cosh(\tfrac{1}{2}\alpha)p_0 a \Phi_0^{-1}[\operatorname{sech}(\pi\tau);\alpha] \\ &= 16p_0 a \frac{\cosh(\tfrac{1}{2}\alpha)}{1 + \cosh\alpha} \\ &= 8p_0 a \operatorname{sech}(\tfrac{1}{2}\alpha).\end{aligned}$$

Now

$$\operatorname{sech}^2 \tfrac{1}{2}\alpha = 1 - \coth^2 \tfrac{1}{2}\alpha = 1 - \rho^2/a^2$$

so that

$$(\psi)_{z=0} = 8p_0\sqrt{(a^2 - \rho^2)}, \qquad 0 \leqslant \rho \leqslant a.$$

7-9 DUAL INTEGRAL EQUATIONS WITH TRIGONOMETRICAL KERNELS

An interesting application of the Mehler–Fock inversion theorem to the solution of dual integral equations with trigonometrical kernels has been given by Babloian (1964). In the next section we shall describe how such equations arise in a natural way in potential theory.

We begin by considering the dual integral equations

$$g(x) \equiv \int_0^\infty f(\tau)\cos(x\tau)\,d\tau = g_1(x), \qquad 0 \leqslant x \leqslant a \tag{7-9-1}$$

$$h(x) \equiv \int_0^\infty \coth(\pi\tau)f(\tau)\sin(x\tau)\,d\tau = h_2(x), \qquad x > a \tag{7-9-2}$$

in which the functions $g_1(x)$ and $h_2(x)$ are prescribed. If we multiply the first of these equations by

$$\frac{\sqrt{2}}{\pi\sqrt{(\cosh\alpha - \cosh x)}}$$

and integrate it with respect to x from 0 to α and use equation (7-4-3), we obtain the equation

$$\int_0^\infty f(\tau)P_{-\frac{1}{2}+i\tau}(\cosh\alpha)\,d\tau = \Omega_1(\alpha), \qquad 0 \leqslant \alpha \leqslant a \tag{7-9-3}$$

where

$$\Omega_1(\alpha) = \frac{\sqrt{2}}{\pi}\int_0^\alpha \frac{g_1(x)\,dx}{\sqrt{(\cosh\alpha - \cosh x)}}. \tag{7-9-4}$$

Similarly, if we multiply the second equation by

$$\frac{\sqrt{2}}{\pi\sqrt{(\cosh x - \cosh\alpha)}}$$

and integrate with respect to x from α to ∞, we find on using equation (7-4-4) that

$$\int_0^\infty f(\tau)P_{-\frac{1}{2}+i\tau}(\cosh\alpha)\,d\tau = \Omega_2(\alpha), \qquad \alpha \geqslant a \tag{7-9-5}$$

where

$$\Omega_2(\alpha) = \frac{\sqrt{2}}{\pi}\int_\alpha^\infty \frac{h_2(x)\,dx}{\sqrt{(\cosh x - \cosh\alpha)}}. \tag{7-9-6}$$

If we now define the function $\Omega(\alpha)$ by the equations

$$\Omega(\alpha) = \begin{cases} \Omega_1(\alpha), & 0 \leqslant \alpha \leqslant a \\ \Omega_2(\alpha), & \alpha > a \end{cases} \tag{7-9-7}$$

we see that equations (7-9-3) and (7-9-5) are equivalent to

$$\Phi_0^{-1}[\tau^{-1}\coth(\pi\tau)f(\tau);\alpha] = \Omega(\alpha)$$

which may be inverted to yield the relation

$$f(\tau) = \tau\tanh(\pi\tau)\,\Phi_0[\Omega(\alpha);\tau] = \tau\tanh(\pi\tau)\Omega_0^*(\tau). \tag{7-9-8}$$

Written out in full this equation takes the form

$$f(\tau) = \tau\tanh(\pi\tau)\int_0^a \Omega_1(\alpha)P_{-\frac{1}{2}+i\tau}(\cosh\alpha)\sinh\alpha\,d\alpha$$
$$+ \tau\tanh(\pi\tau)\int_a^\infty \Omega_2(\alpha)P_{-\frac{1}{2}+i\tau}(\cosh\alpha)\sinh\alpha\,d\alpha. \tag{7-9-9}$$

In many physical problems what we are interested in is the form of $g(x)$ when $x > a$, or the form of $h(x)$ when $0 \leqslant x \leqslant a$. Now from the definition (7-9-1) of $g(x)$ we have the equation

$$\begin{aligned} g(x) &= \int_0^\infty \tau\tanh(\pi\tau)\cos(x\tau)\,d\tau \int_0^\infty \Omega(\alpha)\sinh\alpha\,P_{-\frac{1}{2}+i\tau}(\cosh\alpha)\,d\alpha \\ &= \int_0^\infty \Omega(\alpha)\sinh\alpha\,d\alpha \int_0^\infty \tau\tanh(\pi\tau)\cos(x\tau)P_{-\frac{1}{2}+i\tau}(\cosh\alpha)\,d\tau \\ &= \frac{d}{dx}\int_0^\infty \Omega(\alpha)\sinh\alpha\,d\alpha \int_0^\infty \tanh(\pi\tau)\sin(x\tau)P_{-\frac{1}{2}+i\tau}(\cosh\alpha)\,d\tau. \end{aligned}$$

The inner integral can be evaluated by means of equation (7-4-8) to give

$$g(x) = \frac{1}{\sqrt{2}} \int_0^x \frac{\Omega(\alpha) \sinh \alpha \, d\alpha}{\sqrt{(\cosh x - \cosh \alpha)}}.$$

Hence, when $x > a$, $g(x) = g_2(x)$, where the function $g_2(x)$ is defined by the equation

$$g_2(x) = \frac{1}{\sqrt{2}} \frac{d}{dx} \int_0^a \frac{\Omega_1(\alpha) \sinh \alpha \, d\alpha}{\sqrt{(\cosh x - \cosh \alpha)}} + \frac{1}{\sqrt{2}} \frac{d}{dx} \int_a^x \frac{\Omega_2(\alpha) \sinh \alpha \, d\alpha}{\sqrt{(\cosh x - \cosh \alpha)}}. \tag{7-9-10}$$

Similarly from the definition of $h(x)$ we have the equation

$$h(x) = \int_0^\infty \tau \sin (x\tau) \, d\tau \int_0^\infty \Omega(\alpha) \sinh \alpha \, P_{-\frac{1}{2}+i\tau}(\cosh \alpha) \, d\alpha$$

which can be written as

$$h(x) = -\frac{d}{dx} \int_0^\infty \Omega(\alpha) \sinh \alpha \, d\alpha \int_0^\infty \cos (x\tau) P_{-\frac{1}{2}+i\tau}(\cosh \alpha) \, d\tau$$

and, using equation (7-4-5), we see that this in turn can be written in the form

$$h(x) = -\frac{1}{\sqrt{2}} \frac{d}{dx} \int_x^\infty \frac{\Omega(\alpha) \sinh \alpha \, d\alpha}{\sqrt{(\cosh \alpha - \cosh x)}}.$$

Hence, when $0 \leqslant x \leqslant a$, $h(x) = h_1(x)$ where

$$h_1(x) = -\frac{1}{\sqrt{2}} \frac{d}{dx} \int_x^a \frac{\Omega_1(\alpha) \sinh \alpha \, d\alpha}{\sqrt{(\cosh \alpha - \cosh x)}} - \frac{1}{\sqrt{2}} \frac{d}{dx} \int_a^\infty \frac{\Omega_2(\alpha) \sinh \alpha \, d\alpha}{\sqrt{(\cosh \alpha - \cosh x)}}. \tag{7-9-11}$$

If we have the dual integral equations

$$j(x) \equiv \int_0^\infty \tau^{-1} f(\tau) \sin (x\tau) \, d\tau = j_1(x), \qquad 0 \leqslant x \leqslant a \tag{7-9-12}$$

$$h(x) \equiv \int_0^\infty \coth (\pi\tau) f(\tau) \sin (x\tau) \, d\tau = h_2(x), \qquad x > a \tag{7-9-13}$$

then, differentiating both sides of equation (7-9-12) with respect to x, we obtain the equation

$$j'(x) \equiv \int_0^\infty f(\tau) \cos (x\tau) \, d\tau = j_1'(x), \qquad 0 \leqslant x \leqslant a. \tag{7-9-14}$$

Hence the solution of the pair of equations (7-9-12), (7-9-13) is again given by equation (7-9-9) but now we define $\Omega_1(\alpha)$ by the equation

$$\Omega_1(\alpha) = \frac{\sqrt{2}}{\pi}\int_0^\infty \frac{j_1'(x)\,dx}{\sqrt{(\cosh\alpha - \cosh x)}} \tag{7-9-15}$$

instead of by equation (7-9-4). The function $\Omega_2(\alpha)$ is defined, as previously, by the equation (7-9-6).

From the definition of $j(x)$, we have the equation

$$j(x) = \int_0^\infty \Omega(\alpha)\sinh\alpha\,d\alpha \int_0^\infty \tanh(\pi\tau)\sin(x\tau)P_{-\frac{1}{2}+i\tau}(\cosh\alpha)\,d\tau$$

and evaluating the inner integral by means of the formula (7-4-8), we find that when $x > a$, $j(x) = j_2(x)$, where

$$j_2(x) = \frac{1}{\sqrt{2}}\int_0^a \frac{\Omega_1(\alpha)\sinh\alpha\,d\alpha}{\sqrt{(\cosh x - \cosh\alpha)}} + \frac{1}{\sqrt{2}}\int_a^x \frac{\Omega_2(\alpha)\sinh\alpha\,d\alpha}{\sqrt{(\cosh x - \cosh\alpha)}}. \tag{7-9-16}$$

The value of $h(x)$ for $0 < x < a$ is given again by equation (7-9-11) with—as in equation (7-9-16)—$\Omega_1(\alpha)$ and $\Omega_2(\alpha)$ given by equations (7-9-15) and (7-9-6) respectively.

To obtain the solution of the pair of dual integral equations

$$k(x) \equiv \int_0^\infty \tau f(\tau)\sin(x\tau)\,d\tau = k_1(x), \qquad 0 \leqslant x \leqslant a \tag{7-9-17}$$

$$h(x) \equiv \int_0^\infty \coth(\pi\tau) f(\tau)\sin(x\tau)\,d\tau = h_2(x), \qquad x > a \tag{7-9-18}$$

we integrate the first equation of the pair with respect to x from 0 to x. It then assumes the form (7-9-1) with $g_1(x)$ now defined by the equation

$$g_1(x) = C + \int_0^x k_1(u)\,du \tag{7-9-19}$$

where C is the constant defined by the equation

$$C = \int_0^\infty f(\tau)\,d\tau. \tag{7-9-20}$$

The solution is then given in terms of C by equations (7-9-9), (7-9-4), and (7-9-6) with $g_1(x)$ now defined by equation (7-9-19); it will therefore be of the form

$$f(\tau) = Cf_0(\tau) + f_1(\tau)$$

where $f_1(\tau)$, $f_2(\tau)$ can be computed by the above formulas. The constant C will then, as a consequence of its definition (7-9-20), be determined by the simple equation

$$C\left[1 - \int_0^\infty f_0(\tau)\,d\tau\right] = \int_0^\infty f_1(\tau)\,d\tau.$$

Dual integral equations of the type

$$g(x) \equiv \int_0^\infty f(\tau)\cos(\tau x)\,d\tau = g_1(x), \qquad 0 \leqslant x \leqslant a \tag{7-9-21}$$

$$l(x) \equiv \int_0^\infty \tau^{-1}\coth(\pi\tau)f(\tau)\cos(\tau x)\,d\tau = l_2(x), \qquad x > a \tag{7-9-22}$$

can be treated in a similar way. If we differentiate equation (7-9-22) with respect to x, we find that it becomes

$$\int_0^\infty \coth(\pi\tau)f(\tau)\sin(\tau x)\,d\tau = -l_2'(x), \qquad x > a$$

which is identical with equation (7-9-2) with

$$h_2(x) = -l_2'(x). \tag{7-9-23}$$

The solution of equations (7-9-21) and (7-9-22) is therefore given by equations (7-9-9), (7-9-4), (7-9-6) with $h_2(x)$ now defined by equation (7-9-23).

Similarly the solution of the pair of equations

$$g(x) \equiv \int_0^\infty f(\tau)\cos(\tau x)\,d\tau = g_1(x), \qquad 0 \leqslant x \leqslant a \tag{7-9-24}$$

$$m(x) \equiv \int_0^\infty \tau\coth(\pi\tau)f(\tau)\cos(x\tau)\,d\tau = m_2(x), \qquad x > a \tag{7-9-25}$$

is given by equations (7-9-9), (7-9-4), (7-9-6) with $h_2(x)$ now defined by equation

$$h_2(x) = C - \int_x^\infty m_2(u)\,du. \tag{7-9-26}$$

7-10 MIXED BOUNDARY VALUE PROBLEMS FOR A SEMI-INFINITE STRIP

We shall now show how dual integral equations of the types discussed in the last section arise in the analysis of certain mixed boundary value problems for the plane harmonic equation

$$\frac{\partial^2\psi}{\partial x^2} + \frac{\partial^2\psi}{\partial y^2} = 0, \qquad 0 \leqslant y \leqslant \pi, \qquad x \geqslant 0. \tag{7-10-1}$$

We begin by considering the case (Cf. Fig. 7-5) in which the boundary conditions on the function $\psi(x,y)$ are

$$\begin{aligned}
\psi(0,y) &= 0, && 0 \leqslant y \leqslant \pi \\
\psi(x,0) &= 0, && x \geqslant 0 \\
\psi(x,\pi) &= j_1(x), && 0 \leqslant x \leqslant a \\
\frac{\partial\psi(x,\pi)}{\partial y} &= h_2(x), && x > a.
\end{aligned}$$

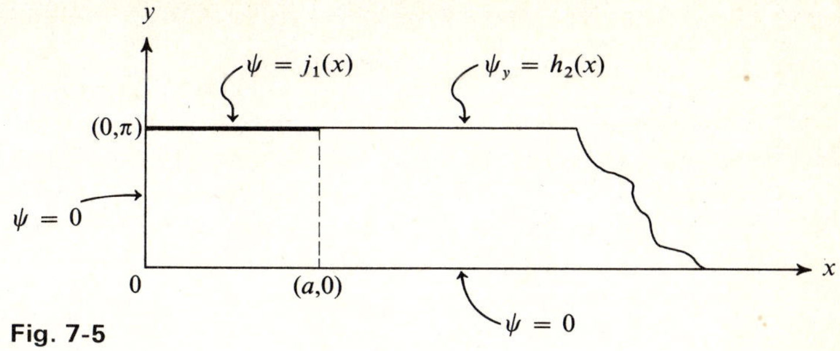

Fig. 7-5

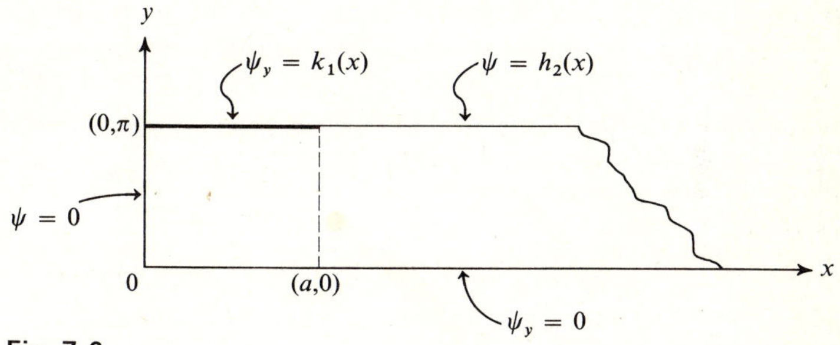

Fig. 7-6

In addition, we require that $\psi(x,y) \to 0$ as $x \to \infty$, $0 \leqslant y \leqslant \pi$. It is easily verified that the solution of this boundary value problem is given by the formula

$$\psi(x,y) = \int_0^\infty \tau^{-1} f(\tau) \frac{\sinh(\tau y)}{\sinh(\pi\tau)} \sin(\tau x)\, d\tau \tag{7-10-2}$$

provided that the function $f(\tau)$ is the solution of the pair of dual integral equations (7-9-12) and (7-9-13).

Similarly the mixed boundary value problem specified by the equations (7-10-1) and

$$\begin{aligned} \psi(0,y) &= 0, & 0 \leqslant y \leqslant \pi \\ \frac{\partial\psi(x,0)}{\partial y} &= 0, & x \geqslant 0 \\ \frac{\partial\psi(x,\pi)}{\partial y} &= k_1(x), & 0 \leqslant x \leqslant a \\ \psi(x,\pi) &= h_2(x), & x > a \end{aligned}$$

(Cf. Fig. 7-6) has solution

$$\psi(x,y) = \int_0^\infty f(\tau) \frac{\cosh(\tau y)}{\sinh(\pi\tau)} \sin(x\tau)\, d\tau \tag{7-10-3}$$

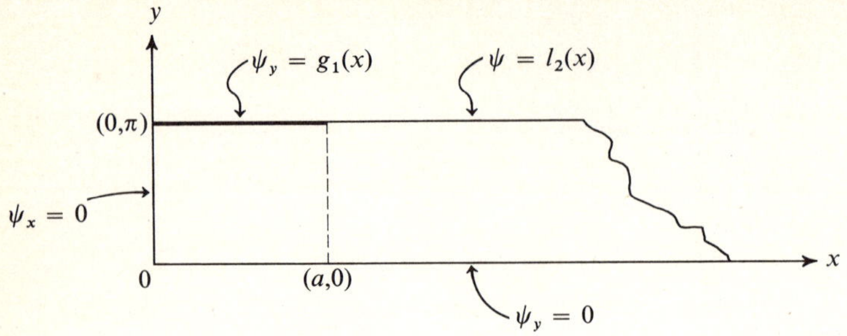

Fig. 7-7

provided that the function $f(\tau)$ is the solution of the pair of dual integral equations (7-9-17) and (7-9-18).

Further, the solution of the mixed boundary value problem described by equations (7-10-1) and

$$\frac{\partial\psi(0,y)}{\partial x} = 0, \qquad 0 \leqslant y \leqslant \pi$$

$$\frac{\partial\psi(x,0)}{\partial y} = 0, \qquad x \geqslant 0$$

$$\frac{\partial\psi(x,\pi)}{\partial y} = g_1(x), \qquad 0 \leqslant x \leqslant a$$

$$\psi(x,\pi) = l_2(x), \qquad x > a$$

(Cf. Fig. 7-7) is given by the equation

$$\psi(x,y) = \int_0^\infty \tau^{-1} f(\tau) \frac{\cosh(\tau y)}{\sinh(\pi\tau)} \cos(\tau x)\, d\tau \tag{7-10-4}$$

where $f(\tau)$ is the solution of the dual integral equations (7-9-21), (7-9-22).

Finally we note that the function $\psi(x,y)$ harmonic in the strip $x \geqslant 0$, $0 \leqslant y \leqslant \pi$ and satisfying the mixed boundary conditions

$$\frac{\partial\psi(0,y)}{\partial x} = 0, \qquad 0 \leqslant y \leqslant \pi$$

$$\psi(x,0) = 0, \qquad x \geqslant 0$$

$$\psi(x,\pi) = g_1(x), \qquad 0 \leqslant x \leqslant a$$

$$\frac{\partial\psi(x,\pi)}{\partial y} = m_2(x), \qquad x \geqslant a$$

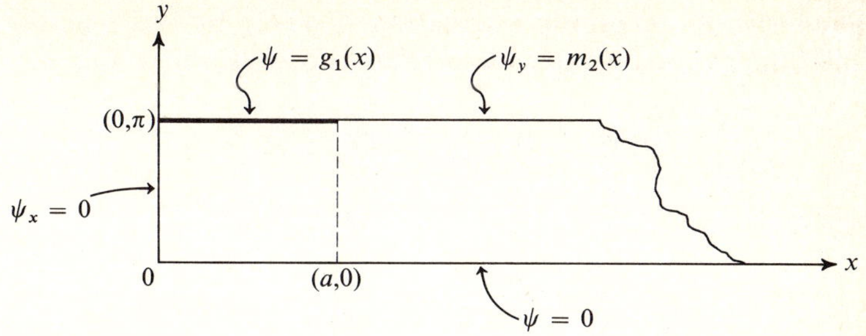

Fig. 7-8

(Cf. Fig. 7-8) is given by the formula

$$\psi(x,y) = \int_0^\infty f(\tau)\frac{\sinh(\tau y)}{\sinh(\pi\tau)}\cos(\tau x)\,d\tau$$

where $f(\tau)$ is the solution of the dual integral equations (7-9-24), (7-9-25).

7-11 DUAL INTEGRAL EQUATIONS WITH LEGENDRE FUNCTION KERNELS

We shall now discuss briefly the methods introduced by Babloian (1964) for the solution of certain dual integral equations with kernel $P_{-\frac{1}{2}+i\tau}(\cosh\alpha)$.

We begin by considering the pair

$$g(\alpha) \equiv \int_0^\infty f(\tau)P_{-\frac{1}{2}+i\tau}(\cosh\alpha)\,d\tau = g_1(\alpha), \qquad 0 \leqslant \alpha \leqslant \alpha_1 \tag{7-11-1}$$

$$h(\alpha) \equiv \int_0^\infty \tau\tanh(\pi\tau)f(\tau)P_{-\frac{1}{2}+i\tau}(\cosh\alpha)\,d\tau = h_2(\alpha), \qquad \alpha > \alpha_1. \tag{7-11-2}$$

Multiplying equation (7-11-1) by $\sinh\alpha(\cosh x - \cosh\alpha)^{-\frac{1}{2}}$, integrating the result with respect to α from 0 to x, and then differentiating with respect to x, we find that if $0 \leqslant x \leqslant \alpha_1$,

$$\frac{1}{\sqrt{\pi}}\int_0^\infty f(\tau)\,d\tau\,\frac{d}{dx}\int_0^x \frac{\sinh\alpha}{\sqrt{(\cosh x - \cosh\alpha)}}P_{-\frac{1}{2}+i\tau}(\cosh\alpha)\,d\alpha = F_1(x)$$

where the function $F_1(x)$ is defined in terms of the prescribed function $g_1(\alpha)$ through the equation

$$F_1(x) = \frac{1}{\sqrt{\pi}}\frac{d}{dx}\int_0^x \frac{g_1(\alpha)\sinh\alpha\,d\alpha}{\sqrt{(\cosh x - \cosh\alpha)}}. \tag{7-11-3}$$

From equation (7-4-6) we see that this is equivalent to the equation

$$\sqrt{\frac{2}{\pi}}\int_0^\infty f(\tau)\cos(\tau x)\,d\tau = F_1(x), \qquad 0 \leqslant x \leqslant \alpha_1. \tag{7-11-4}$$

Similarly, if we multiply equation (7-11-2) by $\sinh\alpha(\cosh\alpha - \cosh x)^{-\frac{1}{2}}$ and integrate with respect to α from x to ∞, we find that

$$\sqrt{\frac{2}{\pi}}\int_0^\infty f(\tau)\cos(\tau x)\,d\tau = F_2(x), \qquad x \geqslant \alpha_1 \tag{7-11-5}$$

where

$$F_2(x) = \frac{1}{\sqrt{\pi}}\int_x^\infty \frac{h_2(\alpha)\sinh\alpha\,d\alpha}{\sqrt{(\cosh\alpha - \cosh x)}}, \qquad x \geqslant \alpha_1. \tag{7-11-6}$$

Hence if we define the function $F(x)$ by the equations

$$F(x) = \begin{cases} F_1(x), & 0 \leqslant x \leqslant \alpha_1 \\ F_2(x) & x > \alpha_1 \end{cases} \tag{7-11-7}$$

we see that

$$\mathscr{F}_c[f(\tau);x] = F(x)$$

and hence that

$$f(\tau) = \mathscr{F}_c[F(x);\tau]. \tag{7-11-8}$$

The solution of the pair of dual integral equations (7-11-1), (7-11-2) is therefore given by equations (7-11-8), (7-11-7), (7-11-3), and (7-11-6).

If we substitute the value of $f(\tau)$ into the equation (7-11-1), we find that if $\alpha > \alpha_1$, $g(\alpha) = g_2(\alpha)$, where

$$g_2(\alpha) = \sqrt{\frac{2}{\pi}}\int_0^{\alpha_1}\frac{F_1(x)\,dx}{\sqrt{(\cosh\alpha - \cosh x)}} + \sqrt{\frac{2}{\pi}}\int_{\alpha_1}^\infty \frac{F_2(x)\,dx}{\sqrt{(\cosh\alpha - \cosh x)}}. \tag{7-11-9}$$

It can also be shown that if $0 < \alpha < \alpha_1$, $h(\alpha) = h_1(\alpha)$, where

$$h_1(\alpha) = \frac{\operatorname{cosech}\alpha}{\sqrt{\pi}}\frac{d}{d\alpha}\int_\alpha^{\alpha_1}\frac{\{F_2(\alpha_1) - F_1(x)\}}{\sqrt{(\cosh x - \cosh\alpha)}}\sinh x\,dx. \tag{7-11-10}$$

In a similar way, we can show that the pair of dual integral equations

$$g(\alpha) \equiv \int_0^\infty \tau f(\tau)P_{-\frac{1}{2}+i\tau}(\cosh\alpha)\,d\tau = g_1(\alpha), \qquad 0 \leqslant \alpha \leqslant \alpha_1 \tag{7-11-11}$$

$$h(\alpha) \equiv \int_0^\infty \tanh(\pi\tau)f(\tau)P_{-\frac{1}{2}+i\tau}(\cosh\alpha)\,d\tau = h_2(\alpha), \qquad \alpha > \alpha_1 \tag{7-11-12}$$

has the solution

$$f(\tau) = \mathscr{F}_s[F(x);\tau] \tag{7-11-13}$$

where now the components of the function $F(x)$ are defined by the pair of equations

$$F_1(x) = \frac{1}{\sqrt{\pi}} \int_0^x \frac{g_1(\alpha) \sinh \alpha \, d\alpha}{\sqrt{(\cosh x - \cosh \alpha)}}, \qquad 0 \leqslant x \leqslant \alpha_1 \tag{7-11-14}$$

$$F_2(x) = -\frac{1}{\sqrt{\pi}} \frac{d}{dx} \int_x^\infty \frac{h_2(\alpha) \sinh \alpha \, d\alpha}{\sqrt{(\cosh \alpha - \cosh x)}}, \qquad x > \alpha_1. \tag{7-11-15}$$

Here we find that, when $\alpha > \alpha_1$, $g(\alpha) = g_2(\alpha)$, where

$$g_2(\alpha) = \frac{\operatorname{cosech} \alpha}{\sqrt{\pi}} \frac{d}{d\alpha} \int_{\alpha_1}^\alpha \frac{\{F_2(x) - F_1(\alpha_1)\}}{\sqrt{(\cosh \alpha - \cosh x)}} \sinh x \, dx \tag{7-11-16}$$

and that, when $0 \leqslant \alpha < \alpha_1$, $h(\alpha) = h_1(\alpha)$, where

$$h_1(\alpha) = \frac{1}{\sqrt{\pi}} \int_\alpha^{\alpha_1} \frac{F_1(x) \, dx}{\sqrt{(\cosh x - \cosh \alpha)}} + \frac{1}{\sqrt{\pi}} \int_{\alpha_1}^\infty \frac{F_2(x) \, dx}{\sqrt{(\cosh x - \cosh \alpha)}}. \tag{7-11-17}$$

As a generalization of the pair of equations (7-11-1), (7-11-2), we could consider the pair

$$\int_0^\infty f(\tau)[1 + N(\tau)] P_{-\frac{1}{2}+i\tau}(\cosh \alpha) \, d\tau = g_1(\alpha), \qquad 0 \leqslant \alpha \leqslant \alpha_1 \tag{7-11-18}$$

$$\int_0^\infty \tau \tanh (\pi\tau) f(P_{-\frac{1}{2}+i\tau}(\cosh \alpha) \, d\tau = 0, \qquad \alpha > \alpha_1 \tag{7-11-19}$$

where the function $N(\tau)$ is prescribed (and is assumed to be absolutely integrable) and it is desired to find the function $f(\tau)$.

If we write equation (7-11-18) in the form

$$\int_0^\infty f(\tau) P_{-\frac{1}{2}+i\tau}(\cosh \alpha) \, d\tau = g_1(\alpha) - \int_0^\infty f(\tau) N(\tau) P_{-\frac{1}{2}+i\tau}(\cosh \alpha) \, d\tau$$

we see that the manipulations which lead to the solution (7-11-8) will, in this case, lead to the relation

$$f(\tau) = \sqrt{\frac{2}{\pi}} \int_0^{\alpha_1} F_1(x) \cos (\tau x) \, dx \tag{7-11-20}$$

where

$$F_1(x) = \frac{1}{\sqrt{\pi}} \frac{d}{dx} \int_0^x \frac{\sinh \alpha \, d\alpha}{\sqrt{(\cosh x - \cosh \alpha)}} \times$$

$$\times \left\{ g_1(\alpha) - \int_0^\infty f(\tau) N(\tau) P_{-\frac{1}{2}+i\tau}(\cosh \alpha) \, d\tau \right\}$$

If we write

$$G_1(x) = \frac{1}{\sqrt{\pi}} \frac{d}{dx} \int_0^x \frac{g_1(\alpha) \sinh \alpha \, d\alpha}{\sqrt{(\cosh x - \cosh \alpha)}}, \qquad 0 \leqslant x \leqslant \alpha_1 \tag{7-11-21}$$

and make use of equation (7-4-6) to write

$$\begin{aligned}\frac{1}{\sqrt{\pi}} \frac{d}{dx} \int_0^x \frac{\sinh \alpha \, d\alpha}{\sqrt{(\cosh x - \cosh \alpha)}} \int_0^\infty f(\tau) N(\tau) P_{-\frac{1}{2}+i\tau}(\cosh \alpha) \, d\tau \\ = \sqrt{\frac{2}{\pi}} \int_0^\infty f(\tau) N(\tau) \cos(\tau x) \, d\tau \\ = \frac{2}{\pi} \int_0^{\alpha_1} F_1(u) \, du \int_0^\infty N(\tau) \cos(\tau x) \cos(\tau u) \, d\tau\end{aligned}$$

we see that the solution of the pair of equations (7-11-18), (7-11-19) can be written in the form (7-11-20), where $F_1(x)$ is the solution of the Fredholm integral equation of the second kind

$$F_1(x) + \int_0^{\alpha_1} F_1(u) K(x,u) \, du = G_1(x), \qquad 0 \leqslant x \leqslant \alpha_1 \tag{7-11-22}$$

with kernel defined by the equation

$$K(x,u) = \frac{1}{\sqrt{(2\pi)}} \{N_c(x + u) + N_c(|x - u|)\},$$
$$N_c(x) = \mathscr{F}_c[N(\tau); x] \tag{7-11-23}$$

and with free term defined by equation (7-11-21).

Similarly, we can show that the dual integral equations

$$\int_0^\infty \tau f(\tau) P_{-\frac{1}{2}+i\tau}(\cosh \alpha) \, d\tau = 0, \qquad 0 \leqslant \alpha \leqslant \alpha_1 \tag{7-11-24}$$

$$\int_0^\infty \tanh(\pi\tau)[1 + N(\tau)] f(\tau) P_{-\frac{1}{2}+i\tau}(\cosh \alpha) \, d\tau = h_2(\alpha), \qquad \alpha > \alpha_1 \tag{7-11-25}$$

have solution

$$f(\tau) = \sqrt{\frac{2}{\pi}} \int_{\alpha_1}^\infty F_2(x) \sin(\tau x) \, dx \tag{7-11-26}$$

where $F_2(x)$ is the solution of the Fredholm integral equation

$$F_2(x) + \int_{\alpha_1}^\infty F_2(u) L(x,u) \, du = H_2(x), \qquad x > \alpha_1 \tag{7-11-27}$$

whose free term is defined by the equation

$$H_2(x) = -\frac{1}{\sqrt{\pi}}\frac{d}{dx}\int_x^\infty \frac{h_2(\alpha)\sinh\alpha\, d\alpha}{\sqrt{(\cosh\alpha - \cosh x)}}, \qquad x > \alpha_1 \tag{7-11-28}$$

and whose kernel is defined by the equation

$$L(x,u) = \frac{1}{\sqrt{(2\pi)}}\{N_c(|x-u|) - N_c(x+u)\} \tag{7-11-29}$$

$N_c(x)$ having the same meaning as before.

7-12 APPLICATION TO A TORSION PROBLEM

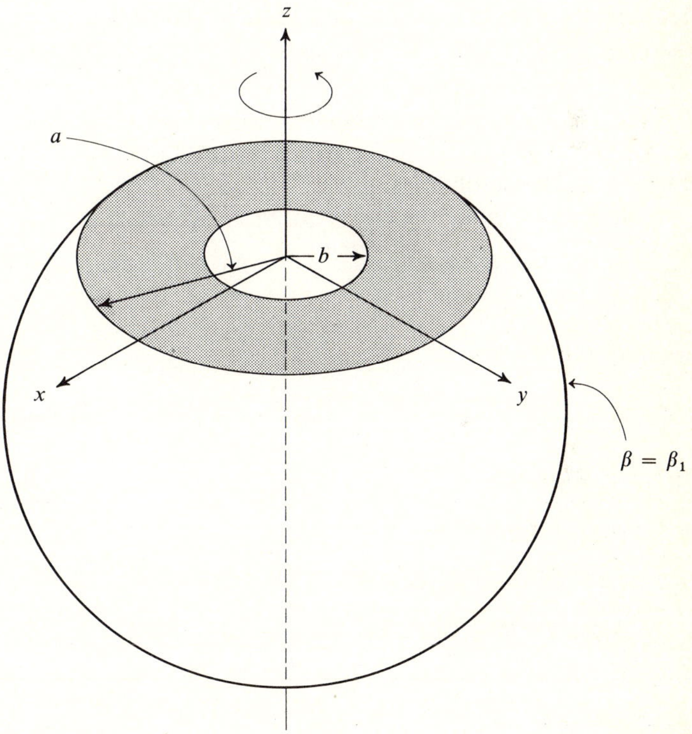

Fig. 7-9

To illustrate the application of the methods of the last section Babloian (1964) considered the problem of the torsion of a spherical segment when the deformation is produced by rotating a circular disk, rigidly attached at the center of the plane section, through a small angle θ. (Cf. Fig. 7-9.) We assume that the radius of the plane is a and that that of the rigid disk is $b < a$. The remaining part of the flat boundary is assumed to be free from external loading, and the spherical surface is assumed to be rigidly fixed.

If we express the torsion equation in terms of toroidal coordinates (α,β,ϕ) then it is easily shown that any displacement vector of the form $(0,0,u_\phi)$ can be characterized by an expression of the form

$$u_\phi = \frac{a \sinh \alpha}{\cosh \alpha + \cos \beta} \psi(\alpha,\beta) \tag{7-12-1}$$

where the function $\psi(\alpha,\beta)$ satisfies the differential equation

$$\frac{\partial^2 \psi}{\partial \alpha^2} + \frac{\partial^2 \psi}{\partial \beta^2} + \frac{3(\cosh \alpha \cos \beta + 1)}{\sinh \alpha (\cosh \alpha + \cos \beta)} \frac{\partial \psi}{\partial \beta} + \frac{3 \sin \beta}{\cosh \alpha + \cos \beta} \frac{\partial \psi}{\partial \beta} = 0. \tag{7-12-2}$$

The only nonvanishing components of the stress tensor are $\sigma_{\alpha\phi}$ and $\sigma_{\beta\phi}$ and these are given by the pair of equations

$$\sigma_{\alpha\phi} = \mu \sinh \alpha \frac{\partial \psi}{\partial \alpha}, \qquad \sigma_{\beta\phi} = \mu \sinh \alpha \frac{\partial \psi}{\partial \beta} \tag{7-12-3}$$

where μ is the rigidity modulus.

If we now make the substitution

$$\psi(\alpha,\beta) = \operatorname{cosech} \alpha (\cosh \alpha + \cos \beta)^{\frac{3}{2}} e^{\pm \tau\beta} v(\mu), \qquad \mu = \cosh \alpha \tag{7-12-4}$$

we find that

$$(1 - \mu^2) \frac{d^2 v}{d\mu^2} - 2\mu \frac{dv}{d\mu} - \left[(\tfrac{1}{4} + \tau^2) + \frac{1}{1 - \mu^2} \right] v(\mu) = 0$$

showing that

$$v(\mu) = P^1_{-\frac{1}{2}+i\tau}(\mu)$$

is a possible form for $v(\mu)$ and hence that

$$\psi(\alpha,\beta) = \operatorname{cosech} \alpha (\cosh \alpha + \cos \beta)^{\frac{3}{2}} \int_0^\infty \psi(\tau) \frac{\sinh(\beta_1 \tau - \beta\tau)}{\cosh(\beta_1 \tau)} P^1_{-\frac{1}{2}+i\tau}(\cosh \alpha)\, d\tau \tag{7-12-5}$$

is a possible form for the function $\psi(\alpha,\beta)$.

Corresponding to the stress function (7-12-5), we have the displacement component

$$u_\phi(\alpha,\beta) = a(\cosh \alpha + \cos \beta)^{\frac{1}{2}} \int_0^\infty \psi(\tau) \frac{\sinh(\beta_1 \tau - \beta\tau)}{\cosh(\beta_1 \tau)} P^1_{-\frac{1}{2}+i\tau}(\cosh \alpha)\, d\tau \tag{7-12-6}$$

and the stress component

$$\sigma_{\beta\phi}(\alpha,\beta) = -\mu(\cosh\alpha + \cos\beta)^{\frac{3}{2}}$$

$$\int_0^\infty \tau\psi(\tau)\frac{\cosh(\beta_1\tau - \beta\tau)}{\cosh(\beta_1\tau)} P^1_{-\frac{1}{2}+i\tau}(\cosh\alpha)\,d\tau \tag{7-12-7}$$

showing that the solution (7-12-5) is such that

$$u_\phi(\alpha,\beta_1) = 0 \tag{7-12-8}$$

$$u_\phi(\alpha,0) = \sqrt{(2)}\,a\cosh(\tfrac{1}{2}\alpha)\int_0^\infty \psi(\tau)\tanh(\beta_1\tau)\,P^1_{-\frac{1}{2}+i\tau}(\cosh\alpha)\,d\tau \tag{7-12-9}$$

$$\sigma_{\beta\phi}(\alpha,0) = -2\sqrt{(2)}\,\mu\cosh^3(\tfrac{1}{2}\alpha)\int_0^\infty \tau\psi(\tau)\,P^1_{-\frac{1}{2}+i\tau}(\cosh\alpha)\,d\tau. \tag{7-12-10}$$

Hence this solution satisfies the condition that the spherical surface is kept fixed and will satisfy the conditions

$$u_\phi = \theta\rho, \qquad 0 \leqslant \rho \leqslant b$$

$$\sigma_{z\phi} = 0, \qquad b \leqslant \rho \leqslant a$$

on the flat face $z = 0$, $0 \leqslant \rho \leqslant a$ provided that we can find a function $\psi(\tau)$ satisfying the dual integral equations

$$\int_0^\infty \psi(\tau)\tanh(\beta_1\tau)\,P^1_{-\frac{1}{2}+i\tau}(\cosh\alpha)\,d\tau = \frac{\theta}{\sqrt{2}}\sinh(\tfrac{1}{2}\alpha)\,\mathrm{sech}^2(\tfrac{1}{2}\alpha), \qquad 0 \leqslant \alpha \leqslant \alpha_1 \tag{7-12-11}$$

$$\int_0^\infty \tau\psi(\tau)\,P^1_{-\frac{1}{2}+i\tau}(\cosh\alpha)\,d\tau = 0, \qquad \alpha > \alpha_1 \tag{7-12-12}$$

where α_1 is defined by the equation

$$b = a\tanh(\tfrac{1}{2}\alpha_1). \tag{7-12-13}$$

Recalling that

$$\frac{d}{d\alpha}P_{-\frac{1}{2}+i\tau}(\cosh\alpha) = P^1_{-\frac{1}{2}+i\tau}(\cosh\alpha)$$

we see that equation (7-12-11) may be replaced by the equation

$$\int_0^\infty \psi(\tau)\tanh(\beta_1\tau)P_{-\frac{1}{2}+i\tau}(\cosh\alpha)\,d\tau = c_1 - \sqrt{(2)}\,\theta\,\mathrm{sech}(\tfrac{1}{2}\alpha), \qquad 0 \leqslant \alpha \leqslant \alpha_1 \tag{7-12-14}$$

where c_1 is a constant, as yet undefined, and that equation (7-12-12) may be replaced by the equation

$$\int_0^\infty \tau\psi(\tau)P_{-\frac{1}{2}+i\tau}(\cosh\alpha)\,d\tau = 0, \qquad \alpha > \alpha_1. \tag{7-12-15}$$

Now equations (7-12-14) and (7-12-15) are equivalent to equations (7-11-18) and (7-11-19) if we write

$$\psi(\tau) = \tanh(\pi\tau)f(\tau),$$

$$N(\tau) = -1 + \tanh(\pi\tau)\tanh(\beta_1\tau) = -\frac{\cosh(\pi\tau - \beta_1\tau)}{\cosh(\pi\tau)\cosh(\beta_1\tau)},$$

so that the solution of the boundary value problem is given by the equation

$$\psi(\tau) = \sqrt{\frac{2}{\pi}}\tanh(\pi\tau)\int_0^{\alpha_1} F_1(x)\cos(\tau x)\,dx \tag{7-12-16}$$

where $F_1(x)$ is the solution of the Fredholm equation

$$F_1(x) - \int_0^{\alpha_1} F_1(u)K(x,u)\,du = \sqrt{2}c_1\coth(\tfrac{1}{2}x) - \tfrac{1}{2}\sqrt{\pi}\theta\sinh x. \tag{7-12-17}$$

In this equation the kernel is defined by the equations

$$K(x,u) = \pi^{-\frac{1}{2}}\{K_c(x+u) + K_c(|x-u|)\} \tag{7-12-18}$$

$$K_c(x) = \mathscr{F}_c\left[\frac{\cosh(\pi\tau - \beta_1\tau)}{\cosh(\pi\tau)\cosh(\beta_1\tau)},\ \tau\to x\right] \tag{7-12-19}$$

and the constant c_1 is arbitrary.

For example, for a hemisphere $\beta = \frac{1}{2}\pi$ and so

$$K_c(x) = \mathscr{F}_c[\operatorname{sech}(\pi\tau);x] = (2\pi)^{-\frac{1}{2}}\operatorname{sech}(\tfrac{1}{2}x).$$

The kernel of the integral equation therefore becomes

$$\frac{2\sqrt{2}}{\pi}\cdot\frac{\cosh(\frac{1}{2}x)\cdot\cosh(\frac{1}{2}u)}{\cosh x + \cosh u}. \tag{7-12-20}$$

The integral equation (7-12-17) may be solved numerically. The value of the constant c_1 is determined by a *physical* condition—namely, by the condition that the sum of the shear stresses which act below the circular disk is bounded. (Cf. Abramian *et al.* 1964.)

7-13 MEHLER–FOCK TRANSFORMS OF ORDER *m*

We shall now consider a transform suitable for the solution of the boundary value problem posed by equations (7-3-29) and (7-3-30); that is, we shall find a form of Φ_m for Φ_m^{-1} defined by equation (7-3-26).

The equation

$$f(\alpha) = (-1)^m \int_0^\infty \tau \tanh(\pi\tau) P_{-\frac{1}{2}+i\tau}{}^m(\cosh\alpha) f_m^*(\tau)\, d\tau \tag{7-13-1}$$

is an obvious generalization of the equation (7-5-2). Writing it in the form

$$f(\cosh^{-1} x) = (-1)^m (x^2 - 1)^{m/2} \frac{d^m}{dx^m} \int_0^\infty \tau \tanh(\pi\tau) P_{-\frac{1}{2}+i\tau}(x) f_m^*(\tau)\, d\tau$$

we see that the analysis which we followed in the derivation of Fock's theorem (sec. 7-5 above) now leads to the equation

$$f(\cosh^{-1} x) = \pi^{-\frac{1}{2}} (-1)^m (x^2 - 1)^{m/2} \frac{d^m}{dx^m} \int_x^\infty \frac{\theta(u)\, du}{\sqrt{(u - x)}}$$

where

$$\mathscr{F}_s[\tau f_m^*(\tau); t] = \theta(\cosh t) \sinh t. \tag{7-13-2}$$

By the methods of sec. 3-16 we can easily show that the solution of this integral equation is

$$\theta(u) = -\frac{1}{\Gamma(m + \frac{1}{2})} \frac{d}{du} \int_u^\infty (x^2 - 1)^{-m/2} f(\cosh^{-1} x)(x - u)^{m-\frac{1}{2}}\, dx$$

so that, by equation (7-13-2), we construct the relation

$$\mathscr{F}_s[\tau f_m^*(\tau); \alpha] = -\frac{1}{\Gamma(m + \frac{1}{2})} \frac{d}{d\alpha} \int_\alpha^\infty (\sinh\xi)^{1-m} f(\xi)(\cosh\xi - \cosh\alpha)^{m-\frac{1}{2}}\, d\xi.$$

Hence we derive the equation

$$\mathscr{F}_c[f_m^*(\tau); \alpha] = \frac{1}{\Gamma(m + \frac{1}{2})} \int_\alpha^\infty (\sinh\xi)^{1-m} f(\xi)(\cosh\xi - \cosh\alpha)^{m-\frac{1}{2}}\, d\xi.$$

Inverting this equation by means of the Fourier cosine inversion theorem, we find that

$$f_m^*(\tau) = \int_0^\infty (\sinh\xi)^{1-m} f(\xi)\, d\xi \frac{2}{\pi\Gamma(m + \frac{1}{2})} \times \\ \times \int_0^\xi \cos(\alpha\tau)(\cosh\xi - \cosh\alpha)^{m-\frac{1}{2}}\, d\alpha.$$

From equation (7-4-1) we see that this reduces to

$$f_m^*(\tau) = \int_0^\infty f(\xi) P_{-\frac{1}{2}+i\tau}{}^{-m}(\cosh\xi) \sinh\xi\, d\xi. \tag{7-13-3}$$

Hence if we define the Mehler–Fock transform[1] of order m of a function $f(\alpha)$ by the equation

$$\Phi_m[f(\alpha);\tau] = f_m^*(\tau) = \int_0^\infty f(\alpha)P_{-\frac{1}{2}+i\tau}^{-m}(\cosh\alpha)\sinh\alpha\,d\alpha, \tag{7-13-4}$$

then the inversion theorem states that

$$\Phi_m^{-1}[f_m^*(\tau);\alpha] = f(\alpha) = (-1)^m\int_0^\infty \tau\tanh(\pi\tau)f_m^*(\tau)P_{-\frac{1}{2}+i\tau}^{m}(\cosh\alpha)\,d\tau. \tag{7-13-5}$$

It would appear that the class of functions f for which this result is valid is not yet clearly defined; Olevskii (1943, 1949) and Vilenkin (1958) impose conditions on the function $f_m^*(\tau)$ but not on $f(\alpha)$. The derivation of Ragab (1965) is purely formal.

Alternatively, if we define

$$\phi_m[f(x);\tau] = \tilde{f}_m(\tau) = \int_1^\infty f(x)P_{-\frac{1}{2}+i\tau}^{m}(x)\,dx \tag{7-13-6}$$

we have the inversion theorem

$$\phi_m^{-1}[\tilde{f}_m(\tau);x] = f(x) = (-1)^m\int_0^\infty \tau\tanh(\pi\tau)f_m^*(\tau)P_{-\frac{1}{2}+i\tau}^{m}(x)\,d\tau. \tag{7-13-7}$$

It is a simple matter to derive a relation of Parseval type:

$$(-1)^m\int_0^\infty \tau\tanh(\pi\tau)\frac{\Gamma(\frac{1}{2}+i\tau+m)}{\Gamma(\frac{1}{2}+i\tau-m)}\tilde{f}_m(\tau)\tilde{g}_m(\tau)\,d\tau$$
$$= (-1)^m\int_0^\infty \tau\tanh(\pi\tau)\frac{\Gamma(\frac{1}{2}+i\tau+m)}{\Gamma(\frac{1}{2}+i\tau-m)}\tilde{f}_m(\tau)\,d\tau\int_1^\infty g(x)P_{-\frac{1}{2}+i\tau}^{-m}(x)\,dx$$

by equation (7-2-9). Hence we have the relation

$$(-1)^m\int_0^\infty \tau\tanh(\pi\tau)\frac{\Gamma(\frac{1}{2}+m+i\tau)}{\Gamma(\frac{1}{2}-m+i\tau)}\tilde{f}_m(\tau)\tilde{g}_m(\tau)\,d\tau = \int_1^\infty f(x)g(x)\,dx.$$

This result can be used in exactly the same way as the Parseval relation (7-7-1) to derive the values of complicated integrals. Examples are to be found in Lowndes (1964).

7-14 APPLICATIONS OF MEHLER–FOCK TRANSFORMS OF ORDER *m*

We shall conclude by considering some applications of the Mehler–Fock transforms of general order.

[1] $f_m^*(\tau)$ in the case $m \neq 0$ is sometimes referred to as a *generalized Mehler transform*.

We begin with the problem of determining a function ψ, harmonic in the half-space $z \geqslant 0$, and satisfying the boundary conditions

$$\psi = F(\rho,\phi), \qquad 0 \leqslant \rho \leqslant a \tag{7-14-1}$$

$$\frac{\partial \psi}{\partial z} = 0, \qquad \rho > a \tag{7-14-2}$$

on the boundary plane $z = 0$. It is easily seen that this is equivalent to the problem of finding a function $\psi(\alpha,\beta,\phi)$ which is harmonic in $0 < \beta < \pi$ and which satisfies

$$\psi(\alpha,0,\phi) = f(\alpha,\phi) \tag{7-14-3}$$

$$\frac{\partial \psi(\alpha,\pi,\phi)}{\partial \beta} = 0 \tag{7-14-4}$$

with the function $f(\alpha,\phi)$ defined by the equation

$$f(\alpha,\phi) = F(a \tanh \tfrac{1}{2}\alpha,\phi). \tag{7-14-5}$$

We have already considered the formal solution of this problem in sec. 7-3; it is given by equations (7-3-32) and (7-3-33).

To illustrate how the method is used in practice we consider the case in which

$$F(\rho,\phi) = c_1 + c_2(\rho/a)\cos\phi \tag{7-14-6}$$

where c_1 and c_2 are constants. The corresponding form for $f(\alpha,\phi)$ is

$$f(\alpha,\phi) = c_1 + c_2 \tanh(\tfrac{1}{2}\alpha)\cos\phi.$$

The function $\psi(\alpha,\beta,\phi)$ is then given by the equation

$$\psi(\alpha,\beta,\phi) = \sqrt{(\cosh\alpha + \cos\beta)}\Phi_0^{-1}\left[A_0(\tau)\frac{\cosh(\pi\tau - \beta\tau)}{\cosh(\pi\tau)};\tau \to \alpha\right]$$
$$+ \cos\phi \cdot \Phi_1^{-1}\left[A_1(\tau)\frac{\cosh(\pi\tau - \beta\tau)}{\cosh(\pi\tau)};\tau \to \alpha\right]$$

where

$$A_0(\tau) = \frac{c_1}{\sqrt{2}}\Phi_0[\operatorname{sech}(\tfrac{1}{2}\alpha);\tau] = \frac{\sqrt{2}c_1}{\tau}\operatorname{cosech}(\pi\tau)$$

$$A_1(\tau) = \frac{c_2}{\sqrt{2}}\Phi_1[\sinh(\tfrac{1}{2}\alpha)\operatorname{sech}^2(\tfrac{1}{2}\alpha);\tau] = \frac{2\sqrt{2}c_2}{\tau}\operatorname{cosech}(\pi\tau)$$

the first transform being found from equation (7-6-10) and the second from Problem 7-8 below.

Hence we have

$$\psi(\alpha,\beta,\phi) = \sqrt{[2(\cosh\alpha + \cos\beta)]}(c_1\Phi_0^{-1} + 2c_2\cos\phi\cdot\Phi_1^{-1})\times$$
$$\times\left[\tau^{-1}\frac{2\cosh(\pi\tau - \beta\tau)}{\sinh(2\pi\tau)};\tau\to\alpha\right]$$

Now it is easily shown (Cf. Problem 7-8) that

$$\Phi_1^{-1}[\chi(\tau);\alpha] = -\frac{\partial}{\partial\alpha}\Phi_0^{-1}[\chi(\tau);\alpha]$$

so that we can write the solution in the form

$$\psi(\alpha,\beta,\phi) = \sqrt{[2(\cosh\alpha + \cos\beta)]}$$
$$\left(c_1 - 2c_2\cos\phi\frac{\partial}{\partial\alpha}\right)\Phi_0^{-1}\left[\frac{2\cosh(\pi\tau - \beta\tau)}{\tau\sinh(2\pi\tau)};\tau\to\alpha\right].$$

We saw in sec. 7-8 that

$$\Phi_0^{-1}\left[\frac{\cosh(\pi\tau - \beta\tau)}{\tau\sinh(2\pi\tau)};\tau\to\alpha\right]$$
$$= 2(\cosh\alpha + \cos\beta)^{-\frac{1}{2}}\tan^{-1}\left[\frac{(\cosh\alpha + \cos\beta)^{\frac{1}{2}}}{2\sin\frac{1}{2}\beta}\right]$$

so that

$$\psi(\alpha,\beta,\phi) = \chi(\alpha,\beta)\left(c_1 - 2c_2\cos\phi\frac{\partial}{\partial\alpha}\right)\times$$
$$\times\ \chi(\alpha,\beta)^{-1}\tan^{-1}[\tfrac{1}{2}\chi(\alpha,\beta)\operatorname{cosec}\tfrac{1}{2}\beta], \qquad (7\text{-}14\text{-}7)$$

where the function $\chi(\alpha,\beta)$ is defined by the equation

$$\chi(\alpha,\beta) = \sqrt{[2(\cosh\alpha + \cos\beta)]}. \qquad (7\text{-}14\text{-}8)$$

In a similar way the boundary conditions

$$\psi = 0, \qquad 0 \leqslant \rho \leqslant a,\ z = 0$$
$$\frac{\partial\psi}{\partial z} = -F(\alpha,\phi), \qquad \rho > a,\ z = 0$$

on a function ψ harmonic in $z \geqslant 0$ are equivalent to the conditions

$$\psi(\alpha,0,\phi) = 0, \qquad 0 \leqslant \phi \leqslant 2\pi$$
$$\frac{\partial\psi(\alpha,\pi,\phi)}{\partial\beta} = f(\alpha,\phi), \qquad 0 \leqslant \phi \leqslant 2\pi$$

where the function $f(\alpha,\phi)$ is defined in terms of the prescribed function $F(\alpha,\phi)$ through the equation

$$f(\alpha,\phi) = \tfrac{1}{2}a\operatorname{cosech}^2(\tfrac{1}{2}\alpha)F(a\coth\tfrac{1}{2}\alpha,\phi). \qquad (7\text{-}14\text{-}9)$$

Suppose that $f(\alpha,\phi)$ has the Fourier expansion

$$f(\alpha,\phi) = \sum_{m=-\infty}^{\infty} f_m(\alpha)e^{im\phi} \tag{7-14-10}$$

then the required solution is

$$\psi(\alpha,\beta,\phi) = \sum_{m=-\infty}^{\infty} e^{im\phi}\Phi_m^{-1}\left[\tau^{-1}A_m(\tau)\frac{\sinh(\beta\tau)}{\cosh(\pi\tau)};\tau \to \alpha\right] \tag{7-14-11}$$

where $A_m(\tau) = f_m^*(\tau) = \Phi_m[f_m(\alpha);\tau]$.

Finally, if ψ is harmonic in the half-space $z \geqslant 0$ and satisfies the conditions

$$\frac{\partial\psi}{\partial z} = -F(\rho,\phi), \qquad 0 \leqslant \rho \leqslant a, 0 \leqslant \phi \leqslant 2\pi$$

$$\psi = 0, \qquad \rho > a, 0 \leqslant \phi \leqslant 2\pi$$

on the boundary plane $z = 0$, then

$$\psi(\alpha,\beta,\phi) = \sum_{m=-\infty}^{\infty} e^{im\phi}\Phi_m^{-1}\left[\tau^{-1}A_m(\tau)\frac{\sinh(\pi\tau - \beta\tau)}{\cosh(\pi\tau)};\tau \to \alpha\right] \tag{7-14-12}$$

where

$$\sum_{m=-\infty}^{\infty} e^{im\phi}\Phi_m^{-1}[A_m(\tau);\alpha] = \tfrac{1}{2}a\,\mathrm{sech}^2(\tfrac{1}{2}\alpha)F(a\tanh\tfrac{1}{2}\alpha,\phi)$$

that is, where $A_m(\tau) = \Phi_m[f_m(\alpha);\tau]$ with

$$f_m(\alpha) = \frac{a}{2\pi}\mathrm{sech}^2(\tfrac{1}{2}\alpha)\int_0^{2\pi} F(a\tanh\tfrac{1}{2}\alpha,\phi)e^{-im\phi}\,d\phi. \tag{7-14-13}$$

PROBLEMS

7-1 Prove that, if $0 < \beta < 2\pi$,

$$\phi_0[(x - \cos\beta)^{-\frac{1}{2}};\tau] = \frac{\sqrt{2}\cosh(\pi\tau - \beta\tau)}{\tau\sinh(\pi\tau)}$$

and deduce that

$$\phi_0[(x - \cos\beta)^{-\frac{1}{2}};\tau] = 2\sqrt{2}\,\mathrm{cosec}\,\beta\,\frac{\sinh(\pi\tau - \beta\tau)}{\sinh(\pi\tau)}.$$

7-2 Prove that, if $-\pi < \beta < \pi$,

$$\phi_0[(\cosh\alpha + \cos\beta)^{-\frac{3}{2}};\tau] = 2\sqrt{2}\operatorname{cosec}\beta\,\frac{\sinh(\beta\tau)}{\sinh(\pi\tau)}$$

$$\phi_0[(\cosh\alpha + \cos\beta)^{-\frac{5}{2}};\tau]$$
$$= \frac{4\sqrt{2}}{3}\operatorname{cosec}^2\beta\left[\tau\,\frac{\cosh(\beta\tau)}{\sinh(\pi\tau)} - \cot\beta\,\frac{\sinh(\beta\tau)}{\sinh(\pi\tau)}\right]$$

and deduce that

$$\phi_0[\operatorname{sech}^3(\tfrac{1}{2}\alpha);\tau] = 8\tau\operatorname{cosech}(\pi\tau)$$

$$\phi_0[\operatorname{sech}^5(\tfrac{1}{2}\alpha);\tau] = \frac{16}{3}\tau^3\operatorname{cosech}(\pi\tau)$$

$$\phi_0[(\operatorname{sech}\alpha)^{\frac{3}{2}};\tau] = \sqrt{2}\operatorname{sech}(\tfrac{1}{2}\pi\tau)$$

$$\phi_0[(\operatorname{sech}\alpha)^{\frac{5}{2}};\tau] = \frac{2\sqrt{2}}{3}\tau\operatorname{cosech}(\tfrac{1}{2}\pi\tau).$$

7-3 Find

$$\phi_0[(1 + x)^{-n/2};\tau]$$

in the three cases $n = 1, 3, 5$ and deduce the corresponding values of the integral

$$\int_0^\infty x^n \operatorname{cosech}(\pi x)\,dx.$$

7-4 If m is a positive integer, show that

$$\phi_0[(x + t)^{-m-1};x \to \tau] = \frac{(-1)^m}{m!}\pi\operatorname{sech}(\pi\tau)(t^2 - 1)^{-m/2}P_{-\frac{1}{2}+i\tau}{}^m(t)$$

and deduce that

$$\phi_m\left[\frac{(x^2 - 1)^{m/2}}{(t + x)^{m+1}};x \to \tau\right] = \frac{\pi\operatorname{sech}(\pi\tau)}{m!}P_{-\frac{1}{2}+i\tau}(t).$$

Show also that

$$\int_0^\infty \tau\,\frac{\sinh(\pi\tau)}{\cosh^3(\pi\tau)}P_{-\frac{1}{2}+i\tau}{}^m(t)P_{-\frac{1}{2}+i\tau}{}^n(t)\,d\tau$$
$$= \frac{(-1)^{m+n}m!\,n!\,(t - 1)^{\frac{1}{2}m+\frac{1}{2}n}}{\pi^2(m + n + 1)(t + 1)^{\frac{1}{2}m+\frac{1}{2}n+1}}.$$

7-5 If m is a positive integer, show that

$$\phi_0[(x+t)^{-m-\frac{1}{2}}; x \to \tau] = \frac{(-1)^m 2^{2m+\frac{1}{2}} m!}{\tau \sinh(\pi\tau)(2m)!} \cdot \frac{d^m}{dt^m} \cosh[\tau \cos^{-1} t].$$

7-6 Prove that the Mehler–Fock transform of order zero of the function

$$g_n(\alpha) = \tfrac{1}{2}\operatorname{cosech}^3(\tfrac{1}{2}\alpha)(\lambda^2 + \coth^2 \tfrac{1}{2}\alpha)^{-n/2}$$

is

$$g^*_{0,n}(\tau) = \mathscr{F}_c[G_n(x); \tau]$$

where the function $G_n(x)$ is defined through the equation

$$G_n(x) = \sqrt{\frac{2}{\pi}} \operatorname{cosech}^2(\tfrac{1}{2}x)\left[\frac{1-k(x)}{1+\lambda^2}\right]^{n/2} {}_2F_1(\tfrac{1}{2}n; \tfrac{1}{2}; \tfrac{3}{2}; k)$$

with the function $k(x)$ given by the equation

$$k(x) = [1 + (\lambda^2 + 1)\sinh^2(\tfrac{1}{2}x)]^{-1}.$$

Deduce that

$$g^*_{0,3}(\tau) = 2\sin^2(\tfrac{1}{2}\xi)\sec(\tfrac{1}{2}\xi)\frac{\sinh(\pi\tau - \xi\tau)}{\sinh(\pi\tau)}, \qquad \cot(\tfrac{1}{2}\xi) = \lambda.$$

Show also that

$$g^*_{0,5}(\tau) = \tfrac{1}{3}\sin^2(\tfrac{1}{2}\xi)g^*_{0,3}(\tau) + \frac{4}{3\pi}\sin^5(\tfrac{1}{2}\xi)\int_0^\infty \frac{\sinh^2(\tfrac{1}{2}s)\cos(s\tau)\,ds}{\{\sinh^2(\tfrac{1}{2}s) + \sin^2(\tfrac{1}{2}\xi)\}^2}$$

and hence deduce that

$$g^*_{0,5}(\tau) = \frac{2}{3}\frac{\sinh(\pi\tau - \xi\tau)}{\sinh(\pi\tau)}$$

$$\sin^4(\tfrac{1}{2}\xi)\sec^3(\tfrac{1}{2}\xi)[1 + \cos^2(\tfrac{1}{2}\xi) - \tau\coth(\pi\tau - \xi\tau)\sin\xi].$$

7-7 The harmonic function ψ satisfies the boundary conditions

$$\psi = 0, \qquad 0 \leqslant \rho \leqslant a$$

$$\frac{\partial\psi}{\partial z} = -k(b^2 + \rho^2)^{-\frac{3}{2}}[1 - 2\eta + 3b^2(b^2 + \rho^2)^{-1}], \qquad \rho > a$$

(where k, η and b are constants) on the boundary $z = 0$ of the half-space $z \geqslant 0$ on which it is defined.

Show that we can write

$$\psi = \sqrt{(\cosh\alpha + \cos\beta)}\Phi_0^{-1}\left[\tau^{-1}A(\tau)\frac{\sinh(\beta\tau)}{\cosh(\pi\tau)}; \tau \to \alpha\right]$$

where, in the notation of Problem 7-6,

$$A(\tau) = \frac{k}{\sqrt{2a^2}} [(1 - 2\eta)g^*_{0,3}(\tau) + 3\lambda^2 g^*_{0,5}(\tau)], \qquad (\lambda = b/a).$$

Hence show that if $\cot(\frac{1}{2}\xi) = b/a$,

$$A(\tau) = \frac{\sqrt{2}k \sin^2(\frac{1}{2}\xi)}{a^2 \sinh(\pi\tau)} [\{2(1 - \eta)\sec(\tfrac{1}{2}\xi) + \cos(\tfrac{1}{2}\xi)\} \sinh(\pi\tau - \xi\tau) - 2\tau \sin(\tfrac{1}{2}\xi)\cosh(\pi\tau - \xi\tau)]$$

and deduce that, if $z = 0, 0 \leqslant \rho \leqslant a$, then

$$\frac{\partial\psi}{\partial z} = \frac{2\sqrt{2}k \sin^2(\frac{1}{2}\xi)}{\pi a^3} \cosh^3(\tfrac{1}{2}\alpha) \left[2(1 - \eta) + \cos\xi + \sin\xi \frac{\partial}{\partial\xi}\right] J$$

where J denotes the integral

$$\int_0^\infty \frac{\sinh x \sinh(\frac{1}{2}x)\, dx}{(\cosh x - \cos\xi)(\cosh\alpha + \cosh x)^{\frac{3}{2}}}.$$

Evaluate J by making the substitution

$$\sinh(\tfrac{1}{2}x) = u \cosh(\tfrac{1}{2}\alpha),$$

and deduce that, if $z = 0, 0 \leqslant \rho \leqslant a$, then

7-8 Prove that

(a) $\Phi_1[g'(\alpha); \tau] = -\Phi_0[g(\alpha); \tau]$

(b) $\Phi_1^{-1}[\chi(\tau); \alpha] = -\dfrac{\partial}{\partial\alpha} \Phi_0^{-1}[\chi(\tau); \alpha].$

Deduce that

$$\Phi_1[\sinh(\tfrac{1}{2}\alpha)\operatorname{sech}^2(\tfrac{1}{2}\alpha); \tau] = 4\tau^{-1} \operatorname{cosech}(\pi\tau).$$

8

Finite Transforms

The method of integral transforms outlined in the previous chapters may, as we have seen, be applied to the solution of boundary value and initial value problems in mathematical physics, in which the range of values which can be assumed by one of the independent variables, x say, is $(0,\infty)$. Making use of an integral transform of the type

$$\int_0^\infty f(x, \ldots)K(x,\xi)\, dx$$

we can often reduce a partial differential equation in n independent variables to one in $n-1$ independent variables, thus reducing the complexity of the problem under discussion. In some instances successive operations of this type will ultimately reduce the problem to a one-point or a two-point boundary value problem for an ordinary differential equation, methods of solution of which have been much more extensively developed.

There are two obvious ways in which this method can be extended. In the first place, by extending the types of kernel K, we might hope to be able to deal with boundary value problems not hitherto tractable to this method. However, the choice of the kernel is largely determined by the form of the differential equation whose solution is sought. The other alternative is to extend the method to *finite* intervals, that is, to employ integral transforms suitable to problems in which the range of variation of the independent variable x is now (a,b), where both a and b are real and finite. One, of course, may be zero. This extension seems to have been first suggested by Doetsch (1935) in the case of trigonometric kernels. Defining the finite Fourier sine transform of a function by the equation

$$\bar{f}_s(n) = \int_0^\pi f(x) \sin(nx)\, dx$$

Doetsch pointed out that the inversion formula

$$f(x) = \frac{2}{\pi} \sum_{n=1}^{\infty} \bar{f}_s(n) \sin(nx)$$

is an immediate consequence of a well-known theorem in the theory of Fourier series.

This method has been developed and generalized by Kneiss (1938), Koschmieder (1941), Roettinger (1947) and others and applied by Brown (1943) to the discussion of the classical boundary value problems of mathematical physics. The use of finite Hankel transforms was suggested by Sneddon (1945, 1946) and a little later Eringen (1954) and Churchill (1955) developed the concept of a finite Sturm–Liouville transform whose properties follow immediately from the theory of Sturm–Liouville problems for ordinary differential equations.

The use of finite transforms does not, of course, solve problems which are incapable of solution by use of the method of separation of variables and the direct application of the theory of orthogonal functions, but it does facilitate the resolution of boundary value problems by providing an "operational" technique which can be used in much the same way as the method of integral transforms developed in the previous chapters.

In this chapter we begin by considering the case of finite Fourier sine and cosine transforms and then go on to a discussion of the general finite Sturm–Liouville transform. The remainder of the chapter is then taken up with special cases of this theory.

8-1 FINITE FOURIER TRANSFORMS

We begin by outlining the simplest case—that in which the kernel $K(\xi,x)$ is a simple cosine or sine—since the theory on which it is based is almost certainly familiar to the reader.

8-1-1 FINITE FOURIER SINE AND COSINE TRANSFORMS

It is a well-known result of the theory of Fourier series[1] that, if $f \in \mathscr{P}^1(0,\pi)$, then the series

$$\frac{1}{\pi} a_0 + \frac{2}{\pi} \sum_{n=1}^{\infty} a_n \cos(nx)$$

in which

$$a_n = \int_0^\pi f(x) \cos(nx)\, dx$$

converges to the value $\frac{1}{2}[f(x+) + f(x-)]$ at each point of the open interval $(0,\pi)$, to the value $f(x)$ at each point of continuity of the interval, and to the value $\frac{1}{2}[f(0) + f(\pi)]$ if $x = 0$ or $x = \pi$.

If, instead of the Fourier coefficient a_n, we introduce the *finite cosine transform* of the function, denoted by $\bar{f}_c(n)$, and defined by the equation

$$\mathscr{f}_c[f(x);n] = \bar{f}_c(n) = \int_0^\pi f(x) \cos(nx)\, dx \tag{8-1-1}$$

we may re-state the above theorem on Fourier series as:

The inversion theorem for finite cosine transforms

If $f(x) \in \mathscr{P}^1(0,\pi)$ and if $\bar{f}_c(n)$ denotes its finite cosine transform, then

$$\mathscr{f}_c^{-1}[\bar{f}_c(n);x] = f(x) = \frac{1}{\pi} \bar{f}_c(0) + \frac{2}{\pi} \sum_{n=1}^{\infty} \bar{f}_c(n) \cos(nx) \tag{8-1-2}$$

at each point of the open interval $(0,\pi)$ at which $f(x)$ is continuous. At any point of the interval at which $f(x)$ has a finite discontinuity the symbol $f(x)$ is taken to mean $\frac{1}{2}[f(x+) + f(x-)]$, and at $x = 0$ or $x = \pi$ it is taken to mean $\frac{1}{2}[f(0) + f(\pi)]$.

In the sequel we shall refer to equation (8-1-2) as *the inversion formula for the finite Fourier cosine transform.*

A similar result holds when the independent variable ranges over the interval $(0,a)$. If we now define the finite cosine transform by the equation

$$\mathscr{f}_c[f(x);n] = \bar{f}_c(n) = \int_0^a f(x) \cos\left(\frac{n\pi x}{a}\right) dx \tag{8-1-3}$$

[1] See, for instance, Sneddon (1961*c*).

a simple change of the variables shows that the inversion formula (8-1-2) is equivalent to the result:

If $f(x) \in \mathscr{P}^1(0,a)$ *and if its finite cosine transform is defined by equation* (8-1-3), *then* $f(x)$ *is given at a point of continuity of* $(0,a)$ *by the formula*

$$\mathscr{f}_c^{-1}[\bar{f}_c(n);x] = f(x) = \frac{1}{a}\bar{f}_c(0) + \frac{2}{a}\sum_{n=1}^{\infty}\bar{f}_c(n)\cos\left(\frac{n\pi x}{a}\right). \tag{8-1-4}$$

There are obvious modifications at a point of finite discontinuity and at the end points of the interval. We shall also refer to (8-1-4) as the *inversion formula for the finite Fourier cosine transform*; this should not lead to any confusion since, in any given problem, it will be clear whether the range of variation of x is the open interval $(0,\pi)$ or the open interval $(0,a)$.

Similar results hold for finite sine transforms. For a function defined on $(0,\pi)$ we define the finite Fourier sine transform $\bar{f}_s(n)$ by the equation

$$\mathscr{f}_s[f(x);n] = \bar{f}_s(n) = \int_0^{\pi} f(x)\sin(nx)\,dx. \tag{8-1-5}$$

The well-known result for half-range Fourier sine series may then be restated as:

The inversion theorem for finite sine transforms *If* $f(x) \in \mathscr{P}^1(0,\pi)$, *and if* $\bar{f}_s(n)$ *denotes its finite Fourier sine transform, then at each point of continuity in the interval* $(0,\pi)$

$$\mathscr{f}_s^{-1}[\bar{f}_s(n);x] = f(x) = \frac{2}{\pi}\sum_{n=1}^{\infty}\bar{f}_s(n)\sin(nx). \tag{8-1-6}$$

(There are obvious modifications at the points of finite discontinuity and at the end-points of the interval.)

We shall refer to equation (8-1-6) or its corollary

$$\mathscr{f}_s^{-1}[\bar{f}_s(n);x] = f(x) = \frac{2}{a}\sum_{n=1}^{\infty}\bar{f}_s(n)\sin\left(\frac{n\pi x}{a}\right) \tag{8-1-7}$$

with

$$\bar{f}_s(n) = \int_0^a f(x)\sin\left(\frac{n\pi x}{a}\right)dx, \qquad f(x) \in \mathscr{P}^1(0,a) \tag{8-1-8}$$

as *the inversion formula for the finite Fourier sine transform*.

In the application of these theorems on finite Fourier transforms to the solution of boundary value problems it is often necessary to express an integral of the type

$$\int_0^a \frac{\partial^r f}{\partial x^r} \sin\left(\frac{n\pi x}{a}\right) dx, \qquad \int_0^a \frac{\partial^r f}{\partial x^r} \cos\left(\frac{n\pi x}{a}\right) dx$$

in terms of the finite Fourier transforms of the function $f(x)$ defined on $(0, a)$.

The basic results are those involving the first derivative of the function. For example, we have, as a result of an integration by parts, that

$$\int_0^a \frac{\partial f}{\partial x} \sin \frac{n\pi x}{a}\, dx = \left[f(x) \sin \frac{n\pi x}{a} \right]_0^a - \frac{n\pi}{a} \int_0^a f(x) \cos \frac{n\pi x}{a}\, dx.$$

Now, since n is an integer, $\sin(n\pi x/a)$ vanishes both at $x = 0$ and at $x = a$, so that if we define $\bar{f}_c(n)$ by equation (8-1-3), we obtain the relation

$$\mathscr{F}_s\left[\frac{\partial f}{\partial x}; x \to n\right] = -\frac{n\pi}{a}\bar{f}_c(n) \tag{8-1-9}$$

whatever the values of $f(0)$ and $f(a)$, provided, of course, that they are finite. In a similar way we can show that

$$\mathscr{F}_c\left[\frac{\partial f}{\partial x}; x \to n\right] = (-1)^n f(a) - f(0) + \frac{n\pi}{a}\bar{f}_s(n) \tag{8-1-10}$$

in which $\bar{f}_s(n)$ denotes the finite sine transform defined by equation (8-1-8). As a special case of this result we see that if

$$f(a) = f(0) = 0$$

then

$$\mathscr{F}_c\left[\frac{\partial f}{\partial x}; x \to n\right] = \frac{n\pi}{a}\bar{f}_s(n) \tag{8-1-11}$$

The results for higher derivatives may be established by the repeated use of the fundamental results (8-1-9) and (8-1-10). For instance, we deduce from (8-1-9) that

$$\mathscr{F}_s\left[\frac{\partial^2 f}{\partial x^2}; x \to n\right] = -\frac{n\pi}{a}\mathscr{F}_c\left[\frac{\partial f}{\partial x}; x \to n\right]$$

and hence by the application of (8-1-10) that

$$\mathscr{F}_s\left[\frac{\partial^2 f}{\partial x^2}; x \to n\right] = -\frac{n^2\pi^2}{a^2}\bar{f}_s(n) + \frac{n\pi}{a}[(-1)^{n+1}f(a) + f(0)]. \tag{8-1-12}$$

In the particular case in which $f(a) = f(0) = 0$ this reduces to the simple form

$$\mathscr{F}_s\left[\frac{\partial^2 f}{\partial x^2}; x \to n\right] = -\frac{n^2\pi^2}{a^2}\bar{f}_s(n). \tag{8-1-13}$$

Similarly it follows from equation (8-1-10) followed by (8-1-11) that

$$\mathscr{F}_c\left[\frac{\partial^2 f}{\partial x^2}; x \to n\right] = (-1)^n f'(a) - f'(0) - \frac{n^2\pi^2}{a^2}\bar{f}_c(n). \tag{8-1-14}$$

If $\partial f/\partial x$ vanishes at the end points $x = 0$ and $x = a$, this equation becomes simply

$$\mathscr{F}_c\left[\frac{\partial^2 f}{\partial x^2}; x \to n\right] = -\frac{n^2\pi^2}{a^2}\bar{f}_c(n). \tag{8-1-15}$$

For derivatives of order greater than the second the results may be obtained by repeated applications of the above results. For instance in the analysis of boundary value problems in elasticity we often need to know the results for derivatives of the fourth order. It follows from equation (8-1-13) that if $\partial^2 f/\partial x^2$ is zero at the end points $x = 0$ and $x = a$ then

$$\mathscr{F}_s\left[\frac{\partial^4 f}{\partial x^4}; x \to n\right] = -\frac{n^2\pi^2}{a^2}\mathscr{F}_s\left[\frac{\partial^2 f}{\partial x^2}; x \to n\right]$$

so that if

$$\left(\frac{\partial^2 f}{\partial x^2}\right)_{x=0} = \left(\frac{\partial^2 f}{\partial x^2}\right)_{x=a} = (f)_{x=0} = (f)_{x=a} = 0 \tag{8-1-16a}$$

then

$$\mathscr{F}_s\left[\frac{\partial^4 f}{\partial x^4}; x \to n\right] = \frac{n^4\pi^4}{a^4}\bar{f}_s(n). \tag{8-1-16b}$$

Similarly if

$$\left(\frac{\partial^3 f}{\partial x^3}\right)_{x=0} = \left(\frac{\partial^3 f}{\partial x^3}\right)_{x=a} = \left(\frac{\partial f}{\partial x}\right)_{x=0} = \left(\frac{\partial f}{\partial x}\right)_{x=a} = 0 \tag{8-1-17a}$$

then

$$\mathscr{F}_c\left[\frac{\partial^4 f}{\partial x^4}; x \to n\right] = \frac{n^4\pi^4}{a^4}\bar{f}_c(n). \tag{8-1-17b}$$

Similar results for the two-dimensional Laplacian operator Δ_2 can be deduced from equations (8-1-12) and (8-1-13).

If we write

$$\bar{u}_s(n,y) = \mathscr{F}_s[u(x,y); x \to n]$$
$$\bar{u}_c(n,y) = \mathscr{F}_c[u(x,y); x \to n] \qquad (8\text{-}1\text{-}18)$$

then it follows from equation (8-1-12) that

$$\mathscr{F}_s[\Delta_2 u(x,y); x \to n]$$
$$= \frac{n\pi}{a}[(-1)^{n+1}u(a,y) + u(0,y)] + \left(\frac{\partial^2}{\partial y^2} - \frac{n^2\pi^2}{a^2}\right)\bar{u}_s(n,y) \qquad (8\text{-}1\text{-}19)$$

and from equation (8-1-13) that

$$\mathscr{F}_c[\Delta_2 u(x,y); x \to n]$$
$$= (-1)^n u_x(a,y) - u_x(0,y) + \left(\frac{\partial^2}{\partial y^2} - \frac{n^2\pi^2}{a^2}\right)\bar{u}_c(n,y). \qquad (8\text{-}1\text{-}20)$$

Similar results can be generated for the biharmonic operator $\Delta_2^2 = \Delta_2\Delta_2$.

8-1-2 CONVOLUTION THEOREMS FOR FINITE FOURIER TRANSFORMS

The convolution theorems for finite Fourier transforms cannot be put into simple, elegant forms of the kind that hold for "ordinary" Fourier transforms. To present the results in a neat form, we introduce two functions $F_1(u)$ and $F_2(u)$, which we call the *odd* and *even periodic extensions* of the function $f(u)$ which is defined in the closed interval $[0,\pi]$. Thus in the interval $[-\pi,\pi]$ we define $F_1(u)$ the odd periodic extension with period 2π, of the function $f(u)$ by the equations

$$F_1(u) = \begin{cases} f(u), & 0 \leqslant u \leqslant \pi \\ -f(-u), & -\pi < u < 0 \end{cases}$$
$$F_1(u + 2r\pi) = F_1(u), \qquad (r = \pm 1, \pm 2, \ldots)$$

Similarly the even periodic extension $F_2(u)$ is defined by the equations

$$F_2(u) = \begin{cases} f(u), & 0 \leqslant u \leqslant \pi \\ f(-u), & -\pi < u < 0 \end{cases}$$
$$F_2(u + 2r\pi) = F_2(u), \qquad (r = \pm 1, \pm 2, \ldots).$$

It follows from the definition of $F_1(u)$ that if $F_1(u)$ and $G_1(u)$ are the odd periodic extensions of $f(u)$ and $g(u)$, respectively, then

$$\int_{-\pi}^{\pi} F_1(x-u)G_1(u)\,du = \int_0^{\pi} f(x-u)g(u)\,du - \int_0^{\pi} f(u-x)g(u)\,du \qquad (8\text{-}1\text{-}21)$$

and also, because F_1 is periodic,

$$\int_{-\pi}^{\pi} F_1(x-u)G_1(u)\,du = \int_0^{\pi} f(-x-u)g(u)\,du - \int_0^{\pi} f(x+u)g(u)\,du. \tag{8-1-22}$$

In the notation introduced by Churchill[1] we write $F_1(x)^*G_1(x)$ for the integral

$$\int_{-\pi}^{\pi} F_1(x-u)G_1(u)\,du$$

and call it the *convolution* of the functions F_1 and G_1.

If we denote the finite Fourier sine transforms of $f(x)$ and $g(x)$ by $\bar{f}_s(n)$ and $\bar{g}_s(n)$ respectively (Cf. equation (8-1-1) above), then

$$\frac{2}{\pi}\sum_{n=1}^{m} \bar{f}_s(n)\bar{g}_s(n)\cos(nx)$$

$$= \frac{2}{\pi}\int_0^{\pi} d\xi \int_0^{\pi} d\eta\, f(\xi)g(\eta) \sum_{n=1}^{m} \cos(nx)\sin(n\xi)\sin(n\eta)$$

on interchanging the orders of summation and integration. Making use of the results

$$4\sin(n\xi)\sin(n\eta)\cos(nx) = \cos(\xi-\eta+x)n + \cos(\xi-\eta-x)n - \cos(\xi+\eta+x)n - \cos(\xi+\eta-x)n$$

$$\sum_{r=1}^{m}\cos(ry) = -\tfrac{1}{2} + \frac{\sin(m-\tfrac{1}{2})y}{2\sin\tfrac{1}{2}y}$$

we find that

$$\frac{2}{\pi}\sum_{r=1}^{m} \bar{f}_s(n)\bar{g}_s(n)\cos(nx)$$

$$= \frac{1}{4\pi}\int_0^{\pi} f(\xi)\,d\xi \int_0^{\pi} g(\eta)\,d\eta \left[\frac{\sin(m-\frac{1}{2})(\xi-\eta+x)}{\sin\frac{1}{2}(\xi-\eta+x)} + \frac{\sin(m-\frac{1}{2})(\xi-\eta-x)}{\sin\frac{1}{2}(\xi-\eta-x)} - \frac{\sin(m-\frac{1}{2})(\xi+\eta+x)}{\sin\frac{1}{2}(\xi+\eta+x)} - \frac{\sin(m-\frac{1}{2})(\xi+\eta-x)}{\sin\frac{1}{2}(\xi+\eta-x)}\right].$$

[1] p. 274 of Churchill, 1944.

If in the first of these four integrals we change the variables of integration from ξ and η to λ and μ where

$$\lambda = \xi - \eta + x, \qquad \mu = \eta$$

we find that

$$\frac{1}{4\pi}\int_0^\pi f(\xi)\,d\xi \int_0^\pi g(\eta)\frac{\sin(m-\frac{1}{2})(\xi-\eta+x)}{\sin\frac{1}{2}(\xi-\eta+x)}\,d\eta$$

$$= \frac{1}{2\pi}\int_0^\pi g(\mu)\,d\mu \int_{x-\mu}^{x+\pi-\mu} f(\lambda+\mu-x)\frac{\sin(m-\frac{1}{2})\lambda}{\lambda}\cdot\frac{\frac{1}{2}\lambda}{\sin\frac{1}{2}\lambda}\,d\lambda$$

which, by the Riemann localization lemma, tends to

$$\frac{1}{4}\int_0^x g(\mu)f(\mu-x)\,d\mu$$

as $m \to \infty$. The other three terms can be evaluated in a similar fashion to give

$$\frac{2}{\pi}\sum_{n=1}^{\infty}\bar{f}_s(n)\bar{g}_s(n)\cos(nx)$$

$$= -\frac{1}{4}\int_0^\pi g(\mu)[f(x-\mu)+f(-x-\mu)-f(\mu-x)-f(\mu+x)]\,d\mu.$$

Hence, by equations (8-1-21) and (8-1-22) we have

$$\frac{2}{\pi}\sum_{n=1}^{\infty}\bar{f}_s(n)\bar{g}_s(n)\cos(nx) = -\tfrac{1}{2}\int_{-\pi}^{\pi}F_1(x-\mu)G_1(\mu)\,d\mu$$

$$= -\tfrac{1}{2}F_1(x)*G_1(x).$$

Using the inversion formula for finite cosine transforms we see that this can be written in the form

$$\mathscr{f}_c[F_1*G_1] = -2\bar{f}_s(n)\bar{g}_s(n) \tag{8-1-23}$$

Similarly by considering the limit of the sum

$$\frac{2}{\pi}\sum_{n=1}^{m}\bar{f}_s(n)\bar{g}_c(n)\sin(nx)$$

as $m \to \infty$ we can show that

$$\mathscr{f}_s[F_1*G_2] = 2\bar{f}_s(n)\bar{g}_c(n). \tag{8-1-24}$$

For an alternative proof of equation (8-1-24) the reader is referred to pp. 274–276 of Churchill (1944).

By either of these methods we can show further that

$$\mathscr{F}_c[F_2{}^*G_2] = 2\bar{f}_c(n)\bar{g}_c(n). \tag{8-1-25}$$

It is also easily verified that the steps in the above analysis are justified if both f and g belong to the class $\mathscr{P}^1(0,\pi)$.

8-1-3 MULTIPLE FINITE FOURIER TRANSFORMS

We now consider the case in which $f(x,y)$ is a function of the two independent variables x and y, defined on the square $0 < x < \pi$, $0 < y < \pi$. Then regarded as a function of x, $f(x,y)$ will, if it satisfies certain wide conditions, possess a finite sine transform. If we denote this transform by $\bar{f}_s(m,y)$, then

$$\bar{f}_s(m,y) = \mathscr{F}_s[f(x,y); x \to m]. \tag{8-1-26}$$

This function will itself have a finite Fourier transform (with respect to y), which we may denote by $\bar{f}_{ss}(m,n)$, defined by the equation

$$\bar{f}_{ss}(m,n) = \mathscr{F}_s[\mathscr{F}_s\{f(x,y); x \to m\}; y \to n]. \tag{8-1-27}$$

If we substitute from (8-1-26) into (8-1-27), we find that the relation between $f(x,y)$ and $\bar{f}_{ss}(m,n)$ written in explicit form is

$$\mathscr{F}_{ss}[f(x,y); (x,y) \to (m,n)] \equiv \bar{f}_{ss}(m,n)$$
$$= \int_0^\pi \int_0^\pi f(x,y) \sin(mx) \sin(ny)\, dxdy. \tag{8-1-28}$$

We shall regard this equation as defining the double finite sine transform of the function $f(x,y)$ defined over the square $0 < x < \pi$, $0 < y < \pi$. To obtain an inversion formula giving $f(x,y)$ in terms of its double finite sine transform $\bar{f}_{ss}(m,n)$, we note that if we apply the inverse theorem for finite sine transforms to equations (8-1-26) and (8-1-27) we obtain the relations

$$\bar{f}_s(m,y) = \frac{2}{\pi} \sum_{n=1}^{\infty} \bar{f}_{ss}(m,n) \sin(ny) \tag{8-1-29}$$

and

$$f(x,y) = \frac{2}{\pi} \sum_{m=1}^{\infty} \bar{f}_s(m,y) \sin(mx). \tag{8-1-30}$$

Substituting into (8-1-30) the series expansion for $\bar{f}_s(m,y)$ given by (8-1-29), we obtain finally the inversion formula

$$\mathscr{F}_{ss}{}^{-1}[\bar{f}_{ss}(m,n); (m,n) \to (x,y)] \equiv f(x,y)$$
$$= \frac{4}{\pi^2} \sum_{m=1} \sum_{n=1} \bar{f}_{ss}(m,n) \sin(mx) \sin(ny) \tag{8-1-31}$$

$(0 < x < \pi,\ 0 < y < \pi)$. We have written the result out in the case in which f is continuous at (x,y); the modification when it has a finite discontinuity is obvious.

The inversion formula given by the pair of equations (8-1-28), (8-1-31) is applicable when the range of variation of the independent variables x and y is given by the inequalities $0 < x < \pi$, $0 < y < \pi$. If we are dealing with a problem in which $f(x,y)$ is defined over the rectangle $0 < x < a$, $0 < y < b$, we use the obvious modification:

If a function $f(x,y) \in \mathscr{P}^1(D)$, where D is the rectangle $\{(x,y): 0 < x < a,\ 0 < y < b\}$, and if its double finite sine transform is defined by the equation

$$\mathscr{f}_{ss}[f(x,y);(x,y) \to (m,n)] \equiv \bar{f}_{ss}(m,n) = \int_0^a \int_0^b f(x,y) \sin\left(\frac{m\pi x}{a}\right) \sin\left(\frac{n\pi y}{b}\right) dxdy \qquad (8\text{-}1\text{-}32)$$

then at points of the rectangle at which the function $f(x,y)$ is continuous

$$\mathscr{f}_{ss}^{-1}[\bar{f}_{ss}(m,n);(m,n) \to (x,y)] \equiv f(x,y) = \frac{4}{ab} \sum_{m=1}^{\infty} \sum_{n=1}^{\infty} \bar{f}_{ss}(m,n) \sin\left(\frac{m\pi x}{a}\right) \sin\left(\frac{n\pi y}{b}\right). \qquad (8\text{-}1\text{-}33)$$

The modification required at a point of finite discontinuity of the function f is obvious.

In a similar way we can define double transforms of the types

$$\bar{f}_{cs}(m,n) \equiv \mathscr{f}_{cs}[f(x,y);(x,y) \to (m,n)] = \mathscr{f}_c[\mathscr{f}_s\{f(x,y);y \to n\};x \to m] \qquad (8\text{-}1\text{-}34)$$

$$\bar{f}_{sc}(m,n) \equiv \mathscr{f}_{sc}[f(x,y);(x,y) \to (m,n)] = \mathscr{f}_s[\mathscr{f}_c\{f(x,y);y \to n\};x \to m] \qquad (8\text{-}1\text{-}35)$$

$$\bar{f}_{cc}(m,n) = \mathscr{f}_{cc}[f(x,y);(x,y) \to (m,n)] = \mathscr{f}_c[\mathscr{f}_c\{f(x,y);y \to n\};x \to m]. \qquad (8\text{-}1\text{-}36)$$

Inversion formulas for double transforms of these types can easily be deduced by successive applications of the inversion formulas for the finite sine and the finite cosine transforms. For example, in the case in which $f(x,y)$ is defined on the square $\{(x,y): 0 < x < \pi,\ 0 < y < \pi\}$, we easily find that

$$\mathscr{f}_{cs}^{-1}[\bar{f}_{cs}(m,n);(m,n) \to (x,y)] \equiv f(x,y) = \frac{2}{\pi^2} \sum_{n=1}^{\infty} \bar{f}_{cs}(0,n) \sin(ny) + \frac{4}{\pi^2} \sum_{m=1}^{\infty} \sum_{n=1}^{\infty} \bar{f}_{cs}(m,n) \cos(mx) \sin(ny). \qquad (8\text{-}1\text{-}37)$$

The double transforms of the partial derivatives of a function may readily be written down. For instance if $f(x, y) \in \mathscr{C}^2(D)$, where

$$D = \{(x,y) : 0 < x < a,\ 0 < y < b\}$$

then

$$\mathcal{F}_s\left[\frac{\partial^2 f}{\partial x^2}; x \to m\right]$$
$$= \frac{m\pi}{a}\{f(0,y) + (-1)^{m+1} f(a,y)\} - \frac{m^2\pi^2}{a^2}\mathcal{F}_s[f(x,y); x \to m]$$

so that

$$\mathcal{F}_{ss}\left[\frac{\partial^2 f}{\partial x^2}; (x,y) \to (m,n)\right] = \Theta_1(m,n) - \frac{m^2\pi^2}{a^2}\bar{f}_{ss}(m,n) \tag{8-1-38a}$$

with

$$\Theta_1(m,n) = \frac{m\pi}{a}\{\mathcal{F}_s[f(0,y); y \to n] + (-1)^{m+1}\mathcal{F}_s[f(a,y); y \to n]\}. \tag{8-1-38b}$$

In particular, if

$$f(0,y) = f(a,y) = 0, \qquad 0 < y < b, \tag{8-1-39a}$$

then

$$\mathcal{F}_{ss}\left[\frac{\partial^2 f}{\partial x^2}; (x,y) \to (m,n)\right] = -\frac{m^2\pi^2}{a^2}\bar{f}_{ss}(m,n). \tag{8-1-39b}$$

A similar result holds for the double finite sine transform of $\partial^2 f/\partial y^2$:

$$\mathcal{F}_{ss}\left[\frac{\partial^2 f}{\partial y^2}; (x,y) \to (m,n)\right] = \Theta_2(m,n) - \frac{n^2\pi^2}{b^2}\bar{f}_{ss}(m,n) \tag{8-1-40a}$$

with

$$\Theta_2(m,n) = \frac{n\pi}{b}[\mathcal{F}_s[f(x,0); x \to m] + (-1)^{n+1}\mathcal{F}_s[f(x,b); x \to m]\}. \tag{8-1-40b}$$

In particular, if

$$f(x,0) = f(x,b) = 0, \qquad 0 < x < a \tag{8-1-41a}$$

then

$$\mathcal{F}_{ss}\left[\frac{\partial^2 f}{\partial y^2}; (x,y) \to (m,n)\right] = -\frac{n^2\pi^2}{b^2}\bar{f}_{ss}(m,n). \tag{8-1-41b}$$

From these results we deduce immediately that

$$\mathscr{f}_{ss}[\Delta_2 f(x,y);(x,y) \to (m,n)]$$
$$= \Theta_1(m,n) + \Theta_2(m,n) - \pi^2\left(\frac{m^2}{a^2} + \frac{n^2}{b^2}\right)\bar{f}_{ss}(m,n). \qquad (8\text{-}1\text{-}42)$$

In particular, if $f(x,y)$ vanishes along the boundary of D,

$$\mathscr{f}_{ss}[\Delta_2 f(x,y);(x,y) \to (m,n)] = -\pi^2\left(\frac{m^2}{a^2} + \frac{n^2}{b^2}\right)\bar{f}_{ss}(m,n). \qquad (8\text{-}1\text{-}43)$$

Similar results can easily be derived for the other operators $\mathscr{f}_{sc}$, $\mathscr{f}_{cs}$, and $\mathscr{f}_{cc}$. We shall not derive them here since the methods are exactly the same as for the operator $\mathscr{f}_{ss}$, and double Fourier sine transforms occur more frequently in applications than do the others.

The theory outlined above can be extended in another way, by considering functions of p variables $x_1,x_2, \ldots ,x_p$. For instance, if $f(x_1, \ldots ,x_n)$ is defined in the region

$$D = \{(x_1,x_2, \ldots ,x_p):0 < x_i < a_i, i = 1,2, \ldots ,p\}$$

and if its multiple sine transform is taken to be defined by

$$\bar{f}_s(n_1, \ldots ,n_p) = \int_0^{a_1}\int_0^{a_2} \cdots \int_0^{a_p} f(x_1, \ldots ,x_p) \sin\frac{n_1\pi x_1}{a_1}$$
$$\ldots \sin\frac{n_p\pi x_p}{a_p}\, dx_1 \ldots dx_p,$$

$f \in \mathscr{P}^1(D)$, then at any point of D at which f is continuous

$$f(x_1,x_2, \ldots ,x_p) = 2^p(a_1a_2 \ldots a_p)^{-1} \sum_{n_1=1}^{\infty} \cdots \sum_{n_p=1}^{\infty}$$
$$\bar{f}_s(n_1, \ldots ,n_p) \sin\frac{n_1\pi x_1}{a_1} \ldots \sin\frac{n_p\pi x_p}{a_p}.$$

Many results from the case in which $p = 2$ go over into the general case. For example, if $f = 0$ on the boundary of D then the transform of the Laplacian of f, $\Delta_p f(x_1, \ldots ,x_p)$ is

$$-\pi^2\bar{f}_s(n_1, \ldots ,n_p) \sum_{j=1}^{p} (n_j^2/a_j^2).$$

8-1-4 APPLICATIONS OF FINITE FOURIER TRANSFORMS

We shall now indicate briefly some applications of finite Fourier transforms to the solution of boundary value and initial value problems.

We begin by considering the solution $u(x,t)$ of the wave equation

$$\frac{\partial^2 u}{\partial x^2} = \frac{1}{c^2}\frac{\partial^2 u}{\partial t^2}, \qquad 0 \leqslant x \leqslant a,\ t > 0 \tag{8-1-44}$$

satisfying the boundary conditions

$$u(0,t) = u(a,t) = 0, \qquad t > 0 \tag{8-1-45}$$

and the initial conditions

$$u(x,0) = \frac{4\varepsilon}{a^2}x(a-x), \qquad \frac{\partial u(x,0)}{\partial t} = 0, \qquad 0 \leqslant x \leqslant a. \tag{8-1-46}$$

Equations (8-1-45) indicate that it is appropriate to introduce the finite sine transform

$$\bar{u}_s(n,t) = \mathscr{f}_s[u(x,t); x \to n].$$

From equation (8-1-13) we see that the wave equation (8-1-44) is equivalent to the differential equation

$$\left(\frac{\partial^2}{\partial t^2} + \frac{c^2 n^2 \pi^2}{a^2}\right)\bar{u}_s(n,t) = 0$$

for the transform $\bar{u}_s(n,t)$. From equations (8-1-46) and Problem 8-1 (*e*), we see that the appropriate initial conditions on $\bar{u}_s(n,t)$ are

$$\bar{u}_s(n,0) = \frac{8\varepsilon a}{\pi^3}\frac{1}{n^3}[1 + (-1)^{n+1}], \qquad \frac{\partial \bar{u}_s(n,0)}{\partial t} = 0$$

so that the solution for the finite sine transform is

$$\bar{u}_s(n,t) = \frac{8\varepsilon a}{\pi^3}\frac{1 + (-1)^{n+1}}{n^3}\cos\left(\frac{n\pi ct}{a}\right).$$

Making use of the inversion formula for the finite sine transform we finally obtain the result

$$u(x,t) = \frac{16\varepsilon}{\pi^3}\sum_{r=0}^{\infty}(2r+1)^{-3}\cos\frac{(2r+1)\pi ct}{a}\sin\frac{(2r+1)\pi x}{a}.$$

As a second example we consider the solution of the equation describing the vibrations of a beam of finite length

$$\frac{\partial^4 u}{\partial x^4} + \frac{1}{c^2}\frac{\partial^2 u}{\partial t^2} = \frac{p(x,t)}{EI}, \qquad 0 \leqslant x \leqslant a,\ t > 0 \tag{8-1-47}$$

when the displacement $u(x,t)$ satisfies the initial conditions

$$u(x,0) = \frac{\partial u(x,0)}{\partial t} = 0, \qquad 0 \leqslant x \leqslant a \tag{8-1-48}$$

and the boundary conditions

$$u(0,t) = \frac{\partial^2 u(0,t)}{\partial x^2} = 0, \qquad u(a,t) = \frac{\partial^2 u(a,t)}{\partial x^2} = 0, \qquad t > 0. \tag{8-1-49}$$

Equations (8-1-49) express the fact that the beam is freely hinged at each of its endpoints—see, for example, p. 151 of Morse (1948).

If we write

$$\bar{u}_s(n,t) = \mathscr{f}_s[u(x,t); x \to n]$$

then it follows from equations (8-1-49) and (8-1-16) that

$$\mathscr{f}_s\left[\frac{\partial^4 u}{\partial x^4}; x \to n\right] = \frac{n^4\pi^4}{a^4}\,\bar{u}_s(n,t)$$

and hence that the transform $\bar{u}_s(n,t)$ satisfies the differential equation

$$\left(\frac{\partial^2}{\partial t^2} + \frac{n^4\pi^4 c^2}{a^4}\right)\bar{u}_s(n,t) = \frac{c^2}{EI}\bar{p}_s(n,t) \tag{8-1-50}$$

where

$$\bar{p}_s(n,t) = \mathscr{f}_s[p(x,t); x \to n]. \tag{8-1-51}$$

The corresponding initial conditions are the transforms of equations (8-1-48)

$$\bar{u}_s(n,0) = \frac{\partial \bar{u}_s(n,0)}{\partial t} = 0\,. \tag{8-1-52}$$

The solution of the initial value problem posed by equations (8-1-50) and (8-1-52) is easily seen to be

$$\bar{u}_s(n,t) = \frac{a^2}{n^2\pi^2 c}\int_0^t \frac{c^2}{EI}\bar{p}_s(n,\tau)\sin\frac{n^2\pi^2 c(t-\tau)}{a^2}\,d\tau$$

so that by the inversion formula for the finite sine transform we have the final result

$$u(x,t) = \frac{2ac}{\pi^2 EI}\sum_{n=1}^{\infty}\frac{1}{n^2}\sin\frac{n\pi x}{a}\int_0^t \bar{p}_s(n,\tau)\sin\frac{n^2\pi^2 c(t-\tau)}{a^2}\,d\tau, \tag{8-1-53}$$

where $\bar{p}_s(n,t)$ is defined by equation (8-1-51).

As an example of the use of the finite Fourier cosine transform we consider the derivation of the solution $u(x,t)$ of the diffusion equation

$$\frac{\partial^2 u}{\partial x^2} = \frac{1}{\kappa}\frac{\partial u}{\partial t}, \qquad 0 \leqslant x \leqslant a,\ t > 0 \tag{8-1-54}$$

satisfying the boundary conditions

$$\frac{\partial u(0,t)}{\partial x} = \frac{\partial u(a,t)}{\partial x} = 0, \qquad t > 0, \tag{8-1-55}$$

and the initial condition

$$u(x,0) = f(x), \qquad 0 \leqslant x \leqslant a. \tag{8-1-56}$$

By equations (8-1-55) and (8-1-15) we see that

$$\mathscr{f}_c\left[\frac{\partial^2 u}{\partial x^2}; x \to n\right] = -\frac{n^2\pi^2}{a^2}\,\bar{u}_c(n,t)$$

where

$$\bar{u}_c(n,t) = \mathscr{f}_c[u(x,t); x \to n]$$

so that equations (8-1-54) and (8-1-56) yield the initial value problem

$$\frac{\partial \bar{u}_c}{\partial t} + \frac{\kappa n^2\pi^2}{a^2}\,\bar{u}_c = 0$$

$$\bar{u}_c(n,0) = \bar{f}_c(n)$$

for the determination of $\bar{u}_c(n,t)$. The solution is easily shown to be

$$\bar{u}_c(n,t) = f_c(n)e^{-\kappa n^2\pi^2 t/a^2}$$

so that by the inversion theorem for the finite Fourier cosine transform

$$u(x,t) = \frac{1}{a}\bar{f}_c(0) + \frac{2}{a}\sum_{n=1}^{\infty} \bar{f}_c(n)e^{-\kappa n^2\pi^2 t/a^2}\cos\frac{n\pi x}{a}.$$

Finally, to illustrate the use of double sine transforms we consider the solution of Poisson's equation

$$\Delta_2 u(x,y) + f(x,y) = 0$$

in the rectangle $D = \{(x,y): 0 \leqslant x \leqslant a,\ 0 \leqslant y \leqslant b\}$, when u vanishes on the boundary of D. From equation (8-1-43) we see that if we introduce the transform

$$\bar{u}_{ss}(m,n) = \mathscr{f}_{ss}[u(x,y); (x,y) \to (m,n)]$$

then

$$\bar{u}_{ss}(m,n) = \omega_{mn}{}^{-1} f_{ss}(m,n)$$

where

$$\omega_{mn} = \pi^2\left(\frac{m^2}{a^2} + \frac{n^2}{b^2}\right). \tag{8-1-57}$$

We therefore obtain the solution

$$u(x,y) = \mathscr{f}_{ss}^{-1}[\omega_{mn}^{-1}\bar{f}_{ss}(m,n);(m,n) \to (x,y)].$$

8-2 FINITE STURM–LIOUVILLE TRANSFORMS

We saw in the last section that the finite Fourier sine and cosine transforms can be used in the analysis of boundary value problems defined in rectangular regions when there are simple boundary conditions. A little reflection shows that the reason that the finite sine transform is appropriate to the solution of the boundary value problem expressed by equations (8-1-44) through (8-1-46) is simply because the functions

$$\sin\left(\frac{n\pi x}{a}\right)$$

are the eigenfunctions of the differential system

$$\frac{d^2z}{dx^2} + \xi^2 z = 0$$

$$z(0) = z(a) = 0$$

(the latter conditions being determined by the boundary conditions (8-1-45)). This encourages us to believe that we could generate finite transforms of a different type by considering the eigenvalues and eigenfunctions of different kinds of differential system. Starting with this idea Eringen (1953) and Churchill (1955) developed a theory of finite transforms of a very general type—which, following Eringen we shall call *finite Sturm–Liouville transforms*—special cases of which yield transforms of use in the solution of boundary value problems in mathematical physics. The advantage of Eringen's method is that it determines the choice of transform appropriate to the analysis of any given boundary value problem.

We suppose that we are concerned with functions $f(x)$ defined on a prescribed finite interval (a,b). We then consider finite transforms of the kind

$$\mathscr{T}[f(x);\xi] = \int_a^b p(x)K(x,\xi)f(x)\,dx \tag{8-2-1}$$

where the prescribed function $p(x)$ is of the nature of a "weight function"—in the sense of the theory of orthogonal polynomials—and the function $K(x,\xi)$ denotes a prescribed function of x on the open interval (a,b) for each value of the parameter ξ, whose domain is also prescribed. It is obvious that $\mathscr{T}$ is a linear operator.

To discover the form of the kernel $K(x,\xi)$ likely to be of value in a particular problem we consider the solutions of the self-adjoint differential equation $\mathbf{L}z = 0$ where $\mathbf{L}$ denotes the linear differential operator

$$\mathbf{L}z = \frac{1}{p(x)}\left\{\frac{d}{dx}\,q(x)\,\frac{dz}{dx}\right\} + r(x)z(x) = 0 \tag{8-2-2}$$

where $q(x) \in \mathscr{C}^2[a,b]$, $r(x) \in \mathscr{C}[a,b]$ and it is assumed that $q(x)/p(x) > 0$ for $x \in [a,b]$. We also introduce boundary operators $\mathbf{M}$ and $\mathbf{N}$ defined by the equations

$$\mathbf{M}f = a_1 f(a) + a_2 f'(a), \qquad \mathbf{N}f = b_1 f(b) + b_2 f'(b) \tag{8-2-3}$$

where a_1, a_2, b_1 and b_2 represent prescribed constants such that at least one of a_1 and a_2 and at least one of b_1 and b_2 is nonzero.

Applying the formula for integration by parts, we see that if $f \in \mathscr{C}^1[a,b]$,

$$\begin{aligned}\mathscr{T}\{\mathbf{L}f(x);\xi\} &= q(a)[K_x(a,\xi)f(a) - K(a,\xi)f'(a)] \\ &\quad - q(b)[K_x(b,\xi)f(b) - K(b,\xi)f'(b)] \\ &\quad + \int_a^b p(x)f(x)\mathbf{L}K(x,\xi)\,dx.\end{aligned} \tag{8-2-4}$$

Now if $a_2 \neq 0$,

$$K_x(a,\xi)f(a) - K(a,\xi)f'(a) = a_2^{-1}[f(a)\mathbf{M}K - K(a,\xi)\mathbf{M}f]$$

and, similarly, if $b_2 \neq 0$ we have a corresponding expression for the second term on the right-hand side of the equation (8-2-4) so that it reduces to

$$\begin{aligned}\mathscr{T}[\mathbf{L}f(x);\xi] &= \int_a^b p(x)f(x)\mathbf{L}K(x,\xi)\,dx \\ &\quad + \frac{q(a)}{a_2}[f(a)\mathbf{M}K - K(a,\xi)\mathbf{M}f] \\ &\quad - \frac{q(b)}{b_2}[f(b)\mathbf{N}K - K(b,\xi)\mathbf{N}f].\end{aligned}$$

If, therefore, we can find $K(x,\xi)$ and ξ such that

$$(\mathbf{L} - \xi)K(x,\xi) = 0, \qquad \mathbf{M}K = \mathbf{N}K = 0 \tag{8-2-5}$$

then we have

$$\mathscr{T}[\mathbf{L}f;\xi] = \xi\mathscr{T}[f;\xi] - a_2^{-1}q(a)K(a,\xi)\mathbf{M}f + b_2^{-1}q(b)K(b,\xi)\mathbf{N}f. \tag{8-2-6}$$

On the other hand, if $a_2 = 0$, $a_1 \neq 0$,

$$K_x(a,\xi)f(a) - K(a,\xi)f'(a) = a_1^{-1}q(a)[K_x(a,\xi)\mathbf{M}f - f'(a)\mathbf{M}K]$$

so that, in this case, the second term on the right-hand side of equation (8-2-6) has to be replaced by

$$a_1^{-1}q(a)K_x(a,\xi)\mathbf{M}f$$

and there is a similar modification to the third term in the case $b_2 = 0$, $b_1 \neq 0$.

The formula (8-2-6) must be modified also if $f \notin \mathscr{C}^1[a,b]$. For instance, suppose that $f \in \mathscr{C}[a,b]$ but that $f'(x)$ is continuous on $[a,b]$ except at the point $x = t$, $a < t < b$, where

$$f'(t+) - f'(t-) = j(t). \tag{8-2-7}$$

The first two terms on the right side of equation (8-2-4) then become

$$q(a)[K_x(a,\xi)f(a) - K(a,\xi)f'(a)] - q(t)[K_x(t,\xi)f(t) - K(t,\xi)f'(t-)]$$
$$+ q(t)[K_x(t,\xi)f(t) - K(t,\xi)f'(t+)] - q(b)[K_x(b,\xi)f(b) - K(b,\xi)f'(b)]$$

and it is easily seen that this reduces to

$$a_2^{-1}q(a)[f(a)\mathbf{M}K - K(a,\xi)\mathbf{M}f]$$
$$- b_2^{-1}q(b)[f(b)\mathbf{N}K - K(b,\xi)\mathbf{N}f] - q(t)K(t,\xi)j(t)$$

and hence that equation (8-2-6) is modified to

$$\mathscr{T}[\mathbf{L}f;\xi] = \xi\mathscr{T}[f;\xi] - a_2^{-1}q(a)K(a,\xi)\mathbf{M}f$$
$$+ b_2^{-1}q(b)K(b,\xi)\mathbf{N}f - q(t)K(t,\xi)j(t). \tag{8-2-8}$$

From the theory of second order linear differential equations, we know that under the conditions stated on the coefficients $p(x)$, $q(x)$, $r(x)$ of the operator $\mathbf{L}$ defined by equation (8-2-2), there is a denumerable infinity of values of the parameter ξ for which the equations (8-2-5) have nontrivial solutions and that this set of eigenfunctions is orthogonal on the interval (a,b) with the weight function $p(x)$.

If, therefore, we denote by $\phi_n(x)$ and ξ_n the (normalized) eigenfunctions and eigenvalues of the Sturm–Liouville problem

$$(\mathbf{L} - \xi)\phi = 0, \qquad \mathbf{M}\phi = \mathbf{N}\phi = 0 \tag{8-2-9}$$

then the finite transform defined by the equation

$$\mathscr{T}[f;n] \equiv \bar{f}_n = \int_a^b p(x)\phi_n(x)f(x)\,dx \tag{8-2-10}$$

has the property that if $f \in \mathscr{C}^1[a,b]$, then

$$\mathscr{T}[\mathbf{L}f;n] = \xi_n\bar{f}_n + \mu_n\mathbf{M}f + \nu_n\mathbf{N}f \tag{8-2-11}$$

where

$$\mu_n = \begin{cases} -a_2^{-1}q(a)\phi_n(a), & (a_2 \neq 0) \\ a_1^{-1}q(a)\phi_n'(a), & (a_2 = 0) \end{cases} \tag{8-2-12}$$

and

$$\nu_n = \begin{cases} b_2^{-1} q(b)\phi_n(b), & (b_2 \neq 0) \\ -b_1^{-1} q(b)\phi'(b), & (b_2 = 0). \end{cases} \tag{8-2-13}$$

The corresponding inversion formula is

$$\mathscr{T}^{-1}[\bar{f}_n;x] \equiv f(x) = \sum_{n=1} \bar{f}_n \phi_n(x) \tag{8-2-14}$$

The formula (8-2-11) has to be modified if $f \notin \mathscr{C}^1[a,b]$. For example, suppose that, as before, $f(x) \in \mathscr{C}[a,b]$ and that $f'(x)$ is continuous everywhere in the same interval except at $t \in (a,b)$ where it has the finite discontinuity (8-2-7). Then

$$\mathscr{T}[\mathbf{L}f;n] = \xi_n \bar{f}_n + \mu_n \mathbf{M}f + \nu_n \mathbf{N}f - q(t)\phi_n(t)j(t). \tag{8-2-15}$$

Other operational properties of the transform $\mathscr{T}$ are easily derived. If we define a function $y(x)$ by the equation

$$\mathscr{T}[y(x);n] = \xi^{-1}\bar{f}_n$$

then $\xi_n y_n$ is the transform of the function $\mathbf{L}y$ if $\mathbf{M}y = \mathbf{N}y = 0$, i.e., $y(x)$ is the solution of the boundary value problem

$$\mathbf{L}y = f(x), \qquad \mathbf{M}y = \mathbf{N}y = 0. \tag{8-2-16}$$

Whenever this two-point boundary value problem has a solution $y \in \mathscr{C}^1[a,b]$,

$$\xi_n y_n = f_n. \tag{8-2-17}$$

If we denote the solution of the problem (8-2-16) by $\mathbf{L}^{-1}f$, we see that the equation (8-2-15) is equivalent to the equation

$$\mathscr{T}[\mathbf{L}^{-1}f;n] = \xi_n^{-1} f_n. \tag{8-2-18}$$

From the theory of ordinary differential equations we know that when $f \in \mathscr{C}[a,b]$, the equation (8-2-16) has a unique solution of class $\mathscr{C}^1[a,b]$ provided that the corresponding homogeneous problem ($f \equiv 0$) has only the trivial solution $y(x) \equiv 0$, $x \in [a,b]$, that is, provided $\xi = 0$ is not an eigenvalue. In this case the solution may be expressed by means of the formula

$$y(x) = \int_a^b p(t)G(x,t)f(t)\,dt \tag{8-2-19}$$

where the Green's function $G(x,t)$ is determined by the equations

$$\left.\begin{aligned} \mathbf{L}G(x,t) = 0, \qquad \mathbf{M}G = \mathbf{N}G = 0, \\ G_x(t+,t) - G_x(t-,t) = \frac{p(t)}{q(t)}. \end{aligned}\right\} \tag{8-2-20}$$

and the continuity condition $G(t+,t) = G(t-,t)$ (Cf. Birkhoff and Rota, 1969, p. 309.)

Now, from equations (8-2-15) and (8-2-20) we have the relation

$$0 = \xi_n \mathscr{T}[G(x,t); x \to n] - \phi_n(t)$$

from which we deduce that if $\xi_n \neq 0$,

$$\mathscr{T}[G(x,t); x \to n] = \xi_n^{-1}\phi_n(t), \qquad (a < t < b). \tag{8-2-21}$$

Applying the inversion formula (8-2-14) to this result we deduce the bilinear formula

$$G(x,t) = \sum_{n=1}^{\infty} \xi_n^{-1}\phi_n(t)\phi_n(x) \tag{8-2-22}$$

familiar from the theory of integral equations.

Also, we have

$$\begin{aligned}\mathscr{T}\left[\int_a^b p(\tau)G(x,\tau)G(\tau,t)\,d\tau; x \to n\right] &= \int_a^b p(\tau)\mathscr{T}[G(x,\tau); x \to n]G(\tau,t)\,d\tau \\ &= \xi_n^{-1}\int_a^b p(\tau)\phi_n(\tau)G(\tau,t)\,d\tau \\ &= \xi_n^{-1}\mathscr{T}[G(\tau,t); \tau \to n]\end{aligned}$$

which shows that

$$\mathscr{T}\left[\int_a^b p(\tau)G(x,\tau)G(\tau,t)\,d\tau; x \to n\right] = \xi_n^{-2}\phi_n(t). \tag{8-2-23}$$

Another interesting result can be derived by putting $f = 1$, $\mathbf{L}f = r(x)$ in equation (8-2-11). We find that

$$\mathscr{T}[r(x); n] = \xi_n \mathscr{T}[1; n] + \mu_n a_1 + \nu_n b_1 \tag{8-2-24}$$

from which we derive the transform of either $r(x)$ or 1 when that of the other is known.

We shall now consider some special cases of the general transform (8-2-10).

8-3 GENERALIZED FINITE FOURIER TRANSFORMS

To illustrate the use of the above procedures we begin by considering the generalized finite Fourier transforms (Cf. Churchill, 1955–6) corresponding to the differential operator

$$\mathbf{L} = \frac{d^2}{dx^2} + \xi^2. \tag{8-3-1}$$

The first of these transforms correspond to the boundary operators

$$\mathbf{M}y = y'(0) \tag{8-3-2}$$

$$\mathbf{N}y = hy(1) + y'(1). \tag{8-3-3}$$

The solution of $\mathbf{L}y = 0$ satisfying $\mathbf{M}y = 0$ is

$$y(x) = \cos(\xi_n x).$$

For this solution

$$hy(x) + y'(x) = h\cos(\xi_n x) - \xi_n \sin(\xi_n x)$$

so that the condition $\mathbf{N}y = 0$ is satisfied if the ξ_n are chosen to be the positive roots of the transcendental equation

$$\xi \tan \xi = h. \tag{8-3-4}$$

If we define a transform $\mathscr{T}_1 f$ by the equation

$$f_1^*(n) \equiv \mathscr{T}_1[f(x);n] = \int_0^1 f(x)\cos(\xi_n x)\,dx \tag{8-3-5}$$

where ξ_n is the n-th positive root of equation (8-3-4) we see that its inversion formula is

$$\mathscr{T}_1^{-1}[f_1^*(n);x] \equiv f(x) = \sum_{n=1}^{\infty} c_n f_1^*(n)\cos(\xi_n x), \qquad 0 < x < 1 \tag{8-3-6}$$

with

$$c_n^{-1} = \int_0^1 \cos^2(\xi_n x)\,dx$$

that is, with

$$c_n = \frac{2h}{h + \sin^2 \xi_n}. \tag{8-3-7}$$

On the other hand, if we replace the conditions (8-3-2) and (8-3-3) by

$$\mathbf{M}y = y(0) \tag{8-3-8}$$

$$\mathbf{N}y = hy'(1) + y''(1) \tag{8-3-9}$$

respectively, we obtain the transform

$$f_2^*(n) \equiv \mathscr{T}_2[f(x);n] = \int_0^1 f(x)\sin(\xi_n x)\,dx \tag{8-3-10}$$

where again ξ_n is the n-th positive root of the transcendental equation (8-3-4), with the corresponding inversion formula

$$\mathscr{T}_2^{-1}[f^*(n);x] \equiv f(x) = \sum_{n=1}^{\infty} d_n f_2^*(n)\sin(\xi_n x) \tag{8-3-11}$$

with

$$d_n = \frac{2h}{h - \sin^2 \xi_n}. \tag{8-3-12}$$

Using the formula for integrating by parts we find that

$$\mathscr{T}_1[f'(x);n] = f(1)\cos\xi_n + \xi_n f_2^*(n) \tag{8-3-13}$$

and that

$$\mathscr{T}_2[f'(x);n] = f(1)\sin\xi_n - \xi_n f_1^*(n). \tag{8-3-14}$$

Applying these formulae in succession we deduce immediately that

$$\mathscr{T}_1[f''(x);n] = [f'(1) + hf(1)]\cos\xi_n - \xi_n{}^2 f_1^*(n) \tag{8-3-15}$$

$$\mathscr{T}_2[f''(x);n] = f'(1)\sin\xi_n - f(1)\xi_n\cos\xi_n - \xi_n{}^2 f_2^*(n). \tag{8-3-16}$$

Also, if in equation (8-3-14) we replace $f(x)$ by

$$\int_x^1 f(t)\,dt$$

we find that

$$\mathscr{T}_2[f(x);n] = \xi_n \mathscr{T}_1\left[\int_x^1 f(t)\,dt;n\right]. \tag{8-3-17}$$

As an example of the use of transforms of this kind we consider the solution $u(x,t)$ of the diffusion equation

$$\frac{\partial^2 u}{\partial x^2} = \frac{1}{\kappa}\frac{\partial u}{\partial t}, \qquad 0 \leqslant x \leqslant 1,\ t \geqslant 0 \tag{8-3-18}$$

satisfying the boundary conditions

$$u(0,t) = 0, \qquad t > 0 \tag{8-3-19}$$

$$\frac{\partial u(1,t)}{\partial x} + hu(1,t) = 0, \qquad t > 0 \tag{8-3-20}$$

and the initial condition

$$u(x,0) = f(x), \qquad 0 \leqslant x \leqslant 1. \tag{8-3-21}$$

From an examination of the boundary conditions (8-3-19), (8-3-20) it is obvious that the appropriate transform to use is

$$u_1^*(n,t) = \mathscr{T}_1[u(x,t);x \to n].$$

It then follows immediately from equation (8-3-15) that the differential equation (8-3-18) is equivalent to the ordinary differential equation

$$\frac{du_1^*}{dt} = -\kappa\xi_n{}^2 u_1^*$$

and that the initial condition (8-3-21) is equivalent to

$$u_1^*(n,0) = f_1^*(n).$$

The solution of this simple initial value problem

$$u_1^*(n,t) = f_1^*(n)e^{-\kappa\xi_n{}^2 t}$$

taken with the inversion theorem (8-3-6) leads immediately to the formula

$$u(x,t) = \sum_{n=1} c_n f_1^*(n)e^{-\kappa\xi_n{}^2 t}\cos(\xi_n x) \tag{8-3-22}$$

with c_n defined by (8-3-7) and $f_1^*(n)$ by (8-3-5), for the solution of the problem.

8-4 FINITE HANKEL TRANSFORMS

The finite transforms corresponding to the linear operator

$$\mathbf{L} = \mathscr{B}_\nu \equiv \frac{d^2}{dx^2} + \frac{1}{x}\frac{d}{dx} - \frac{\nu^2}{x^2} \tag{8-4-1}$$

which is of the form (8-2-1) with

$$p(x) = x, \qquad q(x) = x, \qquad r(x) = -\nu^2 x^{-2} \tag{8-4-2}$$

are called *finite Hankel transforms* (Cf. Sneddon, 1946). If we take $\xi = -\lambda^2$, the first of equations (8-2-9) is simply Bessel's equation

$$\frac{d^2\phi}{dx^2} + \frac{1}{x}\frac{d\phi}{dx} + \left(\lambda^2 - \frac{\nu^2}{x^2}\right) = 0 \tag{8-4-3}$$

which has general solution

$$\phi(\lambda,x) = c_1(\lambda)\mathrm{J}_\nu(\lambda x) + c_2(\lambda)\mathrm{Y}_\nu(\lambda x). \tag{8-4-4}$$

The particular values of λ appropriate to any boundary value problem will, of course, depend on the forms we assume for the operators **M** and **N**. We shall now consider special cases in some detail.

Case (*a*): $0 \leqslant \mathbf{x} \leqslant a, \qquad \mathbf{N}f = f.$

The required solution of equation (8-4-3) in this case is

$$\phi_n(x) = \mathrm{J}_\nu(\xi_n x), \qquad (\nu \geqslant 0) \tag{8-4-5}$$

where $\xi_1, \xi_2, \ldots$ are the positive roots of the transcendental equation

$$\mathrm{J}_\nu(\xi a) = 0.$$

The transform (8-2-10) becomes in this case

$$\bar{f}_{1,\nu}(n) \equiv \mathscr{h}_{1,\nu}[f(x);n] = \int_0^a xf(x)\mathrm{J}_\nu(\xi_n x)\,dx; \tag{8-4-7}$$

we shall call it a *finite Hankel transform of the first kind.* Since the Bessel functions of the first kind satisfy the orthogonality relation

$$\int_0^a x\mathrm{J}_\nu(\xi_m x)\mathrm{J}_\nu(\xi_n x)\,dx = \tfrac{1}{2}a^2\{\mathrm{J}_{\nu+1}(\xi_n a)\}^2\delta_{mn} \tag{8-4-8}$$

it follows that the inversion formula for the operator $\mathscr{h}_{1,\nu}$ is

$$\mathscr{h}_{1,\nu}{}^{-1}[\bar{f}_{1,\nu}(n);x] \equiv f(x) = \frac{2}{a^2}\sum_{n=1}^{\infty}\bar{f}_{1,\nu}(n)\frac{\mathrm{J}_\nu(\xi_n x)}{[\mathrm{J}_{\nu+1}(\xi_n a)]^2}. \tag{8-4-9}$$

This, of course, is merely a restatement of the classical theorem in the theory of Fourier–Bessel series (see, for instance, Chapter XVIII of Watson, 1944) which states that if $f(x) \in \mathscr{P}^1(0,a)$ and if its finite Hankel transform $\bar{f}_{1,\nu}(n)$ is defined by equation (8-4-7) then, at any point x of $(0,a)$, the series

$$\frac{2}{a^2}\sum_{n=1}^{\infty}\bar{f}_{1,\nu}(n)\frac{\mathrm{J}_\nu(\xi_n x)}{[\mathrm{J}_{\nu+1}(\xi_n a)]^2}$$

converges to the sum $\frac{1}{2}[f(x+) + f(x-)]$.

The properties of the operator $\mathscr{h}_{1,\nu}$ follow immediately from those of the Bessel function $\mathrm{J}_\nu(\xi_n x)$.

Integrating by parts we see that

$$\mathscr{h}_{1,\nu}[f'(x);n] = [xf(x)\mathrm{J}_\nu(\xi_n x)]_0^a - \int_0^a f(x)\frac{\partial}{\partial x}\{x\mathrm{J}_\nu(\xi_n x)\}\,dx$$

If we suppose that $\nu \geqslant 0$ then $x\mathrm{J}_\nu(\xi_n x)$ will vanish at $x = 0$ so that if we assume that $f(x)$ is finite at $x = 0$ then the quantity in the square bracket will vanish at the lower limit. It will vanish at the upper limit because of the choice of ξ_n—provided of course that $f(a)$ is finite. Using the recurrence relation

$$\frac{\partial}{\partial x}\{x\mathrm{J}_\nu(\xi_n x)\} = \frac{\nu+1}{2\nu}\xi_n x\mathrm{J}_{\nu-1}(\xi_n x) - \frac{\nu-1}{2\nu}\xi_n x\mathrm{J}_{\nu+1}(\xi_n x), \qquad (\nu \neq 0) \tag{8-4-10}$$

we find that

$$\mathscr{h}_{1,\nu}[f'(x);n] = \frac{\nu+1}{2\nu}\xi_n\bar{f}_{1,\nu-1}(n) - \frac{\nu-1}{2\nu}\xi_n\bar{f}_{1,\nu+1}(n), \qquad (\nu \neq 0). \tag{8-4-11}$$

The case $\nu = 0$ has to be treated separately. Instead of (8-3-10) we use the relation

$$\frac{\partial}{\partial x}\{x\mathrm{J}_0(\xi_n x)\} = \mathrm{J}_0(\xi_n x) - \xi_n x\mathrm{J}_1(\xi_n x)$$

to derive the equation

$$\mathscr{h}_{1,0}[f'(x);n] = \mathscr{h}_{1,0}[x^{-1}f(x);n] - \xi_n \mathscr{h}_{1,1}[f(x);n]. \tag{8-4-12}$$

Similarly from the recurrence relation

$$J_{\nu-1}(x) - \frac{2\nu}{x} J_\nu(x) + J_{\nu+1}(x) = 0, \qquad (\nu \neq 0) \tag{8-4-13}$$

we deduce the relation

$$\mathscr{h}_{1,\nu}[x^{-1}f(x);n] = \frac{\xi_n}{2\nu}\{\bar{f}_{1,\nu-1}(n) + \bar{f}_{1,\nu+1}(n)\}, \qquad (\nu \neq 0). \tag{8-4-14}$$

From equation (8-2-6) we also have the relation

$$\mathscr{h}_{1,\nu}[\mathscr{B}_\nu f;n] = -a\xi_n f(a) J_\nu'(\xi_n a) - \xi_n^2 \bar{f}_{1,\nu}(n)$$

from which, by using the result

$$J_\nu'(\xi_n a) = -J_{\nu+1}(\xi_n a) \tag{8-4-15}$$

we deduce the equation

$$\mathscr{h}_{1,\nu}[\mathscr{B}_\nu f;n] = a\xi_n f(a) J_{\nu+1}(\xi_n a) - \xi_n^2 \bar{f}_{1,\nu}(n). \tag{8-4-16}$$

Also, from the recurrence formula

$$\frac{d}{dx}\{x^\nu J_\nu(x)\} = x^\nu J_{\nu-1}(x) \tag{8-4-17}$$

and the rule for integrating by parts we deduce that

$$\mathscr{h}_{1,\nu}[x^{\nu-1}f'(x);n] = -\xi_n \mathscr{h}_{1,\nu-1}[x^{\nu-1}f(x);n], \qquad (\nu > 0). \tag{8-4-18}$$

In particular, if we take $\nu = 1$ we have

$$\mathscr{h}_{1,1}[f'(x);n] = -\xi_n \bar{f}_{1,0}(n). \tag{8-4-19}$$

To illustrate the use of the finite Hankel transform of the first kind we consider the problem of finding the solution $\theta(r,t)$ of the diffusion equation

$$\frac{\partial^2\theta}{\partial r^2} + \frac{1}{r}\frac{\partial\theta}{\partial r} = \frac{1}{\kappa}\frac{\partial\theta}{\partial t}, \qquad (t \geqslant 0, 0 \leqslant r \leqslant a) \tag{8-4-20}$$

satisfying the boundary condition

$$\theta(a,t) = f(t), \qquad (t \geqslant 0) \tag{8-4-21}$$

and the initial condition

$$\theta(r,0) = 0, \qquad (0 \leqslant r \leqslant a) \tag{8-4-22}$$

which has been considered already in sec. 3-15-3 as an example of the use of the Laplace transform.

We introduce the transform

$$\bar{\theta}_n(t) = \mathscr{h}_{1,0}[\theta(r,t); r \to n]$$

and make use of the result (8-4-16) to show that

$$\mathscr{h}_{1,0}\left[\frac{\partial^2\theta}{\partial r^2} + \frac{1}{r}\frac{\partial\theta}{\partial r}; r \to n\right] = a\xi_n J_1(\xi_n a)f(t) - \xi_n{}^2\bar{\theta}_n(t)$$

and hence that $\bar{\theta}_n(t)$ satisfies the differential equation

$$\left(\frac{\partial}{\partial t} + \kappa\xi_n{}^2\right)\theta_n(t) = \kappa a\xi_n J_1(\xi_n a)f(t).$$

From equation (8-4-22) we see that it also satisfies the initial condition

$$\bar{\theta}_n(0) = 0,$$

from which we immediately deduce that

$$\bar{\theta}_n(t) = \kappa a\xi_n J_1(\xi_n a)c_n(t) \tag{8-4-23}$$

where

$$c_n(t) = \int_0^t f(u)e^{-\kappa\xi_n{}^2(t-u)}\,du.$$

Applying the inversion formula (8-4-9) to equation (8-4-23) we obtain the solution

$$\theta(r,t) = \frac{2}{a^2}\sum_{n=1}^{\infty} c_n(t)\xi_n\frac{J_0(\xi_n r)}{J_1(\xi_n a)}$$

in agreement with equation (3-15-20).

Case (b): $0 \leqslant x \leqslant a, \qquad \mathbf{N}f = f'(a) + hf(a).$

The required solution of (8-4-3) in this case is again $J_\nu(\xi_n x)$, $(\nu \geqslant 0)$, but now $\xi_1, \xi_2, \ldots$ are the positive zeros of the transcendental equation

$$\xi_n J_\nu'(\xi_n a) + hJ_\nu(\xi_n a) = 0 \tag{8-4-24}$$

and the finite Hankel transform of the second kind is defined by the equation

$$\bar{f}_{2,\nu}(n) \equiv \mathscr{h}_{2,\nu}[f(x); n] = \int_0^a xf(x)J_\nu(\xi_n x)\,dx \tag{8-4-25}$$

where ξ_n is the n-th positive root of the equation (8-4-24). Since the Bessel functions now satisfy the orthogonality condition

$$\int_0^a xJ_\nu(\xi_m x)J_\nu(\xi_n x)\,dx = \tfrac{1}{2}a^2\left(1 + \frac{h^2 - \nu^2/a^2}{\xi_n{}^2}\right)\{J_\nu(\xi_n a)\}^2\delta_{mn} \tag{8-4-26}$$

(Cf. Problem 8-14) it follows that the inversion formula for the operator $\mathscr{h}_{2,\nu}$ is

$$\mathscr{h}_{2,\nu}^{-1}[\bar{f}_{2,\nu}(n);x] \equiv f(x)$$

$$\frac{2}{a^2}\sum_{n=1}^{\infty}\bar{f}_{2,\nu}(n)\frac{\xi_n^{\,2}\mathrm{J}_\nu(\xi_n x)}{(\xi_n^{\,2}+h^2-\nu^2/a^2)\{\mathrm{J}_\nu(\xi_n a)\}^2}. \tag{8-4-27}$$

The properties of the operator $\mathscr{h}_{2,\nu}$ can be established in exactly the same way as those of $\mathscr{h}_{1,\nu}$. We obtain the relations:

$$\mathscr{h}_{2,\nu}[f'(x);n] = af(a)\mathrm{J}_\nu(\xi_n a) + \frac{\nu+1}{2\nu}\,\xi_n\bar{f}_{2,\nu-1}(n) - \frac{\nu-1}{2\nu}\,\xi_n\bar{f}_{2,\nu+1}(n) \tag{8-4-28}$$

$$\mathscr{h}_{2,0}[f'(x);n] = af(a)\mathrm{J}_0(\xi_n a) + \mathscr{h}_{2,0}[x^{-1}f(x);n] - \xi_n\bar{f}_{2,1}(n) \tag{8-4-29}$$

$$\mathscr{h}_{2,\nu}[x^{-1}f(x);n] = \frac{\xi_n}{2\nu}\{\bar{f}_{1,\nu-1}(n) + \bar{f}_{1,\nu+1}(n)\}, \qquad (\nu > 0) \tag{8-4-30}$$

$$\mathscr{h}_{2,\nu}[x^{\nu-1}f'(x);n] = a^\nu\mathrm{J}_\nu(\xi_n a)f(a) - \xi_n\mathscr{h}_{2,\nu-1}[x^{\nu-1}f(x);n], \qquad (\nu > 0) \tag{8-4-31}$$

$$\mathscr{h}_{2,\nu}[\mathscr{B}_\nu f;n] = -\xi_n^{\,2}\bar{f}_{2,\nu}(n) + a\mathrm{J}_\nu(\xi_n a)\{f'(a) + hf(a)\}. \tag{8-4-32}$$

To illustrate the use of the finite Hankel transform of the second kind we consider the solution of equation (8-4-20) satisfying the boundary condition

$$\frac{\partial\theta(a,t)}{\partial r} + h\theta(a,t) = 0, \qquad (t \geqslant 0) \tag{8-4-33}$$

$$\theta(r,0) = f(r), \qquad (0 \leqslant r \leqslant a). \tag{8-4-34}$$

If we introduce the transform

$$\bar{\theta}_n(t) = \mathscr{h}_{2,0}[\theta(r,t);r\to n]$$

then from equations (8-4-32) and (8-4-33) we see that

$$\mathscr{h}_{2,0}\left[\frac{\partial^2\theta}{\partial r^2} + \frac{1}{r}\frac{\partial\theta}{\partial r};r\to n\right] = -\xi_n^{\,2}\bar{\theta}_n$$

so that θ_n satisfies

$$\left(\frac{\partial}{\partial t} + \kappa\xi_n^{\,2}\right)\bar{\theta}_n(t) = 0.$$

Also from equation (8-4-34) we see that it satisfies the initial condition

$$\bar{\theta}_n(0) = \bar{f}_{2,0}(n) \equiv \mathscr{h}_{2,0}[f(r);n]$$

and hence that

$$\bar{\theta}_n(t) = \bar{f}_{2,0}(n)e^{-\kappa\xi_n^2 t}.$$

The required solution may therefore be written in the form

$$\theta(r,t) = \mathscr{h}_{2,0}^{-1}[\bar{f}_{2,0}(n)e^{-\kappa\xi_n^2 t};n \to r]$$

and, from equation (8-4-27) with $\nu = 0$, we see that this is equivalent to the series solution

$$\theta(r,t) = \frac{2}{a^2}\sum_{n=1} \frac{\xi_n^2 J_\nu(\xi_n r)}{(\xi_n^2 + h^2)[J_\nu(\xi_n a)]^2} e^{-\kappa\xi_n^2 t} \int_0^a u f(u) J_\nu(\xi_n u)\, du$$

where the summation extends over the positive roots $\xi_1, \xi_2, \ldots$ of equation (8-4-24).

Case (*c*): $a \leqslant x \leqslant b, \qquad \mathbf{M}f = f(a), \qquad \mathbf{N}f = f(b).$

The required solution of equation (8-4-3) in this case is

$$\phi_n(x) = J_\nu(\xi_n x)Y_\nu(\xi_n a) - J_\nu(\xi_n a)Y_\nu(\xi_n x) \tag{8-4-35}$$

where now ξ_n is a root of the transcendental equation

$$J_\nu(\xi_n b)Y_\nu(\xi_n a) - J_\nu(\xi_n a)Y_\nu(\xi_n b) = 0. \tag{8-4-36}$$

We then define the finite Hankel transform of the third kind by the equation

$$\bar{f}_{3,\nu}(n) \equiv \mathscr{h}_{3,\nu}[f(x);n]$$
$$= \int_a^b x f(x)[J_\nu(\xi_n x)Y_\nu(\xi_n a) - J_\nu(\xi_n a)Y_\nu(\xi_n x)]\, dx. \tag{8-4-37}$$

Using the orthogonality condition

$$\int_a^b x\phi_m(x)\phi_n(x)\, dx = \frac{2}{\pi^2\xi_n^2}\frac{\{J_\nu(\xi_n a)\}^2 - \{J_\nu(\xi_n b)\}^2}{\{J_\nu(\xi_n b)\}^2}\delta_{mn} \tag{8-4-38}$$

for the functions $\phi_m(x)$ defined by equation (8-4-35) we see that the inversion theorem for the operator $\mathscr{h}_{3,\nu}$ is

$$\mathscr{h}_{3,\nu}^{-1}[\bar{f}_{3,\nu}(n);x] \equiv f(x) = \tfrac{1}{2}\pi^2 \sum_{n=1}^{\infty} \frac{\xi_n^2 J_\nu^2(\xi_n b)\bar{f}_{3,\nu}(n)}{\{J_\nu(\xi_n a)\}^2 - \{J_\nu(\xi_n b)\}^2} \times$$
$$\times [J_\nu(\xi_n x)Y_\nu(\xi_n a) - J_\nu(\xi_n a)Y_\nu(\xi_n x)] \tag{8-4-39}$$

the sum extending over all the positive roots of equation (8-4-36).

Also, from equation (8-2-6), we have the useful relation

$$\mathscr{h}_{3,\nu}[\mathscr{B}_\nu f(x);n] = \frac{2J_\nu(\xi_n a)}{\pi J_\nu(\xi_n b)} f(b) - \frac{2}{\pi} f(a) - \xi_n^2 \bar{f}_{3,\nu}(n). \tag{8-4-40}$$

For an illustration of the use of this finite Hankel transform of the third kind the reader is referred to Problem 8-24.

Case (*d*): $a \leqslant x \leqslant b, \qquad \mathbf{M}f = f'(a), \qquad \mathbf{N}f = f(b).$

The appropriate solution of (8-4-3) is

$$\phi_n(x) = J_\nu(\xi_n x)Y'_\nu(\xi_n a) - J'_\nu(\xi_n a)Y_\nu(\xi_n x) \tag{8-4-41}$$

with ξ_n a root of the equation

$$J_\nu(\xi_n b)Y'_\nu(\xi_n a) - J'_\nu(\xi_n a)Y_\nu(\xi_n b) = 0. \tag{8-4-42}$$

The functions $\phi_n(x)$ defined by equations (8-4-41) and (8-4-42) are easily shown to satisfy the orthogonality condition

$$\int_a^b x\phi_m(x)\phi_n(x)\,dx = \frac{2}{\pi^2\xi_n^2}\left[\frac{\{J'_\nu(\xi_n a)\}^2}{\{J_\nu(\xi_n b)\}^2} - \left(1 - \frac{\nu^2}{\xi_n^2 a^2}\right)\right]\delta_{mn} \tag{8-4-43}$$

so that the inversion theorem for the operator $\mathscr{h}_{4,\nu}$ of the finite Hankel transform of the fourth kind defined by the equation

$$\mathscr{h}_{4,\nu}[f(x);n] \equiv \bar{f}_{4,\nu}(n) \times \int_a^b xf(x)[J_\nu(\xi_n x)Y'_\nu(\xi_n a) - J'_\nu(\xi_n a)Y_\nu(\xi_n x)]\,dx \tag{8-4-44}$$

with ξ_n the n-th positive root of equation (8-4-42) is

$$f(x) = \mathscr{h}_{4,\nu}^{-1}[\bar{f}_{4,\nu}(n);x] = \tfrac{1}{2}\pi^2 \sum_{n=1} \frac{\xi_n^2\{J_\nu(\xi_n b)\}^2 \bar{f}_{4,\nu}(n)}{\{J'_\nu(\xi_n a)\}^2 - (1 - \nu^2/\xi_n^2 a^2)\{J_\nu(\xi_n b)\}^2} \times \times [J_\nu(\xi_n x)Y'_\nu(\xi_n a) - J'_\nu(\xi_n a)Y_\nu(\xi_n x)] \tag{8-4-45}$$

where the summation extends over all the positive roots of equation (8-4-42). In this case equation (8-2-6) gives

$$\mathscr{h}_{4,\nu}[\mathscr{B}_\nu f(x);n] = -\xi_n^2 \bar{f}_{4,\nu}(n) - \frac{2}{\pi\xi_n} f'(a) + \frac{2}{\pi}\frac{J'_\nu(\xi_n a)}{J_\nu(\xi_n b)} f(b). \tag{8-4-46}$$

An example of the use of this transform is given in Problem 8-25.

Case (*e*): $a \leqslant x \leqslant b, \qquad \mathbf{M}f = f(a), \qquad \mathbf{N}f = f'(b).$

This case is very similar to the last one. It leads to a finite Hankel transform of the fifth kind given by the equation

$$\mathscr{h}_{5,\nu}[f(x);n] \equiv \bar{f}_{5,\nu}(n)$$
$$= \int_a^b xf(x)[\mathrm{J}_\nu(\xi_n x)\mathrm{Y}_\nu(\xi_n a) - \mathrm{J}_\nu(\xi_n a)\mathrm{Y}_\nu(\xi_n x)]\,dx \qquad \text{(8-4-47)}$$

where ξ_n is the n-th positive root of the equation

$$\mathrm{J}_\nu(\xi_n a)\mathrm{Y}'_\nu(\xi_n b) - \mathrm{J}'_\nu(\xi_n b)\mathrm{Y}_\nu(\xi_n a) = 0. \qquad \text{(8-4-48)}$$

The inversion theorem is contained in the equation

$$\mathscr{h}_{5,\nu}^{-1}[\bar{f}_{5,\nu}(n);x] \equiv f(x)$$
$$= \tfrac{1}{2}\pi^2 \sum_{n=1} \frac{\xi_n^{\,2}\{\mathrm{J}_\nu(\xi_n a)\}^2\bar{f}_{5,\nu}(n)}{(1 - \nu^2/\xi_n^{\,2}b^2)\{\mathrm{J}_\nu(\xi_n a)\}^2 - \{\mathrm{J}'_\nu(\xi_n b)\}^2} \times$$
$$\times [\mathrm{J}_\nu(\xi_n x)\mathrm{Y}_\nu(\xi_n a) - \mathrm{J}_\nu(\xi_n a)\mathrm{Y}_\nu(\xi_n x)] \qquad \text{(8-4-49)}$$

the summation extending over all the positive roots of equation (8-4-48). Corresponding to equation (8-4-46) we have the relation

$$\mathscr{h}_{5,\nu}[\mathscr{B}_\nu f(x);n] = -\xi_n^{\,2}\bar{f}_{5,\nu}(n) + \frac{2}{\pi\xi_n}f'(b) - \frac{2\mathrm{J}'_\nu(\xi_n b)}{\pi \mathrm{J}_\nu(\xi_n a)}f(a). \qquad \text{(8-4-50)}$$

Case (f): $a \leqslant x \leqslant b$, $\quad \mathbf{M}f = f'(a)$, $\quad \mathbf{N}f = f'(b)$.

In this case the appropriate eigenfunction is

$$\phi_n(x) = \mathrm{J}_\nu(\xi_n x)\mathrm{Y}'_\nu(\xi_n a) - \mathrm{J}'_\nu(\xi_n a)\mathrm{Y}_\nu(\xi_n x) \qquad \text{(8-4-51)}$$

where ξ_n is the n-th positive root of the equation

$$\mathrm{J}'_\nu(\xi_n b)\mathrm{Y}'_\nu(\xi_n a) - \mathrm{J}'_\nu(\xi_n a)\mathrm{Y}'_\nu(\xi_n b) = 0. \qquad \text{(8-4-52)}$$

Since these functions satisfy the orthogonality relation

$$\int_a^b x\phi_m(x)\phi_n(x)\,dx$$
$$= \frac{2}{\pi^2\xi_n^{\,2}}\delta_{mn}\left\{\left(1 - \frac{\nu^2}{\xi_n^{\,2}b^2}\right)\left[\frac{\mathrm{J}'_\nu(\xi_n a)}{\mathrm{J}'_\nu(\xi_n b)}\right]^2 - \left(1 - \frac{\nu^2}{\xi_n^{\,2}a^2}\right)\right\}$$

we see that the *finite Hankel transform of the sixth kind* defined by the equation

$$\mathscr{h}_{6,\nu}[f(x);n] \equiv \bar{f}_{6,\nu}(n)$$
$$= \int_a^b xf(x)\{\mathrm{J}_\nu(\xi_n x)\mathrm{Y}'_\nu(\xi_n a) - \mathrm{J}'_\nu(\xi_n a)\mathrm{Y}_\nu(\xi_n x)\}\,dx \qquad \text{(8-4-53)}$$

has the inversion formula

$$\mathscr{h}_{6,\nu}^{-1}[\bar{f}_{6,\nu}(n);x] \equiv f(x)$$

$$= \tfrac{1}{2}\pi^2 \sum_{n=1}^{\infty} \frac{\{J'_\nu(\xi_n b)\}^2 \bar{f}_{6,\nu}(n)\phi_n(x)}{(1 - \nu^2/\xi_n{}^2 b^2)\{J'_\nu(\xi_n a)\}^2 - (1 - \nu^2/\xi_n{}^2 a^2)\{J'_\nu(\xi_n b)\}^2} \tag{8-4-54}$$

where $\phi_n(x)$ is defined by equation (8-4-51) and the infinite sum is taken over the positive roots of equation (8-4-52).

In this case equation (8-2-6) takes the form

$$\mathscr{h}_{6,\nu}[\mathscr{B}_\nu f(x);n] = \frac{2}{\pi\xi_n} \cdot \frac{J'_\nu(\xi_n a)}{J'_\nu(\xi_n b)} f'(b) - \frac{2}{\pi\xi_n} f'(a) - \xi_n{}^2 \bar{f}_{6,\nu}(n). \tag{8-4-55}$$

The use of this transform in a typical problem of mathematical physics is shown in Problem 8-26.

The general finite Hankel transform corresponding to the operators

$$\mathbf{L} = \mathscr{B}_\nu, \qquad \mathbf{M}f = f'(a) + hf(a), \qquad \mathbf{N}f = f'(b) + kf(b)$$

has been considered by Cinelli (1966). The formula expressing the inversion theorem for a transform as general as this is very complicated and it is better, in discussion of a problem involving the operator $\mathscr{B}_\nu$, to construct the finite Hankel transform specific to that problem. Further examples of the procedure are provided by Problems 8-27 and 8-28.

8-5 FINITE LEGENDRE TRANSFORMS

Finite transforms corresponding to the differential operator

$$\mathbf{L} = \frac{d}{dx}(1 - x^2)\frac{d}{dx}, \qquad -1 \leqslant x \leqslant 1$$

were first discussed by Tranter (1951) and Churchill (1954). In this case the appropriate eigenfunctions are the Legendre polynomials $P_n(x)$, $(n = 0, 1, \ldots)$, with, in the notation of sec. 8-2, $\xi = -n(n+1)$.

If we define the *finite Legendre transform* by the equation

$$\bar{f}_l(n) \equiv \ell[f(x);n] = \int_{-1}^{1} f(x)P_n(x)\,dx \tag{8-5-1}$$

we can use the procedures of sec. 8-2 to establish an operational calculus of such transforms.

Since $P_0(x) = 1$, $P_1(x) = x$, we have the special cases

$$\bar{f}_l(0) = \int_{-1}^{1} f(x)\,dx \tag{8-5-2}$$

$$\bar{f}_l(1) = \int_{-1}^{1} xf(x)\,dx. \tag{8-5-3}$$

In particular

$$\ell[f'(x);0] = f(1) - f(-1) \tag{8-5-4}$$

$$\ell[f'(x);1] = f(1) + f(-1) - \int_{-1}^{1} f(x)\,dx. \tag{8-5-5}$$

Replacing $f(x)$ by $f'(x)$ in (8-5-5) and using (8-5-4) we find that

$$\ell[f''(x);1] = f'(1) + f'(-1) - f(1) + f(-1). \tag{8-5-6}$$

The well-known orthogonality property for Legendre polynomials

$$\int_{-1}^{1} P_m(x)P_n(x)\,dx = \frac{2}{2n+1}\,\delta_{mn} \tag{8-5-7}$$

can be written in the form

$$\ell[P_m(x);n] = \frac{2}{2n+1}\,\delta_{mn}.$$

In particular, by taking $m = 0$ we see that if c is a constant

$$\left.\begin{aligned} &\ell[c;0] = 2c, \\ &\ell[c;n] = 0, \qquad (n = 1,2,\ldots). \end{aligned}\right\} \tag{8-5-8}$$

The inversion formula for the operator ℓ follows from the Fourier–Legendre expansion theorem—itself a consequence of the orthogonality relation (8-5-7)—which states that if $f(x) \in \mathscr{P}^1[-1,1]$, then the series

$$\sum_{n=0}^{\infty} (n + \tfrac{1}{2})\bar{f}_l(n)P_n(x)$$

converges to $\frac{1}{2}[f(x+) + f(x-)]$ for all x belonging to $(-1,1)$.

We therefore have the inversion formula

$$\ell^{-1}[\bar{f}_l(n);x] = \sum_{n=0} (n + \tfrac{1}{2})\bar{f}_l(n)P_n(x). \tag{8-5-9}$$

Other operational properties of the finite Legendre transform can be derived immediately from elementary properties of the Legendre polynomial.

For example, since

$$\ell[f(-x);n] = \int_{-1}^{1} f(-x)P_n(x)\,dx = \int_{-1}^{1} f(x)P_n(-x)\,dx$$

and $P_n(-x) = (-1)^n P_n(x)$, we see that

$$\ell[f(-x);n] = (-1)^n \bar{f}_l(n)$$

a result which is more often used in the alternative form

$$\ell^{-1}[(-1)^n\bar{f}_l(n);x] = f(-x). \tag{8-5-10}$$

Also, by equation (8-2-6), we have the result

$$\ell\left[\frac{d}{dx}(1-x^2)f'(x);n\right] = -n(n+1)\bar{f}_l(n) \tag{8-5-11}$$

provided that both $(1-x^2)f(x)$ and $(1-x^2)f'(x)$ vanish at the end-points $x = \pm 1$.

From the recurrence relation

$$(2n+1)P_n(x) = P'_{n+1}(x) - P'_{n-1}(x)$$

we deduce that

$$\int_{-1}^{1} g(x)P_n(x)\,dx = (2n+1)^{-1}[g(x)\{P_{n+1}(x) - P_{n-1}(x)\}]_{-1}^{1}$$
$$- (2n+1)^{-1}\int_{-1}^{1} g'(x)\{P_{n+1}(x) - P_{n-1}(x)\}\,dx$$

that is, that

$$\ell[g(x);n] = (2n+1)^{-1}\{\ell[g'(x);n-1] - \ell[g'(x);n+1]\}. \tag{8-5-12}$$

In particular, we have the special case

$$\ell\left[\int_{-1}^{x} f(t)\,dt;n\right] = (2n+1)^{-1}\{\bar{f}_l(n-1) - \bar{f}_l(n+1)\}. \tag{8-5-13}$$

By repeated use of equation (8-5-12) and the equations (8-5-4), (8-5-5) we find that

$$\ell[f'(x);2m] - f(1) - f(-1) - \sum_{r=0}^{m-1}(4m - 4r \quad 1)\bar{f}_l(2m \quad 2r - 1) \tag{8-5-14}$$

$$\ell[f'(x);2m+1] = f(1) + f(-1) - \sum_{r=0}^{m}(4m - 4r + 1)\bar{f}_l(2m - 2r). \tag{8-5-15}$$

From the recurrence relation

$$(n+1)P_{n+1}(x) - (2n+1)xP_n(x) + nP_{n-1}(x) = 0$$

we deduce immediately that

$$\ell[xf(x);n] = (2n+1)^{-1}\{(n+1)\bar{f}_l(n+1) + n\bar{f}_l(n)\}. \tag{8-5-16}$$

Repeating this result we obtain the relation

$$\ell[x^2 f(x);n] = \frac{(n+1)(n+2)}{(2n+1)(2n+3)}\bar{f}_l(n+2) + \frac{2n^2+2n-1}{(2n-1)(2n+3)}\bar{f}_l(n)$$

$$+ \frac{n(n-1)}{(2n-1)(2n+1)}\bar{f}_l(n-2). \qquad (8\text{-}5\text{-}17)$$

In the case where n is a positive integer the second solution of Legendre's equation may be taken to be

$$Q_n(x) = \tfrac{1}{2}\int_{-1}^{1} \frac{P_n(t)\,dt}{x-t}, \qquad |x| > 1$$

(*Neumann's formula* for the Legendre function of the second kind of positive integral order n). This may be interpreted as the form

$$\ell[(a-x)^{-1};n] = 2Q_n(a). \qquad (8\text{-}5\text{-}18)$$

One of the features of the finite Legendre transform which facilitates its use in the discussion of boundary value problems is that Churchill and Dolph (1954) have established a convolution theorem for this transform.

This states that if $f(x)$ and $g(x)$ are bounded integrable functions on the closed interval $[-1,1]$ and $\bar{f}_l(n)$ and $\bar{g}_l(n)$ denote their finite Legendre transforms, then

$$\ell^{-1}[\bar{f}_l(n)\bar{g}_l(n);x] = h(x), \qquad (8\text{-}5\text{-}19)$$

where $h(x)$ is described by any one of the following formulas:

$$h(\cos\theta) = \frac{1}{\pi}\int_0^{\pi} f(\cos\theta')\sin\theta'\,d\theta' \int_0^{\pi} g(\cos\Theta)\,d\phi \qquad (8\text{-}5\text{-}20)$$

with

$$\cos\Theta = \cos\theta\cos\theta' + \sin\theta\sin\theta'\cos\phi \qquad (8\text{-}5\text{-}21)$$

$$h(\cos\theta) = \frac{1}{\pi}\int_0^{\frac{1}{2}\pi} \sin\phi\,d\phi \int_0^{2\pi} f\{\sin\phi\sin(\psi - \tfrac{1}{2}\theta)\} \times$$

$$\times\, g\{\sin\phi\sin(\psi + \tfrac{1}{2}\theta)\}\,d\psi \qquad (8\text{-}5\text{-}22)$$

$$h(x) = \frac{1}{\pi}\iint_{\Omega} f(y)g(z)(1 - x^2 - y^2 - z^2 + 2xyz)^{-\frac{1}{2}}\,dy\,dz \qquad (8\text{-}5\text{-}23)$$

where, for each fixed $x \in (-1,1)$, Ω is the region bounded by the ellipse with equation

$$y^2 + z^2 - 2xyz = 1 - x^2.$$

If we put $x = 1$ in the relation (8-5-19) written out in the form of a Fourier–Legendre series and use $P_n(1) = 1$, we obtain the identity

$$\sum_{n=0}^{\infty} (n + \tfrac{1}{2})\bar{f}_l(n)\bar{g}_l(n) = \int_{-1}^{1} f(x)g(x)\,dx. \tag{8-5-24}$$

Similarly if we put $x = -1$ and use the relation $P_n(-1) = (-1)^n$, we obtain the identity

$$\sum_{n=0}^{\infty} (-1)^n(n + \tfrac{1}{2})\bar{f}_l(n)\bar{g}_l(n) = \int_{-1}^{1} f(x)g(-x)\,dx. \tag{8-5-25}$$

As an illustration of the use of the finite Legendre transform in the solution of boundary value problems we shall derive the solution $\psi(r,\theta)$ of the interior Dirichlet problem

$$\Delta_3\psi(r,0) = 0, \qquad 0 \leqslant r \leqslant a,\ 0 \leqslant \theta \leqslant \pi \tag{8-5-26}$$

$$\psi(a,\theta) = f(\cos\theta), \qquad 0 \leqslant \theta \leqslant \pi. \tag{8-5-27}$$

We have

$$\Delta_3\psi = \frac{1}{r^2}\left(\frac{\partial}{\partial r} r^2 \frac{\partial}{\partial r} + \mathbf{L}\right)\psi \tag{8-5-28}$$

where

$$\mathbf{L} = \frac{d}{dx}(1 - x^2)\frac{d}{dx}, \qquad x = \cos\theta \tag{8-5-29}$$

so that

$$\ell[\Delta_3\psi; x \to n] = \frac{1}{r^2}\left(\frac{\partial}{\partial r} r^2 \frac{\partial}{\partial r} - n(n+1)\right)\bar{\psi}_l(r,n) \tag{8-5-30}$$

with

$$\bar{\psi}_l(r,n) = \ell[\psi(r,x); x \to n].$$

It follows immediately that the transform $\bar{\psi}_l(r,n)$ satisfies the differential equation

$$\left(\frac{\partial}{\partial r} r^2 \frac{\partial}{\partial r} - n(n+1)\right)\bar{\psi}_l(r,n) = 0, \qquad 0 \leqslant r \leqslant a$$

and the boundary condition

$$\bar{\psi}_l(a,n) = \bar{f}_l(n) \equiv \ell[f(x); n].$$

In addition the solution must be such as to be finite everywhere within the sphere of radius a. We therefore have the solution

$$\bar{\psi}_l(r,n) = \bar{f}_l(n)\left(\frac{r}{a}\right)^n$$

so that

$$\psi(r,\theta) = \ell^{-1}\left[\left(\frac{r}{a}\right)^n \bar{f}_1(n); n \to \cos\theta\right].$$

Now by (*c*) of Problem 8-30 we have

$$\ell^{-1}\left[\left(\frac{r}{a}\right)^n; n \to x\right] = \tfrac{1}{2}a(a^2 - r^2)(a^2 - 2arx + r^2)^{-\frac{3}{2}}$$

so that making use of the form of the convolution theorem expressed by equations (8-5-19) and (8-5-20) we obtain the solution

$$\psi(r,\theta) = \frac{a(a^2 - r^2)}{2\pi}\int_0^\pi f(\cos\theta')\sin\theta'\,d\theta'\int_0^\pi \frac{d\phi}{(a^2 - 2ar\cos\Theta + r^2)^{\frac{3}{2}}},$$

where Θ is defined by equation (8-5-21).

Further examples of the use of finite Legendre transforms in the solution of boundary value problems in mathematical physics are given in Problems 8-35 through 8-39.

We can easily adapt the above procedure for use in applications in which the range of the independent variable is (0,1). We do this by considering even and odd extensions of the relevant function over the interval (−1,1).

We define the *odd* Legendre transform by the relation

$$\hat{f}_{2n+1} \equiv \ell_o[f(x); x \to n] = \int_0^1 f(x)\mathrm{P}_{2n+1}(x)\,dx \tag{8-5-31}$$

and the *even* Legendre transform by

$$\hat{f}_{2n} \equiv \ell_e[f(x); x \to n] = \int_0^1 f(x)\mathrm{P}_{2n}(x)\,dx. \tag{8-5-32}$$

The inversion formulas for these transforms are easily seen to be

$$\ell_o^{-1}[\hat{f}_{2n+1}; x] \equiv f(x) = \sum_{n=0}^{\infty}(4n + 3)\hat{f}_{2n+1}\mathrm{P}_{2n+1}(x), \qquad 0 \leqslant x \leqslant 1 \tag{8-5-33}$$

and

$$\ell_e^{-1}[\hat{f}_{2n}; x] \equiv f(x) = \sum_{n=0}^{\infty}(4n + 1)\hat{f}_{2n}\mathrm{P}_{2n}(x), \qquad 0 \leqslant x \leqslant 1 \tag{8-5-34}$$

respectively.

Further, if

$$\mathbf{L} = \frac{\partial}{\partial x}\left[(1 - x^2)\frac{\partial}{\partial x}\right]$$

then

$$\ell_o[\mathbf{L}f(x); x \to n] = (2n+1)\mathrm{P}_{2n}(0)f(0) - (2n+1)(2n+2)\hat{f}_{2n+1} \tag{8-5-35}$$

and

$$\ell_e[\mathbf{L}f(x); x \to n] = -\mathrm{P}_{2n}(0)f'(0) - 2n(2n+1)\hat{f}_{2n} \tag{8-5-36}$$

8-6 TCHEBYCHEFF TRANSFORMS

The eigenfunctions corresponding to the operator

$$\mathbf{L} = (1 - x^2)\frac{d^2}{dx^2} - x\frac{d}{dx} \tag{8-6-1}$$

which is (8-2-2) with $p(x) = (1 - x^2)^{-\frac{1}{2}}$, $q(x) = (1 - x^2)^{\frac{1}{2}}$, $r(x) = 0$, are the Tchebycheff polynomials $\mathrm{T}_n(x)$ defined by the equation

$$\mathrm{T}_n(x) = \cos(n \cos^{-1} x). \tag{8-6-2}$$

In the notation of sec. 8-2, $\xi = -n^2$.

We may therefore introduce a *finite Tchebycheff transform* through the equation

$$t[f(x); n] \equiv f_t^*(n) = \int_{-1}^{1} (1 - x^2)^{-\frac{1}{2}} \mathrm{T}_n(x) f(x)\, dx. \tag{8-6-3}$$

Since the Tchebycheff polynomials satisfy the orthogonality condition

$$\int_{-1}^{1} (1 - x^2)^{-\frac{1}{2}} \mathrm{T}_m(x) \mathrm{T}_n(x)\, dx = \begin{cases} \frac{1}{2}\pi\delta_{mn}, & m \neq 0, n \neq 0 \\ \pi, & m = n = 0 \end{cases}$$

we obtain the inversion formula

$$t^{-1}[f_t^*(n); x] \equiv f(x) = \frac{1}{\pi} f_t^*(0) + \frac{2}{\pi} \sum_{n=1}^{\infty} f_t^*(n) \mathrm{T}_n(x). \tag{8-6-4}$$

With the operator $\mathbf{L}$ defined by equation (8-6-1) we see that equation (8-2-6) yields the relation

$$t[\mathbf{L}f; n] = -n^2 t[f; n] \tag{8-6-5}$$

which is readily generalized to

$$t[\mathbf{L}^r f; n] = (-1)^r n^{2r} f_t^*(n) \tag{8-6-6}$$

for every positive integer r.

It is possible to derive a convolution theorem for the finite Tchebycheff transform.

We have

$$f_t^*(n)g_t^*(n) = \int_{-1}^{1} f(u)\mathrm{T}_n(u)\frac{du}{\sqrt{(1-u^2)}}\int_{-1}^{1} g(v)\mathrm{T}_n(v)\frac{dv}{\sqrt{(1-v^2)}}$$
$$= \int_0^{\pi}\int_0^{\pi} f(\cos\theta)g(\cos\phi)\cos(n\theta)\cos(n\phi)\,d\theta d\phi$$
$$= \tfrac{1}{8}\int_{-\pi}^{\pi}\int_{-\pi}^{\pi} f_e(\cos\theta)g_e(\cos\phi)\ \times$$
$$\times\ \{\cos n(\theta+\phi) + \cos n(\theta-\phi)\}\,d\theta d\phi$$

where $f_e(\cos\theta)$, $g_e(\cos\phi)$ are the *even* extensions of $f(\cos\theta)$, $g(\cos\phi)$ defined in $(0,\pi)$ having period 2π.

The transformation

$$x = \theta + \phi, \qquad y = \phi$$

for which

$$\frac{\partial(x,y)}{\partial(\theta,\phi)} = 1$$

maps the square $\{(\theta,\phi):|\theta| \leqslant \pi, |\phi| \leqslant \pi\}$ onto the parallelogram $\{(x,y): |x-y| \leqslant \pi, |y| \leqslant \pi\}$ and this can in turn, by virtue of the periodicity be replaced by the square $\{(x,y):|x| \leqslant \pi, |y| \leqslant \pi\}$.

Hence

$$\tfrac{1}{8}\int_{-\pi}^{\pi}\int_{-\pi}^{\pi} f_e(\cos\theta)g_e(\cos\phi)\cos n(\theta+\phi)\,d\theta d\phi$$
$$= \tfrac{1}{4}\int_0^{\pi}\cos(nx)\,dx\int_{-\pi}^{\pi} f_e(\cos\overline{x-y})g_e(\cos y)\,dy$$

If we let $x = \cos^{-1} z$ and use

$$\cos(x-y) = z\cos y + \sqrt{(1-z^2)}\sin y$$

we see that this becomes

$$\frac{1}{4}\int_{-1}^{1}\frac{\mathrm{T}_n(z)\,dz}{\sqrt{(1-z^2)}}\int_{-\pi}^{\pi} f_e\{z\cos y + \sqrt{(1-z^2)}\sin y\}g_e(\cos y)\,dy$$
$$= \tfrac{1}{2}\mathscr{t}[(f \times g)(z); z \to n]$$

where we have written

$$(f \times g)(x) = \tfrac{1}{2}\int_{-\pi}^{\pi} f_e(x\cos y + \sqrt{1-x^2}\sin y)g_e(\cos y)\,dy. \tag{8-6-7}$$

We can show similarly that the integral

$$\int_{-\pi}^{\pi}\int_{-\pi}^{\pi} f_e(\cos\theta)g_e(\cos\phi)\cos n(\theta-\phi)\,d\theta d\phi$$

has exactly the same value. Hence

$$\mathcal{T}[(f \times g)(x);n] = f_t^*(n)g_t^*(n) \tag{8-6-8}$$

or, in inverse form,

$$\mathcal{T}^{-1}[f_t^*(n)g_t^*(n);x] = (f \times g)(x). \tag{8-6-9}$$

Further properties of the finite Tchebycheff transform are developed in Problems 8-40 through 8-45.

8-7 FINITE MELLIN TRANSFORMS

We conclude this chapter with an account of a finite transform which is generated in a different way from those considered above—the finite Mellin transform due to Naylor (1963).

We define a finite Mellin transform of the first kind by the equation

$$\mathcal{M}_1[f(x);s] \equiv f_1^*(s) = \int_0^a \left(\frac{a^{2s}}{x^{s+1}} - x^{s-1}\right) f(x)\,dx \tag{8-7-1}$$

and a finite Mellin transform of the second kind by the equation

$$\mathcal{M}_2[f(x);s] \equiv f_2^*(s) = \int_0^a \left(\frac{a^{2s}}{x^{s+1}} + x^{s-1}\right) f(x)\,dx. \tag{8-7-2}$$

From these definitions it follows immediately that

$$f_1^*(-s) = -a^{-2s}f_1^*(s) \tag{8-7-3}$$

and that

$$f^*(-s) = a^{-2s}f_2^*(s). \tag{8-7-4}$$

Replacing $f(x)$ by $xf'(x)$ in the definition (8-7-1) and applying the rule for integration by parts we find that

$$\mathcal{M}_1[xf'(x);s] = sf_2^*(s) \tag{8-7-5}$$

provided that s is such that

$$\lim_{x \to 0} (a^{2s}x^{-s} - x^s)f(x) = 0. \tag{8-7-6}$$

Similarly we can show that

$$\mathcal{M}_2[xf'(x);s] = sf_1^*(s) + 2a^s f(a) \tag{8-7-7}$$

provided that

$$\lim_{x \to 0} (a^{2s}x^{-s} + x^s)f(x) = 0. \tag{8-7-8}$$

If we write

$$\mathbf{L}_1 = x\frac{d}{dx}x\frac{d}{dx} = x^2\frac{d^2}{dx^2} + x\frac{d}{dx} \tag{8-7-9}$$

then, from equations (8-7-5) and (8-7-7) we deduce the relation

$$\mathscr{m}_1[\mathbf{L}_1 f(x); s] = s^2 f_1^*(s) + 2sa^s f(a) \tag{8-7-10}$$

provided that

$$\lim_{x\to 0}\{(a^{2s}x^{-s} - x^s)xf'(x) + s(a^{2s}x^{-s} + x^s)f(x)\} = 0 \tag{8-7-11}$$

and the relation

$$\mathscr{m}_2[\mathbf{L}_1 f(x); s] = s^2 f_2^*(s) + 2a^{s+1}f'(a) \tag{8-7-12}$$

provided that

$$\lim_{x\to 0}\{(a^{2s}x^{-s} + x^s)xf'(x) + s(a^{2s}x^{-s} - x^s)f(x)\} = 0. \tag{8-7-13}$$

These relations are useful in applications to problems involving the two-dimensional Laplacian operator in plane polar coordinates (ρ,ϕ),

$$\Delta_2 = \frac{\partial^2}{\partial\rho^2} + \frac{1}{\rho}\frac{\partial}{\partial\rho} + \frac{1}{\rho^2}\frac{\partial^2}{\partial\phi^2}. \tag{8-7-14}$$

From (8-7-10) we see that provided s is such that

$$\lim_{\rho\to 0}\{(a^{2s}\rho^{-s} - \rho^s)\rho u_\rho(\rho,\phi) + s(a^{2s}\rho^{-s} + \rho^s)u(\rho,\phi)\} = 0 \tag{8-7-15}$$

then

$$\mathscr{m}_1[\rho^2\Delta_2 u(\rho,\phi);\rho\to s] = \left(\frac{\partial^2}{\partial\phi^2} + s^2\right)u_1^*(s,\phi) + 2sa^s u(a,\phi) \tag{8-7-16}$$

where

$$u_1^*(s,\phi) = \mathscr{m}_1[u(\rho,\phi);\rho\to s]. \tag{8-7-17}$$

Similarly from (8-7-12) we can show that if s is such that

$$\lim\{(a^{2s}\rho^{-s} + \rho^s)\rho u_\rho(\rho,\phi) + s(a^{2s}\rho^{-s} - \rho^s)u(\rho,\phi)\} = 0 \tag{8-7-18}$$

then

$$\mathscr{m}_2[\rho^2\Delta_2 u(\rho,\phi);\rho\to s] = \left(\frac{\partial^2}{\partial\phi^2} + s^2\right)u_2^*(s,\phi) + 2a^{s+1}\frac{\partial u(a,\phi)}{\partial\rho} \tag{8-7-19}$$

where

$$u_2^*(s,\phi) = \mathscr{m}_2[u(\rho,\phi);\rho\to s]. \tag{8-7-20}$$

It is obvious from these equations that in potential problems concerning a region $0 \leqslant \rho \leqslant a$, $\alpha \leqslant \phi \leqslant \beta$ we use the m_1-transform when $u(a,\phi)$ is prescribed and the m_2-transform when $\partial u(a,\phi)/\partial \rho$ is prescribed.

To establish the inversion formula for the m_1-transform we consider the integral

$$\int_0^a \left(\frac{a^{2s}}{x^{s+1}} - x^{s-1}\right) f(x)\, dx$$

where the function $f(x)$ is defined by the equation

$$f(x) = \frac{1}{2\pi i}\int_{c-iN}^{c+iN} x^{-s} f_1^*(s)\, ds$$

in which it is assumed that $f_1^*(s)$ is regular in some strip $|\mathrm{Re}\, s| \leqslant \gamma$ of the s-plane and that $0 < c < \gamma$. We then have

$$\begin{aligned}
\int_0^a \left(\frac{a^{2s}}{x^{s+1}} - x^{s-1}\right) f(x)\, dx
&= \frac{1}{2\pi i}\int_{c-iN}^{c+iN} f_1^*(t)\, dt \int_0^a (a^{2s}x^{-t-s-1} - x^{s-t-1})\, dx \\
&= -\frac{1}{2\pi i}\int_{c-iN}^{c+iN} a^{s-t} f_1^*(t)\frac{dt}{t+s} \\
&= +\frac{1}{2\pi i}\int_{c-iN}^{c+iN} a^{s-t} f_1^*(t)\frac{dt}{t-s}.
\end{aligned}$$

Now if we change the variable in the first integral from t to $-t$ and make use of equation (8-7-3) we see that the first of these two integrals becomes

$$\frac{1}{2\pi i}\int_{-c+iN}^{-c-iN} a^{s-t} f_1^*(t)\frac{dt}{t-s}$$

so that

$$\int_0^a \left(\frac{a^{2s}}{x^{s+1}} - x^{s-1}\right) f(x)\, dx = \frac{1}{2\pi i}\left\{\int_{c-iN}^{c+iN} + \int_{-c+iN}^{-c-iN}\right\} a^{s-t} f_1^*(t)\frac{dt}{t-s}.$$

We now close the contour by the two lines $u = \tau \pm iN$, $(-c \leqslant \tau \leqslant c)$, use Cauchy's theorem and then let $N \to \infty$. If we assume that $|f_1^*(t)|$ remains bounded as $|\mathrm{Im}\, t| \to \infty$, $|\mathrm{Re}\, t| \leqslant c$, we see that the integral on the right side of this equation tends to $f_1^*(s)$ as $N \to \infty$. Hence if

$$f(x) = \frac{1}{2\pi i}\int_C x^{-s} f_1^*(s)\, ds$$

where C is the line $|\mathrm{Re}\, s| = c$, $(0 < c < \gamma)$ then

$$\mathscr{M}_1[f(x);s] = f_1^*(s).$$

This gives us the inversion formula for $\mathscr{M}_1$ in the form

$$\mathscr{M}_1^{-1}[f_1^*(s);x] = \frac{1}{2\pi i}\int_C x^{-s} f_1^*(s)\,ds. \tag{8-7-21}$$

By an exactly similar method we can show that the inversion formula for $\mathscr{M}_2$ is

$$\mathscr{M}_2^{-1}[f_1^*(s);x] = \frac{1}{2\pi i}\int_C x^{-s} f_2^*(s)\,ds. \tag{8-7-22}$$

Another pair of finite Mellin transforms—this time suitable for the analysis of boundary value problems in spherical polar coordinates—is defined by the pair of equations

$$\mathscr{M}_3[f(x);s] \equiv f_3^*(s) = \int_0^a \left(\frac{a^{2s+1}}{x^{s+1}} - x^s\right) f(x)\,dx \tag{8-7-23}$$

$$\mathscr{M}_4[f(x);s] = f_4^*(s) = \int_0^a \left(\frac{a^{2s+2}}{x^{s+1}} + x^{s+1}\right) f(x)\,dx. \tag{8-7-24}$$

These transforms have the property that if

$$\mathbf{L}_2 = x^2 \frac{d^2}{dx^2} + 2x\frac{d}{dx}$$

then

$$\mathscr{M}_3[\mathbf{L}_2 f;s] = s(s+1)f_3^*(s) + 2a^{s+1}f(a) \tag{8-7-25}$$

and

$$\mathscr{M}_4[\mathbf{L}_2 f;s] = s(s+1)f_4^*(s) + 2a^{s+3}f'(a). \tag{8-7-26}$$

Using a method similar to that employed in the derivation of the inversion formula for the $\mathscr{M}_1$-transform we can show that the inversion formulae for these transforms are

$$\mathscr{M}_3^{-1}[f_3^*(s);x] = \frac{1}{2\pi i}\int_C x^{-s-1} f_3^*(s)\,ds \tag{8-7-27}$$

$$\mathscr{M}_4^{-1}[f_4^*(s);x] = \frac{1}{2\pi i}\int_C x^{-s-1} f_4^*(s)\,ds \tag{8-7-28}$$

where C has the same meaning as before except that now $|\operatorname{Re} s| \leqslant \gamma$ is the strip of regularity of $f_3^*(s)$ or $f_4^*(s)$.

From equation (8-7-25) we see that if we denote

$$\Delta_3 = \frac{1}{r^2}\frac{\partial}{\partial r} r^2 \frac{\partial}{\partial r} + \frac{1}{\sin\theta}\frac{\partial}{\partial\theta}\sin\theta\frac{\partial}{\partial\theta} \tag{8-7-29}$$

for the axially symmetric Laplace operator in spherical polar coordinates (r,θ,ϕ), then we have

$$\mathscr{m}_3[r^2\Delta_a u(r,\theta);r \to s]$$

$$= \left[\frac{1}{\sin\theta}\frac{\partial}{\partial\theta}\sin\theta\frac{\partial}{\partial\theta} + s(s+1)\right]u_3^*(s,\theta) + (2s+1)a^{s+1}u(a,\theta) \tag{8-7-30}$$

where

$$u_3^*(s,\theta) = \mathscr{m}_3[u(r,\theta);r \to s]. \tag{8-7-31}$$

Similarly from equation (8-7-26) we have the relation

$$\mathscr{m}_4[r^2\Delta_a u(r,\theta);r \to s]$$

$$= \left[\frac{1}{\sin\theta}\frac{\partial}{\partial\theta}\sin\theta\frac{\partial}{\partial\theta} + s(s+1)\right]u_4^*(s,\theta) + 2a^{s+3}u_r(a,\theta) \tag{8-7-32}$$

where

$$u_4^*(s,\theta) = \mathscr{m}_4[u(r,\theta);r \to s]. \tag{8-7-33}$$

As an illustration of the use of these transforms we consider the problem of determining the plane harmonic function $u(\rho,\phi)$ in the sector $0 \leqslant \rho \leqslant a$, $-\alpha \leqslant \phi \leqslant \alpha$ satisfying the boundary conditions

$$u(a,\phi) = 0, \qquad u(\rho;\pm\alpha) = f(\rho).$$

From equation (8-7-16) we deduce that the finite Mellin transform $u_1^*(s,\phi)$ defined by equation (8-7-17) must satisfy the equation

$$\left(\frac{\partial^2}{\partial\phi^2} + s^2\right)u_1^*(s,\phi) = 0$$

and the boundary conditions

$$u_1^*(s,\pm\alpha) = f_1^*(s)$$

where $f_1^*(s)$ is the finite Mellin transform of the first kind of $f(\rho)$. Hence we find that

$$u_1^*(s,\phi) = f_1^*(s)\frac{\cos(s\phi)}{\cos(s\alpha)}.$$

From the inversion formula (8-7-21) we therefore deduce that the required harmonic function is given by the equation

$$u(\rho,\phi) = \frac{1}{2\pi i}\int_C \rho^{-s}f_1^*(s)\frac{\cos(s\phi)}{\cos(s\alpha)}\,ds$$

with

$$f_1^*(s) = \int_0^a (a^{2s}\rho^{-s-1} - \rho^{s-1})f(\rho)\,d\rho.$$

Further illustrations are provided by Problems 8-46 and 8-47.

8-8 A NOTE ON OTHER TYPES OF FINITE TRANSFORM

Reference should be made to other forms of transform which are special cases of the finite Sturm–Liouville transform discussed in sec. 8-2.

Laguerre transforms have been discussed by Debnath (1960), McCully (1960), and Hirschman (1963), *Hermite transforms* by Debnath (1964), *Jacobi transforms* by Scott (1953) and Debnath (1963), and *Gegenbauer transforms* by Conte (1955), Lakshmanarao (1954), and Srivastava (1965).

A transform of a different kind was introduced by Tricomi (1951). This is the *finite Hilbert transform* defined by the equation

$$\bar{f}_h(x) \equiv \mathscr{h}[f(y);x] = \frac{1}{\pi}\int_{-1}^{1} \frac{f(y)\,dy}{y - x}, \tag{8-8-1}$$

where the integral on the right-hand side of the equation denotes a Cauchy principal value. The corresponding inversion formula

$$f(y) \equiv \mathscr{h}^{-1}[\bar{f}_h(x);y] = \frac{1}{\sqrt{(1 - y^2)}}\left[C - \frac{1}{\pi}\int_{-1}^{1} \frac{\sqrt{(1 - x^2)}}{x - y}\,\bar{f}_h(x)\,dx\right] \tag{8-8-2}$$

in which C is an arbitrary constant, was established by Tricomi. (See also Problem 8-50.) This can obviously be written in the alternative form

$$\mathscr{h}^{-1}[\bar{f}_h(x);y] = \frac{1}{\sqrt{(1 - y^2)}}[C - \mathscr{h}\{\sqrt{(1 - x^2)}\bar{f}_h(x);y\}]\,. \tag{8-8-3}$$

Tricomi also established the relation

$$\mathscr{h}[f(y)\bar{g}_h(y) + g(y)\bar{f}_h(y);x] = \bar{f}_h(x)\bar{g}_h(x) - f(x)g(x) \tag{8-8-4}$$

which is the analogue of Parseval's relation for Fourier transforms.

A finite Laplace transform which should prove useful in the analysis of certain physical problems has been discussed by Dunn (1967). He defines

$$\bar{f}(p,T) = \int_0^T f(t)e^{-pt}\,dt, \qquad (\mathrm{Re}\,p \geqslant \sigma) \tag{8-8-5}$$

for $f(t) \in \mathscr{P}^1[0,T]$. This allows us to consider a more general class of functions than does the Laplace transform. For instance, the function e^{t^2} does not possess a Laplace transform but it does have a finite Laplace transform. Also, in determining the response of a dynamical system at time T we need only know the input to the system up to time T. The inversion formula is

$$f(t) = \frac{1}{2\pi i}\int_C \bar{f}(p,T)e^{pt}\,dp$$

where C is an arbitrary contour except that it must terminate at the points $\sigma + i\infty$. The arbitrariness arises from the fact that $\bar{f}(p,T)$ is an entire function of p. Similar results were derived earlier by Amerbaev (1961) in a paper which has been unfortunately neglected.

In recent years several authors have considered integral transforms of the type

$$\mathscr{T}[f(y);x] = \int_x^1 K\left(\frac{x}{y}\right) f(y)\,\frac{dy}{y}$$

The inversion formula for transforms of this general kind is easily derived by the methods of sec. 4-5. We shall not discuss transforms of this kind since they do not appear to have any useful applications and merely refer the interested reader to Sneddon (1968) and the papers cited there.

PROBLEMS

8-1 Prove that

$$(a)\quad \mathscr{f}_s[1;n] = \frac{a}{n\pi}[1 + (-1)^{n+1}]$$

$$(b)\quad \mathscr{f}_s[x;n] = (-1)^{n+1}\frac{a^2}{\pi n}$$

$$(c)\quad \mathscr{f}_s^{-1}[n^{-1};x] = \frac{\pi}{a}\left(1 - \frac{x}{a}\right)$$

$$(d)\quad \mathscr{f}_s[x^2;n] = (-1)^{n+1}\frac{a^3}{n\pi} - \frac{2a^3}{n^3\pi^3}[1 + (-1)^{n+1}]$$

$$(e)\quad \mathscr{f}_s[x(a - x);n] = \frac{2a^3}{n^3\pi^3}[1 + (-1)^{n+1}].$$

8-2 Calculate $\mathscr{f}_s[e^{kx};x \to n]$ and $\mathscr{f}_c[e^{kx};x \to n]$ and deduce that

$$\mathscr{f}_s^{-1}\left[\frac{n}{k^2 + n^2\pi^2/a^2};n \to x\right] = \frac{a}{\pi}\frac{\sinh k(a - x)}{\sinh ka}, \qquad 0 \leqslant \mathbf{x} \leqslant a.$$

8-3 Show that the solution $u(x,t)$ of the diffusion equation

$$\frac{\partial^2 u}{\partial x^2} = \frac{1}{\kappa}\frac{\partial u}{\partial t}, \qquad 0 \leqslant x \leqslant a, t > 0$$

satisfying the boundary conditions

$$u(0,t) = u(a,t) = 0, \qquad t > 0$$

and the initial condition

$$u(x,0) = f(x), \qquad 0 \leqslant x \leqslant a$$

can be expressed by the formula

$$u(x,t) = \frac{2}{\pi}\sum_{n=1}^{\infty} \bar{f}_s(n) e^{-\kappa n^2 \pi^2 t/a^2} \sin\left(\frac{n\pi x}{a}\right).$$

8-4 Find the solution in $D = \{(x,y):0 \leqslant x \leqslant a,\ 0 \leqslant y \leqslant b\}$, of the diffusion equation

$$\kappa \Delta_2 u(x,y,t) = \frac{\partial u}{\partial t}, \qquad t > 0,$$

when u vanishes on the boundary of D and

$$u(x,y,0) = f(x,y), \qquad (x,y) \in D.$$

8-5 The function $u(x,y)$ satisfies $\Delta_2 u = 0$ in the region D defined in the previous problem, and the boundary conditions

$$u(0,y) = u(a,y) = 0, \qquad 0 \leqslant y \leqslant b$$

$$u(x,0) = f(x), \qquad u(x,b) = 0, \qquad 0 \leqslant x \leqslant a.$$

Show that

$$u(x,y) = \mathscr{f}_s^{-1}\left[\bar{f}_s(m)\frac{\sin hm\pi/a(b - y)}{\sin hm\pi b/a}; m \rightarrow x\right]$$

where $\bar{f}_s(m) = \mathscr{f}_s[f(x); x \rightarrow m]$.

Show also that we may write

$$u(x,y) = \frac{\pi}{b}\,\mathscr{f}_{ss}^{-1}[n\omega_{mn}^{-1}\bar{f}_s(m); (m,n) \rightarrow (x,y)]$$

where ω_{mn} is defined by equation (8-1-57).

Deduce the first form of solution from the second.

8-6 The function $u(x,y)$ satisfies $\Delta_2 u = 0$ in the semi-infinite strip $\{(x,y): 0 \leqslant x \leqslant \infty, 0 \leqslant y \leqslant b\}$, and the boundary conditions

$$u(x,0) = 0, \qquad u(x,b) = 0, \qquad x \geqslant 0$$
$$u(0,y) = f(y), \qquad 0 \leqslant y \leqslant b$$
$$u(x,y) \to 0 \text{ as } x \to \infty, \qquad 0 \leqslant y \leqslant b.$$

Show that

$$u(x,y) = \frac{2}{b} \sum_{n=1}^{\infty} \bar{f}_s(n) e^{-n\pi x/b} \cos \frac{n\pi y}{b}.$$

8-7 Show that if the normal derivative of u vanishes at all points on the boundary, ∂D, of the region D defined in Problem 8-4, then

$$\mathcal{F}_{cc}[\Delta_2 u(x,y);(x,y) \to (m,n)] = -\omega_{mn} \bar{u}_{cc}(m,n)$$

where ω_{mn} is again defined by equation (8-1-57).

If $u(x,y,t)$ satisfies the diffusion equation

$$\Delta_2 u(x,y,t) = \frac{\partial u}{\partial t}, \qquad (x,y) \in D,\ t > 0$$

if its normal derivative vanishes on ∂D, and if

$$u(x,y,0) = f(x,y), \qquad (x,y) \in D$$

show that

$$u(x,y,t) = \frac{1}{ab} \bar{f}_{cc}(0,0) + \frac{2}{ab} \sum_{m=1}^{\infty} \bar{f}_{cc}(m,0) e^{-\omega_{m0} t} \cos \frac{m\pi x}{a}$$
$$+ \frac{2}{ab} \sum_{n=1}^{\infty} \bar{f}_{cc}(0,n) e^{\ \omega_{0n} t} \cos \frac{n\pi y}{b}$$
$$+ \frac{4}{ab} \sum_{m=1}^{\infty} \sum_{n=1}^{\infty} \bar{f}_{cc}(m,n) e^{-\omega_{mn} t} \cos\left(\frac{m\pi x}{a}\right) \cos\left(\frac{n\pi y}{b}\right)$$

where $\bar{f}_{cc}(m,n) = \mathcal{F}_{cc}[f(x,y);(m,n)]$.

8-8 The function $u(x,y)$ satisfies $\Delta_2 u = 0$ in the rectangle D of Problem 8-4, and the boundary conditions

$$\frac{\partial u(0,y)}{\partial x} = \frac{\partial u(a,y)}{\partial x} = 0, \qquad 0 \leqslant y \leqslant b$$

$$u(x,0) = f(x), \qquad u(x,b) = 0, \qquad 0 \leqslant x \leqslant a.$$

By considering $\mathscr{f}_c[u(x,y);x \to m]$ show that

$$u(x,y) = \mathscr{f}_c^{-1}\left[\bar{f}_c(m)\frac{\sinh\{m\pi(b-y)/a\}}{\sinh(m\pi b/a)};m \to x\right]$$

and by considering $\mathscr{f}_{cs}[u(x,y);(x,y) \to (m,n)]$ show that

$$u(x,y) = \frac{\pi}{b}\,\mathscr{f}_{cs}^{-1}[n\omega_{mn}^{-1}\bar{f}_c(m);(m,n) \to (x,y)]$$

where ω_{mn} is defined by equation (8-1-57).

Show that the two results are equivalent.

8-9 Prove that $\mathscr{F}$

(a) $\mathscr{T}_1[1;n] = \xi_n^{-1}\sin\xi_n$

(b) $\mathscr{T}_1[\mathrm{H}(t-x);x \to n] = \xi_n^{-1}\sin(\xi_n t), \qquad 0 < t < 1$

(c) $\mathscr{T}_1[x;n] = \xi_n^{-2}\{(1+h)\cos\xi_n - 1\}$

(d) $\mathscr{T}_1[1-x;n] = \xi_n^{-2}(1-\cos\xi_n)$.

8-10 Show that if $0 < t < 1$ and h is the constant occurring in equation (8-3-4),

$$\mathscr{T}_1^{-1}[\xi_n^{-2}\cos(\xi_n t);x] = 1 + h^{-1} - t - (x-t)\mathrm{H}(x-t)$$

where H is the Heaviside unit function.

8-11 Show that

$$\mathscr{T}_1[f(1-x);n] = \cos\xi_n\mathscr{T}_1\left[f(x) + h\int_x^1 f(t)\,dt;n\right]$$

and deduce that

$$\mathscr{T}_1^{-1}[\cos\xi_n f_1^*(n);x] = f(1-x) - h\int_0^x e^{-ht}f(1-t)\,dt, \qquad (0 < x < 1).$$

8-12 Show that

$$\mathscr{T}_1^{-1}[\xi_n^{-2}f_1^*(n);x] = (1-x)\int_0^x f(t)\,dt + \int_x^1 (1-t)f(t)\,dt + h^{-1}\int_0^1 f(t)\,dt, \qquad (0 < x < 1).$$

8-13 Show that the integral transform $\mathscr{T}_3$ defined by the operators

$$\mathbf{L} = \frac{d^2}{dx^2}, \qquad \mathbf{M}f = hf(0) - f'(0), \qquad \mathbf{N}f = f'(1)$$

has the property that

$$\mathscr{T}_3[f(x);n] = \mathscr{T}_1[f(1-x);n].$$

8-14 If w_ν and W_ν are any two solutions of Bessel's equation show that, if $k \neq m$,

$$(m^2 - k^2)\int_\alpha^\beta x w_\nu(kx)W_\nu(mx)\,dx = [x\{kw_\nu'(kx)W_\nu(mx) - mw_\nu(kx)W_\nu'(mx)\}]_\alpha^\beta.$$

Putting $m = k + \varepsilon$ and letting ε tend to zero deduce that

$$\int_\alpha^\beta x w_\nu(kx)W_\nu(kx)\,dx = \tfrac{1}{2}\left[x^2\left\{w_\nu'(kx)W_\nu'(kx) + \left(1 - \frac{\nu^2}{k^2x^2}\right)w_\nu(kx)W_\nu(kx)\right\}\right]_\alpha^\beta.$$

Deduce the orthogonality relations (8-4-8), (8-4-26), (8-4-38), (8-4-43), (8-4-54).

8-15 Prove that

(a) $\mathscr{h}_{1,\nu}[x^\nu;n] = a^{\nu+1}\xi_n^{-1}\mathrm{J}_{\nu+1}(\xi_n a)$

(b) $\mathscr{h}_{1,\nu}\left[\dfrac{\mathrm{J}_\nu(kx)}{\mathrm{J}_\nu(ka)};n\right] = a\xi_n(\xi_n^2 - k^2)^{-1}\mathrm{J}_{\nu+1}(\xi_n a)$

where k is not a zero of $\mathrm{J}_\nu(\xi a)$.

Deduce the value of

$$\mathscr{h}_{1,\nu}^{-1}[\xi_n^{-1}(\xi_n^2 - k^2)^{-1}\mathrm{J}_{\nu+1}(\xi_n a);x].$$

8-16 Show that

$$\mathscr{h}_{1,0}^{-1}[\xi_n^{-3}\mathrm{J}_1(\xi_n a);x] = \frac{a^2 - x^2}{4a}, \qquad 0 < x < a.$$

8-17 Show that if $\nu \geqslant 0$

$$\mathscr{h}_{1,\nu}[x^{-\nu-1}f'(x);n] = \xi_n \mathscr{h}_{1,\nu+1}[x^{-\nu-1}f(x);n] - \frac{\xi_n f(0)}{2^\nu\Gamma(\nu + 1)}.$$

8-18 Show that the equation

$$\sigma\frac{\partial^2 y}{\partial t^2} = g\sigma\frac{\partial}{\partial x}\left(x\frac{\partial y}{\partial x}\right) + p(x,t), \qquad 0 < x < a,\ t > 0$$

governing the horizontal deflection $y(x,t)$ of a heavy chain of uniform line density σ, fixed at $x = a$, and under the influence of an external transverse force of intensity $p(x,t)$ can be transformed to

$$\sigma\frac{\partial^2 y}{\partial t^2} = \frac{g\sigma}{4a}\frac{\partial^2 y}{\partial z^2} + \frac{1}{z}\frac{\partial y}{\partial z} + p(az^2,t), \qquad 0 \leqslant z \leqslant 1,\ t \geqslant 0$$

by the substitution $x = az^2$.

Hence show that if initially the chain is at rest under gravity the horizontal deflection at any subsequent time is given by

$$y(x,t) = \frac{4}{\sigma}\sqrt{\frac{a}{g}} \sum_{n=1}^{\infty} \frac{J_0(\xi_n\sqrt{x/a})}{\{J_1(\xi_n)\}^2} \int_0^t \bar{p}_n(t-\tau)\sin\left(\tfrac{1}{2}\xi_n\tau\sqrt{\frac{g}{a}}\right)d\tau$$

where $\xi_1, \xi_2, \ldots$ are the positive zeros of $J_0(\xi)$ and

$$\bar{p}_n(t) = \frac{1}{2a}\int_0^a p(x,t)J_0(\xi_n\sqrt{x/a})\,dx.$$

8-19 If the chain described in the last problem is released from rest in the position $y = \varepsilon(1 - x/a)$, $0 \leqslant x \leqslant a$, and swings freely, show that at a subsequent time the horizontal deflection is given by

$$y(x,t) = 8\varepsilon \sum_{n=1} \frac{J_0(\xi_n\sqrt{x/a})}{\xi_n{}^3 J_1(\xi_n)} \cos\left(\tfrac{1}{2}\xi_n t\sqrt{\frac{g}{a}}\right)$$

where $\xi_1, \xi_2, \ldots$ are the positive zeros of $J_0(\xi)$.

8-20 The symmetrical forced vibrations of a circular membrane of radius a and density σ, under a tension T and subjected to an applied transverse pressure $p(r,t)$ may be described by the inhomogeneous wave equation

$$\frac{\partial^2 z}{\partial r^2} + \frac{1}{r}\frac{\partial z}{\partial r} = \frac{1}{c^2}\frac{\partial^2 z}{\partial t^2} - \frac{p(r,t)}{T}, \qquad 0 \leqslant r \leqslant a,\ t \geqslant 0$$

($c^2 = T/\sigma$). If initially the membrane is at rest in the equilibrium position show that

$$z(r,t) = \frac{2}{c\sigma a^2} \sum_{n=1}^{\infty} \frac{J_0(\xi_n r)}{\xi_n[J_1(\xi_n a)]^2} \int_0^t \bar{p}_n(\tau)\sin\{c\xi_n(t-\tau)\}\,d\tau$$

where the summation extends over the positive zeros of $J_0(\xi a)$ and

$$\bar{p}_n(t) = \mathscr{h}_{1,0}[p(r,t); r \to n].$$

If $p(r,t) = P_0\psi(t)H(b - r)$, $0 < b \leqslant a$, calculate $\bar{p}_n(t)$ and examine the two special cases:

(*a*) $b = a$;

(*b*) $b \to 0$, $P_0 \to \infty$ in such a way that $\pi b^2 P_0 \to F$.

8-21 If, in the notation of the last problem, we take

$$p(r,t) = P_0 \sin(\omega t), \qquad 0 \leqslant r \leqslant a$$

show that

$$z(r,t) = \frac{P_0}{\sigma\omega^2}\left[\frac{\mathrm{J}_0(\omega r/c)}{\mathrm{J}_0(\omega a/c)} - 1\right]\sin(\omega t) +$$

$$+ \frac{2P_0}{c\sigma a}\sum_{n=1}^{\infty}\frac{\sin(\xi_n t)\mathrm{J}_0(\xi_n r)}{\xi_n^{\ 2}(\omega^2 - c^2\xi_n^{\ 2})\{\mathrm{J}_1(\xi_n a)\}^2}.$$

8-22 The free symmetrical vibrations of a thin circular plate of radius a held along its circumference in such a way that the displacement of the plate is zero and the mean curvature vanishes there also, satisfies the partial differential equation

$$b^2\left(\frac{\partial r}{\partial r^2} + \frac{1}{r}\frac{\partial}{\partial r}\right)^2 w + \frac{\partial^2 w}{\partial t^2} = 0, \qquad 0 \leqslant r \leqslant a,\ t \geqslant 0$$

where c is a constant, and the boundary conditions

$$w(a,t) = 0, \qquad \frac{\partial^2 w(a,t)}{\partial r^2} + \frac{1}{a}\frac{\partial w(a,t)}{\partial r} = 0, \qquad (t \geqslant 0).$$

If the plate is deformed to the shape

$$w(r,0) = \varepsilon(1 - r^2/a^2), \qquad 0 \leqslant r \leqslant a$$

and then released from rest show that

$$w(r,t) = \frac{8\varepsilon}{a^3}\sum_{n=1}^{\infty}\frac{\cos(c\xi_n t)\mathrm{J}_0(\xi_n r)}{\xi_n^{\ 3}\mathrm{J}_1(\xi_n a)}$$

where the summation is over the positive zeros of $\mathrm{J}_0(\xi a)$.

8-23 Show that

(a) $\mathscr{h}_{2,\nu}^{-1}[\xi_n^{-2}\mathrm{J}_\nu(\xi_n a); x] = (\nu + ha)^{-1}(x/a)^\nu$;

(b) $\mathscr{h}_{2,\nu}^{-1}[(\xi_n^{\ 2} - k^2)^{-1}\mathrm{J}_\nu(\xi_n a); x] = \dfrac{\mathrm{J}_\nu(kx)}{a[h\mathrm{J}_\nu(ka) + k\mathrm{J}'_\nu(ka)]}$

8-24 The diffusion of heat in an infinite hollow cylinder bounded by $r = a$, $r = b$ $(b > a)$ when these surfaces are kept at zero temperature is governed by the equation

$$\frac{\partial^2\theta}{\partial r^2} + \frac{1}{r}\frac{\partial\theta}{\partial r} = \frac{1}{\kappa}\frac{\partial\theta}{\partial t}, \qquad a \leqslant r \leqslant b,\ t \geqslant 0$$

and the boundary conditions

$$(a)\ \theta(a,t) = 0, \qquad (b)\ \theta(b,t) = 0,\ (t \geqslant 0).$$

If, initially,

$$\theta(r,0) = f(r), \qquad a \leqslant r \leqslant b$$

show that

$$\theta(r,t) = \tfrac{1}{2}\pi^2 \sum_{n=1}^{\infty} \frac{\xi_n^{\,2}\{J_0(\xi_n b)\}^2}{\{J_0(\xi_n a)\}^2 - \{J_0(\xi_n b)\}^2} \bar{f}_n e^{-\kappa\xi_n^2 t}[J_0(\xi_n r)Y_0(\xi_n a) - J_0(\xi_n a)Y_0(\xi_n r)]$$

where the sum is taken over the positive roots of equation (8-4-36) and

$$\bar{f}_n = \int_a^b rf(r)\{J_0(\xi_n r)Y_0(\xi_n a) - J_0(\xi_n a)Y_0(\xi_n r)\}\, dr.$$

8-25 If, instead of being kept at zero temperature, the surface $r = a$ is insulated against the flow of heat the equations of Problem 8-24 remain unaltered except that (a) is replaced by

$$\frac{\partial\theta(a,t)}{\partial r} = 0, \qquad (t \geqslant 0).$$

Show that

$$\theta(r,t) = \tfrac{1}{2}\pi^2 \sum_{n=1}^{\infty} \frac{\xi_n^{\,2}\{J_0(\xi_n b)\}^2 \hat{f}_n e^{-\kappa\xi_n^2 t}}{\{J_1(\xi_n a)\}^2 - \{J_1(\xi_n b)\}^2} [J_1(\xi_n a)Y_0(\xi_n r) - Y_1(\xi_n a)J_0(\xi_n r)]$$

where the sum is taken over the positive roots of the equation

$$J_0(\xi_n b)Y_1(\xi_n a) - J_1(\xi_n a)Y_0(\xi_n b) = 0$$

and

$$\hat{f}_n = \int_a^b rf(r)[J_1(\xi_n a)Y_0(\xi_n r) - J_0(\xi_n r)Y_1(\xi_n a)]\, dr.$$

8-26 The surfaces $r = a$, $r = b$ of an infinite hollow cylinder are insulated against the flow of heat. The temperature field $\theta(r,t)$ is due to a distributed source of density $\Theta(r,t)$ so that

$$\frac{\partial\theta}{\partial t} = \kappa\left(\frac{\partial^2\theta}{\partial r^2} + \frac{1}{r}\frac{\partial\theta}{\partial r}\right) + \Theta(r,t), \qquad a \leqslant r \leqslant b,\ t \geqslant 0$$

$$\frac{\partial\theta(a,t)}{\partial r} = \frac{\partial\theta(b,t)}{\partial r} = 0, \qquad (t \geqslant 0).$$

If, initially, the temperature at every point of the cylinder is the reference temperature so that $\theta(r,0) = 0$, $(a \leqslant r \leqslant b)$, show that

$$\theta(r,t) = \tfrac{1}{2}\pi^2 \sum_{n=1} \frac{\{J_1(\xi_n b)\}^2 \phi_n(r)}{\{J_1(\xi_n a)\}^2 - \{J_1(\xi_n b)\}^2} \int_0^t \bar{\Theta}_n(\tau) e^{-\kappa\xi_n^2(t-\tau)}\, d\tau$$

where $\xi_1, \xi_2, \ldots$ are the positive roots of the equation

$$J_1(\xi_n b)Y_1(\xi_n a) - J_1(\xi_n a)Y_1(\xi_n b) = 0$$

and the functions $\phi_n(r)$ and $\Theta_n(t)$ are defined by the equations

$$\phi_n(r) = J_1(\xi_n a)Y_0(\xi_n r) - Y_1(\xi_n a)J_0(\xi_n r)$$

$$\bar{\Theta}_n(t) = \int_a^b r\Theta(r,t)\phi_n(r)\, dr.$$

8-27 Show that the eigenfunctions corresponding to the operators

$$\mathbf{L} = \mathscr{B}_\nu, \qquad \mathbf{M}f = f'(a) + hf(a), \qquad \mathbf{N}f = f(b)$$

are

$$\phi_n(x) = J_\nu(\xi_n x)Y_\nu(\xi_n b) - J_\nu(\xi_n b)Y_\nu(\xi_n x), \qquad (a \leqslant x \leqslant b). \qquad (i)$$

where ξ_n is a root of the equation

$$[\xi_n J_\nu'(\xi_n a) + hJ_\nu(\xi_n a)]Y_\nu(\xi_n b) - [\xi_n Y_\nu'(\xi_n a) + hY_\nu(\xi_n a)]J_\nu(\xi_n b) = 0. \qquad (ii)$$

If a finite transform is defined by the equation

$$\mathscr{h}_{7,\nu}[f(x);n] \equiv \bar{f}_{7,\nu}(n) = \int_a^b xf(x)\phi_n(x)\, dx$$

with $\phi_n(x)$ defined by equations (*i*) and (*ii*), show that its inversion formula is

$$\mathscr{h}_{7,\nu}^{-1}[\bar{f}_{7,\nu}(n)] = \tfrac{1}{2}\pi^2 \sum_{n=1} \frac{\xi_n^2\{\xi_n J_\nu'(\xi_n a) + hJ_\nu(\xi_n a)\}^2 \bar{f}_{7,\nu}(n)}{\{\xi_n J_\nu'(\xi_n a) + hJ_\nu(\xi_n a)\}^2 - (h^2 + \xi_n^2 - \nu^2/a^2)\{J_\nu(\xi_n b)\}^2}\, \phi_n(x).$$

Write down the expression for $\mathscr{h}_{7,\nu}[\mathscr{B}_\nu f(x);n]$ in terms of the boundary values of $f(x)$ and its transform $\bar{f}_{7,\nu}(n)$.

8-28 Show that the eigenfunctions corresponding to the operators

$$\mathbf{L} = \mathscr{B}_\nu, \qquad \mathbf{M}f = f'(a) + hf(a), \qquad \mathbf{N}f = f'(b)$$

are

$$\phi_n(x) = J_\nu(\xi_n x)Y_\nu'(\xi_n b) - J_\nu'(\xi_n b)Y_\nu(\xi_n x), \qquad (a \leqslant x \leqslant b) \qquad (i)$$

where ξ_n is a positive root of the equation

$$[\xi_n \mathrm{Y}'_\nu(\xi_n a) + h\mathrm{Y}_\nu(\xi_n a)]\mathrm{J}'_\nu(\xi_n b) - [\xi_n \mathrm{J}'_\nu(\xi_n a) + h\mathrm{J}_\nu(\xi_n a)]\mathrm{Y}'_\nu(\xi_n b) = 0. \qquad (ii)$$

If a finite transform is defined by the equation

$$\mathscr{h}_{8,\nu}[f(x);n] \equiv \bar{f}_{8,\nu}(n) = \int_a^b x\phi_n(x)f(x)\,dx$$

where $\phi_n(x)$ is defined by equations (*i*) and (*ii*), show that its inversion formula is

$$f(x) = \tfrac{1}{2}\pi^2 \sum_{n=1}^{\infty} \frac{\xi_n^2\{\xi_n \mathrm{J}'_\nu(\xi_n a) + h\mathrm{J}_\nu(\xi_n a)\}^2 \bar{f}_{8,\nu}(n)\phi_n(x)}{(1 - \nu^2/\xi_n^2 b^2)\{\xi_n \mathrm{J}'_\nu(\xi_n a) + h\mathrm{J}_\nu(\xi_n a)\}^2 - (h^2 + \xi_n^2 - \nu^2/a^2)\{\mathrm{J}'_\nu(\xi_n b)\}^2}$$

and write down the formula for $\mathscr{h}_{8,\nu}[\mathscr{B}_\nu f(x);n]$.

8-29 Prove that

$$\ell[\mathrm{H}(x);n] = (2n+1)^{-1}\{P_{n-1}(0) - \mathrm{P}_{n+1}(0)\}, \qquad (n \geqslant 1)$$

and deduce that if n is even $\ell[\mathrm{H}(x);n] = 0$, while if n is odd,

$$\ell[\mathrm{H}(x);n] = \frac{(-1)^{\frac{1}{2}(n-1)}(n-1)!}{2^n(\frac{1}{2}n - \frac{1}{2})!(\frac{1}{2}n + \frac{1}{2})!}.$$

8-30 Show that if $|a| \leqslant 1$,

$$(a)\ \ell^{-1}\left[\frac{2a^n}{2n+1};x\right] = (1 - 2ax + a^2)^{-\frac{1}{2}}$$

and deduce that

$$(b)\ \ell^{-1}\left[\frac{a^{n+1}}{(2n+1)(n+1)};x\right] = \log\frac{a - x + \sqrt{(1 - 2ax + a^2)}}{1 - x}$$

$$(c)\ \ell^{-1}[a^n;x] = \tfrac{1}{2}(1 - a^2)(1 - 2ax + a^2)^{-\frac{3}{2}}.$$

$$(d)\ \ell^{-1}\left[\frac{2a^{n+k}}{(2n+1)(n+k)};x\right] = \int_0^a t^{k-1}(1 - 2xt + t^2)^{-\frac{1}{2}}\,dt.$$

8-31 Show that if $|t| \leqslant 1$

$$\ell[t^{-1}\{(1 - 2xt + t^2)^{-\frac{1}{2}} - 1\};x \to n] = \begin{cases} \dfrac{2t^{n-1}}{2n+1}, & n = 1,2,\ldots \\ 0, & n = 0, \end{cases}$$

and deduce that if n is a positive integer

$$\ell^{-1}\left[\frac{2a^n}{n(2n+1)};x\right]$$
$$= \log 2 - \log\{1 - ax + \sqrt{(a^2 - 2ax + 1)}\}, \qquad |a| \leqslant 1.$$

8-32 If

$$\mathbf{L} = \frac{d}{dx}(1 - x^2)\frac{d}{dx}$$

show that

$$\mathbf{L}\log(1 - x) = -1,$$

and that

$$\ell[\mathbf{L}\log(1 - x);n] = -2 + \int_{-1}^{1}(1 + x)\mathrm{P}_n'(x)\,dx.$$

Writing $1 + x$ as $(1 - x^2)/(1 - x)$ and integrating by parts in the integral on the right-hand side of this equation show that if $n = 1,2,\ldots$

$$\ell[\log(1 - x);n] = -\frac{2}{n(n+1)}.$$

Find $\ell[\log(1 - x);0]$ and deduce $\ell[\log(1 + x);n]$, $(n = 0,1,2,\ldots)$.

8-33 If $g \in \mathscr{P}^1[-1,1]$ is such that

$$\int_{-1}^{1} g(x)\,dx = 0,$$

show that

$$\ell^{-1}\left[\frac{\bar{g}_l(n)}{n(n+1)};x\right] = c - \int_0^x \frac{ds}{1 - s^2}\int_{-1}^{s} g(t)\,dt$$

where c is an arbitrary constant.

8-34 Prove that

$$(a)\ \ell[f(x);n] = \frac{1}{2^n n!}\int_{-1}^{1}(1 - x^2)^n f^{(n)}(x)\,dx$$

and deduce that

$$(b)\ \ell[e^{ax};n] = \sqrt{\frac{2\pi}{a}}\,\mathrm{I}_{n+\frac{1}{2}}(a)$$

$$(c)\ \ell[\cos(ax);n] = \begin{cases} 0, & \text{if } n \text{ is odd,} \\ (-1)^{-n}\sqrt{\dfrac{2\pi}{a}}\,J_{n+\frac{1}{2}}(a), & \text{if } n \text{ is even;} \end{cases}$$

$$(d)\ \ell[\sin(ax);n] = \begin{cases} (-1)^{-(n-1)}\sqrt{\dfrac{2\pi}{a}}\,J_{n+\frac{1}{2}}(a), & \text{if } n \text{ is odd,} \\ 0, & \text{if } n \text{ is even.} \end{cases}$$

8-35 $\chi(r,\theta)$ is the solution of the Dirichlet problem

$$\Delta_3\chi(r,\theta) = 0, \qquad r \geqslant a$$
$$\chi(a,\theta) = f(\cos\theta)$$
$$\chi(r,\theta) \to 0 \qquad \text{as } r \to \infty.$$

Show that

$$\chi(r,\theta) = \frac{a}{r}\psi\left(\frac{a^2}{r},\theta\right), \qquad r \geqslant a$$

where $\psi(r,\theta)$ is the solution of the interior Dirichlet problem specified by equations (8-5-26) and (8-5-27).

8-36 Show that the solution of the interior Robin problem

$$\Delta_3\phi(r,\theta) = 0, \qquad 0 \leqslant r \leqslant a$$
$$\frac{\partial\phi(a,0)}{\partial r} + h\phi(a,0) = f(\cos\theta), \qquad 0 \leqslant \theta \leqslant \pi \tag{I}$$

(h a positive constant), can be written in the form

$$\phi(r,\theta) = a\int_0^1 \psi(rt,\theta)t^{ha-1}\,dt$$

where $\psi(r,\theta)$ is the solution of the interior Dirichlet problem specified by equations (8-5-26) and (8-5-27).

Show also that the solution of the corresponding problem in $r \geqslant a$ with (I) unaltered but h how a negative constant, is given by

$$\phi(r,\theta) = -\frac{a^2}{r}\int_0^1 \psi\left(\frac{a^2 t}{r},\theta\right)t^{-ha}\,dt.$$

8-37 Show that the solution $\psi(r,\theta)$ of Poisson's equation

$$\Delta_3\psi = f(r,\cos\theta), \qquad 0 \leqslant r \leqslant a, 0 \leqslant \theta \leqslant \pi$$

satisfying the boundary condition

$$\psi(a,0) = 0$$

can be written in the form

$$\psi(r,\theta) = \frac{1}{a}\int_0^a H(rs/a^2, s, \cos\theta)s^2\,ds - \frac{1}{r}\int_0^r H(s/r, s, \cos\theta)s^2\,ds - \int_r^a H(r/s, s, \cos\theta)\,ds$$

where

$$H(h,s,\cos\theta) = \frac{1}{2\pi}\int_0^\pi f(s,\cos\theta')\sin\theta'\,d\theta'\int_0^\pi \frac{d\phi}{\sqrt{(1 - 2h\cos\Theta + h^2)}},$$

$$\cos\Theta = \cos\theta\cos\theta' + \sin\theta\sin\theta'\cos\phi.$$

8-38 The function $\psi(r,\theta)$ satisfies the Poisson equation

$$\Delta_3\psi(r,\theta) = r^{-\alpha}, \qquad (\alpha > 3)$$

in the region $r \geqslant a$. If $\psi(r,\theta) \to 0$ as $r \to \infty$ and

$$\psi(a,\theta) = (c - \cos\theta)^{-1}, \qquad |c| > 1$$

show that

$$\psi(r,\theta) = \frac{a}{r}\mathrm{Q}_0(c) + \frac{1}{(\alpha - 2)(\alpha - 3)}(r^{2-\alpha} - a^{2-\alpha}) + \sum_{n=1}^{\infty}(2n+1)\left(\frac{a}{r}\right)^{n+1}\mathrm{P}_n(\cos\theta)\mathrm{Q}_n(c).$$

8-39 In terms of oblate spheroidal coordinates (μ,ζ) the problem of determining the potential $\psi(\mu,\zeta)$ due to a circular disk of unit radius charged to unit potential can be reduced to that of solving

$$\frac{\partial}{\partial\mu}(1-\mu^2)\frac{\partial\psi}{\partial\mu} + \frac{\partial}{\partial\zeta}(1+\zeta^2)\frac{\partial\psi}{\partial\zeta} = 0, \qquad |\mu| \leqslant 1, 0 < \zeta < \infty$$

subject to the boundary conditions

$$\psi(\mu,0) = 1, \qquad \frac{\partial\psi(0,\zeta)}{\partial\mu} = 0, \qquad \psi(\mu,\zeta) \to 0 \quad \text{as } \zeta \to \infty.$$

Show that

$$\psi(\mu,\zeta) = \frac{2}{\pi}\cot^{-1}\zeta.$$

8-40 Find the Tchebycheff transform of the Heaviside function $\mathrm{H}(x)$.

8-41 Prove that

$$T_m(x)T_n(x) = \tfrac{1}{2}\{T_{m+n}(x) + T_{|m-n|}(x)\}$$

and deduce that

(*a*) $\mathcal{t}[T_m(x)f(x);n] = \tfrac{1}{2}\{f_t^*(m+n) + f_t^*(|m-n|)\}$

(*b*) $\mathcal{t}[xf(x);n] = \tfrac{1}{2}\{f_t^*(n+1) + f_t^*(n-1)\}$

(*c*) $\mathcal{t}[x^r f(x);n] = 2^{-r}\sum_{s=0}^{r}\binom{r}{s} f_t^*(n+2s-r).$

Evaluate

$$\mathcal{t}[x^r;n].$$

8-42 Prove that

(*a*) $2n\mathcal{t}[f(x);n] = \mathcal{t}[f'(x);n-1] - \mathcal{t}[f'(x);n+1];$

(*b*) $\mathcal{t}[(1-x^2)f'(x);n] = \tfrac{1}{2}(n+1)f_t^*(n+1) - \tfrac{1}{2}(n-1)f_t^*(n-1).$

Deduce that, if

$$F(x) = \int_{-1}^{x} \frac{f(t)\,dt}{1-t^2},$$

(*c*) $F_t^*(2m) = m^{-1}\sum_{r=1}^{m} f_t^*(2r-1);$

(*d*) $F_t^*(2m+1) = \dfrac{2}{2m+1}\left[\sum_{r=1}^{m} f_t^*(2r) + \tfrac{1}{2}F^*(1)\right].$

8-43 The function $F(x)$ is defined in terms of a function $f(t)$ which satisfies

$$\int_{-1}^{1} \frac{f(t)\,dt}{\sqrt{(1-t^2)}} = 0$$

through the equation

$$F(x) = (1-x^2)^{-\frac{1}{2}} \int_{-1}^{x} \frac{f(t)\,dt}{\sqrt{(1-t^2)}}, \qquad |x| < 1.$$

Show that

$$F_t^*(n+1) - F_t^*(n-1) = \frac{2}{n} f_t^*(n),$$

and deduce that if m is a positive integer,

$$(a)\ F_t^*(2m) = F_t^*(0) + 2\sum_{r=1}^{m} (2r-1)^{-1} f_t^*(2r-1)$$

$$(b)\ F^*(2m+1) = F^*(1) + \sum_{r=1}^{m} r^{-1} f_t^*(2r).$$

8-44 Show that

$$t[e^{ax};n] = \pi \mathrm{I}_n(a)$$

and deduce the expressions for

$$t[\cos ax;n], \qquad t[\sin ax;n].$$

8-45 Prove that

$$t^{-1}[n^{-2} f_t^*(n);x] = C - \int_{-1}^{x} \frac{du}{\sqrt{(1-u^2)}} \int_{-1}^{u} \frac{dv}{\sqrt{(1-v^2)}}$$

where C is an arbitrary constant.

8-46 The function $u(\rho,\phi)$ is harmonic in the sector $0 \leqslant \rho \leqslant a$, $0 \leqslant \phi \leqslant \alpha$ and satisfies the boundary conditions

$$u_\rho(a,\phi) = 0, \qquad (0 \leqslant \phi \leqslant \alpha)$$
$$u(\rho,0) = 0, \qquad (0 \leqslant \rho \leqslant a)$$
$$u(\rho,\alpha) = f(\rho), \qquad (0 \leqslant \rho \leqslant a).$$

Show that

$$u(\rho,\phi) = \frac{1}{2\pi i} \int_C \rho^{-s} U(s) \frac{\sin(s\phi)}{\sin(s\alpha)}\, ds$$

where

$$U(s) = \int_0^a (a^{2s}\rho^{-s-1} + \rho^{s-1}) f(\rho)\, d\rho.$$

8-47 The axisymmetric function $u(r,\theta)$ is harmonic in the conical sector $0 \leqslant r \leqslant a$, $0 \leqslant \theta \leqslant \alpha$, and satisfies the boundary conditions

$$u(a,\theta) = 0, \qquad 0 \leqslant \theta \leqslant \alpha$$
$$u(r,\alpha) = f(r), \qquad 0 \leqslant r \leqslant a.$$

Show that

$$u(r,\theta) = \frac{1}{2\pi i} \int_C r^{-s-1} U(s) \frac{P_s(\cos\theta)}{P_s(\cos\alpha)}\, ds$$

where

$$U(s) = \int_0^a \left(\frac{a^{2s+1}}{r^{s+1}} - r^s \right) f(r)\, dr.$$

8-48 Prove that

(*a*) $\mathscr{h}[(1 - y^2)^{-\frac{1}{2}}; x] = 0$
(*b*) $\mathscr{h}[(1 - y^2)^{\frac{1}{2}}; x] = -x.$

8-49 Prove that

$$\mathscr{h}[yf(y); x] = x\bar{f}_h(x) + \frac{1}{\pi} \int_{-1}^{1} f(y)\, dy.$$

8-50 Prove that

$$\mathscr{h}[\sqrt{(1 - y^2)}\bar{f}_h(y); x] = \frac{1}{\pi} \int_{-1}^{1} f(y)\, dy - \sqrt{(1 - x^2)} f(x).$$

Deduce the inversion formula (8-8-2) and the identity

$$\mathscr{h}[\sqrt{(1 - y^2)}\bar{f}_h(y) - yf(y); x] = -\sqrt{(1 - x^2)} f(x) - x\bar{f}_h(x).$$

Calculate

$$\mathscr{h}[y\sqrt{(1 - y^2)}; x].$$

9
Generalized Functions

We shall conclude with a *brief* account of some of the problems encountered in applied mathematics when transform methods are applied to analyze physical situations in which impulsive forces, or point sources, are involved. Formal calculations are much simplified by the introduction of the Dirac delta "function" but the rules for its manipulation do not follow naturally from the methods of classical analysis. This led to the introduction of the concept of a "generalized" function. The first ideas of such an approach are to be found in the work of Bochner (1932) and Sobolev (1936) but it was the work of Schwartz in the early 1940's, culminating in the publication of his treatise on the theory of distributions (Schwartz, 1950–51) which put the theory of generalized functions on a firm foundation.

Only the barest outline of these ideas will be given here. A reader wishing to undertake the study of these ideas is advised to continue with reading Zemanian's books—Zemanian, (1965) and Zemanian, (1969).

Since we are only considering the elements of the theory we shall restrict our attention to the one-dimensional case.

9-1 THE DIRAC DELTA FUNCTION

In the solution of problems in theoretical mechanics we often encounter situations in which the applied force is of an impulsive nature, that is in which the relevant disturbance is of very short duration but has a high peak value. By the application of an *impulse* of magnitude I at time $t = 0$ we mean the application of a force of magnitude $F(t)$ which acts for a very short time τ and which satisfies the relation

$$\int_0^\tau F(t)\,dt = I. \tag{9-1-1}$$

Indeed in many high school texts on mechanics it is just assumed that $F(t)$ has a constant value F_0, say, so that $F_0\tau = I$. Such a force could therefore be thought of as being of the form

$$F(t) = I\delta(t), \tag{9-1-2}$$

where $\delta(t)$ is the limit as ε tends to zero of the function $\delta_\varepsilon(t)$, which has the properties

$$\delta_\varepsilon(t) = 0, \qquad \text{if } t \notin (0,\varepsilon) \tag{9-1-3}$$

$$\int_{-\infty}^{\infty} \delta_\varepsilon(t)\,dt = 1. \tag{9-1-4}$$

If, then, we formally let ε tend to zero we should expect $\delta(t)$ to have the corresponding properties

$$\delta(t) = 0, \qquad \text{if } t \neq 0 \tag{9-1-5}$$

$$\int_{-\infty}^{\infty} \delta(t)\,dt = 1. \tag{9-1-6}$$

This concept was introduced into quantum mechanics by Dirac (see, for instance, p. 58 of Dirac (1947)) to deal with a similar problem of representing certain physical quantities in that theory. The function $\delta(t)$ defined in this way is not of course a function in the sense in which that word is used in classical analysis; it was called by Dirac an "improper function" and he recommended its use in analysis only when it is obvious that no inconsistency will follow from this use. Initially it was used merely to derive formal solutions of physical problems; once such solutions have been obtained they should then be subjected to the processes of classical analysis to establish rigorously that they do in fact satisfy all the conditions posed in the original formulation of the problem.

For the moment we shall proceed in a purely formal way and set up equations concerning the Dirac delta function which we would not claim necessarily held rigorously. The equations (9-1-3) and (9-1-4) do not specify the variation of $\delta_\varepsilon(t)$ in the interval $(0,\varepsilon)$. Suppose that we assume that it is a step function of the form:

$$\delta_\varepsilon(t) = \begin{cases} 0, & t < 0, \\ \varepsilon^{-1}, & 0 < t < \varepsilon, \\ 0, & t > \varepsilon. \end{cases}$$

Then if the function $f(t)$ is continuous in an interval containing the origin and the sub-interval $(0,\varepsilon)$,

$$\int_{-\infty}^{\infty} \delta_\varepsilon(t) f(t)\, dt = \varepsilon^{-1} \int_0^\varepsilon f(t)\, dt.$$

Now by the mean value theorem of the integral calculus

$$\int_0^\varepsilon f(t)\, dt = \varepsilon f(\theta\varepsilon), \qquad 0 \leqslant \theta \leqslant 1$$

so that we have the equation

$$\int_{-\infty}^{\infty} \delta_\varepsilon(t) f(t)\, dt = f(\theta\varepsilon), \qquad 0 \leqslant \theta \leqslant 1.$$

Letting ε tend to zero we should then obtain the relation

$$\int_{-\infty}^{\infty} \delta(t) f(t)\, dt = f(0) \tag{9-1-7}$$

for all functions f which are continuous in a neighborhood of the origin. (This property was named the *sifting property* of the delta function by van der Pol.)

Changing the variable in the integral from t to x, where $x = t + a$, we obtain the equivalent relation

$$\int_{-\infty}^{\infty} \delta(x - a) f(x)\, dx = f(a) \tag{9-1-8}$$

valid for all functions f continuous in a neighborhood of $x = a$. We can rewrite this equation as

$$\delta(x - a) f(x) = \delta(x - a) f(a) \tag{9-1-9}$$

the meaning of such an equation being that its two sides give equivalent results as factors in an integrand. With a similar interpretation we may write

$$x\delta(x) = 0. \tag{9-1-10}$$

Continuing this formal procedure we consider the integral

$$\int_{-\infty}^{\infty} \mathrm{H}'(x) f(x)\, dx$$

where $\mathrm{H}'(x)$ denotes the "derivative" of Heaviside's unit function which it will be recalled is defined by the equations

$$\mathrm{H}(x) = \begin{cases} 0, & x < 0 \\ 1, & x > 0. \end{cases}$$

If we apply the rule for integrating by parts we find that

$$\int_{-\infty}^{\infty} \mathrm{H}'(x)f(x)\,dx = [\mathrm{H}(x)f(x)]_{-\infty}^{\infty} - \int_{-\infty}^{\infty} \mathrm{H}(x)f'(x)\,dx$$

$$= -\int_{0}^{\infty} f'(x)\,dx$$

$$= f(0)$$

if we assume that $f(x) \to 0$ as $x \to \infty$ (as we must if the integral on the left-hand side of this equation is to converge). We may now use equation (9-1-7) to rewrite this last equation in the form

$$\int_{-\infty}^{\infty} \mathrm{H}'(x)f(x)\,dx = \int_{-\infty}^{\infty} \delta(x)f(x)\,dx$$

or, in the symbolic form

$$\delta(x) = \mathrm{H}'(x) \tag{9-1-11}$$

To anyone without a background in rigorous mathematical analysis it may not be clear why there is anything wrong with the analysis up to this point. The reason is that we have treated the Dirac delta "function" (or, what is the same thing the derivative of the Heaviside function) as though it were a function in the sense of classical analysis. It is obviously *not*, since it is defined to be zero everywhere except at the origin, i.e. to be zero almost everywhere, and yet to have a finite integral. It is not admissible, therefore to apply the standard theorems of calculus, such as the formula for integration by parts, to such "functions." When we attempt to give a rigorous interpretation to these equations we have to generalize the whole concept of a function; there are several such generalizations which reproduce the essential features of the Dirac delta function.

9-1-1 THE MIKUSINSKI–TEMPLE THEORY OF GENERALIZED FUNCTIONS

The simplest attempt at such a generalization (though by no means the first) is due to Mikusinski (1948) and developed by Temple (1953, 1955). Basically this starts from a function $p(t)$ which satisfies the conditions

$$p(t) = 0, \qquad |t| > 1 \tag{9-1-12}$$

$$\int_{-\infty}^{\infty} p(t)\,dt = 1 \tag{9-1-13}$$

from which is defined a sequence $\{p_n(t)\}$ by the rule

$$p_n(t) = np(nt). \tag{9-1-14}$$

The sifting relation is then replaced by

$$\lim_{n\to\infty} \int_{-\infty}^{\infty} p_n(t)f(t)\,dt = f(0)$$

for continuous functions f, and the δ-function is regarded as the limit of the sequence $\{p_n\}$.

Other generalized functions can be defined in a similar way. The method defines generalized functions as classes of equivalent fundamental sequences of continuous functions in a way similar to that used when real numbers are introduced with the help of fundamental sequences of rational numbers. We shall outline the method in sec. 9-2.

9-1-2 SCHWARTZ'S THEORY OF DISTRIBUTIONS

In the method of Schwartz, a full account of which is given in Schwartz (1950, 1951), a generalized function or distribution is defined as a linear functional[1] on a class of "test" functions. In this interpretation the sifting relation

$$\delta\langle\phi\rangle = \phi(0)$$

defines the Dirac delta distribution.

A brief account of this approach is given in sec. 9-3.

9-1-3 GENERALIZED DERIVATIVES OF LOCALLY INTEGRABLE FUNCTIONS

This method which may be considered to be inspired by the relation (9-1-11) is based on the idea of identifying distributions, on finite intervals, with generalized derivatives of locally integrable functions. It was discussed briefly by Halperin (1952) and developed by König (1953) and Sauer (1958).

We shall not discuss this method here since it is one which is more suited to the development of the theory of boundary value problems for partial differential equations than for that of integral transforms.

9-1-4 MIKUSINSKI'S OPERATIONAL CALCULUS

Another approach to the problem of constructing a consistent theory of generalized functions is also due to Mikusinski; (see, for instance, Mikusinski, 1953). Unlike his first theory this one is algebraic in nature. It closely mirrors the process by which the concept of number is extended from integers to rational numbers, and provides a natural approach to operational calculus as well as to generalized functions. Its greatest drawback is that although it has been successful with functions defined on the positive real line—and has been extended to functions of several such variables—it is unable to deal with functions of unrestricted real variables or with functions defined on an arbitrary region of a space of n dimensions ($n \geqslant 2$).

[1] It will be recalled that a linear functional is a linear mapping of the space of functions into the field of real (or complex) numbers.

Mikusinski shows that the set $\mathscr{C}[0,\infty)$ with addition and multiplication by scalars defined in an obvious way and multiplication of two functions a,b of the set defined by the convolution $a*b$ forms a commutative ring which is called the *convolution ring*. As a result of Titchmarsh's theorem which states that $a*b = \theta$, the function which is identically zero, if and only if $a = \theta$ or $b = \theta$ or both we see that in this ring *division* is a meaningful operation. This is similar to the situation we find when we introduce the idea of division of integers, and there we ensure the feasibility of division by extending the concept of number from integers to rational numbers in terms of classes of equivalent ordered pairs of integers.

In the present situation we consider ordered pairs of elements of $\mathscr{C}[0,\infty)$ and consider (a,b) to be equivalent to (c,d) if $a*d = c*b$. The class of all ordered pairs of continuous functions equivalent to (a,b) is denoted by a/b which is called a *convolution quotient*. It is a simple matter to define the operations of addition, multiplication by a scalar and multiplication in $\mathscr{Q}$, the set of all convolution quotients, and to show that the embedding of $\mathscr{C}[0,\infty)$ in $\mathscr{Q}$ preserves all these operations. For this reason, we may write $(a*f)/a$ as f for any pair of continuous functions a and f and $(\lambda a)/a$ as λ for any scalar λ. The unit element e, in $\mathscr{Q}$, may be written a/a and it is easy to show that multiplication by e reproduces f. Hence we may identify the unit element in $\mathscr{Q}$ with the Dirac delta function.

The elements of $\mathscr{Q}$ comprise numbers, functions, the delta function, and its derivatives so that they may be regarded as generalized functions. The elements of $\mathscr{Q}$ may also be thought of as operators thereby providing a convenient approach to Heaviside's operational calculus.

We shall not develop these ideas and methods here since Mikusinski's method is much more suitable to a text on operational calculus than to one on integral transforms. It is drawn to the reader's attention because it provides a viable alternative to the method of the Laplace transform as a justification of Heaviside's methods. The interested reader should consult Mikusinski (1953), or its English translation (Mikusinski, 1959); an excellent brief account is given also by Erdelyi (1962) and by pp. 9–24 of Erdelyi (1961).

An alternative approach to generalized functions has been developed by Weston (1957, 1959) where they are operators acting on a class of "perfect" functions rather than convolution quotients of elements of $\mathscr{C}[0,\infty)$.

9-2 THE MIKUSINSKI–TEMPLE THEORY OF GENERALIZED FUNCTIONS

In this section we shall outline the theory of generalized functions due to Mikusinski and Temple as simplified by Lighthill (1958). The spaces $\mathscr{N}$ and $\mathscr{S}$ introduced in sec. 1-8 play an important part in this theory so the reader should refer back to that section now.

9-2-1 DEFINITION OF GENERALIZED FUNCTIONS

From the definitions of $\mathcal{N}$ and $\mathcal{S}$ we deduce immediately the following results:

(*a*) $f \in \mathcal{S} \Rightarrow f' \in \mathcal{S}$;

(*b*) $f, g \in \mathcal{S} \Rightarrow f + g \in \mathcal{S}$;

(*c*) $f \in \mathcal{S}, g \in \mathcal{N} \Rightarrow fg \in \mathcal{S}$;

and we have shown in sec. 2-3 that

(*d*) $f \in \mathcal{S} \Rightarrow \mathcal{F}[f] \in \mathcal{S}$

(*e*) $f, g \in \mathcal{S}$ and $F(\xi) = \mathcal{F}[f(x); \xi]$, $G(\xi) = \mathcal{F}[g(x);\ \xi] \Rightarrow$

$$\int_{-\infty}^{\infty} F(\xi)G(\xi)\, d\xi = \int_{-\infty}^{\infty} f(-x)g(x)\, dx.$$

A sequence $\{\phi_r(x)\}_{r=1}^{\infty}$, $\phi_r(x) \in \mathcal{S}$ is called a *regular $\mathcal{S}$-sequence* if, for any $f \in \mathcal{S}$, the limit

$$\lim_{r\to\infty} \int_{-\infty}^{\infty} \phi_r(x)f(x)\, dx \tag{9-2-1}$$

exists. Two regular $\mathcal{S}$-sequences $\{\phi_r(x)\}_{r=1}^{\infty}$ and $\{\psi_r(x)\}_{r=1}^{\infty}$ are said to be *equivalent* if, for any $f \in \mathcal{S}$,

$$\lim_{r\to\infty} \int_{-\infty}^{\infty} \phi_r(x)f(x)\, dx = \lim_{r\to\infty} \int_{-\infty}^{\infty} \psi_r(x)f(x)\, dx.$$

For example, the sequences

$$\{e^{-x^2/r^2}\}_{r=1}^{\infty},\ \{e^{-x^2/4r^2}\}_{r=1}^{\infty},\ \{e^{-x^4/r^4}\}_{r=1}^{\infty}$$

are equivalent $\mathcal{S}$-sequences leading to the limit

$$\int_{-\infty}^{\infty} f(x)\, dx.$$

We now define generalized functions in terms of these sequences: *Each class of equivalent regular $\mathcal{S}$-sequences defines a generalized function.*

We shall, as a matter of notation, write $\phi(x)$ for the generalized function defined by the equivalence class of which the sequence $\{\phi_r(x)\}_{r=1}^{\infty}$ is a member, and we define the integral

$$\int_{-\infty}^{\infty} \phi(x)f(x)\, dx, \qquad f \in \mathcal{S}$$

of the product of a generalized function $\phi(x)$ and a function of rapid descent $f(x)$, to be the limit

$$\lim_{r\to\infty} \int_{-\infty}^{\infty} \phi_r(x)f(x)\, dx.$$

We can define the *sum* of two generalized functions in an obvious way:

If $\phi(x)$ and $\psi(x)$ are two generalized functions defined by the regular $\mathscr{S}$-sequences $\{\phi_r(x)\}_{r=1}^{\infty}$ and $\{\psi_r(x)\}_{r=1}^{\infty}$, respectively, we define $\phi(x) + \psi(x)$ to be the generalized function defined by the class equivalent to the regular $\mathscr{S}$-sequence

$$\{\phi_r(x) + \psi_r(x)\}_{r=1}^{\infty}.$$

That this sequence is in fact a regular $\mathscr{S}$-sequence follows from the result (*b*) above, and since

$$\lim_{r\to\infty} \int_{-\infty}^{\infty} \{\phi_r(x) + \psi_r(x)\} f(x)\,dx = \lim_{r\to\infty} \int_{-\infty}^{\infty} \phi_r(x) f(x)\,dx + \lim_{r\to\infty} \int_{-\infty}^{\infty} \psi_r(x) f(x)\,dx$$

it follows immediately that the relevant limit exists.

From (*a*) above we see that if $\phi_r(x) \in \mathscr{S}$, then so does $\phi_r'(x)$. Also since both $\phi_r'(x)$ and $f(x)$ belong to $\mathscr{S}$ we can apply the rule for integrating by parts to obtain the equation

$$\int_{-\infty}^{\infty} \phi_r'(x) f(x)\,dx = -\int_{-\infty}^{\infty} \phi_r(x) f'(x)\,dx.$$

Since—again by (*a*)—we know that $f' \in \mathscr{S}$ we find on letting $r \to \infty$ that the limit

$$\lim_{r\to\infty} \int_{-\infty}^{\infty} \phi_r'(x) f(x)\,dx$$

exists and hence that $\{\phi_r'(x)\}_{r=1}^{\infty}$ is a regular $\mathscr{S}$-sequence. It therefore defines a generalized function. We shall denote this generalized function by $\phi'(x)$ and we shall call it the *derivative* of the generalized function $\phi(x)$. We shall sometimes also use the notation

$$\frac{d}{dx}\phi(x)$$

to denote the derivative of the generalized function $\phi(x)$.

We notice also that

$$\int_{-\infty}^{\infty} \phi'(x) f(x)\,dx = -\int_{-\infty}^{\infty} \phi(x) f'(x)\,dx, \qquad f \in \mathscr{S}. \tag{9-2-2}$$

By continuing this process we can define the k-th derivative of the generalized function $\phi(x)$—which we denote by $\phi^{(k)}(x)$—by the regular $\mathscr{S}$-sequence $\{\phi_r^{(k)}(x)\}_{r=1}^{\infty}$. The relation corresponding to (9-2-2) is then

$$\int_{-\infty}^{\infty} \phi^{(k)}(x) f(x)\,dx = (-1)^k \int_{-\infty}^{\infty} \phi(x) f^{(k)}(x)\,dx, \qquad f \in \mathscr{S} \tag{9-2-3}$$

where k is any positive integer. We see from its method of definition that a generalized function possesses derivatives of all orders.

If $\{\phi_r(x)\}_{r=1}^{\infty}$ is a regular $\mathscr{S}$-sequence, then it is easily shown that so is the sequence $\{\phi_r(\lambda x + \mu)\}_{r=1}^{\infty}$, λ, $\mu \in \mathbf{R}$ and hence that it defines a generalized function which we denote by $\phi(\lambda x + \mu)$. Also, since

$$\int_{-\infty}^{\infty} \phi_r(\lambda x + \mu) f(x)\, dx = |\lambda|^{-1} \int_{-\infty}^{\infty} \phi_r(x) f\left(\frac{x - \mu}{\lambda}\right) dx,$$

it follows that

$$\int_{-\infty}^{\infty} \phi(\lambda x + \mu) f(x)\, dx$$
$$= |\lambda|^{-1} \int_{-\infty}^{\infty} \phi(x) f\left(\frac{x - \mu}{\lambda}\right) dx, \qquad \lambda \neq 0, \qquad f \in \mathscr{S}. \tag{9-2-4}$$

Finally, if $a(x) \in \mathscr{N}$ then by (c) $af \in \mathscr{S}$ if $f \in \mathscr{S}$. Hence if $\phi(x)$ is the generalized function defined by the class of sequences equivalent to the regular $\mathscr{S}$-sequence $\{\phi_r(x)\}_{r=1}^{\infty}$, it is easily shown that $\{a(x)\phi_r(x)\}_{r=1}$ is a regular $\mathscr{S}$-sequence. We denote the generalized function defined by this sequence by $a(x)\phi(x)$.

9-2-2 ORDINARY FUNCTIONS AS GENERALIZED FUNCTIONS

If $\phi(x)$ is a function of x in the ordinary sense of classical analysis and is such that there exists a positive number m such that $(1 + x^2)^{-m}\phi(x) \in \mathscr{A}_1(\mathbf{R})$, i.e., such that

$$\int_{-\infty}^{\infty} (1 + x^2)^{-m} |\phi(x)|\, dx < \infty, \tag{9-2-5}$$

then it is possible[1] to construct a regular $\mathscr{S}$-sequence $\{\phi_r(x)\}_{r=1}^{\infty}$ with the property that

$$\lim_{r\to\infty} \int_{-\infty}^{\infty} \phi_r(x) f(x)\, dx = \int_{-\infty}^{\infty} \phi(x) f(x)\, dx$$

for all $f \in \mathscr{S}$.

By this procedure we can greatly increase the number of generalized functions available to us, by using not only ordinary functions such that (9-2-5) is finite but also the new generalized functions which can be obtained by differentiation in accordance with the rule (9-2-3).

It is also a simple matter to show that if $\phi(x)$ is an ordinary differentiable function such that both $\phi(x)$ and $\phi'(x)$ satisfy (9-2-5), then the derivative of the generalized function generated by $\phi(x)$ is identical with the generalized function generated by $\phi'(x)$.

[1] See, for example, Problem 9-1.

Among the ordinary functions which can be interpreted as generalized functions are the generalized powers $|x|^\alpha$. Whatever the value of m the integral

$$\int_{-\infty}^{\infty} (1 + x^2)^{-m} |x|^\alpha \, dx$$

is convergent only if $\alpha > -1$, and if $\alpha > -1$ this integral is convergent as long as we take $m > \frac{1}{2}(\alpha + 1)$. Hence we can define a generalized function $|x|^\alpha$ if $\alpha > -1$. If we take derivatives of both sides of the equation

$$\log |x|^\alpha = \alpha \log |x|$$

and recall that

$$\frac{d}{dx} \log |x| = |x|^{-1} \operatorname{sgn} x$$

we find that

$$\frac{d}{dx} |x|^\alpha = \alpha |x|^{\alpha - 1} \operatorname{sgn} x \tag{9-2-6}$$

provided that now $\alpha > 0$ so that $|x|^{\alpha-1}$ as well as $|x|^\alpha$ is defined.

When $\alpha < -1$ and is *not* an integer we define $|x|^\alpha$ to be the generalized function

$$|x|^\alpha = \frac{1}{(\alpha + 1)_n} \frac{d^n}{dx^n} \{|x|^{\alpha+n} (\operatorname{sgn} x)^n\}, \tag{9-2-7}$$

where $(a)_n$ denotes the Pochhammer symbol

$$(a)_n = a(a + 1) \ . \ . \ . \ (a + n - 1),$$

and we have chosen n to be a positive integer such that $\alpha + n > -1$. It has to be verified that this definition is in fact independent of the value chosen for n. This is simple enough—it is Problem 9-9 below—but it depends on a property of the generalized function $\delta(x)$ defined in sec. 9-2-3.

In a similar way we define the generalized function $|x|^\alpha \operatorname{sgn} x$ by the equation

$$|x|^\alpha \operatorname{sgn} x = \frac{1}{(\alpha + 1)_n} \frac{d^n}{dx^n} \{|x|^{\alpha+n} (\operatorname{sgn} x)^{n+1}\} \, (\alpha + n > -1) \tag{9-2-8}$$

and the generalized function $|x|^\alpha \mathrm{H}(x)$ by the equation

$$|x|^\alpha \mathrm{H}(x) = \frac{1}{(\alpha + 1)_n} \frac{d^n}{dx^n} \{x^{\alpha+n} \mathrm{H}(x)\} \tag{9-2-9}$$

with the same restriction on n.

9-2-3 THE DIRAC DELTA FUNCTION

We now consider the sequence $\{\phi_r(x)\}_{r=1}^{\infty}$, where

$$\phi_r(x) = \frac{r}{\sqrt{\pi}} e^{-r^2x^2} \tag{9-2-10}$$

for which it is obvious that $\phi_r \in \mathscr{S}$ for all r. We now consider

$$\lim_{r\to\infty} \int_{-\infty}^{\infty} \phi_r(x)f(x)\,dx, \qquad f \in \mathscr{S}.$$

If $f(x)$ can be expanded in a power series

$$f(x) = \sum_{s=0}^{\infty} \frac{f^{(s)}(0)}{s!} x^s$$

valid for all values of x, then

$$\begin{aligned}\int_{-\infty}^{\infty} \phi_r(x)f(x)\,dx &= \sum_{s=0}^{\infty} \frac{f^{(s)}(0)}{s!} \int_{-\infty}^{\infty} x^s \phi_r(x)\,dx \\ &= \sum_{s=0}^{\infty} \frac{f^{(s)}(0)}{n^{2s}} \frac{(\frac{1}{2})_s}{(2s)!}\end{aligned}$$

which *indicates* that

$$\lim_{r\to\infty} \int_{-\infty}^{\infty} \phi_r(x)f(x)\,dx = f(0). \tag{9-2-11}$$

To *prove* that this is indeed so, we consider

$$\left| \int_{\infty}^{\infty} \phi_r(x)f(x)\,dx - f(0) \right|, \qquad f \in \mathscr{S}$$

noting that since

$$\int_{-\infty}^{\infty} \phi_r(x)\,dx = 1,$$

this can be written

$$\left| \int_{-\infty}^{\infty} \phi_r(x)\{f(x) - f(0)\}\,dx \right| \leqslant M \int_{-\infty}^{\infty} \phi_r(x)\,|x|\,dx$$

where $M = \max f'(x)$.

$$\int_{-\infty}^{\infty} |x|\,\phi_r(x)\,dx = \frac{2r}{\sqrt{\pi}} \int_0^{\infty} x e^{-r^2x^2}\,dx = \frac{1}{r\sqrt{\pi}}$$

so that it follows that

$$\left| \int_{-\infty}^{\infty} \phi_r(x) f(x)\, dx - f(0) \right| \leqslant M\pi^{-\frac{1}{2}} r^{-1}$$

and hence that (9-2-11) holds for all $f \in \mathscr{S}$.

The sequence $\{\phi_r(x)\}_{r=1}^{\infty}$ is therefore a regular $\mathscr{S}$-sequence which converges to a generalized function $\delta(x)$ with the property

$$\int_{-\infty}^{\infty} \delta(x) f(x)\, dx = f(0), \qquad f \in \mathscr{S}. \tag{9-2-12}$$

For an obvious reason we call this generalized function *the Dirac delta function.*

From equation (9-2-3) we see that the k-th derivative of the generalized function $\delta(x)$, defined by the regular $\mathscr{S}$-sequence $\{\phi_r^{(k)}(x)\}_{r=1}^{\infty}$ has the property

$$\int_{-\infty}^{\infty} f(x) \delta^{(k)}(x)\, dx = (-1)^k \int_{-\infty}^{\infty} f^{(k)}(x) \delta(x)\, dx, \qquad f \in \mathscr{S}$$

which is equivalent to

$$\int_{-\infty}^{\infty} f(x) \delta^{(k)}(x)\, dx = (-1)^k f^{(k)}(0), \qquad f \in \mathscr{S}. \tag{9-2-13}$$

We now consider the proof of the equation (9-1-11). Since $H(x)$ satisfies the condition (9-2-5) with $m = 1$, it is a generalized function and for any $f \in \mathscr{S}$,

$$\int_{-\infty}^{\infty} \mathrm{H}'(x) f(x)\, dx = -\int_{-\infty}^{\infty} \mathrm{H}(x) f'(x)\, dx = -\int_{0}^{\infty} f'(x)\, dx = f(0)$$

which proves the result.

Other properties of the Dirac delta function are collected together in Problems 9-5 through 9-8 which the reader should solve for himself.

9-2-4 FOURIER TRANSFORMS OF GENERALIZED FUNCTIONS

We know that if $\phi_r(x) \in \mathscr{S}$ then its Fourier transform $\Phi_r(\xi) = \mathscr{F}[\phi_r(x); \xi]$ also belongs to $\mathscr{S}$. Similarly, if $f \in \mathscr{S}$, then $F = \mathscr{F}[f] \in \mathscr{S}$ and by Parseval's theorem

$$\int_{-\infty}^{\infty} \Phi_r(x) F(x)\, dx = \int_{-\infty}^{\infty} \phi_r(x) f(-x)\, dx, \qquad f \in \mathscr{S}$$

so that if the limit

$$\lim_{r \to \infty} \int_{-\infty}^{\infty} \phi_r(x) f(-x)\, dx$$

exists for an arbitrary member of the set $\mathscr{S}$, then

$$\lim_{r \to \infty} \int_{-\infty}^{\infty} \Phi_r(x) F(x)\, dx$$

also exists for any $F \in \mathscr{S}$. In other words, if $\{\phi_r(x)\}_{r=1}^{\infty}$ is a regular $\mathscr{S}$-sequence, then so also is the sequence $\{\Phi_r(x)\}_{r=1}^{\infty}$.

If the generalized function $\phi(x)$ is defined by the regular $\mathscr{S}$-sequence $\{\phi_r(x)\}_{r=1}^{\infty}$, we call the generalized function $\Phi(\xi)$ defined by the regular $\mathscr{S}$-sequence $\{\Phi_r(\xi)\}_{r=1}^{\infty}$ with $\Phi_r(\xi) = \mathscr{F}[\phi_r(x);\xi]$ the Fourier transform of ϕ and we denote it by $\mathscr{F}[\phi]$ or by $\mathscr{F}[\phi(x);\xi]$.

For example, if we take

$$\delta_r(x) = r\pi^{-\frac{1}{2}}e^{-r^2x^2},$$

then from equation (2-7-10) we see that its Fourier transform is

$$\Delta_r(x) = (2\pi)^{-\frac{1}{2}}e^{-x^2/4r^2}.$$

Now the regular sequence $\{\delta_r(x)\}_{r=1}^{\infty}$ defines the Dirac delta function and the sequence $\{e^{-x^2/4r^2}\}_{r=1}^{\infty}$ defines the generalized function 1 so that

$$\mathscr{F}[\delta] = (2\pi)^{-\frac{1}{2}} \tag{9-2-14}$$

where the right-hand side denotes the generalized function 1 multiplied by the constant $(2\pi)^{-\frac{1}{2}}$.

Alternatively, if we had taken $\{e^{-x^2/4r^2}\}_{r=1}^{\infty}$ as our initial $\mathscr{S}$-sequence we should have obtained the result

$$\mathscr{F}[1] = (2\pi)^{\frac{1}{2}}\,\delta. \tag{9-2-15}$$

This is, of course, a particular case of the *Fourier inversion theorem for generalized functions* which can readily be established from that for functions of the class $\mathscr{S}$.

Suppose that the regular $\mathscr{S}$-sequence $\{\phi_r(x)\}_{r=1}^{\infty}$ defines the generalized function $\phi(x)$, then the sequence of Fourier transforms $\{\Phi_r(\xi)_{r=1}^{\infty}$ defines the Fourier transform $\Phi(\xi) = \mathscr{F}[\phi(x);\xi]$. Hence the sequence $\{\Phi_r(-\xi)\}_{r=1}^{\infty}$ defines $\Phi(-\xi)$. But since, for $\phi_r \in \mathscr{S}$, we find that

$$\mathscr{F}[\Phi_r(-\xi);x] = \phi_r(x)$$

we see that

$$\mathscr{F}[\phi(x);\xi] = \Phi(\xi) \tag{9-2-16}$$

implies that

$$\mathscr{F}[\Phi(-\xi);x] = \phi(x) \tag{9-2-17}$$

which proves the Fourier inversion theorem for generalized functions.

Further, since $\{\phi_r^{(k)}(x)\}_{r=1}^{\infty}$ defines $\phi^{(k)}(x)$ and

$$\mathscr{F}[\phi_r^{(k)}(x);\xi] = (-i\xi)^k\Phi_r(\xi),$$

we deduce that for any generalized function

$$\mathscr{F}[\phi^{(k)}(x);\xi] = (-i\xi)^k\Phi(\xi). \tag{9-2-18}$$

In particular

$$\mathscr{F}[\delta^{(k)}(x);\xi] = (-i\xi)^k(2\pi)^{-\frac{1}{2}} \tag{9-2-19}$$

which in its inverse form can be written

$$\mathscr{F}[x^k;\xi] = (2\pi)^{\frac{1}{2}} i^{-k}\delta^{(k)}(\xi). \tag{9-2-20}$$

Similarly we can show that if λ and μ are real

$$\mathscr{F}[\delta(\lambda x + \mu);\xi] = \frac{1}{|\lambda|\sqrt{(2\pi)}}\, e^{-i\mu\xi/\lambda}, \qquad (\lambda \neq 0) \tag{9-2-21}$$

or, in its inverse form

$$\mathscr{F}[e^{i\alpha x};\xi] = \sqrt{(2\pi)}\,\delta(\xi + \alpha), \qquad (\alpha \in \mathbf{R}). \tag{9-2-22}$$

9-3 SCHWARTZ'S THEORY OF DISTRIBUTIONS

We conclude this chapter with a brief account of Schwartz's theory of distribution.

9-3-1 DEFINITION OF A DISTRIBUTION

As indicated in subsection 9-1-2 we define a *distribution* to be a linear functional on the space $\mathscr{D}$. If the value of a distribution T on a test function ϕ is denoted by $T(\phi)$ or $\langle T,\phi\rangle$, then T is mapping of $\mathscr{D}$ into $\mathbf{C}$ satisfying the conditions:

$$T(\lambda_1\phi_1 + \lambda_2\phi_2) = \lambda_1 T(\phi_1) + \lambda_2 T(\phi_2) \tag{9-3-1}$$

where $\phi_1,\phi_2 \in \mathscr{D}$ and $\lambda_1,\lambda_2 \in \mathbf{C}$, and

$$\phi_j \to \phi \text{ in } \mathscr{D} \text{ as } j \to \infty \quad \Rightarrow \quad T(\phi_j) \to T(\phi) \text{ as } j \to \infty. \tag{9-3-2}$$

It is easily seen that the distributions form a vector space $\mathscr{D}'$, a subspace of the dual space $\mathscr{D}^*$. Thus $\langle T,\phi\rangle$ is a bilinear form.

The simplest example of a distribution is provided by a locally integrable function f. For such a function we define a distribution T_f through the equation

$$\langle T_f,\phi\rangle = \int_{\mathbf{R}} f(x)\phi(x)\,dx, \qquad \phi \in \mathscr{D}. \tag{9-3-3}$$

From the definition we see that two functions f and g define the same distribution T_f if and only if they are equal almost everywhere.

It is also obvious that for such a function f the integral

$$\int_{\mathbf{R}} f(x)\phi^{(r)}(x)\,dx, \qquad \phi \in \mathscr{D}$$

also defines a distribution.

Any distribution of the type T_f is called *regular*; all other distributions are called *singular*.

Although a distribution does not have a value at a point it is possible to speak of local properties of distributions in a certain sense.

If Ω is an *open* subset of $\mathbf{R}$, we say that a distribution T is *zero on* Ω if $T(\phi) = 0$, for all test functions ϕ such that $\operatorname{supp}\phi \subset \Omega$, and that two distributions are equal on Ω if their difference is zero on Ω. A distribution is *zero in the neighborhood of a point a* if it is zero on some open set containing a; we also say that a distribution is *regular on* Ω if it is equal to a regular distribution on Ω. By a *distribution on an open set* Ω we mean a continuous linear functional on $\mathscr{D}(\Omega)$ the space of infinitely smooth functions with compact supports contained in Ω; these distributions in turn form a vector space which is denoted by $\mathscr{D}'_\Omega$.

The question which naturally arises is: To what extent do such "local" distributions determine a "global" distribution? This is answered by the theorem:

If $\{\Omega_1,\Omega_2, \ldots\}$ *is a finite or infinite collection of open sets whose union is* Ω, *and* $\{T_1, T_2, \ldots\}$ *is a collection of distributions each defined on the respective element* Ω_i *with the property that*

$$T_i = T_j \qquad \text{on} \qquad \Omega_i \cap \Omega_j \neq \emptyset$$

then there exists a unique distribution T *on* Ω *such that*

$$T_i = T \qquad \text{on} \qquad \Omega_i (i = 1,2, \ldots).$$

In particular, a distribution which is zero on each set of a collection of open sets is zero in their union, so that it is reasonable to define the support of a distribution T, denoted by $\operatorname{supp} T$, to be the complement of the union of all the open sets on which T is zero.

As an example we see that if T is the distribution associated with a locally integrable function then $\operatorname{supp} T = \operatorname{supp} f$.

9-3-2 THE DERIVATIVES OF A DISTRIBUTION

If f is a continuously differentiable function on R

$$\int_{\mathbf{R}} f'(x)\phi(x)\,dx = -\int_{\mathbf{R}} f(x)\phi'(x)\,dx$$

so that the distributions T_f and $T_{f'}$ defined by such a function f have the property

$$\langle T_{f'},\phi\rangle = -\langle T_f,\phi'\rangle, \qquad \phi \in \mathscr{D}. \tag{9-3-4}$$

If, therefore, we wish a regular distribution to have the property that

$$(T_f)' = T_{f'}$$

we may define the *derivative of a distribution T* by the equation

$$\langle T',\phi\rangle = -\langle T,\phi'\rangle, \qquad \phi \in \mathscr{D}. \tag{9-3-5}$$

The higher derivatives are defined in an obvious way:

$$\langle T^{(r)},\phi\rangle = (-1)^r\langle T,\phi^{(r)}\rangle, \qquad \phi \in \mathscr{D}. \tag{9-3-6}$$

It follows immediately from this definition that every distribution possesses derivatives of all orders.

In particular a regular distribution T_f has derivatives of all orders but these are not necessarily functions; if $f'(x)$ is not locally integrable there need be no connection between f' and $(T_f)'$.

9-3-3 THE PRODUCT OF A DISTRIBUTION AND A SMOOTH FUNCTION

From the definitions of the classes $\mathscr{N}$ and $\mathscr{D}$ (Cf. sec. 1-8) we see immediately that if $\psi \in \mathscr{N}$ and $\phi \in \mathscr{D}$, then their product $\psi\phi \in \mathscr{D}$. Hence if T_f is a regular distribution, we have the identity

$$\langle \psi T_f,\phi\rangle = \langle T_f,\psi\phi\rangle, \qquad \phi \in \mathscr{D}.$$

This suggests that we define the product ψT of a distribution T and a function $\psi \in \mathscr{N}$ by the equation

$$\langle \psi T,\phi\rangle = \langle T,\psi\phi\rangle, \qquad \phi \in \mathscr{D} \tag{9-3-7}$$

Combining this with the definition of the last subsection we get an important formula concerning the linear differential operator **L** defined by the equation

$$\mathbf{L} = \sum_{r=0}^{n} a_r(x)\frac{d^r}{dx^r}, \qquad a_r(x) \in \mathscr{N}, \tag{9-3-8}$$

and its adjoint $\mathbf{L}^*$ defined by

$$\mathbf{L}^*u = \sum_{r=0}^{n} (-1)^r \frac{d^r}{dx^r}\{a_r(x)u(x)\}. \tag{9-3-9}$$

If we define $\mathbf{L}T$ by the equation

$$\mathbf{L}T = \sum_{r=0}^{n} a_r T^{(r)} \tag{9-3-10}$$

then we can easily show that

$$\langle \mathbf{L}T,\phi\rangle = \langle T,\mathbf{L}^*\phi\rangle, \qquad \phi \in \mathscr{D}. \tag{9-3-11}$$

Similarly, if $T(x)$ is a distribution defined over **R**, and $a \in \mathbf{R}$, we may define the *shifted distribution* $T(x - a)$ by the equation

$$\langle T(x-a),\phi(x)\rangle = \langle T(x),\phi(x+a)\rangle, \qquad \phi \in \mathscr{D} \tag{9-3-12}$$

since this is obviously the analogue of the formula

$$\int_{-\infty}^{\infty} f(x-a)\phi(x)\,dx = \int_{-\infty}^{\infty} f(x)\phi(x+a)\,dx.$$

In a similar way the formula

$$\int_{-\infty}^{\infty} f(-x)\phi(x)\,dx = \int_{-\infty}^{\infty} f(x)\phi(-x)\,dx$$

suggests that we define the *transpose* $T(-x)$ of a distribution by the equation

$$\langle T(-x),\phi(x)\rangle = \langle T(x),\phi(-x)\rangle, \qquad \phi \in \mathscr{D}. \tag{9-3-13}$$

We say that a distribution T is *even* if $T(x) = T(-x)$ and *odd* if $T(x) = -T(-x)$.

The definition of $T(ax)$, $a > 0$, can be framed by keeping similar points in mind. We take the definition

$$\langle T(ax),\phi(x)\rangle = \langle T(x),a^{-1}\phi(x)\rangle, \qquad (a > 0, \phi \in \mathscr{D}). \tag{9-3-14}$$

Combining equations (9-3-12) and (9-3-14) we see that if a and b are real,

$$\langle T(ax-b),\phi(x)\rangle = |a|^{-1}\langle T(x),\phi(x/a+b)\rangle \tag{9-3-15}$$

for all $\phi \in \mathscr{D}$.

9-3-4 THE DELTA DISTRIBUTION

As already stated in subsection 9-1-2 we define the *delta distribution* δ, by the equation

$$\langle \delta(x);\phi(x)\rangle = \phi(0), \qquad \phi \in \mathscr{D}. \tag{9-3-16}$$

From equation (9-3-12) we deduce that

$$\langle \delta(x-a),\phi(x)\rangle = \phi(a), \qquad \phi \in \mathscr{D} \tag{9-3-17}$$

and from equation (9-3-6) that

$$\langle \delta^{(r)}(x-a),\phi(x)\rangle = (-1)^r\phi^{(r)}(a). \tag{9-3-18}$$

Also, from equations (9-3-7) and (9-3-16) we deduce that

$$\langle x\delta(x),\phi(x)\rangle = 0 \qquad \phi \in \mathscr{D}$$

a result which may be written as

$$x\delta(x) = 0. \qquad \phi \in \mathscr{D}$$

9-3-5 FOURIER TRANSFORMS OF DISTRIBUTIONS

If in the relation

$$\int_{-\infty}^{\infty} f(x)g(-x)\,dx = \int_{-\infty}^{\infty} F(t)G(t)\,dt$$

connecting the functions f,g with their Fourier transforms F,G, we take

$$G(t) = \phi(t), \qquad g(x) = \mathscr{F}^{-1}[\phi(t);x] = \mathscr{F}[\phi(t);-x]$$

we see that

$$\int_{-\infty}^{\infty} F(t)\phi(t)\,dt = \int_{-\infty}^{\infty} f(x)\Phi(x)\,dx$$

or

$$\langle \mathscr{F}f,\phi\rangle = \langle f,\mathscr{F}\phi\rangle.$$

Similarly we could show that

$$\langle \mathscr{F}^*f,\phi\rangle = \langle f,\mathscr{F}^*\phi\rangle.$$

This suggests that we might define $\mathscr{F}T$ and $\mathscr{F}^*T$ by means of the equations

$$\langle \mathscr{F}T,\phi\rangle = \langle T,\mathscr{F}\phi\rangle$$
$$\langle \mathscr{F}^*T,\phi\rangle = \langle T,\mathscr{F}^*\phi\rangle$$

respectively, and this would be alright provided that if $\phi \in \mathscr{D}$ then so did $\mathscr{F}\phi$. But we know that this is not so, so that we cannot use these equations to define the Fourier transform of an arbitrary distribution.

To overcome this difficulty we extend the class of test functions from $\mathscr{D}$ to $\mathscr{S}$.

We say that a distribution is *tempered*, or *of slow growth*, if it can be extended to a continuous linear functional on $\mathscr{S}$.

The set of tempered distributions is the dual of $\mathscr{S}$, and is denoted by $\mathscr{S}'$. We have the relations

$$\mathscr{D} \subset \mathscr{S} \subset \mathscr{S}' \subset \mathscr{D}'.$$

Since $\mathscr{S}'$ is a subspace of $\mathscr{D}'$ all the operations that are defined for distributions also apply to tempered distributions.

In addition it can be shown that

(*a*) If f is an integrable function, T_f is tempered;

(*b*) If f is bounded, T_f is tempered;

(*c*) If $f \in \mathscr{N}$ is locally integrable, T_f is tempered;

(*d*) If T has compact support it is tempered;

(*e*) If T is tempered, so are $T^{(r)}$, $(r = 1,2,3, \ldots)$;

(*f*) If T is tempered and $\psi \in \mathscr{N}$ then ψT is tempered.

(See Chap. 4 of Zemanian, 1965.)

If T is a tempered distribution we define its Fourier transform $\mathscr{F}T$

$$\langle \mathscr{F}T,\phi\rangle = \langle T,\mathscr{F}\phi\rangle, \qquad \phi \in \mathscr{S}. \tag{9-3-19}$$

Similarly we define $\mathscr{F}^*T$ by the equation

$$\langle \mathscr{F}^*T,\phi \rangle = \langle T,\mathscr{F}^*\phi \rangle, \qquad \phi \in \mathscr{S} \tag{9-3-20}$$

and it is easily shown that $\mathscr{F}T$ and $\mathscr{F}^*T$ are tempered distributions.

To prove the Fourier inversion theorem for distributions we put $\hat{T} = \mathscr{F}T$ in (9-3-19) to obtain

$$\langle \hat{T},\phi \rangle = \langle T,\mathscr{F}\phi \rangle.$$

Writing $\Phi = \mathscr{F}\phi$, $\phi = \mathscr{F}^*\Phi$ we see that this is equivalent to the relation

$$\langle T,\Phi \rangle = \langle \hat{T},\mathscr{F}^*\Phi \rangle, \qquad \Phi \in \mathscr{S}$$

and by (9-3-20) we see that this in turn is equivalent to

$$\langle T,\Phi \rangle = \langle \mathscr{F}^*\hat{T},\Phi \rangle, \qquad \Phi \in \mathscr{S}$$

showing that

$$\hat{T} = \mathscr{F}T \quad \Rightarrow \quad T = \mathscr{F}^*\hat{T}. \tag{9-3-21}$$

This is Fourier's inversion theorem for distributions.

From the above definitions we have that

$$\begin{aligned}
\langle (\mathscr{F}T)',\phi \rangle &= -\langle \mathscr{F}T,\phi' \rangle \\
&= -\langle T,\mathscr{F}\phi' \rangle \\
&= \langle T,i\xi\Phi(\xi) \rangle \\
&= \langle i\xi T,\Phi(\xi) \rangle \\
&= \langle i\xi T,\mathscr{F}\phi \rangle \\
&= \langle \mathscr{F}(ixT),\phi(x) \rangle
\end{aligned}$$

showing that

$$(\mathscr{F}T)' = i\mathscr{F}(xT). \tag{9-3-22}$$

On the other hand

$$\begin{aligned}
\langle \mathscr{F}T',\phi \rangle &= \langle T',\mathscr{F}\phi \rangle \\
&= \langle T',\Phi \rangle \\
&= -\langle T,\Phi' \rangle \\
&= -\langle T,\mathscr{F}(ix\phi) \rangle \\
&= -i\langle \mathscr{F}T,x\phi \rangle \\
&= -i\langle x\mathscr{F}T,\phi \rangle
\end{aligned}$$

showing that

$$\mathscr{F}T' = -ix\mathscr{F}T. \tag{9-3-23}$$

These results can easily be generalized to

$$(\mathscr{F}T)^{(r)} = i^r\mathscr{F}(x^rT), \tag{9-3-24}$$

$$\mathscr{F}T^{(r)} = (-ix)^r\mathscr{F}T. \tag{9-3-25}$$

In particular

$$\langle \mathscr{F}\delta(x - a),\phi(x)\rangle = \langle \delta(x - a),\Phi(x)\rangle = \Phi(a).$$

But

$$\Phi(a) = \langle (2\pi)^{-\frac{1}{2}}e^{i\xi a},\phi(\xi)\rangle$$

so that we may write

$$\mathscr{F}[\delta(x - a);\xi] = (2\pi)^{-\frac{1}{2}}e^{i\xi a}. \tag{9-3-26}$$

Putting $a = 0$ we have the special case

$$\mathscr{F}[\delta(x);\xi] = (2\pi)^{-\frac{1}{2}}. \tag{9-3-27}$$

By (9-3-25)

$$\mathscr{F}[\delta^{(r)}(x);\xi] = (2\pi)^{-\frac{1}{2}}(-i\xi)^r. \tag{9-3-28}$$

Using the inversion theorem we can rewrite this in the form

$$(2\pi)^{\frac{1}{2}}\,\delta^{(r)}(x) = \mathscr{F}^*[(-i\xi)^r;x] = \mathscr{F}[(i\xi)^r;x]$$

from which we deduce that

$$\mathscr{F}[x^r;\xi] = (-i)^r(2\pi)^{\frac{1}{2}}\,\delta^{(r)}(\xi). \tag{9-3-29}$$

There is a convolution theorem for the Fourier transforms of distributions the nature of which we shall only outline here. Given two distributions T_1 and T_2 we define their *tensor product* $T_1 \times T_2$ in terms of test functions which are of the class $\mathscr{S}$ on the Euclidean plane by means of the equation

$$\langle T_1 \times T_2,\phi\rangle = \langle T_1(x),\langle T_2(y),\phi(x,y)\rangle\rangle.$$

We then define the convolution $T_1 * T_2$ by the equation

$$\langle T_1 * T_2,\phi\rangle = \langle T_1 \times T_2,\phi(x + y)\rangle.$$

With this definition it can then be shown that if T_1 and T_2 are distributions with bounded supports whose Fourier transforms are denoted respectively by $\hat{T}_1$ and $\hat{T}_2$, then

$$\mathscr{F}(T_1 * T_2) = \hat{T}_1 \cdot \hat{T}_2$$

For the details of the proof the reader is referred to Chap. 7 of Zemanian, (1965).

9-3-6 INTEGRAL TRANSFORMS OF DISTRIBUTIONS

Other integral transforms—such as the Laplace, Mellin, Hankel, etc.,—of distributions can be defined in a similar way. As in the case of the Fourier transform the basic difficulty is defining a suitable linear space of smooth functions in terms of which to frame the definition of the transform.

A complete account of recent developments in this area is given by Zemanian (1969), to which the reader is referred.

PROBLEMS

9-1 If $\phi(x)$ is a function of x in the ordinary sense of classical analysis such that there exists a positive integer m for which $(1 + x^2)^{-m}\phi(x) \in \mathscr{A}_1(\mathbf{R})$, show that the sequence $\{\phi_r(x)\}_{r=1}^{\infty}$ defined by

$$\phi_r(x) = \int_{-\infty}^{\infty} \phi(t)p_r(t - x)e^{-t^2/r^2}\, dt,$$

where $p_r(t)$ is a positive function of class $\mathscr{S}$ satisfying equations (9-1-12) through (9-1-14), is a regular $\mathscr{S}$-sequence.

Show further that it has the property

$$\lim_{r\to\infty} \int_{-\infty}^{\infty} \phi_r(x)f(x)\, dx = \int_{-\infty}^{\infty} \phi(x)f(x)\, dx, \qquad f \in \mathscr{S}.$$

9-2 Determine the generalized functions determined by $\{\phi_r(x)\}_{r=1}$ when

(*a*) $\phi_r(x) = \dfrac{\sin(rx)}{\pi x}$;

(*b*) $\phi_r(x) = r\,\text{sech}^2(rx)$.

9-3 Show that the function $\operatorname{sgn} x$ is a generalized function and find its derivative.

9-4 If $\phi(x)$ and $\psi(x)$ are generalized functions show that the derivative of their product is the generalized function $\phi\psi' + \phi'\psi$.

9-5 If the ordinary function $g(x)$ has only simple zeros $x_1, x_2, \ldots x_n$, show that

$$\delta\{g(x)\} = \sum_{i=1}^{n} \frac{\delta(x - x_i)}{|g'(x_i)|}.$$

Deduce that

$$\delta(x^2 - a^2) = \frac{1}{2a}\{\delta(x + a) + \delta(x - a)\}.$$

9-6 Prove that, if n is a positive integer, and $r > 0$,

(*a*) $\delta^{(n)}(x) = (-1)^n x^{-n} n! \delta(x)$

(*b*) $x^{n+r}\delta(x) = 0$.

If m is a positive integer ($m < n$),

(*c*) $x^{n-m}\delta^{(n)}(x) = (-1)^{n-m}\dfrac{n!}{m!}\delta^{(m)}(x)$.

Deduce that

(*d*) $e^{-\lambda x}\delta^{(n)}(x) = \displaystyle\sum_{m=0}^{n}\binom{n}{m}\lambda^{n-m}\delta^{(m)}(x)$

(*e*) $g(x)\delta^{(n)}(x) = \displaystyle\sum_{m=0}^{n}(-1)^m\binom{n}{m}g^{(n-m)}(0)\delta^{(m)}(x), \qquad g \in \mathcal{N}$.

9-7 (*a*) If $f(x) \in \mathcal{S}$, show that

$$\int_{-\infty}^{x} f(t)\,dt \in \mathcal{S}$$

if and only if

$$\int_{-\infty}^{\infty} f(t)\,dt = 0.$$

By considering

$$f_1(x) = \int_{-\infty}^{x}\left\{f(t) - \pi^{-\frac{1}{2}}e^{-t^2}\int_{-\infty}^{\infty} f(u)\,du\right\}dt$$

show that if $\phi(x)$ is a generalized function whose derivative is zero, then $\phi(x)$ is equal to a constant multiple of the generalized function 1.

(*b*) Prove that $f(x) \in \mathcal{S} \quad \Rightarrow \quad f_1(x) = x^{-1}\{f(x) - e^{-x^2}f(0)\} \in \mathcal{S}$.

By considering the integral of $x\phi(x)f_1(x)$ show that if $\phi(x)$ is a generalized function satisfying $x\phi(x) = 0$, then $\phi(x)$ is a constant multiple of $\delta(x)$.

9-8 Show that the general solution of the equation

$$x^3\phi(x) = 0$$

is

$$\phi(x) = c_1\delta(x) + c_2\delta'(x) + c_3\delta''(x)$$

where c_1, c_2, and c_3 are arbitrary constants.

9-9 Show that the definition (9-2-7) of $|x|^\alpha$ is independent of the value chosen for n, provided that the condition $\alpha + n > -1$ is satisfied.

9-10 Show that if $\alpha < -1$ but is not a negative integer,

$$(a)\ \mathscr{F}[x^\alpha H(x);\xi] = \frac{\Gamma(\alpha+1)}{\sqrt{(2\pi)}}|\xi|^{-\alpha-1}e^{-\pi i(\alpha+1)\operatorname{sgn}\xi};$$

$$(b)\qquad \mathscr{F}[|x|^\alpha;\xi] = \sqrt{\frac{2}{\pi}}\,\Gamma(\alpha+1)|\xi|^{-\alpha-1}\cos\{\tfrac{1}{2}\pi(\alpha+1)\};$$

$$(c)\ \mathscr{F}[|x|^\alpha \operatorname{sgn} x;\xi] = i\operatorname{sgn}\xi\sqrt{\frac{2}{\pi}}\,\Gamma(\alpha+1)|\xi|^{-\alpha-1}\sin\{\tfrac{1}{2}\pi(\alpha+1)\}.$$

9-11 The generalized function $\phi(x)$ is said to be *odd* (*even*) if

$$\int_{-\infty}^{\infty} f(x)\phi(x)\,dx = 0$$

for all even (odd) functions $f(x) \in \mathscr{S}$.

Prove that

(*a*) $\delta(x)$ is even;

(*b*) if $\phi(x)$ is odd (even), then $\phi'(x)$ is even (odd);

(*c*) $\delta^{(k)}(x)$ is even if k is even and odd if k is odd;

(*d*) if $\phi(x)$ is even (odd), then $\mathscr{F}[\phi]$ is even (odd).

9-12 We define the generalized function x^{-1} to be the *odd* generalized function satisfying the equation

$$x\phi(x) = 1$$

and x^{-m}, m a positive integer, by the relation

$$x^{-m} = \frac{(-1)^{m-1}}{(m-1)!}\frac{d^{m-1}}{dx^{m-1}}(x^{-1}).$$

Prove that

$$\mathscr{F}[x^{-m};\xi] = i\operatorname{sgn}\xi\sqrt{\frac{\pi}{2}}\frac{(i\xi)^{m-1}}{(m-1)!}.$$

9-13 Show that, if n is a positive integer,

$$(a)\ \mathscr{F}[x^n \operatorname{sgn} x;\xi] = \sqrt{\frac{2}{\pi}}(-i\xi)^{-n-1}n!$$

and deduce that

$$(b)\ \mathscr{F}[x^n H(x);\xi] = \sqrt{(2\pi)}i^n\left[\tfrac{1}{2}\delta^{(n)}(\xi) + \frac{(-1)^n n!}{2\pi i\xi^{n+1}}\right].$$

9-14 Show that if $f(x)$ has a power series valid for all values of x, then

$$i\mathscr{F}_s[f(x);\xi] = \sum_{r=0}^{\infty} \frac{f^{(2r)}(0)}{(2r)!} \mathscr{F}[x^{2r}\operatorname{sgn} x;\xi]$$

$$\mathscr{F}_c[f(x);\xi] = \sum_{r=0}^{\infty} \frac{f^{(2r+1)}(0)}{(2r+1)!} \mathscr{F}[x^{2r+1}\operatorname{sgn} x;\xi]$$

and deduce that

$$\mathscr{F}_s[f(x);\xi] = \sqrt{\frac{2}{\pi}}\left[\frac{f(0)}{\xi} - \frac{f''(0)}{\xi^3} + \frac{f^{iv}(0)}{\xi^5} - \cdots\right]$$

$$\mathscr{F}_c[f(x);\xi] = -\sqrt{\frac{2}{\pi}}\left[\frac{f'(0)}{\xi^2} - \frac{f'''(0)}{\xi^4} + \frac{f^{v}(0)}{\xi^6} - \cdots\right]$$

(That this is truly an asymptotic expansion for all $f \in \mathscr{S}$ is shown on p. 56 of Lighthill, 1958.)

9-15 Find the Fourier transforms of the functions

$(a)\ \dfrac{1}{x^2(1+x^2)}$, $\quad (b)\ \dfrac{x^3}{x^2+4}$, $\quad (c)\ \dfrac{x^3 - x + 2}{x^2 - 1}$

9-16 If (q_1, q_2, q_3) are arbitrary orthogonal curvilinear coordinates with Lame coefficients h_1, h_2, h_3, show that

$$\delta(\mathbf{r} - \mathbf{r}^\circ) = (h_1 h_2 h_3)^{-1}\delta(q_1 - q_1^\circ)\delta(q_2 - q_2^\circ)\delta(q_3 - q_3^\circ).$$

Write down the expression for $\delta(\mathbf{r} - \mathbf{r}^\circ)$ in:

(*a*) cylindrical coordinates (ρ, ϕ, z)

(*b*) spherical polar coordinates (r, θ, ϕ)

and the corresponding expressions for $\delta(\mathbf{r})$.

9-17 Show that a line source of uniform density σ distributed along the z-axis can be represented by the density function $(2\pi\rho)^{-1}\sigma\delta(\rho)$, and that a line source of uniform density σ distributed along the *positive* z-axis can be represented by the density function $(2\pi r^2 \sin\theta)^{-1}\sigma\delta(\theta)$.

State the corresponding result for a source distributed along the *negative* z-axis.

appendix A
Bessel Functions

We shall derive here some of the properties of Bessel functions which are of most frequent use in the theory of integral transforms. Thc subjcct cannot be adequately treated in so short a space so that only the most important properties can be mentioned even briefly. For a fuller discussion of these topics and of further properties of Bessel functions the reader is referred to the standard treatise on the subject:

> G. N. Watson, "The Theory of Bessel Functions" second edition, (Cambridge Univ. Press, London, 1944).

A-1 BESSEL'S DIFFERENTIAL EQUATION

If we seek solutions of Laplace's equation in cylindrical coordinates (ρ,ϕ,z)

$$\frac{\partial^2 u}{\partial \rho^2} + \frac{1}{\rho}\frac{\partial u}{\partial \rho} + \frac{1}{\rho^2}\frac{\partial^2 u}{\partial \phi^2} + \frac{\partial^2 u}{\partial z^2} = 0 \qquad \text{(A-1)}$$

of the form

$$u(\rho,\phi,z) = R(\rho)e^{-\xi z + iv\phi} \tag{A-2}$$

where $R(\rho)$ is a function of ρ only we find, on substituting from (A-2) into (A-1) that

$$\frac{d^2R}{d\rho^2} + \frac{1}{\rho}\frac{dR}{d\rho} + \left(\xi^2 - \frac{v^2}{\rho^2}\right)R = 0.$$

If now we change the independent variable in this equation from ρ to $t = \xi\rho$, we find that R satisfies the ordinary differential equation

$$\frac{d^2R}{dt^2} + \frac{1}{t}\frac{dR}{dt} + \left(1 - \frac{v^2}{t^2}\right)R = 0, \qquad t = \xi\rho. \tag{A-3}$$

The equation (A-3) is called *Bessel's differential equation.*

One of the standard ways of solving a linear ordinary differential equation of this type is to assume a power series solution of the type

$$R = \sum_{r=0}^{\infty} a_r t^{\rho+r}, \qquad (a_0 \neq 0)$$

and substitute in equation (A-3) to obtain, by equating coefficients of $t^{\rho+r-2}$, the recurrence relation

$$\{(\rho + r)^2 - v^2\}a_r + a_{r-2} = 0$$

and, by equating coefficients of $t^{\rho-2}$, the indicial equation

$$\rho^2 = v^2$$

giving possible values of ρ.

Taking $\rho = v$ we then find that the function defined by the infinite series

$$\mathrm{J}_v(t) = \sum_{r=0}^{\infty} \frac{(-1)^r(\frac{1}{2}t)^{v+2r}}{r!\,\Gamma(v + r + 1)} \tag{A-4}$$

is a solution of Bessel's equation (A-3).

If v is neither zero nor an integer the function $\mathrm{J}_{-v}(t)$ will also be solution of Bessel's equation so that we may take the general solution to be

$$R = c_1\mathrm{J}_v(t) + c_2\mathrm{J}_{-v}(t) \tag{A-5}$$

where c_1 and c_2 are arbitrary constants.

Where v is zero or an integer the functions $\mathrm{J}_v(t)$ and $\mathrm{J}_{-v}(t)$ are not linearly independent. We therefore take as the second solution of Bessel's equation the function defined by the equations

$$Y_n(t) = \frac{2}{\pi}\{\gamma + \log(\tfrac{1}{2}t)\}\mathrm{J}_n(t) - \frac{1}{\pi}\sum_{r=0}^{n-1}\frac{(n-r-1)!}{r!}\left(\frac{2}{t}\right)^{n-2r}$$

$$-\frac{1}{\pi}\sum_{r=0}^{\infty}\frac{(-1)^r(\tfrac{1}{2}t)^{n+2r}}{r!(n+r)!}\{\phi(r+n)+\phi(r)\} \qquad \text{(A-6)}$$

where γ denotes Euler's constant and

$$\phi(r) = \sum_{s=1}^{r}\frac{1}{s}. \qquad \text{(A-7)}$$

The function $Y_n(x)$ is called *Weber's Bessel function of the second kind of order* n; it is sometimes denoted by $N_n(x)$.

In other words if $v = n$, a positive integer or zero, Bessel's differential equation has general solution

$$R = c_1\mathrm{J}_n(t) + c_2 Y_n(t). \qquad \text{(A-8)}$$

A-2 RECURRENCE RELATIONS

If we differentiate both sides of the equation

$$t^v\mathrm{J}_v(t) = \sum_{r=0}^{\infty}\frac{(-1)^r(\tfrac{1}{2})^{v+2r}t^{2v+2r}}{r!\,\Gamma(v+r+1)}$$

and make use of the identity

$$\Gamma(v+r+1) = (v+r)\Gamma(v+r)$$

we obtain the equation

$$\frac{d}{dt}\{t^v\mathrm{J}_v(t)\} = \sum_{r=0}^{\infty}\frac{(-1)^r(\tfrac{1}{2})^{v+2r-1}t^{2v+2r-1}}{r!\,\Gamma(v+r)}$$

which is easily seen to be equivalent to the recurrence relation

$$\frac{d}{dt}\{t^v\mathrm{J}_v(t)\} = t^v\mathrm{J}_{v-1}(t), \qquad (v>0). \qquad \text{(A-9)}$$

In a similar way we can establish the relation

$$\frac{d}{dt}\{t^{-v}\mathrm{J}_v(t)\} = -t^{-v}\mathrm{J}_{v+1}(t), \qquad (v>-1). \qquad \text{(A-10)}$$

Carrying out the differentiations we find that these equations are equivalent to the equations

$$t\mathrm{J}_v'(t) = t\mathrm{J}_{v-1}(t) - v\mathrm{J}_v(t) \qquad \text{(A-11)}$$

$$t\mathrm{J}_v'(t) = v\mathrm{J}_v(t) - t\mathrm{J}_{v+1}(t) \qquad \text{(A-12)}$$

respectively. Eliminating $J'_\nu(t)$ between these equations we obtain the further recurrence relation

$$J_{\nu-1}(t) + J_{\nu+1}(t) = \frac{2\nu}{t} J_\nu(t). \tag{A-13}$$

A-3 INTEGRAL EXPRESSIONS FOR $J_\nu(t)$

If we replace the cosine term by its series expansion and interchange the order of summation and integration we find that

$$\int_0^{2\pi} \cos(t\cos\theta)\, d\theta = \sum_{r=0}^{\infty} \frac{(-1)^r}{(2r)!} t^{2r} \int_0^{2\pi} \cos^{2r}\theta\, d\theta.$$

Now

$$\frac{1}{(2r)!}\int_0^{2\pi} \cos^{2r}\theta\, d\theta = 2\pi(\tfrac{1}{2})^{2r}\frac{1}{(r!)^2}$$

so that

$$\int_0^{2\pi} \cos(t\cos\theta)\, d\theta = 2\pi \sum_{r=0}^{\infty} \frac{(-1)^r(\frac{1}{2}t)^{2r}}{(r!)^2}.$$

In other words we have shown that

$$J_0(t) = \frac{1}{2\pi}\int_0^{2\pi} \cos(t\cos\theta)\, d\theta; \tag{A-14}$$

since

$$\int_0^{2\pi} \sin(t\cos\theta)\, d\theta = 0$$

we can write this last equation in the alternative form

$$J_0(t) = \frac{1}{2\pi}\int_0^{2\pi} e^{it\cos\theta}\, d\theta. \tag{A-15}$$

Another form:

$$J_0(t) = \frac{2}{\pi}\int_0^{\frac{1}{2}\pi} \cos(t\cos\theta)\, d\theta \tag{A-16}$$

is known as Poisson's integral for $J_0(t)$.

These results may readily be generalized by considering the integral

$$\int_0^{\pi} \sin^{2\nu}\theta \cos(t\cos\theta)\, d\theta, \qquad (\nu > -\tfrac{1}{2}).$$

Expanding the cosine factor in ascending powers of $t\cos\theta$ and making use of the *duplication formula* for the gamma function

$$\Gamma(\tfrac{1}{2})\Gamma(2z) = 2^{2z-1}\Gamma(z)\Gamma(z+\tfrac{1}{2}) \tag{A-17}$$

in the form

$$\frac{1}{(2r)!} = \frac{\Gamma(\frac{1}{2})}{2^{2r}r!\,\Gamma(r+\frac{1}{2})}$$

we readily show that

$$\mathrm{J}_\nu(t) = \frac{(\frac{1}{2}t)^\nu}{\Gamma(\frac{1}{2})\Gamma(\nu+\frac{1}{2})}\int_0^\pi \sin^{2\nu}\theta\cos(t\cos\theta)\,d\theta, \qquad (\nu > -\tfrac{1}{2}). \tag{A-18}$$

If we put $v = 0$ and recall that $\Gamma(\frac{1}{2}) = \sqrt{\pi}$, we may recover (A-16) from (A-18). We also note that

$$\mathrm{J}_\nu(t) = \frac{(\frac{1}{2})^\nu}{\Gamma(\frac{1}{2})\Gamma(\nu+\frac{1}{2})}\int_{-1}^1 e^{izt}(1-z^2)^{\nu-\frac{1}{2}}\,dz, \qquad (\nu > -\tfrac{1}{2}) \tag{A-19}$$

is equivalent to (A-18).

A second integral representation for $\mathrm{J}_\nu(t)$ is

$$\mathrm{J}_\nu(t) = \frac{1}{\pi}\int_0^\pi \cos(t\sin\theta - \nu\theta)\,d\theta - \sin(\nu\pi)\int_1^\infty u^{-\nu-1}\exp\left\{-\tfrac{1}{2}t\left(u-\frac{1}{u}\right)\right\}du. \tag{A-20}$$

In particular if n is a positive integer

$$\mathrm{J}_n(t) = \frac{1}{\pi}\int_0^\pi \cos(t\sin\theta - n\theta)\,d\theta. \tag{A-21}$$

A-4 RELATION BETWEEN THE BESSEL FUNCTIONS AND CIRCULAR FUNCTIONS

If we put $\nu = \frac{1}{2}$ in (A-4) and make use of the duplication formula (A-17) for the gamma function, we obtain

$$(\tfrac{1}{2}\pi t)^{\frac{1}{2}}\mathrm{J}_{\frac{1}{2}}(t) = \sum_{r=0}^\infty (-1)^r\frac{t^{2r+1}}{(2r+1)!} = \sin t$$

which is equivalent to the equation

$$\mathrm{J}_{\frac{1}{2}}(t) = \left(\frac{2}{\pi t}\right)^{\frac{1}{2}}\sin t. \tag{A-22}$$

Again, if we put $\nu = -\frac{1}{2}$ in (A-4) we obtain the relation

$$J_{-\frac{1}{2}}(t) = \left(\frac{2}{\pi t}\right)^{\frac{1}{2}} \cos t. \tag{A-23}$$

A-5 DEFINITE INTEGRALS INVOLVING BESSEL FUNCTIONS

It is easily seen that equations (A-9) and (A-10) are equivalent to the equations

$$\frac{\partial}{\partial x}\{x^{\nu} J_{\nu}(\xi x) = \xi x^{\nu} J_{\nu-1}(\xi x) \tag{A-24}$$

$$\frac{\partial}{\partial x}\{x^{-\nu} J_{\nu}(\xi x)\} = -\xi x^{-\nu} J_{\nu+1}(\xi x). \tag{A-25}$$

Integrating both sides of (A-24) with respect to x from 0 to a, replacing ν by $\nu + 1$ and using the fact that $x^{\nu+1} J_{\nu+1}(x) \to 0$ as $x \to 0$ if $\nu > 0$ we obtain the definite integral

$$\int_0^a x^{\nu+1} J_{\nu}(\xi x)\, dx = \frac{a^{\nu+1}}{\xi} J_{\nu+1}(\xi a), \qquad (\nu > 0). \tag{A-26}$$

Putting $\nu = 0$ in this result we obtain the integral

$$\int_0^a x J_0(\xi x)\, dx = \frac{a}{\xi} J_1(\xi a) \tag{A-27}$$

which is used repeatedly in the application of the theory of Hankel transforms to special problems.

Further we can write

$$\int_0^a x^3 J_0(\xi x)\, dx = \int_0^a x^2 \frac{1}{\xi}\frac{\partial}{\partial x}\{x J_1(\xi x)\}\, dx$$

$$= \left[\frac{x^3}{\xi} J_1(\xi x)\right]_0^a - \frac{2}{\xi}\int_0^a x^2 J_1(\xi x)\, dx.$$

Now we can evaluate the integral on the right side of this equation merely by putting $\nu = 1$ in equation (A-26). Substituting this value and making use of (A-13) with $\nu = 1$ we obtain the equation

$$\int_0^a x^3 J_0(\xi x)\, dx = \frac{a^2}{\xi^2}\left[2J_0(\xi a) + \left(a\xi - \frac{4}{a\xi}\right) J_1(\xi a)\right]. \tag{A-28}$$

Combining equations (A-27) and (A-28) we obtain

$$\int_0^a x(a^2 - x^2) J_0(\xi x)\, dx = \frac{4a}{\xi^3} J_1(\xi a) - \frac{2a^2}{\xi^2} J_0(\xi a). \tag{A-29}$$

A-6 INFINITE INTEGRALS INVOLVING BESSEL FUNCTIONS

If we use Poisson's formula (A-16) we find that

$$\int_0^\infty e^{-pt}\mathrm{J}_0(at)\,dt = \frac{2}{\pi}\int_0^\infty e^{-pt}\,dt\int_0^\infty \cos(at\cos\theta)\,d\theta.$$

Interchanging the order of the integrations and making use of the result

$$\int_0^\infty e^{-pt}\cos(at\cos\theta)\,dt = \frac{p}{p^2+a^2\cos^2\theta}, \qquad (\mathrm{Re}\,p > 0)$$

we find that

$$\int_0^\infty e^{-pt}\mathrm{J}_0(at)\,dt = \frac{2p}{\pi}\int_0^{\frac{1}{2}\pi}\frac{d\theta}{p^2+a^2\cos^2\theta}$$

Writing the integral on the right as

$$\int_0^{\frac{1}{2}\pi}\frac{\sec^2\theta\,d\theta}{(p^2+a^2)+p^2\tan^2\theta}$$

and making the substitution $u = \tan\theta$, we see that it has the value

$$\frac{1}{p^2}\int_0^\infty\frac{du}{u^2+k^2} = \frac{\pi}{2p^2k}$$

where $k^2 = (p^2+a^2)/p^2$, and hence that

$$\int_0^\infty e^{-pt}\mathrm{J}_0(at)\,dt = (a^2+p^2)^{-\frac{1}{2}}, \qquad (\mathrm{Re}\,p > 0). \tag{A-30}$$

Differentiating both sides of this equation with respect to p we obtain the equation

$$\int_0^\infty te^{-pt}\mathrm{J}_0(at)\,dt = p(a^2+p^2)^{\frac{3}{2}}. \tag{A-31}$$

On the other hand, differentiating with respect to a and using $\mathrm{J}_0'(z) = -\mathrm{J}_1(z)$, we obtain

$$\int_0^\infty te^{-pt}\mathrm{J}_1(at)\,dt = a(a^2+p^2)^{-\frac{3}{2}}. \tag{A-32}$$

These results are particular cases of the general formula

$$\int_0^\infty t^\mu e^{-pt}\mathrm{J}_\nu(at)\,dt = \frac{2^\mu a^\nu}{(a^2+p^2)^{\frac{1}{2}\mu+\frac{1}{2}\nu+\frac{1}{2}}}\,\frac{\Gamma(\frac{1}{2}\mu+\frac{1}{2}\nu+\frac{1}{2})\Gamma(\frac{1}{2}\mu+\frac{1}{2}\nu+1)}{\Gamma(\nu+1)\Gamma(\frac{1}{2})}\times$$

$$\times\,{}_2F_1\left(\tfrac{1}{2}\mu+\tfrac{1}{2}\nu+\tfrac{1}{2},\tfrac{1}{2}\nu-\tfrac{1}{2}\mu;1+\nu;\frac{a^2}{a^2+p^2}\right) \tag{A-33}$$

where, as is usual, the hypergeometric function ${}_2F_1$ is defined by the equation

$$ {}_2F_1(a,b;c;x) = \sum_{r=0}^{\infty} \frac{(a)_r(b)_r}{r!(c)_r} x^r, \qquad (a)_r = a(a+1)\ldots(a+r-1) $$

letting p tend to zero and using Gauss' formula

$$ {}_2F_1(a,b;c;1) = \frac{\Gamma(c)\Gamma(c-a-b)}{\Gamma(c-a)\Gamma(c-b)}, \qquad c > a + b $$

we find that

$$ \int_0^{\infty} t^{\mu} J_{\nu}(at)\, dt = \frac{2^{\mu}\Gamma(\frac{1}{2} + \frac{1}{2}\mu + \frac{1}{2}\nu)}{a^{\mu+1}\Gamma(\frac{1}{2} - \frac{1}{2}\mu + \frac{1}{2}\nu)}. \tag{A-34} $$

Equation (A-33) can also be rewritten in terms of the associated Legendre function P_{μ}^{ν}:

$$ \int_0^{\infty} t^{\mu} e^{-pt} J_{\nu}(at)\, dt = \frac{\Gamma(\mu - \nu + 1)}{(a^2 + p^2)^{\frac{1}{2}\mu + \frac{1}{2}}} P_{\mu}^{\nu}\left\{\frac{p}{(a^2 + p^2)^{\frac{1}{2}}}\right\}. \tag{A-35} $$

In addition, we have the formula

$$ \int_0^{\infty} t^{\mu-1} e^{-pt^2} J_{\nu}(at)\, dt $$

$$ = \frac{a^{\nu}\Gamma(\frac{1}{2}\mu + \frac{1}{2}\nu)}{2^{\nu+1}p^{\frac{1}{2}\mu+\frac{1}{2}\nu}\Gamma(1+\nu)}\, {}_1F_1\left(\tfrac{1}{2}\mu + \tfrac{1}{2}\nu; \nu + 1; -\frac{a^2}{4p}\right) \tag{A-36} $$

where $\nu > -1$, $\mu + \nu > 0$, $\operatorname{Re} p > 0$ and

$$ {}_1F_1(a;b;x) = \sum_{r=0}^{\infty} \frac{(a)_r}{r!(b)_r} x^r $$

of which an important special case is

$$ \int_0^{\infty} t^{\nu+1} e^{-pt^2} J_{\nu}(at)\, dt = \frac{a^{\nu}}{(2p)^{\nu+1}} e^{-a^2/4p}. \tag{A-37} $$

appendix B
Tables of Integral Transforms

We collect together here the main formulae derived either in the text or in the problems to provide short tables of the integral transforms in most common use.

Unless otherwise stated, a, b are assumed to be positive constants and k, m, and n are assumed to be positive integers.

The most complete set of integral transforms is given in Erdelyi *et al.* (1954). Other collections which should be consulted are those of Campbell and Foster (1948), Ditkin and Prudnikov (1961), McLachlan and Humbert (1950), and Oberhettinger and Higgins (1961).

B-1 FOURIER TRANSFORMS

	$f(t)$	$\mathscr{F}[f(t);\xi]$	*Reference*
F1	1	$\sqrt{(2\pi)}\delta(\xi)$	(9-2-15)
F2	t^{-1}	$\sqrt{(\frac{1}{2}\pi)}i\,\mathrm{sgn}\,\xi$	(2-7-5)
F3	$(a^2+t^2)^{-1}$, $\mathrm{Re}(a)>0$	$\sqrt{(\frac{1}{2}\pi)}a^{-1}e^{-a\|\xi\|}$	(2-7-3)
F4	$t(a^2+t^2)^{-1}$, $\mathrm{Re}(a)>0$	$-\frac{1}{2}i\sqrt{(\frac{1}{2}\pi)}a^{-1}\xi e^{-a\|\xi\|}$	F4 & (2-3-16)
F5	$(1-t)^{a-1}(1+t)^{b-1}\mathrm{H}(1-\|t\|)$	$2^{a+b-\frac{3}{2}}\xi^{-\frac{1}{2}}B(a,b)\,{}_1F_1\,(b;a+b;2i\xi)$	Prob. 2-13
F6	$P_n(t)\mathrm{H}(1-\|t\|)$	$i^n\xi^{-\frac{1}{2}}\mathrm{J}_{n+\frac{1}{2}}(\xi)$	Prob. 2-14
F7	$e^{i\alpha t}$, $(\alpha\in\mathbf{R})$	$\sqrt{(2\pi)}\delta(\xi+\alpha)$	(9-2-22)
F8	$e^{-a\|t\|}$	$\sqrt{(2/\pi)}a(a^2+\xi^2)^{-1}$	(2-7-3)
F9	$te^{-a\|t\|}$	$\sqrt{(2/\pi)}2ai\xi(a^2+\xi^2)^{-2}$	(2-7-8)
F10	$\|t\|e^{-a\|t\|}$	$\sqrt{(2/\pi)}(a^2-\xi^2)(a^2+\xi^2)^{-2}$	(2-7-9)
F11	$e^{-a^2t^2}$	$2^{-\frac{1}{2}}a^{-1}e^{-\frac{1}{4}\xi^2/a^2}$	(2-7-13)
F12	$t^{-\frac{1}{2}}\mathrm{J}_{n+\frac{1}{2}}(t)$	$i^nP_n(\xi)\mathrm{H}(1-\|\xi\|)$	Prob. 2-14
F13	$t\,\mathrm{cosech}\,t$	$\sqrt{(2\pi^3)}e^{\pi\xi}(1+e^{\pi\xi})^{-2}$	Prob. 2-20
F14	$\sinh(\alpha t)\,\mathrm{cosech}\,(\beta t)$, $0<\alpha<\beta$	$\sqrt{(\frac{1}{2}\pi)}\beta^{-1}\sin(\pi\alpha/\beta)\times$ $\times\{\cosh(\pi\xi/\beta)+\cos(\pi\alpha/\beta)\}^{-1}$	Prob. 2-21
F15	$\cosh(\alpha t)\,\mathrm{cosech}\,(\beta t)$, $0<\alpha<\beta$	$i\sqrt{(\frac{1}{2}\pi)}\beta^{-1}\sinh(\pi\xi/\beta)\times$ $\times\{\cosh(\pi\xi/\beta)+\cos(\pi\alpha/\beta)\}^{-1}$	
F16	$\delta(t)$	$(2\pi)^{-\frac{1}{2}}$	(9-2-14)
F17	$\delta(\lambda t+\mu)$, $\lambda\neq 0$	$\|\lambda\|^{-1}(2\pi)^{-\frac{1}{2}}e^{-i\mu\xi/\lambda}$	(9-2-21)
F18	$\delta^{(k)}(t)$	$(-i\xi)^k(2\pi)^{-\frac{1}{2}}$	(9-2-19)
F19	t^k	$(2\pi)^{\frac{1}{2}}i^{-k}\delta^{(k)}(\xi)$	(9-2-20)
F20	$t_+^{\alpha}\equiv t^{\alpha}\mathrm{H}(t)$ ($\alpha<-1$ but not an integer)	$\dfrac{\Gamma(\alpha+1)}{\sqrt{(2\pi)}}\|\xi\|^{-\alpha-1}e^{-\pi i(\alpha+1)\mathrm{sgn}\xi}$	Prob. 9-10
F21	$\|t\|^{\alpha}$ ($\alpha<-1$ but not an integer)	$\sqrt{(2/\pi)}\Gamma(\alpha+1)^{-\alpha-1}\times$ $\times\cos\{\frac{1}{2}\pi(\alpha+1)\}$	Prob. 9-10
F22	$\|t\|^{\alpha}\,\mathrm{sgn}\,t$ ($\alpha<-1$) but not an integer)	$i\,\mathrm{sgn}\,\xi\sqrt{(2/\pi)}\Gamma(\alpha+1)\,\|\xi\|^{-\alpha-1}\times$ $\times\sin\{\frac{1}{2}\pi(\alpha+1)\}$	Prob. 9-10
F23	$t^k\,\mathrm{sgn}\,t$	$\sqrt{(2/\pi)}(-i\xi)^{-k-1}k!$	Prob. 9-13
F24	$t^k\mathrm{H}(t)$	$\sqrt{(2\pi)}i^k\left[\frac{1}{2}\delta^{(k)}(\xi)+\dfrac{(-1)^kk!}{2\pi i\xi^{k+1}}\right]$	Prob. 9-13

B-2 FOURIER COSINE TRANSFORMS

	$f(t)$	$\mathscr{F}_c[f(t);\xi]$	*Reference*
C1	t^{p-1}, $0<p<1$	$\sqrt{(2/\pi)}\xi^{-p}\Gamma(p)\cos(\frac{1}{2}p\pi)$	(2-7-21)
C2	$(a^2+t^2)^{-1}$, $\mathrm{Re}(a)>0$	$\sqrt{(\frac{1}{2}\pi)}a^{-1}e^{-a\xi}$	(2-7-1)
C3	$(a^2+t^2)^{-2}$, $\mathrm{Re}(a)>0$	$\frac{1}{2}\sqrt{(\frac{1}{2}\pi)}a^{-3}e^{-a\xi}(1+a\xi)$	(2-7-6)

(continued overleaf)

B-2 FOURIER COSINE TRANSFORMS (continued)

	$f(t)$	$\mathscr{F}_c[f(t);\xi]$	*Reference*
C4	e^{-at}	$\sqrt{(2/\pi)}a(a^2+\xi^2)^{-1}$	(2-7-1)
C5	te^{-at}	$\sqrt{(2/\pi)}(a^2-\xi^2)(a^2+\xi^2)^{-2}$	(2-7-6)
C6	$t^{n-1}e^{-at}$	$\sqrt{(2/\pi)}\cdot(n-1)!r^{-n}\cos(n\theta)$, $r=\sqrt{(a^2+\xi^2)}$, $\theta=\tan^{-1}(\xi/a)$	Prob. 2-10
C7	$e^{-a^2t^2}$	$2^{-\frac{1}{2}}a^{-1}e^{-\frac{1}{4}\xi^2/a^2}$	(2-7-13)
C8	$t^{-1}(e^{-bt}-e^{-at})$, $(0<b<a)$	$(2\pi)^{-\frac{1}{2}}\log\dfrac{a^2+\xi^2}{b^2+\xi^2}$	Prob. 2-26
C9	$\cos(\frac{1}{2}t^2)$	$1/(\sqrt{2})\{\cos(\frac{1}{2}\xi^2)+\sin(\frac{1}{2}\xi^2)\}$	(2-7-15)
C10	$\sin(\frac{1}{2}t^2)$	$1/(\sqrt{2})\{\cos(\frac{1}{2}\xi^2)-\sin(\frac{1}{2}\xi^2)\}$	(2-7-16)
C11	$(a^2-t^2)^{\nu-\frac{1}{2}}\mathrm{H}(a-t)$, $(\nu>-\frac{1}{2})$	$2^{\nu-\frac{1}{2}}\Gamma(\nu+\frac{1}{2})a^\nu\xi^{-\nu}\mathrm{J}_\nu(a\xi)$	(2-7-19)
C12	$\dfrac{\sinh(\alpha t)}{\sinh(\beta t)}$, $0<\alpha<\beta$	$\sqrt{(\frac{1}{2}\pi)}\beta^{-1}\dfrac{\sin(\pi\alpha/\beta)}{\cosh(\pi\xi/\beta)+\cos(\pi\alpha/\beta)}$	Prob. 2-21
C13	$\dfrac{\cosh(\alpha t)}{\cosh(\beta t)}$, $0<\alpha<\beta$	$\sqrt{(2\pi)}\beta^{-1}\dfrac{\cos(\pi\alpha/2\beta)\cosh(\pi\xi/2\beta)}{\cosh(\pi\xi/\beta)+\cos(\pi\alpha/\beta)}$	Prob. 2-22
C14	$(b^2+t^2)^{-1}\mathrm{J}_0(at)$	$\sqrt{(\frac{1}{2}\pi)}b^{-1}e^{-b\xi}I_0(ab)$, $\xi>a$	Prob. 2-34
C15	$e^{-a\cosh t}$	$\sqrt{(2/\pi)}K_{i\xi}(a)$	(6-1-8)
C16	$t^{-\nu}\mathrm{J}_\nu(at)$, $\nu>-\frac{1}{2}$	$\dfrac{(a^2-\xi^2)^{\nu-\frac{1}{2}}H(a-\xi)}{2^{\nu-\frac{1}{2}}a^\nu\Gamma(\nu+\frac{1}{2})}$	(2-7-19)
C17	$K_0(at)$	$\sqrt{(\frac{1}{2}\pi)}(a^2+\xi^2)^{\frac{1}{2}}$	(6-1-5)
C18	$K_{it}(a)$	$\sqrt{(\frac{1}{2}\pi)}e^{-a\cosh\xi}$	(6-1-9)

B-3 FOURIER SINE TRANSFORMS

	$f(t)$	$\mathscr{F}_s[f(t);\xi]$	*Reference*
S1	t^{-1}	$\sqrt{(\frac{1}{2}\pi)}\,\mathrm{sgn}\,\xi$	(2-7-4)
S2	t^{p-1}, $(0<p<1)$	$\sqrt{(2/\pi)}\xi^{-p}\Gamma(p)\sin(\frac{1}{2}\pi p)$	(2-7-22)
S3	$t^{-\frac{1}{2}}$	$\xi^{-\frac{1}{2}}$	(2-7-23)
S4	$t(a^2+t^2)^{-1}$	$\sqrt{(\frac{1}{2}\pi)}e^{-a\xi}$	(2-7-2)
S5	$t(a^2+t^2)^{-2}$	$(2\pi)^{-\frac{1}{2}}a^{-1}\xi e^{-a\xi}$, $(\xi>0)$	(2-7-7)
S6	$t^{-1}(a^2+t^2)^{-1}$	$\sqrt{(\frac{1}{2}\pi)}a^{-2}(1-e^{-a\xi})$, $(\xi>0)$	(2-8-4)
S7	e^{-at}	$\sqrt{(2/\pi)}\xi(a^2+\xi^2)^{-1}$	(2-7-1)
S8	te^{-at}	$\sqrt{(2/\pi)}2a\xi(a^2+\xi^2)^{-2}$	(2-7-7)
S9	$t^{n-1}e^{-at}$	$\sqrt{(2/\pi)}(n-1)!r^{-n}\sin(n\theta)$, $r=\sqrt{(a^2+\xi^2)}$, $\theta=\tan^{-1}(\xi/a)$	Prob. 2-10
S10	$te^{-a^2t^2}$	$2^{-\frac{3}{2}}a^{-3}\xi e^{-\frac{1}{4}\xi^2/a^2}$	(2-7-14)
S11	$t^{-1}e^{-at}$	$\sqrt{(2/\pi)}\tan^{-1}(\xi/a)$	Prob. 2-26
S12	$t(a^2-t^2)^{\nu-\frac{3}{2}}\mathrm{H}(a-t)$, $\nu>\frac{1}{2}$	$2^{\nu-\frac{3}{2}}a^\nu\xi^{1-\nu}\Gamma(\nu-\frac{1}{2})\mathrm{J}_\nu(\xi a)$	(2-7-20)

(continued)

	$f(t)$	$\mathscr{L}[f(t);p]$	*Reference*
S13	$\dfrac{\cosh(\alpha t)}{\sinh(\beta t)}$, $0 < \alpha < \beta$	$\sqrt{(\frac{1}{2}\pi)}\beta^{-1}\dfrac{\sinh(\pi\xi/\beta)}{\cosh(\pi\xi/\beta) + \cos(\pi\alpha/\beta)}$	Prob. 2-21
S14	$\dfrac{\sinh(\alpha t)}{\cosh(\beta t)}$, $0 < \alpha < \beta$	$\sqrt{(2\pi)}\beta^{-1}\dfrac{\sin(\pi\alpha/2\beta)\sinh(\pi\xi/2\beta)}{\cosh(\pi\xi/\beta) + \cos(\pi\alpha/\beta)}$	Prob. 2-22
S15	$t^{1-\nu}J_\nu(at)$, $\nu > \frac{1}{2}$	$\dfrac{\xi(a^2 - \xi^2)^{\nu-\frac{3}{2}}H(a - \xi)}{2^{\nu-\frac{3}{2}}a^\nu\Gamma(\nu - \frac{1}{2})}$	(2-7-20)
S16	$t^{-1}J_0(at)$	$\begin{cases}\sqrt{(\frac{1}{2}\pi)}, & 0 < a < \xi;\\ \sqrt{(2/\pi)}\sin^{-1}(\xi/a), & 0 < \xi < a\end{cases}$	Prob. 2-33
S17	$t(b^2 + t^2)^{-1}J_0(at)$	$\sqrt{(\frac{1}{2}\pi)}e^{-b\xi}I_0(ab)$, $\xi > a$	Prob. 2-34
S18	$t^{-\nu}\mathbf{H}_\nu(t)$	$\dfrac{2^{\frac{1}{2}-\nu}}{\Gamma(\nu + \frac{1}{2})}(1 - \xi^2)^{\nu-\frac{1}{2}}H(1 - \xi)$	Prob. 2-38

B-4 LAPLACE TRANSFORMS

	$f(t)$	$\mathscr{L}[f(t);p]$	*Reference*
L1	1	p^{-1}	(3-2-1)
L2	t^ν, $(\nu > -1)$	$p^{-\nu-1}\Gamma(\nu + 1)$	(3-2-3)
L3	e^{at}	$(p - a)^{-1}$, $\operatorname{Re} p > a$	(3-2-6)
L4	$\cosh at$	$p(p^2 - a^2)^{-1}$, $\operatorname{Re} p > a$	(3-2-7)
L5	$\sinh at$	$a(p^2 - a^2)^{-1}$, $\operatorname{Re} p > a$	(3-2-8)
L6	$t\cosh at$	$(p^2 + a^2)(p^2 - a^2)^{-2}$, $\operatorname{Re} p > a$	L5 & (3-3-12)
L7	$t\sinh at$	$2ap(p^2 - a^2)^{-2}$, $\operatorname{Re} p > a$	L4 & (3-3-12)
L8	$\frac{1}{2}at\cosh at - \frac{1}{2}\sinh at$	$a^3(p^2 - a^2)^{-2}$, $\operatorname{Re} p > a$	L5, L6
L9	$\cosh at + \frac{1}{2}at\sinh at$	$p^3(p^2 - a^2)^{-2}$, $\operatorname{Re} p > a$	L4, L7
L10	$\cos at$	$p(p^2 + a^2)^{-1}$	(3-2-5)
L11	$\sin at$	$a(p^2 + a^2)^{-1}$	(3-2-4)
L12	$t\cos at$	$(p^2 - a^2)(p^2 + a^2)^{-2}$	(3-3-14)
L13	$t\sin at$	$2ap(p^2 + a^2)^{-2}$	(3-3-15)
L14	$\frac{1}{2}\sin at - \frac{1}{2}at\cos at$	$a^3(p^2 + a^2)^{-2}$	L11, L12
L15	$\frac{1}{2}\sin at + \frac{1}{2}at\cos at$	$ap^2(p^2 + a^2)^{-2}$	L11, L12
L16	$\cos at - \frac{1}{2}at\sin at$	$p^3(p^2 + a^2)^{-2}$	L10, L13
L17	$t^{-1}\sin(at)$	$\cot^{-1}(p/a)$	(3-3-22)
L18	$\operatorname{Si}(t)$	$p^{-1}\cot^{-1}p$	(3-5-5)
L19	$\operatorname{Ci}(t)$	$-(1/2p)\log(1 + p^2)$	(3-5-10)
L20	$e^{-\frac{1}{4}b^2t^2}$	$\pi^{\frac{1}{2}}b^{-1}e^{p^2/b^2}\operatorname{Erfc}(p/b)$	(3-6-8)
L21	$\operatorname{Erf}(t/2a)$	$p^{-1}e^{a^2p^2}\operatorname{Erfc}(ap)$	(3-6-10)
L22	$\operatorname{Erf}(a\sqrt{t})$	$ap^{-1}(p + a^2)^{-\frac{1}{2}}$	(3-6-12)
L23	$\operatorname{Erfc}(a\sqrt{t})$	$p^{-1}[1 - a(p + a^2)^{-\frac{1}{2}}]$	(3-6-12)

(*continued overleaf*)

B-4 LAPLACE TRANSFORMS (continued)

	$f(t)$	$\mathscr{L}[f(t);p]$	*Reference*
L24	$e^{a^2t}\,\mathrm{Erfc}\,(a\sqrt{t})$	$p^{-\frac{1}{2}}(p^{\frac{1}{2}}+a)^{-1}$	(3-6-16)
L25	$\mathrm{Erfc}\,(a/\sqrt{t})$	$p^{-1}e^{-2a\sqrt{p}}$, $\mathrm{Re}\,p>0$	(3-6-17)
L26	$t^{-\frac{1}{2}}e^{-c/t}$	$(\pi/p)^{\frac{1}{2}}e^{-2\sqrt{(cp)}}$, $\mathrm{Re}\,p>0$	(3-2-13)
L27	$t^{-\frac{3}{2}}e^{-c/t}$	$(\pi/c)^{\frac{1}{2}}e^{-2\sqrt{(cp)}}$, $\mathrm{Re}\,p>0$	(3-2-14)
L28	$t^{\nu}\mathrm{J}_{\nu}(at)$, $(\nu>-\frac{1}{2})$	$(2a)^{\nu}\pi^{-\frac{1}{2}}\Gamma(\nu+\frac{1}{2})(p^2+a^2)^{-\nu-\frac{1}{2}}$, $\mathrm{Re}\,p>0$	(3-7-3)
L29	$t^{\nu+1}\mathrm{J}_{\nu}(at)$, $(\nu>-\frac{1}{2})$	$\pi^{-\frac{1}{2}}2^{\nu+1}a^{\nu}p\Gamma(\nu+\frac{3}{2})(p^2+a^2)^{-\nu-\frac{3}{2}}$, $\mathrm{Re}\,p>0$	(3-7-6)
L30	$t^{n}\mathrm{J}_{n}(at)$	$\dfrac{(2n)!}{2^n n!}a^n(p^2+a^2)^{-n-\frac{1}{2}}$, $\mathrm{Re}\,p>0$	Prob. 3-12
L31	$t^{n+1}\mathrm{J}_{n}(at)$	$\dfrac{(2n+1)!}{2^n n!}a^{n}(p^2+a^2)^{-n-\frac{3}{2}}$, $\mathrm{Re}\,p>0$	Prob. 3-12
L32	$\mathrm{J}_{\nu}(t)$ $(\nu>-\frac{1}{2})$	$(p^2+1)^{-\frac{1}{2}}\{p+\sqrt{(p^2+1)}\}^{-\nu}$, $\mathrm{Re}\,p>0$	Prob. 3-17
L33	$t\mathrm{J}_{\nu}(t)$ $(\nu>-\frac{1}{2})$	$\{p+\nu\sqrt{(p^2+1)}\}(p^2+1)^{-\frac{3}{2}}\times$ $\times\{p+\sqrt{(p^2+1)}\}^{-\nu}$, $\mathrm{Re}\,p>0$	Prob. 3-17
L34	$t^{-1}\mathrm{J}_{\nu}(t)$ $(\nu>0)$	$\nu^{-1}\{p+\sqrt{(p^2+1)}\}^{-\nu}$, $\mathrm{Re}\,p>0$	Prob. 3-17
L35	$t^{\frac{1}{2}\nu}\mathrm{J}_{\nu}(2\sqrt{at})$, $(\nu>-\frac{1}{2})$	$a^{\frac{1}{2}\nu}p^{-\nu-1}e^{-a/p}$, $\mathrm{Re}\,p>0$	(3-7-8)
L36	$t^{-\frac{1}{2}}\mathrm{J}_{\nu}(2a\sqrt{t})$, $(\nu>-\frac{1}{2})$	$\sqrt{(\pi/p)}e^{-a^2/2p}\mathrm{I}_{\frac{1}{2}\nu}\left(\dfrac{a^2}{2p}\right)$, $\mathrm{Re}\,p>0$	Prob. 3-18
L37	$e^{t\cos\theta}\mathrm{J}_0(t\sin\theta)$	$(1-2p\cos\theta+p^2)^{-\frac{1}{2}}$, $\mathrm{Re}\,p>0$	Prob. 3-14
L38	$t^{\nu}\mathrm{I}_{\nu}(at)$, $(\nu>-\frac{1}{2})$	$(2a)^{\nu}\pi^{-\frac{1}{2}}\Gamma(\nu+\frac{1}{2})(p^2-a^2)^{-\nu-\frac{1}{2}}$, $\mathrm{Re}\,p>a>0$	(3-7-11)
L39	$t^{\nu+1}\mathrm{I}_{\nu}(at)$, $(\nu>-\frac{1}{2})$	$2^{\nu+1}a^{\nu}\pi^{-\frac{1}{2}}p\Gamma(\nu+\frac{3}{2})(p^2-a^2)^{-\nu-\frac{3}{2}}$, $\mathrm{Re}\,p>a>0$	(3-7-12)
L40	$t^{\frac{1}{2}\nu}\mathrm{I}_{\nu}(2\sqrt{at})$	$a^{\frac{1}{2}\nu}p^{-\nu-1}e^{a/p}$, $\mathrm{Re}\,p>0$	(3-7-13)
L41	$t^{\lambda}e^{at}\mathrm{J}_{\lambda}{}^{\mu}(t^{\mu}x)$	$(p-a)^{-\lambda-1}\exp\{-x(p-a)^{-\mu}\}$, $\mathrm{Re}\,p>a>0$	(3-7-23)
L42	$K_{i\tau}(t)$, $(\tau\in\mathbf{R})$	$\dfrac{\pi\sin(\alpha\tau)}{\sinh\alpha\cdot\sinh(\pi\tau)}$, $(p=\cosh\alpha)$	(6-1-34)
L43	$t^{-\frac{1}{2}}\cos(a\sqrt{t})$	$\sqrt{(\pi/p)}e^{-a^2/4p}$, $\mathrm{Re}\,p>0$	Prob. 3-9
L44	$t^{-1}\sin(a\sqrt{t})$	$\mathrm{Erf}\,(\frac{1}{2}ap^{-\frac{1}{2}})$, $\mathrm{Re}\,p>0$	Prob. 3-9
L45	$\sin(a\sqrt{t})$	$\frac{1}{2}a\pi^{\frac{1}{2}}p^{-\frac{3}{2}}e^{-a^2/4p}$, $\mathrm{Re}\,p>0$	Prob. 3-9
L46	$S_{\tau}(t)$	$p^{-1}\tanh(\frac{1}{2}p\tau)$, $\mathrm{Re}\,p>0$	(3-8-4)
L47	$L_n(t)$	$p^{-n-1}(p-1)^n$, $\mathrm{Re}\,p>0$	sec. 3-12
L48	$e^{-bt}\displaystyle\sum_{n=0}^{\infty}a_nL_n(2bt)$	$(p+b)^{-1}\displaystyle\sum_{n=0}^{\infty}a_n\left(\dfrac{p-b}{p+b}\right)^n$ $\mathrm{Re}\,p>b>0$	(3-12-4)

(*continued*)

	$f(t)$	$\mathscr{L}[f(t);p]$	*Reference*
L49	$E_1(t)$	$p^{-1}\log(p+1)$, $\operatorname{Re} p > 0$	Prob. 3-8
L50	$\vartheta(t)$	$p^{-\frac{1}{2}}\coth(p^{\frac{1}{2}})$, $\operatorname{Re} p > 0$	Prob. 3-25

B-5 MELLIN TRANSFORMS

	$f(x)$	$\mathscr{M}[f(x);s]$	*Reference*
M1	$(1-x)^{-1}$	$\pi\cot(\pi s)$, $0 < \operatorname{Re} s < 1$	(4-1-22)
M2	$(1+x)^{-1}$	$\pi\operatorname{cosec}(\pi s)$, $0 < \operatorname{Re} s < 1$	(4-1-13)
M3	$(1+x^a)^{-b}$	$\dfrac{\Gamma(s/a)\Gamma(b-s/a)}{a\Gamma(b)}$, $0 < \operatorname{Re} s < ab$	(4-1-14)
M4	$\dfrac{T_n(x)\mathrm{H}(1-x)}{\sqrt{(1-x^2)}}$	$\dfrac{2^{-s}\pi\Gamma(s)}{\Gamma(\frac{1}{2}+\frac{1}{2}s+\frac{1}{2}n)\Gamma(\frac{1}{2}+\frac{1}{2}s-\frac{1}{2}n)}$, $\operatorname{Re} s > 0$	Prob. 4-26
M5	$\dfrac{T_n(x^{-1})\mathrm{H}(1-x)}{\sqrt{(1-x^2)}}$	$\dfrac{2^{s-2}\Gamma(\frac{1}{2}n+\frac{1}{2}s)\Gamma(\frac{1}{2}s-\frac{1}{2}n)}{\Gamma(s)}$, $\operatorname{Re} s > n$	Prob. 4-27
M6	$P_n(x)\mathrm{H}(1-x)$	$\dfrac{\Gamma(\frac{1}{2}s)\Gamma(\frac{1}{2}s+\frac{1}{2})}{2\Gamma(\frac{1}{2}s-\frac{1}{2}n+\frac{1}{2})\Gamma(\frac{1}{2}s+\frac{1}{2}n+1)}$, $\operatorname{Re} s > 0$	Prob. 4-6
M7	$P_n(x^{-1})\mathrm{H}(1-x)$	$\dfrac{2^{s-1}\Gamma(\frac{1}{2}s+\frac{1}{2}n+\frac{1}{2})\Gamma(\frac{1}{2}s-\frac{1}{2}n)}{\Gamma(\frac{1}{2})\Gamma(s+1)}$, $\operatorname{Re} s > n$	Prob. 4-5
M8	$(1-x^2)^{\frac{1}{2}\lambda}P_v^{-\lambda}(x)\mathrm{H}(1-x)$	$\dfrac{2^{-\lambda-s}\Gamma(\frac{1}{2})\Gamma(s)}{\Gamma(\frac{1}{2}\lambda-\frac{1}{2}v+\frac{1}{2}s+\frac{1}{2})\Gamma(\frac{1}{2}\lambda+\frac{1}{2}v+\frac{1}{2}s+1)}$, $\operatorname{Re} s > \min(0, v-\lambda-1)$	Prob. 4-9
M9	$x^{n-1}(1-x^2)^{\mu-\frac{1}{2}}C_n^{\mu}(x) \times$ $\times\, \mathrm{H}(1-x)$, $(\mu > 0)$	$\dfrac{2^{1-n-s}\Gamma(\frac{1}{2})\Gamma(\mu+\frac{1}{2})\Gamma(2\mu+n)\Gamma(s+n-1)}{n!\,\Gamma(2\mu)\Gamma(\frac{1}{2}s)\Gamma(\frac{1}{2}s+n+\mu)}$,	Prob. 4-8
		$\operatorname{Re} s > 0$	Prob. 4-8
M10	$(1-x^2)^{\lambda-\frac{1}{2}}C_n^{\lambda}(x^{-1}) \times$ $\times\, \mathrm{H}(1-x)$, $(\lambda > 0)$	$\dfrac{2^{s-1}\Gamma(n+2\lambda)\Gamma(\frac{1}{2}s-\frac{1}{2}n)\Gamma(\frac{1}{2}s+\lambda+\frac{1}{2}n)}{n!\,\Gamma(\lambda)\Gamma(s+2\lambda)}$, $\operatorname{Re} s > n$	Prob. 4-7
M11	$\dfrac{1+x\cos\phi}{1-2x\cos\phi+x^2}$, $-\pi < \phi < \pi$	$\dfrac{\pi\cos(s\phi)}{\sin(s\pi)}$, $0 < \operatorname{Re} s < 1$	Prob. 4-4
M12	$\dfrac{x\sin\phi}{1-2x\cos\phi+x^2}$, $-\pi < \phi < \pi$	$\dfrac{\pi\sin(s\phi)}{\sin(s\pi)}$, $0 < \operatorname{Re} s < 1$	Prob. 4-5
M13	e^{-x}	$\Gamma(s)$, $\operatorname{Re} s > 0$	(4-1-11)
M14	e^{-x^2}	$\frac{1}{2}\Gamma(\frac{1}{2}s)$, $\operatorname{Re} s > 0$	M13
M15	$\cos x$	$\Gamma(s)\cos(\frac{1}{2}\pi s)$, $0 < \operatorname{Re} s < 1$	(4-1-7)
M16	$\sin x$	$\Gamma(s)\sin(\frac{1}{2}\pi s)$, $0 < \operatorname{Re} s < 1$	(4-1-8)

(*continued overleaf*)

B-5 MELLIN TRANSFORMS (continued)

	$f(x)$	$\mathscr{M}[f(x);s]$	*Reference*
M17	$e^{-x\cos\phi}\cos(x\sin\phi)$, $-\frac{1}{2}\pi < \phi < \frac{1}{2}\pi$	$\Gamma(s)\cos(s\phi)$, $\mathrm{Re}\, s > 0$	Prob. 4-4
M18	$e^{-x\sin\phi}\sin(x\sin\phi)$, $-\frac{1}{2}\pi < \phi < \frac{1}{2}\pi$	$\Gamma(s)\sin(s\phi)$, $\mathrm{Re}\, s > -1$	Prob. 4-4
M19	$x^{-\nu}J_\nu(x)$, $\nu > -\frac{1}{2}$	$\dfrac{2^{s-\nu-1}\Gamma(\frac{1}{2}s)}{\Gamma(\nu - \frac{1}{2}s + 1)}$, $0 < \mathrm{Re}\, s < 1$	(4-1-9)
M20	$Y_\nu(x)$, $\nu \in \mathbf{R}$	$-2^{s-1}\pi^{-1}\Gamma(\frac{1}{2}s + \frac{1}{2}\nu)\Gamma(\frac{1}{2}s - \frac{1}{2}\nu) \times$ $\times \cos(\frac{1}{2}s - \frac{1}{2}\nu)\pi$, $\lvert\nu\rvert < \mathrm{Re}\, s < \frac{3}{2}$	Prob. 2-37
M21	$K_\nu(x)$, $\nu \in \mathbf{R}$	$2^{s-2}\Gamma(\frac{1}{2}s + \frac{1}{2}\nu)\Gamma(\frac{1}{2}s - \frac{1}{2}\nu)$, $\mathrm{Re}\, s > \nu > 0$	Prob. 4-22
M22	$\mathbf{H}_\nu(x)$, $\nu \in \mathbf{R}$	$\dfrac{2^{s-1}\tan(\frac{1}{2}\pi s + \frac{1}{2}\pi\nu)\Gamma(\frac{1}{2}s + \frac{1}{2}\nu)}{\Gamma(\frac{1}{2}\nu - \frac{1}{2}s + 1)}$ $-1 - \nu < \mathrm{Re}\, s < \min(\frac{3}{2}, 1 - \nu)$	Prob. 2-38

B-6 HANKEL TRANSFORMS

	$f(x)$	$\mathscr{H}_\nu[f(x);\xi]$	*Reference*
H1	$x^\nu H(a - x)$	$\xi^{-1}a^{\nu+1}J_{\nu+1}(a\xi)$	(5-5-4)
H2	x^{s-1}, $(-\nu - 1 < \mathrm{Re}\, s < \nu + 1)$	$\dfrac{2^s\Gamma(\frac{1}{2}s + \frac{1}{2}\nu + \frac{1}{2})}{\xi^{s+1}\Gamma(\frac{1}{2}\nu - \frac{1}{2}s + \frac{1}{2})}$	(2-10-7)
H3	$x^\nu(a^2 - x^2)^{\mu-\nu-1}H(a - x)$, $(\mu > \nu > 0)$	$2^{\mu-\nu-1}\Gamma(\mu - \nu)a^\mu\xi^{\nu-\mu}J_\mu(\xi a)$	(5-5-1)
H4	$x^{\nu-1}e^{-ax}$	$(2\xi)^\nu\pi^{-\frac{1}{2}}\Gamma(\nu + \frac{1}{2})(a^2 + \xi^2)^{-\nu-\frac{1}{2}}$	(5-5-14)
H5	$x^\nu e^{-ax}$	$2^{\nu+1}\xi^\nu a\pi^{-\frac{1}{2}}\Gamma(\nu + \frac{3}{2})(a^2 + \xi^2)^{-\nu-\frac{3}{2}}$	(5-5-15)
H6	$x^{-1}e^{-ax}$	$\xi^\nu(a^2 + \xi^2)^{-\frac{1}{2}}\{a + \sqrt{(a^2 + \xi^2)}\}^{-\nu}$	(5-5-16)
H7	e^{-ax}	$\xi^\nu(a^2 + \xi^2)^{-\frac{3}{2}}\{a + \nu\sqrt{(a^2 + \xi^2)}\}\times$ $\times\{a + \sqrt{(a^2 + \xi^2)}\}^{-\nu}$	(5-5-17)
H8	$x^\nu e^{-x^2/a^2}$	$(\frac{1}{2}a^2)^{\nu+1}\xi^\nu e^{-\frac{1}{4}\xi^2a^2}$	(5-5-19)
H9	$x^\mu e^{-x^2/a^2}$	$\dfrac{\xi^\nu a^{\mu+\nu+2}}{2^{\nu+1}\Gamma(1 + \nu)} \times \Gamma(1 + \frac{1}{2}\mu + \frac{1}{2}\nu) \times$ $\times\, {}_1F_1(1 + \frac{1}{2}\mu + \frac{1}{2}\nu; \nu + 1; -\frac{1}{4}a^2\xi^2)$	Prob. 5-1

B-7 KONTOROVICH–LEBEDEV TRANSFORMS

	$f(x)$	$\mathscr{K}[f(x);\tau]$	*Reference*
KL1	1	$\frac{1}{2}\pi\tau^{-1}\operatorname{cosech}(\frac{1}{2}\pi\tau)$	Prob. 6-3 (*b*)
KL2	x	$\frac{1}{2}\pi\operatorname{sech}(\frac{1}{2}\pi\tau)$	Prob. 6-2 (*b*)
KL3	e^{-x}	$\pi\tau^{-1}\operatorname{cosech}(\pi\tau)$	Prob. 6-3 (*c*)
KL4	e^x	$\pi\tau^{-1}\coth(\pi\tau)$	Prob. 6-3 (*d*)

(*continued*)

	$f(x)$	$\mathscr{K}[f(x);\tau]$	*Reference*
KL5	xe^{-x}	$\pi\tau \operatorname{cosech}(\pi\tau)$	Prob. 6-2 (*c*)
KL6	$e^{-x\cos t}$	$\pi\tau^{-1}\cosh(\tau t)\operatorname{cosech}(\pi\tau)$	Prob. 6-3 (*a*)
KL7	$xe^{-x\cos t}$	$\pi\sinh(\tau t)\operatorname{cosec} t \operatorname{cosech}(\pi\tau)$	Prob. 6-2 (*a*)
KL8	$e^{-x\cosh t}$	$\pi\tau^{-1}\cos(\tau t)\operatorname{cosech}(\pi\tau)$	Prob. 6-4 (*a*)
KL9	$xe^{-x\cosh t}$	$\pi\sin(\tau t)\operatorname{cosech} t \operatorname{cosech}(\pi\tau)$	Prob. 6-4 (*b*)

B-8 MEHLER-FOCK TRANSFORMS OF ZERO ORDER

	$f(x)$	$\phi_0[f(x);\tau]$	*Reference*
MF1	$x^{-\frac{3}{2}}$	$\sqrt{2}\operatorname{sech}(\frac{1}{2}\pi\tau)$	(7-6-23)
MF2	$x^{-\frac{5}{2}}$	$\frac{2}{3}\sqrt{(2)}\tau\operatorname{cosech}(\frac{1}{2}\pi\tau)$	(7-6-24)
MF3	$x^{-2n-\frac{1}{2}}, (n \geqslant 2)$	$\dfrac{2^{2n-\frac{5}{2}}\pi^{\frac{1}{2}}\tau\operatorname{cosech}(\frac{1}{2}\pi\tau)}{\Gamma(2n+\frac{1}{2})}\prod_{r=1}^{n-1}(r^2+\frac{1}{4}\tau^2)$	(7-6-25)
MF4	$x^{-2n-\frac{3}{2}}, (n \geqslant 1)$	$\dfrac{2^{2n-\frac{1}{2}}\pi^{\frac{1}{2}}\operatorname{sech}(\frac{1}{2}\pi\tau)}{\Gamma(2n+\frac{3}{2})}\prod_{r=1}^{n}[(r-\frac{1}{2})^2+\frac{1}{4}\tau^2]$	(7-6-26)
MF5	$(x+t)^{-1}$	$\pi\operatorname{sech}(\pi\tau)\mathrm{P}_{-\frac{1}{2}+i\tau}(t)$	(7-6-27)
MF6	$(x+t)^{-m-1}$	$\dfrac{(-1)^m}{m!}\pi\operatorname{sech}(\pi\tau)(t^2-1)^{-\frac{1}{2}m}\mathrm{P}_{-\frac{1}{2}+i\tau}{}^{m}(t)$	Prob. 7-4
MF7	$(x+t)^{-\frac{1}{2}}$	$\dfrac{\sqrt{2}\cosh\{\tau\cos^{-1}t\}}{\tau\sinh(\pi\tau)}$	(7-6-9)
MF8	$(x+t)^{-m-\frac{1}{2}}$	$\dfrac{(-1)^m 2^{2m+\frac{1}{2}}m!}{\tau\sinh(\pi\tau)(2m)!}\cdot\dfrac{d^m}{dt^m}\cosh\{\tau\cos^{-1}t\}$	Prob. 7-5
MF9	$(x+1)^{-\frac{3}{2}}$	$2\sqrt{(2)}\tau\operatorname{cosech}(\pi\tau)$	Prob. 7-3
MF10	$(x+1)^{-\frac{5}{2}}$	$\frac{2}{3}\sqrt{(2)}\tau^3\operatorname{cosech}(\pi\tau)$	Prob. 7-3
MF11	$(x-\cos\beta)^{-\frac{1}{2}}$	$\sqrt{2}\dfrac{\cosh(\pi\tau-\beta\tau)}{\tau\sinh(\pi\tau)}$	Prob. 7-1
MF12	$(x-\cos\beta)^{-\frac{3}{2}}$	$2\sqrt{2}\operatorname{cosec}\beta\dfrac{\sinh(\pi\tau-\beta\tau)}{\sinh(\pi\tau)}$	Prob. 7-1
MF13	e^{-ax}	$\sqrt{(2/\pi a)}K_{i\tau}(a)$	(7-6-22)
MF14	$e^{ax}ei(a+ax)$	$\sqrt{(2/\pi a)}\operatorname{sech}(\pi\tau)K_{i\tau}(a)$	(7-7-3)

Bibliography

Abramian, B. L., Arutiunian, N. Kh., and Babloian, A. A., 1964: On two contact problems for the elastic sphere (in Russian), *Prik. Mat. i Mekh.*, **28,** 622: Engl. Trans. *App. Math. & Mech.*, **28,** 769.

Amerbaev, V. M., 1961: A finite Laplace transform (Russian), *Inž-Fiž. Ž.*, **4,** 109.

——— Babloian, A. A., 1964: The solution of some dual integral equations, *Prik. Mat. i. Mekh.*, **28,** 105: *Engl. Trans. App. Math. & Mech.*, **28,** 1227.

Bartels, R. F. C., and Churchill, R. V., 1942: Resolution of boundary problems by the use of a generalized convolution, *Bull. Amer. Math. Soc.*, **48,** 276.

Beckenbach, E. F. (ed.), 1961: "Modern Mathematics for the Engineer, Second Series," McGraw-Hill, New York.

Berg, L., 1967: "Introduction to the Operational Calculus," North Holland Pub. Co., Amsterdam.

Birkhoff, G., and Rota, G-C., 1969: "Ordinary Differential Equations," Blaisdell Pub. Co., Waltham, Mass.

Bochner, S., 1932: "Vorlesungen über Fourierische Integrale," Akademie-Verlag, Berlin.

1959: *Lectures on Fourier Integrals* (*Ann. Math. Studies No.* 42), Princeton University Press.

——— and Chandrasekharan, K., 1949: Fourier Transforms (*Ann. Math. Studies No.* 19), Princeton University Press.

Brown, H. K., 1943: The resolution of boundary value problems by means of the finite Fourier transformation, *J. Appl. Phys.*, **14,** 609.

——— Bruyn de, N. G., 1967; Uncertainty principles in Fourier analysis, in Oshiva "Inequalities," Academic Press, New York.

Campbell, G. A., and Foster, R. M., 1948: "Fourier Integrals for Practical Application," John Wiley & Sons, New York.

Carslaw, H. S., 1930: "Fourier Series and Integrals," 3rd edition, Macmillan, London.

——— and Jaeger, J. C., 1953: "Operational Methods in Applied Mathematics," Oxford University Press.

Chadwick, P., and Powdrill, B., 1964: Application of the Laplace transform method to wave motions involving strong discontinuities, *Proc. Cambridge Philos. Soc.*, **60,** 313.

Churchill, R. V., 1944: "Modern Operational Mathematics in Engineering," McGraw-Hill, New York.

1952: Integral transforms and boundary value problems, *Amer. Math. Monthly*, **59,** 149.

1954: The operational calculus of Legendre transforms, *J. Math. Phys.*, **33,** 165.

1955: Extensions of operational mathematics, *Proc. Conf. Diff. Eqns. Univ. of Maryland*, University of Maryland Bookstore.

1955–6: Generalized finite Fourier cosine transforms, *Mich. Math. Journ.*, **3,** 85.

1958: "Operational Mathematics," McGraw-Hill, New York—2nd ed. of Churchill, 1944.

——— and Dolph, C. L., 1954: Inverse transforms of products of Legendre transforms, *Proc. Amer. Math. Soc.*, **5,** 93.

Cinelli, G., 1965: An extension of the finite Hankel transform and applications, *Int. J. Engng. Sci.*, **3,** 539.

Conte, S. D., 1955: Gegenbauer transforms, *Quart. J. Math. Oxford* (2), **6,** 48.

Coulson, C. A., 1941*a*: Momentum distribution in molecular systems: I The single bond, *Proc. Cambridge Philos. Soc.*, **37,** 55.

1941*b*: Momentum distribution in molecular systems: III Bonds of higher order, *Proc. Cambridge Philos. Soc.*, **37,** 74.

——— and Duncanson, W. E., 1941: Momentum distribution in molecular systems: II Carbon and the C—H bond, *Proc. Cambridge Philos. Soc.*, **37,** 67.

1942: Momentum distribution in molecular systems, *Proc. Cambridge Philos. Soc.*, **38,** 100.

Debnath, L., 1960: On Laguerre transform, *Bull. Calcutta Math. Soc.*, **52,** 69.

1963: On Jacobi transform, *Bull. Calcutta Math. Soc.*, **55,** 113.

1964: On Hermite transform, *Mat. Vesnik*, **1,** 285.

Dirac, P. A. M., 1947: "The Principles of Quantum Mechanics," Clarendon Press, Oxford.

Ditkin, V. A., and Prudnikov, A. P., 1961: "Integral transforms and Operational Calculus," Moscow: Engl. Trans.—Pergamon Press, Oxford, 1965.

1968: Operational Calculus (pp. 1–74 of vol. 1 of Gamkrelidze, 1968).

Doetsch, G., 1937: "Theorie und Anwendung der Laplace-Transformation," Springer-Verlag, Berlin.

1935: Integration von Differentialgleichungen vermittels der endlichen Fourier-Transformation, *Math. Ann.*, **112,** 52.

1950: "Handbuch der Laplace-Transformation," Band 1, Verlag Birkhauser, Basel.

1955: The same, Band 2.

Duncanson, W. E., 1941: Momentum distribution in molecular systems: IV The hydrogen molecular ion H_2^+, *Proc. Cambridge Philos. Soc.*, **37,** 397.

——— and Coulson, C. A., 1945: Theoretical shape of the Compton profile for atoms from H to Ne. *Proc. Phys. Soc. London*, **57,** 190.

1948: Electron momenta in atoms, *Proc. Phys. Soc. London*, **60,** 175.

Dunn, H. S., 1967: A generalization of the Laplace transform, *Proc. Cambridge Philos. Soc.*, **63,** 155.

Eddington, A. S., 1913: On a formula for correcting statistics for the effects of a known probable error of observation, *Mon. Not. Roy. Astron. Soc.*, **73,** 359.

Erdelyi, A., 1961: From delta functions to distributions (in pp. 5–50 of Beckenbach, 1961).
1962: "Operational Calculus and Generalized Functions," Holt, Rinehart & Winston, Inc., New York).
——— (ed.) with Magnus, W., Oberhettinger, F., and Tricomi, F. G., 1954: "Tables of Integral Transforms," 2 vols., McGraw-Hill, New York.
Eringen, A. C., 1954: The finite Sturm-Liouville transform, *Quart. J. Math. Oxford*, (2), **5,** 120.
Fock, V., 1943: On the representation of an arbitrary function by an integral involving Legendre's function with a complex index, *Dokl. Akad. Nauk. SSSR*, **39,** 253.
Friedmann, B., 1956: "Principles and Techniques of Applied Mathematics," John Wiley & Sons, New York.
Gamkrelidze, R. V. (ed.), 1968: "Progress in Mathematics," 2 vols., Engl. Trans.—Plenum Press, New York.
Gelfand, I. M., and Shilov, G. E., 1964–8: "Generalized Functions," vols. 1, 2, 3, Academic Press, New York.
——— and Vilenkin, 1964: "Generalized Functions," vol. 4, Academic Press, New York.
Goldberg, R. R., 1961: "Fourier Transforms" (Cambridge Tract No. 52), Cambridge University Press.
Goodman, J. W., 1968: "Introduction to Fourier Optics," McGraw-Hill, New York.
Halperin, I., 1952: "Introduction to the Theory of Distributions based on lectures by Laurent Schwartz," University of Toronto Press.
Haskell, N. A., 1935: Motion of a viscous fluid under a surface load, *Physics*, **6,** 265.
1936: Motion of a viscous fluid under a surface load: part II, *Physics*, **7,** 56.
Hirschman, I. I., 1963: Laguerre transforms, *Duke Math. J.*, **30,** 495.
Jones, D. S., 1965: Some remarks on Hilbert transforms, *J. Inst. Math. Applics.*, **1,** 226.
1966*a*: "Generalized Functions," McGraw-Hill, New York.
1966*b*: Fourier transforms and the method of stationary phase, *J. Inst. Math. Applics.*, **2,** 197.
Jury, E. I., 1964: "Theory and Application of the Z-Transform," John Wiley & Sons, New York.
Kniess, H., 1938: Lösung von Randwertproblemen bei System gewöhnlichler Differentialgleichungen vermittels der endlichen Fourier Transformation, *Math. Z.*, **44,** 266.
Kober, H., 1943*a*: A note on Hilbert transforms, *Quart. J. Math. Oxford.*, **14,** 49.
1943*b*: A note on Hilbert transforms, *J. London Math. Soc.*, **18,** 66.
1944: A note on Fourier transforms, *J. London Math. Soc.*, **19,** 144.
König, H., 1953: Neue Begrundung der theorie der "Distributionen" von L. Schwartz, *Math. Nachr.*, **9,** 129.
Koschmieder, L., 1941, Die endliche Fouriersche Abbildung, *Deut. Math.* **5,** 521.
Kumar, R., 1952: A pair of functions which are Hankel transforms of each other, *Ganita*, **3,** 79.
Kurth, R., 1965: On Eddington's solution of the convolution integral equation, *Rend. Circ. Mat. Palermo*, (2), **14,** 76.
Lagrange, R., 1939: "Polynomes et Fonctions de Legendre," *Mem. des Sci. Math. Fasic. XCVII*, Gauthier-Villars, Paris.
Lakshmanarao, S. K., 1954: Gegenbauer transforms, *Math. Student*, **22,** 161.
Lavoine, J., 1959: "Calcul Symbolique, Distributions et Pseudo-Fonctions," C.N.R.S., Paris.
Lebedev, N. N., 1949: Parseval's formula for the Mehler–Fock transform (Russian), *Dokl. Akad. Nauk. SSSR*, **68,** 445.
1950: Some integral representations for products of sphere functions (Russian), *Dokl. Akad. Nauk. SSSR*, **73,** 449.
Lighthill, M. J., 1958: "Introduction to Fourier Analysis and Generalised Functions," Cambridge University Press.

Loeve, M., 1960: "Probability Theory," 2nd edn., D. van Nostrand, Princeton.

Lowndes, J. S., 1959: An application of the Kontorovich–Lebedev transform, *Proc. Edinburgh Math. Soc.*, **11,** 135.

1961: Note on an integral transform, *SIAM Review*, **3,** 162.

1962: Parseval relations for Kontorovich–Lebedev transforms, *Proc. Edinburgh Math. Soc.*, (*ii*), **13,** 5.

1964: Note on the generalized Mehler transform, *Proc. Cambridge Philos. Soc.*, **60,** 57.

Lukacs, E., 1960: "Characteristic Functions," Griffin, London.

Macauley-Owen, P., 1939: Parseval's theorem for Hankel transforms, *Proc. London Math. Soc.*, **45,** 458.

McCully, J., 1960: The Laguerre transform, *SIAM Review*, **2,** 185.

MacFarlane, G. G., 1949: The application of Mellin transforms to the summation of slowly convergent series, *Phil. Mag.*, (*vii*), **40,** 188.

MacRobert, T. M., 1931: Fourier integrals, *Proc. Roy. Soc. Edinburgh*, **51,** 116.

McLachlan, N. W., and Humbert, P., 1950: "Formulaire pour le calcul symbolique," Gauthier-Villars, Paris.

Mehler, F. G., 1881: Ueber eine mit den Kugel- und Cylinderfunctionen verwandte Function und ihre Anwendung in der Theorie der Elektricitatsvertheilung, *Math. Ann.*, **18,** 161.

Mercer, A. McD., 1963: On integral transform pairs arising from differential equations of any even order, *Quart. J. Math. Oxford*, (2), **14,** 9.

Mikusinski, J. G., 1948: Sur la méthode de généralisation de M. Laurent Schwartz et sur la convergence faible, *Fund. Math.*, **35,** 235.

1953: "Rachunek Operatów" (Polskie Towarzystwo, Warsaw), Engl. Trans.—"The Calculus of Operators," Pergamon Press, Oxford, 1959.

Miles, J. W., 1961: Integral transforms (in pp. 68–99 of Beckenbach, 1961).

Morse, P. M., 1948: "Sound and Vibration," McGraw-Hill, New York.

——— and Feshbach, H., 1953: "Methods of Theoretical Physics," McGraw-Hill, New York.

Mott, N. F., and Sneddon, I. N., 1948: "Wave Mechanics and its Applications," Clarendon Press, Oxford—reprint by Dover Publ. Inc.

Murnaghan, F. D., 1962: "The Laplace Transform," Spartan Books, Washington.

Muskhelishvili, N. I., 1953, "Singular Integral Equations," Noordhoff, Groningen.

Naylor, D., 1963*a*: On a Mellin type integral transform, *J. Math. Mech.*, **12,** 265.

1963*b*: On a finite Lebedev transform, *J. Math. Mech.*, **12,** 375.

1964: On a Lebedev theorem, *J. Math. Mech.*, **13,** 353.

1966: On a finite Lebedev transform, *J. Math. Mech.*, **15,** 455.

Noble, B., 1958: "Methods based on the Wiener–Hopf Technique," Pergamon Press, Oxford.

Oberhettinger, F., and Higgins, T. P., 1961: Tables of Lebedev, Mehler and generalized Mehler transforms, *Boeing Sci. Res. Lab. Math. Note* 246, Seattle.

Olevskii, M. N., 1943: On a generalization of Bessel functions, *Dokl. Akad. Nauk. SSSR*, **40,** 5.

1949: On the representation of an arbitrary function in the form of an integral with a kernel containing a hypergeometric function (Russian), *Dokl. Akad. Nauk. SSSR*, **69,** 11.

Palama, G., 1937: La trasformazione di Laplace e i polinomi di Laguerre e di Hermite, *Rend. Inst. Lombardo di Sci. e Let.*, **70,** Fasc., 3.

1938: Sulla trasformazione di Laplace di alcune note voli funzioni e su alcuni noti sviluppi in serie, *Math. Z.*, **44,** 347.

Paley, R. E. A. C., and Wiener, N., 1934: "Fourier Transforms in the Complex Domain." AMS Colloquium Publ., XV, New York.

Papoulis, A., 1963: "The Fourier Integral and its Applications," McGraw-Hill, New York.

Pauling, L., and Wilson, E. B., 1935: "Introduction to Quantum Mechanics—with Applications to Chemistry," McGraw-Hill, New York.

Podolsky, B., and Pauling, L., 1929: Momentum distribution in hydrogen-like atoms, *Phys. Rev.*, **34,** 109.

van der Pol, B., and Bremmer, H., 1950: "Operational Calculus based on the Two-Sided Integral," Cambridge University Press, London.

Post, E. L., 1930: Generalized differentiation, *Trans. Amer. Math. Soc.*, **32,** 373.

Powzner, A., 1948: On differential equations of Sturm–Liouville type on a half-axis (Russian), *Mat. Sbornik*, **23,** 3, Engl. Trans.—*Amer. Math. Soc. Translation No.* 5 (1950).

Ragab, F. M., 1965: E-Transforms, *Math. Res. Center Univ. of Wisconsin Rpt. No.* 563.

Riesz, F. and Sz.-Nagy, B., 1955, "Functional Analysis", Ungar Publ. Co. New York.

Roettinger, I., 1947: A generalization of the finite Fourier transformation and applications, *Quart. Appl. Math.*, **5,** 298.

Sauer, R., 1958: "Anfangswertprobleme bei partiellen Differentialgleichungen," 2nd edn., Springer-Verlag, Berlin.

Schwartz, L., 1950–1: "Théorie des Distributions," Tomes 1 & 2, Hermann & Cie, Paris.

1961: "Méthodes Mathématiques pour les Sciences Physiques," Hermann & Cie, Paris.

1966: "Mathematics for the Physical Sciences," Addison-Wesley, Reading, Mass. (Engl. Trans. of previous item.)

Scott, E. J., 1953: Jacobi transforms, *Quart. J. Math. Oxford*, **4,** 36.

Slater, L. J., 1966: "Generalized Hypergeometric Functions," Cambridge University Press.

Sneddon, 1946: Finite Hankel transforms, *Phil. Mag.*, **37,** 17.

1951: "Fourier Transforms," McGraw-Hill, New York.

1960: The elementary solution of dual integral equations, *Proc. Glasgow Math. Assoc.*, **4,** 108.

1961*a*: A note on some relations between Fourier and Hankel transforms, *Bull. Acad. Polon. Sci. Ser. Sci. Math. Astronom. Phys.*, **9,** 799.

1961*b*: "The Special Functions of Mathematical Physics and Chemistry," 2nd edn., Oliver & Boyd, Edinburgh—Interscience, New York.

1961*c*: "Fourier Series," Routledge & Kegan Paul, London.

1965: A relation involving Hankel transforms with applications to boundary value problems in potential theory, *J. Math. Mech.*, **14,** 33.

1966: "Mixed Boundary Value Problems in Potential Theory," North Holland Publ. Co., Amsterdam.

1968: A procedure for deriving inversion formulae for integral transform pairs of a general kind, *Glasgow Math. J.*, **9,** 67.

1969: The inversion of Hankel transforms of order zero and unity, *Glasgow Math. J.*, **10,** 156.

Sneddon, I. N., and Lowangrub, M., 1969: "Crack Problems in the Mathematical Theory of Elasticity," John Wiley & Sons, New York.

Sobolev, S. L., 1936: Methode nouvelle a resoudre le probleme de Cauchy pour les equations lineaires hyperboliques normales, *Mat. Sbornik*, **1,** 39.

Sölingen, H., 1939: Die Lösungen der Integralgleichungen $g(x) = \dfrac{1}{2\pi}\displaystyle\int_{-a}^{a}\frac{f(\xi)\,d\xi}{x-\xi}$ und deren Anwendung in der Tragflügeltheorie, *Math. Z.*, **45,** 245.

1953: Algebraisierung der endlichen Hilbert-Transformation, *Z. Angew. Math. Mech.*, **33,** 280.

1954: Zur Theorie der endlichen Hilbert Transformation, *Math. Z.*, **60,** 31.

Srivistav, R. P., 1963: A note on the solution of an integral equation, *Proc. Edinburgh Math. Soc.*, **13,** 271.

Srivastava, K. N., 1965: On Gegenbauer transforms, *Math. Student*, **33,** 129.

Sterling, R. C., 1960: Kernels for integral transforms: I. Chebyshev transforms of the first kind, *Rept. Math. Dept. Northampton Coll., London.*

Svartholm, N., 1945: The binding energies of the lightest atomic nuclei (*Thesis, Univ. of Lund*).
1947: The binding energies of the lightest atomic nuclei, *Arkiv. Mat. Astron. Fysik*, **35A**, Nos. 7 & 8.
Tamarkin, J. D., and Shohat, J. A., 1943: "The Problem of Moments," Amer. Math. Soc., New York.
Temple, G., 1953*a*: Theories and applications of generalized functions, *J. London Math. Soc.*, **28,** 134.
1953*b*: La théorie de la convergence généralisée et les fonctions généralisées et leur application à la physique mathématique, *Rend. Mat. e Appl.*, (*v*), **11,** 113.
1955: The theory of generalized functions, *Proc. Roy. Soc. London*, **228,** 175.
Titchmarsh, E. C., 1948: "An Introduction to the Theory of Fourier Integrals," 2nd edn., Oxford University Press.
Todhunter, I., and Pearson, K., 1893: "A History of the Theory of Elasticity," Cambridge University Press.
Tranter, C. J., 1950: Legendre transforms, *Quart. J. Math., Oxford*, (2), **1,** 1.
1951: "Integral Transforms in Mathematical Physics," Methuen, London; John Wiley & Sons, New York.
Tricomi, F. G., 1935: Trasformazione di Laplace e polinomi di Laguerre, *Rend. Accad. Lincei*, (*vi*), **21,** 235.
1951: On the finite Hilbert transformation, *Quart. J. Math. Oxford*, **2,** 199.
Uflyand, Ya. S., 1963: "Survey of Articles on the Applications of Integral Transforms in the Theory of Elasticity" (Russian), Leningrad.
Vilenkin, N. Ya., 1958: The matrix elements of irreducible unitary representations of a group of Lobachevsky space motions and the generalized Fock-Mehler transformations (Russian), *Dokl. Akad. Nauk. SSSR*, **118,** 219.
Ward, E. E., 1954: The calculation of transients in dynamical systems, *Proc. Cambridge Philos. Soc.*, **50,** 49.
Watson, G. N., 1944: "A Treatise on the Theory of Bessel Functions," 2nd edn., Cambridge University Press.
Weston, J. D., 1957: An extension of the Laplace-transform calculus, *Rend. Circ. Mat. Palermo*, (*ii*), **6,** 1.
1959: Operational calculus and generalized functions, *Proc. Roy. Soc. London A*, **250,** 460.
Wheelson, A. D., 1954: On the summation of infinite series in closed form, *J. Appl. Phys.*, **25,** 113.
1955: Note on the summation of infinite series, *J. Math. Phys.*, **34,** 182.
Whittaker, E. T., and Watson, G. N., 1927: "A Course of Modern Analysis," 4th edn., Cambridge University Press.
Widder, D. V., 1934: The inversion of the Laplace integral and the related moment problem, *Trans. Amer. Math. Soc.*, **36,** 107.
1937: The successive iterates of the Stieltjes kernel expressed in terms of the elementary functions, *Bull. Amer. Math. Soc.*, **43,** 813.
1938: The Stieltjes transform, *Trans. Amer. Math. Soc.*, **43,** 7.
1941: "The Laplace Transform," Princeton University Press.
1961: "Advanced Calculus," 2nd edn., Prentice-Hall, Inc.
Wiener, N., 1932: "The Fourier Integral," Cambridge University Press.
Wintner, A., 1947: "Fourier Transforms of Probability Distributions," (mimeographed notes: Princeton University Press).
Wright, E. M., 1935: The asymptotic expansion of the generalized Bessel function, *Proc. London Math. Soc.*, **38,** 257.
Zemanian, A. H., 1965: "Distribution Theory and Transform Analysis," McGraw-Hill, New York.
1969: "Generalized Integral Transformations," John Wiley & Sons, Inc., New York.

Index of Symbols

Author Index

Subject Index